The science and
practice of welding

The science and practice of welding

A. C. DAVIES

B.Sc (*London Hons. and Liverpool*), C.Eng., M.I.E.E.
Fellow of the Welding Institute

SEVENTH EDITION

CAMBRIDGE UNIVERSITY PRESS
CAMBRIDGE
LONDON · NEW YORK · MELBOURNE

Published by the Syndics of the Cambridge University Press
The Pitt Building, Trumpington Street, Cambridge CB2 1RP
Bentley House, 200 Euston Road, London NW1 2DB
32 East 57th Street, New York, NY 10022, USA
296 Beaconsfield Parade, Middle Park, Melbourne 3206,
Australia

First published 1941
Second edition 1943
Third edition 1945
Reprinted 1947, 1950
Fourth edition 1955
Reprinted 1959
Fifth edition 1963
Reprinted 1966, 1969, 1971
Sixth edition 1972
Reprinted 1975
Seventh edition 1977

Phototypeset by Western Printing Services Ltd, Bristol
and printed in Great Britain by the Pitman Press, Bath

Library of Congress cataloguing in publication data

Davies, Arthur Cyril
The science and practice of welding

Includes index
1. Welding   I. Title
TS227.D22 1977   671.5'2   77-71408
ISBN 0 521 21557 9
(ISBN 0 521 08443 1: 6th edition)

TS
227
D22
1977

Contents

Preface

The book has been extensively revised with new sections on submerged arc, stud (arc and capacitor), explosive and gravity processes. A new chapter has been added on resistance welding and many sections have been brought up to date including the new processes in iron and steel production, and additional information is included in the chapters on TIG and MIG processes. The new electrode classification to BS 639 1976 has been included together with impact machines and testing.

My thanks are due to each and all of the following firms who have helped me in every way by offering advice and supplying information and photographs as indicated.

A.I. Welders Ltd: flash butt welding technology and photograph.

Air Products Cryogenic Division: the welding of aluminium alloys and stainless steel, with illustrations and details of impact tests.

Avery-Denison Ltd: impact testing machines and photographs.

British Oxygen Co. Ltd: oxy-acetylene welding equipment and photographs, industrial gases and diagrams, manual metal arc welding electrodes and filler wires, manual metal arc, TIG, MIG and plasma welding plant and plasma cutting with diagrams and photographs.

Copper Development Association: the welding of copper and its alloys.

Crompton Parkinson Ltd: stud welding with diagrams.

ESAB Ltd: manual metal arc welding electrodes and filler wires, manual metal arc, TIG, MIG, submerged arc, gravity, and electroslag welding equipment, positioners, and robot welding with illustrations and photographs.

G.K.N. Lincoln Ltd: submerged arc welding equipment, wire electrodes and fluxes.

British Railways Board: details of flash butt and thermit welding of rails and diagrams of 'adjustment switch'.

British Steel Corporation, Library and Information Services of the Sheffield Laboratories: modern blast furnaces, direct reduction of iron ores, basic oxygen steel, electric arc steel with illustrations.

KSM Stud Welding Ltd: stud welding with diagrams.

Pirelli General Cable Co.: welding cables.

The Welding Institute: information on the classification of electrodes.

Cooperheat Ltd: pre- and post-heating equipment with photographs.

Sciaky Ltd: spot, seam projection and other types of resistance welding with photographs, laser beam welding with photograph.

Henry Wiggin Ltd: the welding of nickel and nickel alloys.

Union Carbide UK Ltd: TIG and MIG technology, plasma welding, cutting and surfacing technology with diagrams.

Yorkshire Imperial Metals Ltd: explosive welding with diagrams.

Birlec Ltd: induction furnace photograph.

Rockweld Ltd: photographs of CO_2 welding equipment.

Gamma-Rays Ltd: information on non-destructive testing with photographs of radiographic equipment.

I would again like to express my thanks to the City and Guilds of London Institute for permission to reproduce, with some amendments, examination questions set in recent years and to Mr D. G. J. Brunt, T.ENG (C.E.I), F.I.T.E., ASSOC. MEM.I.E.E., M.WELD.I., for help in the reading of proofs, and to Mr M. S. Wilson, B.SC., M.MET., for help in the revised sections in metallurgy.

Abstracts of British Standards are included by permission of the British Standards Institution, 2 Park Street, London, from whom copies of the latest complete standards may be obtained.

The terms TIG, MIG and CO_2 have been retained for these welding processes, pending revision of BS 499, as they are so widely used. The use of gas mixtures of inert and active gases (argon–oxygen, argon–oxygen–CO_2, argon–hydrogen, etc.) as the shielding gas together with the pulsed, modulated feed, modulated arc length and flux cored processes, etc. have made the present terminology rather inadequate.

Oswestry
1977   A. C. Davies

Preface to the sixth edition

I would like to express my thanks to the following firms for their co-operation and help during the writing of the previous five editions of this book:

Air Products Ltd; Associated Electrical Industries; Aluminium Federation; B.E.A.M.A.; British Aluminium Co. Ltd; British Insulated Callenders Cables Ltd; British Oxygen Co. Ltd; British Welding Research Association; Buck and Hickman Ltd; Copper Development Association; Deloro Stellite Ltd; English Electric Co. Ltd; English Steel Corporation; Firth Vickers Stainless Steels; Samuel Fox and Co. Ltd; Fusarc Co. Ltd; Gamma Rays Ltd; Kelvin Hughes and Co. Ltd; Laurence Scott and Electromotors; Lincoln Electric Co. Ltd; Magnesium Elektron Ltd; Magnesium Industry Council; Metalectric Furnaces; Mond Nickel Co.; Murex-Quasi-Arc Ltd; Rockweld Ltd; Siemens UK Ltd; Thorn and Hoddle Ltd; Weldcraft Ltd.

The change from the Imperial to the SI system of units is already well under way in the welding industry and it was felt that an SI edition of this book would be acceptable to students of welding.

The book has been extensively revised with new chapters on TIG, plasma arc, gas shielded metal arc (MIG, CO_2 and mixed gases) and additional sections have been included on electrical and welding technology with a view to the needs of the welding technician.

My thanks are due to all of the following firms who have been helpful in every way by supplying information as indicated.

Air Products Ltd: techniques used in TIG, MIG and pulse fabrications in aluminium alloys and stainless steel.

A.I. Welders Ltd: friction welding and photographs.

Aluminium Federation: the weldability of aluminium and its alloys.

British Aluminium Co.: the weldability of aluminium and its alloys.

British Oxygen Co. Ltd: basic technology of TIG, plasma, MIG, pulse arc processes and their applications, metal-arc welding and metal-arc welding electrodes, industrial gases, and the provision of many diagrams and photographs.

Copper Development Association: the weldability of copper and its alloys.

Distillers Co. (Carbon Dioxide) Ltd: production of carbon dioxide.

Firth Vickers Stainless Steels Ltd: the metallurgy and weldability of stainless and heat-resistant steels.

G.K.N. Lincoln Ltd: the CO_2 welding process, and photographs.

Magnesium Elektron Ltd: the weldability of the magnesium alloys.

Pirelli General Cable Co. Ltd: welding cables and recent developments.

Sandvik UK Ltd: stainless steel welding wire details.

Union Carbide Ltd: TIG, MIG, and plasma welding technology and diagrams.

Henry Wiggin Ltd: the metallurgy and weldability of monel and nickel alloys.

I would also like to express my thanks to the City and Guilds of London Institute for permission to reproduce, with some amendments, examination questions set in recent years and to Mr D. G. J. Brunt, M.I.T.E., ASSOC. MEM.I.E.E., M.WELD.I., for help in the reading of proofs.

Extracts from British Standards are reproduced by permission of the British Standards Institution, 2 Park Street, London, from whom copies of the latest complete standards may be obtained.

Oswestry
1972   A. C. Davies

The metric system and the use of SI units

The metric system was first used in France after the French Revolution and has since been adopted for general measurements by all countries of the world except the United States. For scientific measurements it is generally used universally.

It is a decimal system, based on multiples of ten, the following multiples and sub-multiples being added, as required, as a prefix to the basic unit.

Prefix	Symbol	Multiplying factor		
pico-	p	0.000 000 000 001	or	10^{-12}
nano-	n	0.000 000 001	or	10^{-9}
micro-	μ	0.000 001	or	10^{-6}
milli-	m	0.001	or	10^{-3}
centi-	c	0.01	or	10^{-2}
deci-	d	0.1	or	10^{-1}
deca-	da	10	or	10^{1}
hecto-	h	100	or	10^{2}
kilo-	k	1 000	or	10^{3}
mega-	M	1 000 000	or	10^{6}
giga-	G	1 000 000 000	or	10^{9}
tera-	T	1 000 000 000 000	or	10^{12}

Examples of the use of these multiples of the basic unit are: hectobar, milliampere, meganewton, kilowatt.

In past years, the CGS system, using the centimetre, gram and second as the basic units, has been used for scientific measurements. It was later modified to the MKS system, with the metre, kilogram and second as the basic units, giving many advantages, for example in the field of electrical technology.

Note on the use of indices
A velocity measured in metres per second may be written m/s, indicating

that the second is the denominator, thus: $\dfrac{\text{metre}}{\text{second}}$ or $\dfrac{\text{m}}{\text{s}}$. Since $\dfrac{1}{a^n} = a^{-n}$, the velocity can also be expressed as metre second^{-1} or m s^{-1}. This method of expression is often used in scientific and engineering articles. Other examples are, pressure and stress: newton per square metre or pascal (N/m^2 or Nm^{-2}); density: kilograms per cubic metre (kg/m^3 or kg m^{-3}).

SI units (Système Internationale d'Unités)

To rationalize and simplify the metric system the Système Internationale d'Unités was adopted by the ISO (International Organization for Standardization). In this system there are six primary units, thus:

Quantity	Basic SI unit	Symbol
length	metre	m
mass	kilogram	kg
time	second	s
electric current	ampere	A
temperature	kelvin	K
luminous intensity	candela	cd

In addition there are derived and supplementary units, thus:

Quantity	Unit	Symbol
plane angle	radian	rad
area	square metre	m^2
volume[1]	cubic metre	m^3
velocity	metre per second	ms
angular velocity	radian per second	rad/s
acceleration	metre per second squared	m/s^2
frequency	hertz	Hz
density	kilogram per cubic metre	kg/m^3
force	newton	N
moment of force	newton metre	Nm
pressure, stress	newton per square metre	N/m^2 (or pascal, Pa)
surface tension	newton per metre	N/m
work, energy, quantity of heat	joule	J (Nm)

Quantity	Unit	Symbol
power, rate of heat flow	watt	W (J/s)
impact strength	joule per square metre	J/m²
temperature	degree Celsius	°C
thermal coefficient of linear expansion	reciprocal degree Celsius or kelvin	°C⁻¹, K⁻¹
thermal conductivity	watt per metre degree C	W/m°C
coefficient of heat transfer	watt per square metre degree C	W/m²°C
heat capacity	joule per degree C	J/°C
specific heat capacity	joule per kilogram degree C	J/kg°C
specific latent heat	joule per kilogram	J/kg
quantity of electricity	coulomb	C (As)
electric tension, potential difference, electromotive force	volt	V (W/A)
electric resistance	ohm	Ω (V/A)
electric capacitance	farad	F
magnetic flux	weber	Wb
inductance	henry	H
magnetic flux density	tesla	T (Wb/m²)
magnetic field strength	ampere per metre	A/m
magnetomotive force	ampere	A
luminous flux	lumen	lm
luminance	candela per square metre	cd/m²
illumination	lux	lx

[1] *Note*. N m³ is the same as m³ at normal temperature and pressure, i.e. 0 °C and 760 mm Hg (NTP or STP).

The litre is used instead of the cubic decimetre (1 litre = 1 dm³) and is used in the welding industry to express the volume of a gas.

Pressure and stress many also be expressed in bar (b) or hectobar (hbar) instead of newton per square metre.

Conversion factors from British units to SI units are given in the appendix.

1 metric tonne = 1000 kg.

1

Welding science

HEAT

Solids, liquids and gases: atomic structure

Substances such as copper, iron, oxygen and argon which cannot be broken down into any simpler substances are called elements; there are at the present time over 100 known elements. A substance which can be broken down into two or more elements is known as a compound.

An *atom* is the smallest particle of an element which can take part in a chemical reaction. It consists of a number of negatively charged particles termed electrons surrounding a massive positively charged centre termed the nucleus. Since like electric charges repel and unlike charges attract, the electrons experience an attraction due to the positive charge on the nucleus. Chemical compounds are composed of atoms, the nature of the compound depending upon the number, nature and arrangement of the atoms.

A molecule is the smallest part of a substance which can exist in the free state and yet exhibit all the properties of the substance. Molecules of elements such as copper, iron and aluminium contain only one atom and are mon-atomic. Molecules of oxygen, nitrogen and hydrogen contain two atoms and are di-atomic. A molecule of a compound such as carbon dioxide contains three atoms and complicated compounds contain many atoms.

An atom is made up of three elementary particles: (1) protons, (2) electrons, (3) neutrons.

The *proton* is a positively charged particle and its charge is equal and opposite to the charge on an electron. It is a constituent of the nucleus of all atoms and the simplest nucleus is that of the hydrogen atom which contains one proton.

The *electron* is 1/1836 of the mass of a proton and has a negative charge equal and opposite to the charge on the proton. The electrons form a cloud around the nucleus moving within the electric field of the positive charge and around which they are arranged in shells.

The *neutron* is a particle which carries no electric charge but has a mass equal to that of the proton and is a constituent of the nuclei of all atoms except hydrogen. The atomic number of an element indicates the number of protons in its nucleus and because an atom in its normal state exhibits no external charge, it is the same as the number of electrons in the shells.

Isotopes are forms of an element which differ in their atomic mass but not in some of their chemical properties. The atomic weight of an isotope is known as its mass number. For example, an atom of carbon has 6 protons and 6 neutrons in its nucleus so that its atomic number is 6. Other carbon atoms exist, however, which have 7 neutrons and 8 neutrons in the nucleus. These are termed isotopes and their mass numbers are 13 and 14 respectively, compared with 12 for the normal carbon atom. One isotope of hydrogen called heavy hydrogen or deuterium has a mass number 2 so that it has one proton and one neutron in its nucleus.

Electron shells. The classical laws of mechanics as expounded by Newton do not apply to the extremely minute world of the atom and the density, energy and position of the electrons in the shells are evaluated by quantum or wave mechanics. Since an atom in its normal state is electrically neutral, if it loses one or more electrons it is left positively charged and is known as a *positive ion*; if the atom gains one or more electrons it becomes a *negative ion*. It is the electrons which are displaced from their shells, the nucleus is unaffected, and if the electrons drift from shell to shell in an organized way in a completed circuit this constitutes an electric current.

In the *periodic classification*, the elements are arranged in order of their mass numbers, horizontal rows ending in the inert gases and vertical columns having families of related elements.

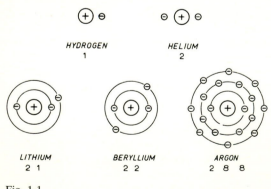

HYDROGEN
1

HELIUM
2

LITHIUM
2 1

BERYLLIUM
2 2

ARGON
2 8 8

Fig. 1.1

The lightest element, hydrogen, has one electron in an inner shell and the following element in the table, helium, has two electrons in the inner shell. This shell is now complete so that for lithium, which has three electrons, two occupy the inner shell and one is in the next outer shell. With succeeding elements this shell is filled with electrons until it is complete with the inert gas neon which has two electrons in the inner shell and eight in the outer shell, ten electrons in all. Sodium has eleven electrons, two in the inner, eight in the second and one in a further outer shell. Electrons now fill this shell with succeeding elements until with argon it is temporarily filled with eight electrons so that argon has eighteen electrons in all. This is illustrated in Fig. 1.1 and this brief study will suffice to indicate how atoms of the elements differ from each other. Succeeding elements in the table have increasing numbers of electrons which fill more shells until the table is, at the present time, complete with just over 100 elements.

hydrogen	helium						
1	2						
lithium	beryllium	boron	carbon	nitrogen	oxygen	fluorine	neon
2	2	2	2	2	2	2	2
1	2	3	4	5	6	7	8
sodium	magnesium	aluminium	silicon	phosphorus	sulphur	chlorine	argon
2	2	2	2	2	2	2	2
8	8	8	8	8	8	8	8
1	2	3	4	5	6	7	8

The shells are then filled up thus:

```
2   8.
2   8    8.
2   8   18.
2   8   18    8.
2   8   18   18.
2   8   18   18    8.
2   8   18   32   18.
2  18   18   32   18    8.
```

The electrons in their shells possess a level of energy and with any change in this energy light is given out or absorbed. The elements with completed or temporarily completed shells are the inactive or inert gases helium, neon, argon, xenon and radon, whereas when a shell is nearly complete (oxygen, fluorine) or has only one or two electrons in a shell (sodium, magnesium), the element is very reactive, so that the characteristics of an element are greatly influenced by its electron structure. When a metal filament such as tungsten is heated in a vacuum it emits electrons, and if a positively charged plate (anode) with an aperture in it is put in front near the filament, the electrons stream through the aperture attracted by the positive charge and form an electron beam.

This beam can be focused and guided and is used in the television tube, while a beam of higher energy can be used for welding by the electron beam process (see p. 508).

If the atoms in a substance are not grouped in any definite pattern the substance is said to be amorphous, while if the pattern is definite the substance is crystalline. Solids owe their rigidity to the fact that the atoms are closely packed in geometrical patterns called space lattices which, in metals, are usually a simple pattern such as a cube. The positions which atoms occupy to make up a lattice can be observed by X-rays.

Atoms vibrate about their mean position in the lattice, and when a solid is heated the heat energy supplied increases the energy of vibration of the atoms until their mutual attraction can no longer hold them in position in the lattice so that the lattice collapses, the solid melts and turns into a liquid which is amorphous. If we continue heating the liquid, the energy of the atoms increases until those having the greatest energy and thus velocity, and lying near the surface, escape from the attraction of neighbouring atoms and become a vapour or gas. Eventually when the vapour pressure of the liquid equals atmospheric pressure (or the pressure above the liquid) the atoms escape wholesale throughout the mass of the liquid which changes into a gaseous state and the liquid boils.

Suppose we now enclose the gas in a closed vessel and continue heating. The atoms are receiving more energy and their velocity continues to increase so that they will bombard the walls of the vessel, causing the pressure in the vessel to increase.

Atoms are grouped into molecules which may be defined as the smallest particles which can exist freely and yet exhibit the chemical properties of the original substance. If an atom of sulphur, two atoms of hydrogen, and four atoms of oxygen combine, they form a molecule of sulphuric acid. This molecule is the smallest particle of the acid which can exist since if we split it up we are back to the original atoms which combined to form it.

From the foregoing, it can be seen that the three states of matter – solids, liquids and gases – are very closely related, and that by giving or taking away heat we can change from one state to the other. Ice, water and steam give an everyday example of this change of state.

Metals require considerable heat to liquefy or melt them, as for example, the large furnaces necessary to melt iron and steel.

We see examples of metals in the gaseous state when certain metals are heated in the flame. The flame becomes coloured by the gas of the metal, giving it a characteristic colour, and this colour indicates what metal is being heated. For example, sodium gives a yellow coloration and copper a green coloration.

This change of state is of great importance to the welder, since he is concerned with the joining together of metals in the liquid state (termed fusion welding) and he has to supply the heat to cause the solid metal to be converted into the liquid state to obtain correct fusion.

Temperature: thermometers and pyrometers

The temperature of a body determines whether it will give heat to, or receive heat from, its surroundings.

Our sense of determining hotness by touch is extremely inaccurate, since iron will always feel colder than wood, for example, even when actually at the same temperature.

Instruments to measure temperature are termed thermometers and pyrometers. Thermometers measure comparatively low temperatures, while pyrometers are used for measuring the high temperatures as, for example, in the melting of metals.

In the thermometer, use is made of the fact that some liquids expand by a great amount when heated. Mercury and alcohol are the usual liquids used. Mercury boils at 357°C and thus can be used for measuring temperatures up to about 330°C.

Mercury is contained in a glass bulb which connects into a very fine bore glass tube called a capillary tube and up which the liquid expands (Fig. 1.2).

The whole is exhausted of air and sealed off. The fixed points on a thermometer are taken as the melting point of ice and the steam from pure water at boiling point at standard pressure (760 mm mercury).

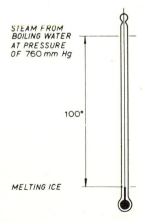

STEAM FROM
BOILING WATER
AT PRESSURE
OF 760 mm Hg

100°

MELTING ICE

Fig. 1.2. Celsius graduations.

In the Celsius thermometer the freezing point is marked 0 and boiling point 100; thus there are 100 divisions, called degrees and shown thus °. The Kelvin scale (K) has its zero at the absolute zero of temperature which is -273.16 °C. To convert approximately from °C to °K add 273 to the Celsius figure.

To measure temperatures higher than those measurable with an ordinary thermometer we can employ:

(1) Temperature cones.
(2) Temperature paints or crayons.
(3) Pyrometers: (*a*) Electrical resistance.
(*b*) Thermo-electric.
(*c*) Radiation.
(*d*) Optical.

(1) Temperature cones (Seger cones) are triangular pyramids made of a mixture of china clay, lime, quartz, iron oxide, magnesia, and boric acid in varying proportions so that they melt at different temperatures and can be used to measure temperatures between 600°C and 2000°C. They are numbered according to their melting points and are generally used in threes, numbered consecutively, of approximately the temperature required. When the temperature reaches that of the lowest melting point cone it bends over until its apex touches the floor. The next cone bends slightly out of the vertical while the third cone remains unaffected. The temperature of the furnace is that of the cone which has melted over.

(2) Indicating paints and crayons either melt or change colour or appearance at definite temperatures. Temperature indicators are available as crayons (sticks), pellets or in liquid form and operate on the melting principle and not colour change. They are available in a range from 30°C to 1650°C and each crayon has a calibrated melting point To use the crayon, one of the temperature range required is stroked on the work as the temperature rises and leaves a dry opaque mark until at the calibrated temperature it leaves a liquid smear which on cooling solidifies to a translucent or transparent appearance. Up to 700°C a mark can be made on the work piece before heating and liquefies at the temperature of the stick. Similarly a pellet of the required temperature is placed on the work and melts at the appropriate temperature while the liquid is sprayed on to the surface such as polished metal (or glass) which is difficult to mark with a crayon, and dries to a dull opaque appearance. It liquefies sharply at its calibrated temperature and remains glossy and transparent upon cooling.

(3) Pyrometers. (*a*) Electrical resistance pyrometers. Pure metals increase in resistance fairly uniformly as the temperature increases. A

platinum wire is wound on a mica former and is placed in a refractory sheath, and the unit placed in the furnace. The resistance of the platinum wire is measured (in a Wheatstone's bridge network) by passing a current through it. As the temperature of the furnace increases the resistance of the platinum increases and this increase is measured and the temperature read from a chart.

(b) Thermo-electric (thermo-couple). When two dissimilar metals are connected together at each end and one pair (or junction) of ends is heated while the other pair is kept cold, an electromotive force (e.m.f.) or voltage is set up in the circuit (Peltier Effect). The magnitude of this e.m.f. depends upon (a) the metals used and (b) the difference in temperature between the hot and cold junctions. In practice the hot junction is placed in a refractory sheath while the other ends (the cold junction) are connected usually by means of compensating leads to a millivoltmeter which measures the e.m.f. produced in the circuit and which is calibrated to read the temperature directly on its scale. The temperature of the cold junction must be kept steady and since this is difficult, compensating leads are used. These are made of wires having the same thermo-electric characteristics as those of the thermo-couple but are much cheaper and they get rid of the thermo-electric effect of the junction between the thermo-couple wires and the leads to the mil-livoltmeter, when the temperature of the cold junction varies. The couples generally used are copper–constantan (60% Cu, 40% Ni) used up to 300°C; chromel (90% Ni, 10% Cr); alumel (95% Ni, 3% Mn, 2% Al) up to 1200°C; and platinum–platinum-rhodium (10% Rh) up to 1500°C.

(c) Radiation pyrometers. These pyrometers measure the radiation emitted from a hot body. A 'black body' surface is one that absorbs all radiation falling upon it and reflects none, and conversely will emit all radiations. For a body of this kind, E, the heat energy radiated is proportional to the fourth power of the absolute temperature, i.e. $E \propto T^4$ (Stefan–Boltzmann Law) so that $E = kT^4$. If a body is however radiating heat in the open, the ratio of the heat which it radiates to the heat that a black body would radiate at the same temperature is termed the emis-sivity, e, and this varies with the nature, colour, and temperature of the body. Knowing the emissivity of a substance we can calculate the true temperature of it when radiating heat in the open from the equation:

$$(\text{True temperature})^4 = \frac{(\text{Apparent temperature})^4}{\text{emissivity}}$$

(temperatures are on the absolute scale).

In an actual radiation pyrometer the radiated heat from the hot source is focused on to a thermo-couple by means of a mirror (the focusing can be either fixed or adjustable) and the image of the hot body must cover the whole of the thermo-couple. The e.m.f. generated in the thermo-couple circuit is measured as previously described on a millivoltmeter.

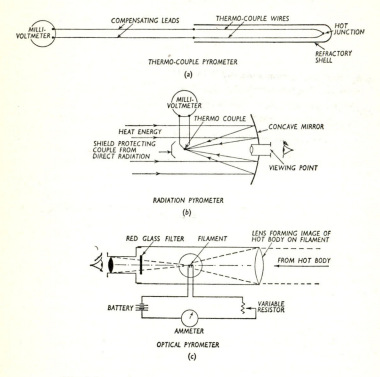

Fig. 1.3

(*d*) Optical pyrometers. The disappearing filament type is an example of this class of pyrometer. A filament contained in an evacuated bulb like an electric light bulb is viewed against the hot body as a background. By means of a control resistor the colour of the filament can be varied by varying the current passing through it until the filament can no longer be seen, hot body and filament then being at the same temperature. An ammeter measures the current taken by the filament and can be calibrated to read the temperature of the filament directly.

The judging of temperatures by colour is usually very inaccurate. If steel is heated, it undergoes a colour change varying from dull red to brilliant white. After considerable experience it is possible to estimate roughly the temperature by this means, but no reliance can be placed on it.

Degrees C	Colour
500	Red (visible in daylight)
800	Cherry red
1000	Bright red
1200	Reddish yellow
1400	White welding heat

Expansion and contraction

When a solid is heated, the atoms of which it is composed vibrate about their mean position in the lattice more and more. This causes them to take up more room and thus the solid expands.

Most substances expand when heated and contract again when cooled, as the atoms settle back into their normal state of vibration.

Metals expand by a much greater amount than other solid substances, and there are many practical examples of this expansion in everyday life.

Gaps are left between lengths of railway lines, since they expand and contract with atmospheric temperature changes. Fig. 1.4a shows the expansion joint used by British Rail. With modern methods of track construction only the last 100 m of rail is allowed to expand or contract longitudinally irrespective of the total continuous length of welded rail, and this movement is well within the capacity of the expansion joint or adjustment switch.

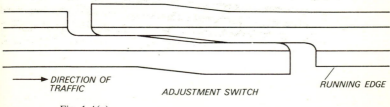

DIRECTION OF TRAFFIC
ADJUSTMENT SWITCH
RUNNING EDGE

Fig. 1.4(a)

Iron tyres are made smaller than the wheel they are to fit. They are heated and expand to the size of the wheel and are fitted when hot. On being quickly cooled, they contract and grip the wheel firmly.

Large bridges are mounted on rollers fitted on the supporting pillars to allow the bridge to expand.

In welding, this expansion and contraction is of the greatest importance. Suppose we have two pieces of steel bar about 1 m long. If these are set together at an angle of 90°, as shown, and then welded and allowed to cool, we find that they have curled or bent up in the direction of the weld (Fig. 1.4*b*).

PLATES ARE
STRAIGHT WHEN WELDED

SHAPE ON COOLING
DUE TO CONTRACTION

Fig. 1.4(b)

The hot weld metal, on contracting, has caused the bar to bend up as shown, and it is evident that considerable force has been exerted to do this.

A well-known example of the use to which these forces, exerted during expansion and contraction, are put is the use of iron bars to pull in or strengthen defective walls of buildings.

Plates or **S** pieces are placed on the threaded ends of the bar, which projects through the walls which need pulling in. The bar is heated to redness and nuts on each end are drawn up tight against the plates on the walls. As the bar cools, gradually the walls are pulled in (Fig. 1.5).

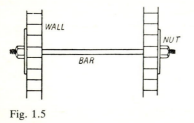

WALL

NUT

BAR

Fig. 1.5

Different metals expand by different amounts. This may be shown by riveting together a bar of copper and a bar of iron about 0.5 m long and 25 mm wide. If this straight composite bar is heated it will become bent, with the copper on the outside of the bend, showing that the copper expands more than the iron (Fig. 1.6). This composite bar is known as a bi-metal strip and is used in engineering for automatic control of temperature.

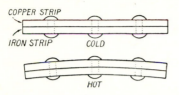

Fig. 1.6

Coefficient of linear expansion

The fraction of its length which a bar will expand when heated through one degree rise in temperature is termed its *coefficient of linear expansion*. (This also applies to contraction when the bar is cooled.) This fraction is very small; for example, for iron it is

$$\frac{12}{1\ 000\ 000}$$

That is, a bar of iron length l would expand by $l \times \dfrac{12}{1\ 000\ 000}$ for every degree rise in temperature. Hence, if the rise was $t°$, the expansion would be $\dfrac{12}{1\ 000\ 000} \times l \times t$.

The fraction $\dfrac{12}{1\ 000\ 000}$ is usually denoted by the letter a. Thus the increase in length of a bar of original length l, made of material whose coefficient of linear expansion is a, when heated through $t°$ is lat.

Thus, the final length of a bar when heated equals it original length plus its expansion, that is:

$$L \quad = \quad l \quad + \ lat$$

Final length = original length + expansion.

This can also be written: $L = l\,(1 + at)$.

l	lat

Length after being
heated through $t°C$

Example

Given that the coefficient of linear expansion of copper is $\dfrac{17}{1\ 000\ 000}$ or 0.000 017 per degree C, find the final length of a bar of copper whose original length was 75 mm, when heated through 50 °C.

Final length = original length + expansion, i.e.:

$$\text{Final length} = 75 + \left(75 \times \frac{17}{1\ 000\ 000} \times 50 \right)$$

$$= 75 + \frac{63\ 750}{1\ 000\ 000} = 75 + \frac{6375}{100\ 000} = 75.06 \text{ mm.}$$

The above is equally true for calculating the contraction of a bar when cooled.

Table of coefficients of linear expansion of metals per degree C

Metal	a	Metal	a
Lead	0.000 027	Zinc	0.000 026
Tin	0.000 021	Cast iron	0.000 010
Aluminium	0.000 025	Nickel	0.000 013
Copper	0.000 017	Wrought iron	0.000 012
Brass,	0.000 020	Mild steel	0.000 012
60% copper,			
40% zinc			

Invar, a nickel steel alloy containing 36% nickel, has a coefficient of linear expansion of only 0.000 000 9, that is, only $\frac{1}{13}$ of that of mild steel, and thus we can say that invar has practically no expansion when heated.

The expansion and contraction of metal is of great importance to the welder, because, as we have previously shown, large forces or stresses are called into play when it takes place. If the metal that is being welded is fairly elastic, it will stretch, or give, to these forces, and this is a great help, although stresses may be set up as a result in the welded metal. Some metals, however, like cast iron, are very brittle and will snap rather than give or show any elasticity when any force is applied. As a result, the greatest care has to be taken in applying heat to cast iron and in welding it lest we introduce into the metal, when expanding and contracting, any forces which will cause it to break. This will be again discussed at a later stage.

Coefficient of cubical expansion

If we imagine a solid being heated, it is evident that its volume will increase, because each side undergoes linear expansion.

A cube, for example, has three dimensions, and each will expand according to the previous rule for linear expansion. Suppose each face of the cube was originally length l and final length L after being heated through $t\,°C$. Let the coefficient of linear expansion be a per degree C.

The original volume was $l \times l \times l = l^3$.

Each edge will have expanded, and for each edge we have:

Final length $L = l\,(1 + at)$ as before (Fig. 1.7).

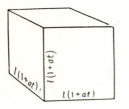

Fig. 1.7

Thus the new volume $= l(1 + at) \times l(1 + at) \times l(1 + at)$
$$= l^3(1 + 3at) \text{ approximately.}$$

Thus, final volume $=$ original volume $(1 + 3at)$.

That is, the *coefficient of cubical expansion* may be taken as being three times the coefficient of linear expansion.

Example

A brass cube has a volume of $0.006\ \text{m}^3$ ($6 \times 10^6\ \text{mm}^3$) and is heated through a $65\,°C$ rise in temperature. Find its final volume, given that the coefficient of linear expansion of brass $= 0.000\,02$ per degree C.

Final volume $=$ original volume $(1 + 3at)$
$$V = 0.006(1 + 3 \times 0.000\,02 \times 65)$$
$$= 0.006(1.0039)$$
$$= 0.006\,023\,4\text{m}^3.$$

The joule and newton

Heat is a form of energy and the unit of energy is the joule (J). A joule may be defined as the energy expended when a force of 1 newton (N) moves through a distance of 1 metre (m). (Note: a newton is that force which, acting on a mass of 1 kilogram (kg) gives it an acceleration of 1 metre per second per second (1 m/sec²). The gravitational force on a mass of 1 kg equals 9.81 N so that for practical purposes, to convert from kilograms force to newtons, multiply by 10.)

Specific heat capacity

This is the quantity of heat required to raise a mass of a substance through 1 degree in temperature. It is expressed in joules per degree C (J/°C).

The specific heat capacity is the quantity of heat required to raise unit mass of a substance through 1 °C rise in temperature and the unit is joules per kilogram °C (joules/kg °C). Thus heat capacity = mass × specific heat capacity.

If a body of mass m kg is heated through t °C,

heat supplied = mass × specific heat capacity × rise in temperature.

Example

Find the heat gained by a mass of 20 kg of cast iron which is raised through a temperature of 30 °C, given that the specific heat capacity of cast iron is 0.5×10^3 J/kg°C.

Heat gained = mass × specific heat capacity × rise in temperature
= $20 \times 0.5 \times 10^3 \times 30$ joules
= 3×10^5 joules or 300 kJ.

Substance	Specific heat capacity	Substance	Specific heat capacity
Water	$4.2 \ \times 10^3$	Mild steel	0.45×10^3
Aluminium	0.91×10^3	Wrought iron	0.47×10^3
Tin	0.24×10^3	Zinc	$0.4 \ \times 10^3$
Lead	0.13×10^3	Cast iron	0.55×10^3
Copper	0.39×10^3	Nickel	0.46×10^3
Brass	0.38×10^3		

The above values are approximately 4.2×10^3 as great as the values when specific heat capacities were expressed in calories per gram degree C.

Melting point

The melting point of a substance is the temperature at which the change of state from solid to liquid occurs, and this is usually the same temperature at which the liquid will change back to solid form or freeze.

Substances which expand on solidifying have their freezing point lowered by increase of pressure while others which contract on freezing have their freezing point raised by pressure increase.

The melting point of a solid with a fairly low melting point can be determined by attaching a small glass tube, with open end containing some of the solid, to the bulb of a thermometer. The thermometer is then placed in a container holding a liquid, whose boiling point is above the melting point of the solid, and fitted with a cover, as shown in Fig. 1.8, and a stirrer is also included. The container is heated and the temperature at which the solid melts is observed. The apparatus is now allowed to cool and the temperature at which the substance solidifies is noted. The mean of these two readings gives the melting point of the solid. By using mercury, which boils at 357 °C, as the liquid in the container, the melting point of solids which melt between 100 and 300°C could be obtained.

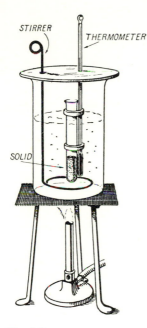

Fig. 1.8

Determination of the melting point by method of cooling

The solid, of which the melting point is required, is placed in a suitable container, fitted with a cork or stopper through which a thermometer is inserted (Fig. 1.9). A hole in the stopper prevents pressure rise. The container is heated until the solid melts, and heating is continued until the temperature is raised well above this point. The liquid is now allowed to cool and solidify and the temperature is taken every quarter or half

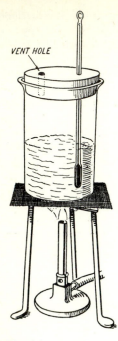

VENT HOLE

Fig. 1.9

minute. This temperature is plotted on a graph against the time, and the shape of the graph should be as shown in Fig. 1.10.

If the melting point of a metal is required, the metal is placed in a fireclay or graphite crucible and heated by means of a furnace, and the temperature is measured, at the same intervals, by a pyrometer. The

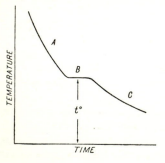

Fig. 1.10

metal, on cooling, begins to solidify and form crystals in exactly the same way as any other solid. The portion A shows the fall in temperature of the liquid or molten metal. The portion B indicates the steady temperature while solidification is taking place, and portion C shows the further fall in temperature as the solid loses heat. The temperature $t°$ of the portion B of the curve is the melting point of the solid.

In practice we may find that the temperature falls below the dotted line, as shown, that is, below the solidifying temperature. This is due to the difficulty which the liquid may experience in commencing to form crystals, and is called 'super-cooling'. It then rises again to the true solidifying point and cooling then takes place as before (Fig. 1.11).

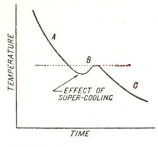

Fig. 1.11

This method of determination of the melting point is much used in finding the melting point of alloys and in observing the behaviour of the constituents of the alloys when melting and solidifying.

The melting point of a metal is of great importance in welding, since, together with the capacity for heat of the metal, it determines how much heat is necessary for fusion. The addition of other substances or metals to a given metal (thus forming an alloy) will affect its melting point.

Specific latent heat

If a block of ice is placed in a vessel with a thermometer and heat is applied, the temperature remains steady at $0°C$ ($273°K$) until the whole of the ice has been melted and then the temperature begins to rise. The heat given to the ice has not caused any rise in temperature but a change of state from solid to liquid and is called the specific latent heat of fusion. When the change of state is from liquid to gas it is termed the specific latent heat of vaporization (or evaporation) and is expressed in joules or kilojoules (kJ) per kilogram (J/kg).

Specific latent heat of fusion in kJ/kg

Aluminium	393	Nickel	273
Copper	180	Tin	58
Iron	205	Ice	333

Specific latent heat of fusion is more important in welding than specific latent heat of vaporization, because a comparison of these figures gives an indication of the relative amounts of heat required to change the solid metal into the liquid state before fusion.

Since heat must be given to a solid to convert it to a liquid, it follows that heat will be given out by the liquid when solidifying. This has already been demonstrated when determining the melting point of a liquid by the method of cooling. When the change of state from liquid to solid takes place (*B* on the curve) heat is given out and the temperature remains steady until solidification is complete, when it again begins to fall.

Transfer of heat

Heat can be transferred in three ways: conduction, convection, radiation.

Conduction. If the end of a short piece of metal rod is heated in a flame, it rapidly gets too hot to hold (Fig. 1.12). Heat has been transferred by conduction from atom to atom through the metal from the flame to the hand. If a rod of copper and one of steel are placed in the flame, the copper rod gets hotter more quickly than the steel one, showing that the heat has been conducted by the copper more quickly than the steel. If the rods are held in a cork and the cork gripped in the hand, they can now be held comfortably. The cork is a bad conductor of heat. All metals are good conductors, but some are better than others, and the rate at which heat is conducted is termed the *thermal conductivity* and is measured in watts per metre degree (W/m °C).

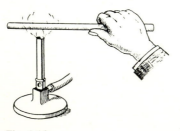

Fig. 1.12

The conductivity depends on the purity of the metal, its structure and the temperature.

As the temperature rises the conductivity decreases and impurities in a metal greatly reduce the conductivity.

The thermal conductivity is closely allied to the electrical conductivity, that is, the ease with which an electric current is carried by a metal. It is interesting to compare the second and third columns in the table. From these we see that in general the better a metal conducts heat, the better it conducts electricity.

Table of comparative conductivities (*taking copper as* 100)

	Thermal conductivity	Electrical conductivity
Silver	106	108
Copper	100	100
Aluminium	62	56
Zinc	29	29
Nickel	25	15
Iron	17	17
Steel	13–17	13–17
Tin	15	17
Lead	8	9

The effect of conductivity of heat on welding practice can clearly be seen from the calculations in Fig. 1.13, where a block of copper and one of steel of equal mass are to be welded. It is seen that if the two blocks were to be each brought up throughout their mass to melting point, the steel would take *a much greater quantity of heat* than the copper would.

When the heat is applied at one spot, copper being such a good conductor, heat is rapidly transferred from this spot throughout its mass, and we find that the spot where the heat is applied will not melt until the whole mass of the copper has been raised to a very high temperature indeed.

With the mild steel block, on the other hand, the heat conductivity is only about $\frac{1}{6}$ (from the table) that of the copper, that is, the heat is conducted away at only $\frac{1}{6}$ the rate. Hence we find that the spot where the heat is applied will be raised to melting point long before the rest of the block has become very hot.

Because of this high conductivity of copper, it is usual to employ greater heat than when welding the same thickness of steel or iron.

For this reason also, when welding copper, whether by arc or oxy-acetylene, it is always advisable to heat the work up to a high temperature over a large area around the area to be welded. In this way the heat will not be conducted to colder regions so rapidly and better fusion in the weld itself can be obtained.

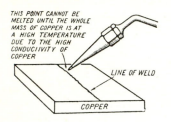 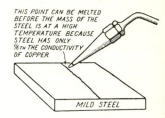

Fig. 1.13

Mass 1 kg
Melting point of copper 1083 °C
Specific heat capacity 0.39 ×
     10^3 J/kg °C
Heat required to raise copper
     to melting point
     = 1 × 1083 × 0.39 × 10^3 J
     = 422 370 J
     = 0.422 × 10^6 J.

Mass 1 kg
Melting point of steel 1400 °C
Specific heat capacity 0.45 ×
     10^3 J/kg °C
Heat required to raise steel to
     melting point
     = 1 × 1400 × 0.45 × 10^3 J
     = 630 000 J
     = 0.63 × 10^6 J.

Cast iron is a comparatively poor conductor of heat compared with copper. If we heat a casting in one spot, therefore, heat will only be transferred away slowly. The part being heated thus expands more quickly than the surrounding parts and, since expansion is irregular, great forces, as before explained, are set up and, since cast iron is brittle and has very small elasticity, the casting fractures. The welding of cast iron is thus a study of expansion and contraction and conduction of heat and, to weld cast iron successfully, care must be taken that the temperature of the whole casting is raised and lowered equally throughout its mass. This will be discussed at a later stage.

Convection. When heat is transferred from one place to another by motion of heated particles, this is termed convection. For example, in the hot water system of a house, heat from the fire heats the water and hot water being less dense than colder water rises in the pipes, forming convection currents and transferring heat to the storage tank.

In the heat treatment of steel it is often necessary to cool the steel slightly more quickly than if it cooled naturally, in order to harden it. It is cooled, therefore, in an air blast, the heat being transferred thus by convection.

Radiation. Heat is transferred by radiation as pulses of energy termed quanta through the intervening space. We sit in front of a fire and it feels warm. There is no physical contact between our bodies and the fire. The heat is being transferred by radiation. Heat transferred in this manner travels according to the laws of light and is reflected and bent in the same way.

The sun's heat is transmitted by radiation to our planet but the method by which the heat travels through space is not fully understood. Metal, if allowed to cool in a still atmosphere, loses its heat by radiation and any other bodies in the neighbourhood will become warmed.

It is evident that the outside of the hot metal will lose heat more quickly than the interior, and we find, for example, that the surface of cast iron is much harder than below the surface, because it has lost heat more quickly.

Chills are strips or blocks of metal placed adjacent to the line of weld during the welding operation in order to dissipate heat and reduce the area affected by the input of heat, the heat-affected zone, HAZ. Heat is removed by conduction, convection and radiation and copper is often used because of its good heat conducting properties. Heat control can be effected by moving the chills nearer or further from the weld.

BEHAVIOUR OF METALS UNDER LOADS

Stress, strain and elasticity

When a force, or load, is applied to a solid body it tends to alter the shape of the body, or deform it.

The atoms of the body, owing to their great attraction for each other, resist, up to a certain point, the attempt to alter their position and there is only a slight distortion of the crystal lattice.

If the applied force is removed before this point is reached, the body will regain its original shape.

This property, which most substances possess, of regaining their original shape upon removal of the applied load is termed *elasticity*.

Should the applied load be large enough, however, the resistance of the atoms will be overcome and they will move and take up new positions in the lattice. If the load is now removed, the body will no longer return to its former shape. It has become permanently distorted (Fig. 1.14).

The point at which a body ceases to be elastic and becomes permanently distorted or set is termed the yield point, and the load which is applied to cause this is the yield-point load. The body is then said to have undergone plastic deformation of flow.

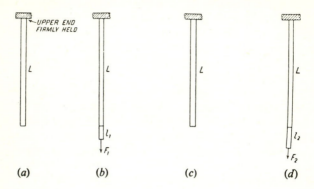

Fig. 1.14. (*a*) Original length of specimen. (*b*) Extension produced = l_1. Elastic limit not reached. F_1 = applied force. (*c*) Force removed, specimen recovers its original dimensions. (*d*) Extension produced = l_2. Elastic limit exceeded by application of force. Specimen now remains permanently distorted or set, and does not recover its original dimensions when force is removed. F_2 = applied force.

Whenever a change of dimensions of a body occurs, from whatever cause, a state of *strain* is set up in that body. Strain is usually measured (for calculation purposes) by the ratio or fraction:

$$\frac{\text{change of dimensions in direction of applied load}}{\text{original dimensions in that direction}}.$$

Example

A bar is 100 mm long and is stretched $\frac{1}{4}$ mm by an applied load along its length. Find the strain.

$$\text{Strain} = \frac{\text{change in length}}{\text{original length}} = \frac{\frac{1}{4}}{100} = \frac{1}{400}.$$

The magnitude of the force or load on unit area of cross-section of the body producing the strain is termed the *stress*.

Stress = force or load per unit area.

Hooke's Law states that for an elastic body strain is proportional to stress.

The mass of a body is the quantity of matter which it contains so that it is dependent upon the number of atoms in its structure. Mass is measured in kilograms (kg) and 1000 grams (g) equal 1 kg. Note 1 lb = 0.4536 kg and 1 kg = 2.2 lb. Newton's Universal Law of Gravitation states that every particle of matter attracts every other particle of matter with a force (F) which is proportional to the product of the masses (m_1 and m_2) of the two particles and inversely proportional to the square of the distance (d) between them $F \propto m_1m_2/d^2$. The weight of a body is the force by which it is attracted to the earth (the force of gravity), but because the earth is a flattened sphere, this force and hence the weight of the body vary somewhat according to its position on the earth's surface. On the surface of the moon, which has about one-sixth of the mass of the earth, a mass of one kilogram would weigh about one-sixth of a kilogram. To distinguish the mass of a kilogram from a force of one kilogram which is the force of attraction due to the gravitational pull of the earth, the letter f is added thus, kgf.

The unit of force termed the newton avoids the distinction between mass and weight and is defined as 'that force which will give an acceleration of 1 metre per second per second to a mass of 1 kilogram'.

Units of stress or pressure
The following multiples of units are used:

tera- (T)	= one million million	10^{12}
giga- (G)	= one thousand million	10^9
mega- (M)	= one million	10^6
kilo- (k)	= one thousand	10^3
hecto- (h)	= one hundred	10^2.

The SI unit of stress or pressure is the newton per square metre (N/m^2) which is also known as the pascal (Pa), and 1 N/m^2 = 1 Pa. This is a small unit and when using it to express tensile strengths of materials large numbers are involved with the use of the mega-newton per square metre (MN/m^2) or megapascal (MPa).

If, however, the newton per square millimetre (N/mm^2) is used, as in this book, large figures are avoided and the change to the SI unit is easily made since 1 N/mm^2 = 1 MN/m^2 or 1 MPa.

The bar (b) and its multiple the hectobar (hbar) are also used as units of pressure and stress. 1 bar is equal to the pressure of a vertical column of mercury 750 mm high and for conversion purposes it can be taken to equal 15 lbf[1] per square inch. It should be noted that 1 bar = 10^5 N/m^2 or 10^5 Pa, and 1 hbar = 10 N/mm^2.

---

[1] 1 bar   = 14.508 lbf/in².
   1 lbf/in² = 0.0689 bar.

Gauges for cylinders of compressed gases can be calibrated in bar, a cylinder pressure of 2500 lbf/in² being 172 bar.

Tensile strength can be expressed in hbar. A specimen of aluminium may have a tensile strength of 12 hbar which is equal to 120 N/mm².

If stress is stated in tonf/in² or kgf/mm² the following conversions can be used. (A full list of conversion factors is given in the appendix.)

Tonf/in² to MN/m², or N/mm², multiply by 15.5; MN/m² or N/mm² to tonf/in², multiply by 0.0647; kgf/m² to N/m² multiply by 9.8; and approximately 1 hbar = 1 kgf/mm².

If a stress is applied to a body and it changes its shape within its elastic limits, the ratio stress/strain is termed the modulus of elasticity or Young's modulus (E) of the material. The unit is N/m² or Pa, and a typical value for a specimen of aluminium is 69×10^3 MN/m² or MPa or N/mm².

There are three kinds of simple stress: (1) tensile, (2) compression, (3) shear.

Tensile stress

If one end of a metal rod is fixed firmly and a force is applied to the other end to pull the rod, it stretches. A tensile force has been applied to the rod and when it is measured on unit cross-sectional area it is termed a tensile stress.

Example

A force of 0.5 MN is applied so as to stretch a bar of cross-sectional area 400 mm². Find the tensile stress.

$$\text{Tensile stress} = \frac{\text{load}}{\text{area of cross-section}} = \frac{500\ 000}{400} = 1250\ \text{N/mm}^2.$$

A machine known as a tensile strength testing machine, which will be described later (Chapter 11), is used for determining the tensile strength of materials and welded joints.

The specimen under test is clamped between two sets of jaws, one fixed and one moving, and the force can be increased until the specimen breaks.

Suppose a piece of mild steel is placed in the machine. As the tensile stress is increased, the bar becomes only very slightly longer for each increase of force. Then a point is reached when, for a very small increase of force, the bar becomes much longer. This is the yield point and the bar has been stretched beyond its elastic limit, and is now deforming plastically.

If the applied load had been reduced before this point was reached, the bar would have recovered its normal size, but will not do so when the yield point has been passed.

As the load is increased beyond the yield point, the elongation of the bar for the same increase of loading becomes much greater, until a point is reached when the bar begins to get reduced in cross-sectional area and forms a waist, as shown. Less load is now required to extend the bar, since the load is now applied on a smaller area, the waist becomes smaller and the bar breaks. The accompanying diagram (Fig. 1.15) will make this clear.

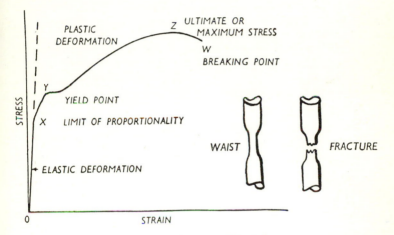

Fig. 1.15. Stress–strain diagram, mild steel.

When the stress is first applied, the extension of the bar is very small and needs accurate measurement, but it is proportional to the load so that the graph of stress/strain is a straight line. At the point X the graph deviates a little from the straight line OX, so that after X the strain is no longer proportional to the stress. The point X is the *limit of proportionality* and Hooke's Law is no longer obeyed. At Y the extension suddenly becomes much greater than before for an equal increase in load and Y is termed the yield point, the stress at this point being the yield-point stress.

Increase of load produces progressive increase of length to the point Z. At this point the waist forms; Z is the maximum load. Breakage occurs at W under a smaller load than at Z. A substance which has a fair elongation during the plastic stage is called ductile, while if the elongation is very small it is said to be brittle.

Given a table of tensile strengths of various metals, we can calculate the maximum force or stress that any given section will stand.

Table of tensile strengths

(The tensile strength of a metal depends upon its condition, whether cast, annealed, work-hardened, heat-treated, etc.)

	N/mm²	hbar	tonsf/in²		N/mm²	hbar	tonsf/in²
Lead	12–22	1.2–2.2	0.8–1.4	Brass	220–340	22–34	14–22
Zinc	30–45	3–4.5	2–3	Cast iron	220–300	22–30	14–20
Tin	30	3	2	Wrought iron	250–300	25–30	16–20
Aluminium	60–90	6–9	4–6	Mild steel	380–450	38–45	25–30
Copper	220–300	20–30	14–20	High tensile steel	600–800	60–80	40–50

Example

A certain grade of steel has a tensile strength of 450 N/mm². What tensile force in newtons will be required to break a specimen of this steel of cross-section 25 mm × 20 mm?

Area of cross-section = 500 mm²

$$\text{Force required} = \text{tensile strength} \times \text{area of cross-section}$$
$$F = 450 \times 500 \text{ newtons}$$
$$= 225\,000 \text{ newtons}$$
$$= 225 \times 10^3 \text{ newtons} = 0.225 \text{ MN.}$$

The tensile strength of a metal depends largely upon the way it has been worked (hammered, rolled, drawn, etc.) during manufacture, its actual composition and the presence of impurities (see Fig. 1.16).

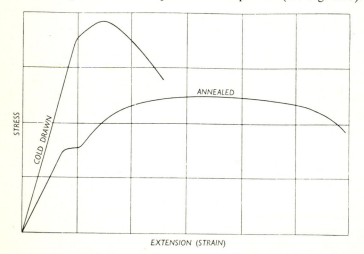

Fig. 1.16. Stress–strain diagram for a steel in (1) annealed, (2) cold drawn condition.

From the tensile test we can obtain:

(1)   Yield point $= \dfrac{\text{yield stress}}{\text{original area of cross-section}}$ (N/mm² or hbar).

(2)   Ultimate tensile stress (UTS)

$= \dfrac{\text{maximum stress}}{\text{original area of cross-section}}$ (N/mm² or hbar).

(3)   Percentage elongation on length between gauge marks

$= \dfrac{\text{extension}}{\text{original length between gauge marks}} \times 100.$

The distance between the gauge marks can be 50 mm or 5 × diameter of the specimen. Standard areas of cross-section can be 75 mm² or 150 mm².

(4)   Percentage reduction of area (R of A)

$= \dfrac{\text{reduction of area at the fracture}}{\text{original area}} \times 100.$

A typical example for one particular grade of weld metal is: Composition: 0.07% C, 0.4% Si, 0.68% Mn, remainder Fe. Yield stress 479 N/mm², ultimate tensile stress 556 N/mm², elongation on gauge length of 5 × D, 26%; reduction of area, 58%.

The elongation will depend on the gauge length. The shorter this is the greater the percentage elongation, since the greatest elongation occurs in the short length where 'waisting' or 'necking' has occurred. Reduction in area and elongation are an indication of the ductility of a metal.

As temperature rises there is usually a decrease in tensile strength and an increase in elongation, and the limit of proportionality is reduced so that at red heat application of stress produces plastic deformation. A fall in temperature usually produces the opposite effect. Internal stresses which have been left in a welded structure can be relieved by heating the members and lowering the limit of proportionality. The stresses then produce plastic deformation, and are relieved. This stress relief, however, may cause distortion.

Proof stress

Non-ferrous metals, such as aluminium and copper, etc., and also very hard steels, do not show a definite yield point, as just explained, and

load–extension curves are shown in Fig. 1.17. A force which will produce a definite permanent extension of 0.1% or 0.2% of the gauge length
is known as the proof stress (Fig. 1.18) and is measured in N/mm² or
hbar.

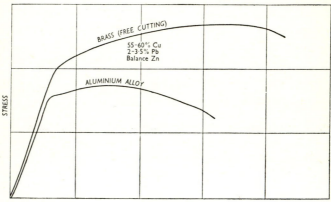

Fig. 1.17

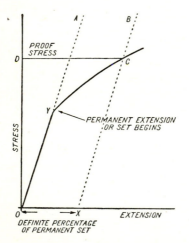

Fig. 1.18. Load–extension curve of hard steels and non-ferrous metals
illustrating proof stress.

Compressive stress

If the forces applied in the previous experiments on tensile strength are reversed, the body is placed under compression.

Compressive tests are usually performed on specimens having a short length compared with their diameter to prevent buckling when the load is applied. Ductile metals increase in diameter to a barrel shape and cracking round the periphery is some indication of the ductility of the specimen. For practical purposes E, the Young's modulus, can be assumed to be the same for compression and tension. Compressive

$$\text{stress} = \frac{\text{compressive load (N)}}{\text{area of cross-section (mm}^2)} \text{ N/mm}^2.$$

A good example of compressive stress is found in building and structural work. All foundations, concrete, bricks and steel columns are under compressive stress, and in the making or fabrication of welded columns and supports, the strength of welded joints in compression is of great importance.

Shearing stress

If a cube has its face fixed to the table on which it stands and a force is applied parallel to the table on one of the upper edges, this force per unit area is termed a shearing stress and it will deform the cube, as indicated by the dotted line (Fig. 1.19). The angle θ through which the cube is deformed is a measure of the shearing strain, while the shearing stress will be in N/mm².

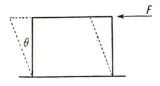

Fig. 1.19

This is a very common type of stress in welded construction. For example, if two plates are lapped over each other and welded, as indicated, then a load applied to the plates as shown puts the welds under a shearing stress. If the load is known and also the shearing strength of the metal of the weld, then sufficient metal can be deposited to withstand the load.

A welded structure should be designed to ensure that there is sufficient area of weld metal in the joint to withstand safely the load required.

Mechanical properties of metals and the effect of heat on these properties

Plasticity may be defined as the ease with which a metal may be bent or moulded into a given shape. At ordinary temperatures, lead is one of the most plastic metals. The plasticity usually increases as temperature rises. Iron and steel are difficult to bend and shape when cold, but it becomes easy to do this when heated above red heat. Wrought iron, however, because of impurities in it, sometimes breaks when we attempt to bend it when hot (called hot shortness), and thus increase of temperatures is not always accompanied by an increase in plasticity.

Brittleness is the opposite of plasticity and denotes lack of elasticity. A brittle metal will break when a force is applied. Cast iron and high carbon steel are examples of brittle metals. The wrought iron in the above paragraph has become brittle through heating. Copper becomes brittle near its melting point, but most metals become less brittle when heat is applied. Carbon steel is an example; when cold it is extremely brittle, but can easily be bent and worked when hot. Brittle metals require care when welding them, due to the lack of elasticity.

Malleability is the property possessed by a metal of becoming permanently flattened or stretched by hammering or rolling. The more malleable a metal is, the thinner the sheets into which it can be hammered. Gold is the most malleable metal (the gold in a sovereign can be hammered into 4 m^2 of gold leaf, less than 0.0025 mm thick).

Copper is very malleable, except near its melting point, while zinc is only malleable between 140 and 160 °C. Metals such as iron and steel become much more malleable as the temperature rises and are readily hammered and forged.

The presence of any impurities greatly reduces the malleability, as we find that the metal cracks when it stretches.

Order of malleability when cold

(1) Gold	(3) Aluminium	(5) Tin	(7) Zinc
(2) Silver	(4) Copper	(6) Lead	(8) Iron.

Ductility is the property possessed by a substance of being drawn out into a wire and it is a property possessed in the greatest degree by certain metals. Like malleability this property enables a metal to be deformed mechanically. Metals are usually more ductile when cold, and thus wire drawing and tube drawing are often done cold, but not always.

In the wire-drawing operation, wire is drawn through a succession of tapered holes called *dies*, each operation reducing the diameter and increasing the deformation of the lattice structure. The brittleness thus increases and the wire must be softened again by a process termed annealing.

Order of ductility

(1) Gold	(3) Iron	(5) Aluminium	(7) Tin
(2) Silver	(4) Copper	(6) Zinc	(8) Lead.

Tenacity is another name for tensile strength. The addition of various substances to a metal may increase or decrease its tensile strength. Sulphur reduces the tenacity of steel while carbon increases it (see section on Tensile Strength).

Hardness is the property possessed by a metal of resisting scratching or indentation. It is measured on various scales, the most common of which are: (1) Brinell, (2) Rockwell, (3) Vickers.

Table of comparative hardness

Material	Brinell	Vickers	Material	Brinell	Vickers
Lead	6	6	Brass 70/30,		
Tin	14	15	annealed	60	64
Aluminium			rolled	150	162
pure			Cast iron	150–250	160–265
annealed	19	20	Mild steel	100–120	108–130
Zinc	45	48	Stainless steel	150–165	160–180
Copper, cast	40–45	42–48			
cold worked	80–100	85–108			

Hardness decreases with rise in temperature. The addition of carbon to steel greatly increases its hardness after heat treatment, and the operations of rolling, drawing, pressing and hammering greatly affect it

It will be noted that there is considerable latitude in the higher figures. Copper, for example, varies from 40 to 100 according to the way it is prepared. Copper is hardened by cold working, that is drawing, pressing and hammering, and this also decreases its ductility.

The tensile strength of steels can be approximately determined in N/mm^2 by multiplying their Brinell hardness figure by 3.25 for hard steels and by 3.56 for those in the soft or annealed condition.

Creep

This is the term applied to the gradual change in dimensions which occurs when a load (tensile, compressive, bending, etc.) is applied to a specimen for a long period of time. Creep generally refers to the extension which occurs in a specimen to which a steady tensile load is applied over a period of weeks and months. In these tests it is generally found that the specimen shows greater extension for a given load over a long period than for a short period and may fracture at a load much less than its usual tensile load. The effect of creep is greater at elevated temperatures and is important as for example in pipes carrying high-pressure steam at high superheat temperatures. In creep testing, the specimen is surrounded by a heating coil fitted with a pyrometer. The specimen is heated to a given temperature, the load is applied and readings taken of the extension that occurs over a period of weeks, a graph of the results being made. The test is repeated for various loads and at various temperatures.

Special electrodes usually containing molybdenum are supplied for welding 'creep-resisting' steels, that is, steels which have a high resistance to elongation when stresses are applied for long periods of time at either ordinary or elevated temperatures.

Fatigue

Fatigue is the tendency which a metal has to fail under a rapidly alternating load, that is a load which acts first in one direction, decreases to zero and then rises to a maximum in the opposite direction, this cycle of reversals being repeated a very great number of times. If the stress is plotted against the number of stress reversals, the curve first falls steadily and then runs almost parallel to the stress reversal axis. The stress at which the curve becomes horizontal is the fatigue limit (Fig. 1.20). The load causing failure is generally much less than would cause failure if it

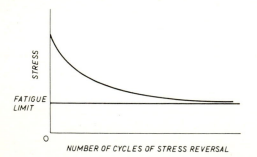

Fig. 1.20. Number of cycles of stress reversal–stress curve.

was applied as a steady load. Many factors, such as the frequency of the applied stress, temperature, internal stresses, variation in section, and sharp corners leading to stress concentration, affect the fatigue limit. Methods of fatigue testing are given in Chapter 11.

CHEMISTRY APPLIED TO WELDING

Elements, compounds and mixtures

All substances can be divided into two classes: (1) elements, (2) compounds.

An element is a simple substance which cannot be split up into anything simpler. For example, aluminium (Al), copper (Cu), iron (Fe), tin (Sn), zinc (Zn), sulphur (S), silicon (Si), hydrogen (H), oxygen (O) are all *elements*.

A table of the elements is given in the appendix, together with their chemical symbols.

A compound is formed by the chemical combination of two or more elements, and the property of the compound differs in all respects from the elements of which it is composed.

We have already mentioned the occurrence of matter in the form of molecules, and now it will be well to consider how these molecules are arranged among themselves and how they are made up.

If a mixture of iron filings and sand is made, we can see the grains of sand among the filings with the naked eye. This mixture can easily be separated by means of a magnet, which will attract the iron filings and leave the sand. Similarly, a mixture of sand and salt can be separated by using the fact that salt will dissolve in water, leaving the sand. In the case of mixtures, we can always separate the components by such simple means as this (called mechanical means).

Similarly, a mixture of iron filings and powdered sulphur can be separated, either by using a magnet or by dissolving the sulphur in a liquid such as carbon disulphide, in which it dissolves readily.

Now suppose we heat this mixture. We find that it first becomes black and then, even after removing the flame, it glows like a coal fire and much heat is given off. After cooling, we find that the magnet will no longer attract the black substance which is left, neither will the liquid carbon disulphide dissolve it. The black substance is, therefore, totally different in character from the iron filings or the sulphur. It can be shown by chemical means that the iron and sulphur are still there, contained in the black substance. This substance is termed a chemical compound and is called iron sulphide. It has properties quite different from those of iron and sulphur.

Previously it has been stated that molecules can be sub-divided. Molecules are themselves composed of atoms, and the number of atoms contained in each molecule depends upon the substance.

For example, a molecule of the black iron sulphide has been formed by the combination of one atom of iron and one atom of sulphur joined together in a chemical bond. This may be written:

$$\text{\textcircled{Fe}} + \text{\textcircled{S}} = \text{\textcircled{Fe}} — \text{\textcircled{S}}$$

iron + sulphur → iron sulphide (FeS).

The molecules of some elements contain more than one atom. A molecule of hydrogen contains two atoms, so this is written: $H_2 = O \underset{H \quad H}{——} O$. Similarly, a molecule of oxygen contains two atoms, thus: $O_2 = O \underset{O \quad O}{——} O$.

A molecule of copper contains only one atom, thus: $Cu = \dfrac{O}{Cu}$.

The atmosphere

Let us now study the composition of the atmosphere, since it is of primary importance in welding.

Suppose we float a lighted candle, fastened on a cork, in a bowl of water and then invert a glass jar over the candle, as shown in Fig. 1.21. We find that the water will gradually rise in the jar, until eventually the candle goes out. By measurement, we find that the water has risen up the jar $\frac{1}{5}$ of the way, that is, $\frac{1}{5}$ of the air has been used up by the burning of the candle, while the remaining $\frac{4}{5}$ of the air still in the jar will not enable the candle to continue burning. The gas remaining in the jar is nitrogen (Fig. 1.22). It has no smell, no taste, will not burn and does not support burning. The gas which has been used up by the burning candle is oxygen.

Evidently, then, air consists of four parts by volume of nitrogen to one part of oxygen. That oxygen is necessary for burning is very evident. Sand thrown on to a fire excludes the air, and thus the oxygen, and the fire is extinguished. If a person's clothes catch fire, rolling him in a blanket or mat will exclude the oxygen and put out the flames. In addition to oxygen and nitrogen the atmosphere contains a small percentage of carbon dioxide and also small percentages of the inert gases first discovered and isolated by Rayleigh and Ramsay. These gases are argon, neon, krypton and xenon. An inert gas is colourless, odourless, and tasteless, it is not combustible neither does it support combustion and it does not enter into chemical combination with other elements. Argon, which is present in greater proportion than the other inert gases is used as the gaseous shield in 'inert gas welding' because it forms a protective shield around the arc and prevents the molten metal from combining with the oxygen and nitrogen of the atmosphere. Helium, which is the lightest of the inert gases occurs only about 1 part in 2 000 in the atmosphere but occurs in association with other natural gases in large quantities especially in the United States, where it is often used instead of argon. The various gases of the atmosphere are extracted by fractionation of liquid air.

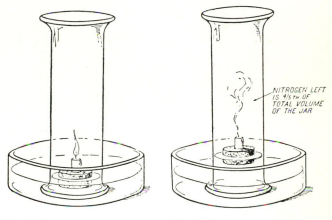

Fig. 1.21   Fig. 1.22

The percentage composition, by volume, of dry air of the Earth's atmosphere is: Nitrogen 78.1, Oxygen 20.9, Argon 0.93, Neon 0.0018, Krypton 0.000 14, Xenon 0.000 008 6, Helium 0.000 52, Carbon dioxide 0.03, Hydrogen 0.000 05, Ozone 0.000 04, Methane 0.000 15, Nitrous oxide 0.000 05.

Nitrogen

Nitrogen is a colourless, odourless, tasteless gas, boiling point -195.8 °C, which does not burn or support combustion. It is di-atomic with an atomic weight of 14, and dissociates in the heat of the arc to form iron nitride which reduces the ductility of a steel weld. For this reason it is not used as a shielding gas to any extent. It also forms nitrogen dioxide NO_2 and nitric oxide NO, which are toxic. It is widely dispersed in compound form in nitrates, ammonia and ammonium salts. It is produced by the liquefaction of air, and the considerable volumes produced by plants such as those supplying tonnage oxygen to steel plants can be used as the top pressure gas in blast furnaces and for the displacement of air in tanks, pipelines, etc.

Nitrogen is supplied in compressed form in steel cylinders of 6.2 and 4.6 m³ (220 and 165 ft³) capacity at a pressure of 13.5 N/mm², and in liquid form by bulk tankers to an evaporator which in turn feeds gas into a pipeline. (See liquid oxygen.)

Argon

This mon-atomic gas, chemical symbol Ar and atomic weight 18, is present in the atmosphere to the extent of about 1% and is obtained by fractional distillation from liquid air. It has no taste, no smell, is non-toxic, colourless and neither burns nor supports combustion. It does not form chemical compounds and has special electrical properties. It is extensively used in welding either on its own or mixed with carbon dioxide or hydrogen in the welding of aluminium, magnesium, titanium, copper, stainless steel and nickel by the TIG and MIG processes and in plasma welding of stainless steel, nickel and titanium, etc. Argon is used for the inert gas filling of electric lamps and valves, with nitrogen, and in metal refining and heat treatment, for inert atmospheres. It is supplied in compressed form in steel cylinders of 8.5 m³ or 68 m³ capacity at a pressure of 17.2 N/mm² and in liquid form delivered by road tankers which pump it directly into vacuum-insulated storage vessels as for liquid oxygen (q.v.).

Helium

Helium is an inert gas present in the atmosphere to an extent of 0.000 052%. It is obtained from underground sources in the USA and is very much more expensive in this country than argon. It is mon-atomic with an atomic weight 4, and is thus lighter than argon. Like the other inert gases it is colourless, odourless and tasteless, does not burn nor support combustion, is non-toxic and does not form chemical compounds. Because of its lightness, a flow rate of 2 to $2\frac{1}{2}$ times that of argon is required to provide an efficient gas shield in inert gas welding processes.

Carbon dioxide CO_2

Carbon dioxide is now extensively used as a shielding gas in the gas shielded metal arc welding process. It is a non-flammable gas of molecular weight of 44.01, with a slightly pungent smell and is about $1\frac{1}{2}$ times as heavy as air (specific gravity relative to air is 1.53). It is soluble in water, giving carbonic acid H_2CO_3, and it can be readily liquefied, the liquid being colourless; the critical temperature (that is the temperature above which it is impossible to liquefy a gas by increasing the pressure) is 31.02

°C. Because its heat of formation is high it is a stable compound, enabling it to be used as a protective shield around the arc to protect the molten metal from contamination by the atmosphere, and it can be mixed with argon for the same purpose. During the CO_2 shielded metal arc process some of the molecules will be broken down or dissociated to form small quantities of carbon monoxide and oxygen. The carbon monoxide recombines with oxygen from the atmosphere to form CO_2 again and only very small quantities (the generally accepted threshold is 50 p.p.m.) escape into the atmosphere and the oxygen is removed by powerful deoxidizers in the welding wire. The gas is very much cheaper than argon; it is not an inert gas.

Carbon dioxide is formed when limestone is heated strongly in the lime kiln and also by the action of hydrochloric acid on limestone. It may be obtained as a by-product in the production of nitrogen and hydrogen in the synthesis of ammonia and also as a by-product in the fermentation process when yeast acts on sugar or starch to produce alcohol and carbon dioxide.

Large supplies for industrial use may be obtained by burning oil, coke or coal in a boiler. The steam generated can be used for driving prime movers for electricity generation and the flue gases, consisting of CO_2, nitrogen and other impurities, are passed into a washer where the impurities are removed and then into an absorber where the CO_2 is absorbed and the nitrogen thus separated. The absorber containing the CO_2 passes into a stripping column where the CO_2 is removed and water vapour set free, and is removed by a condenser. The CO_2 is then stored in a gas-holder from which it passes through a further purifying process, is then compressed in a compressor, passed through a drier and a condenser and stored in the liquid state at a pressure of 2 N/mm^2 at a temperature of -18 °C, the storage tank being well insulated. The liquid CO_2 is then pumped into the cylinders used for welding purposes or into bulk supply tanks, or it may be further converted into the solid state (Cardice) which has a surface temperature of -78.4 °C.

The use of CO_2 as a shielding gas is fully discussed in the chapter on this process and, in addition to this, the following are the main uses of the gas at the present time: in nuclear power stations, where it can be used for transference of heat from the reactor to the electricity generating unit; for the CO_2–silicate process in the foundry for core and mould making; for the soft drink trade where the gas is dissolved under pressure in the water of the mineral water or beer and gives a sparkle to the drink when the pressure is released; and in the solid state for refrigerated transport, the perishable foodstuffs being packed in heavily insulated containers with the solid CO_2 which evaporates to the gaseous state and leaves no residue.

Oxygen

In view of the importance of oxygen to the welder, it will be useful to prepare some oxygen and investigate some of its properties.

Place a small quantity of potassium chlorate in a hard glass tube (test tube) and heat by means of a gas flame. The substance melts, accompanied by crackling noises. Now place a glowing splinter in the mouth of the tube. The splinter bursts into flame and burns violently (Fig. 1.23). Oxygen is being given off by the potassium chlorate and causes this violent burning. The glowing splinter test should *not* be used for testing for an escape of oxygen from welding plant.

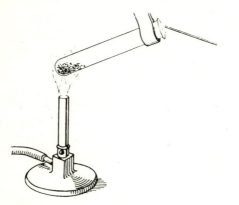

Fig. 1.23. Test for oxygen, glowing splinter bursts into flame.

Oxygen is prepared on a commercial scale by one of two methods: (1) liquefaction of air, (2) electrolysis of water. In the first method air is liquefied by reducing its temperature to about $-140\,°C$ and then compressing it to a pressure of 40 bar (4 N/mm^2). The pressure is then reduced and the nitrogen boils off first, leaving the liquid oxygen behind. This is then allowed to boil off into its gaseous form and is compressed into steel cylinders at 175 bar (17.5 N/mm^2).

The second method is generally used when there is a plentiful supply of cheap water power for generating electricity. An electric current is passed through large vats containing water, the current entering at the anode (positive) and leaving at the cathode (negative). The passage of the current splits up the water into hydrogen and oxygen. The hydrogen is collected from the cathode and the oxygen from the anode, there being twice the volume of hydrogen evolved as oxygen. (This operation is known as electrolysis.) The gases are then dried, compressed and stored in steel containers, the hydrogen being compressed to 172 bar (17.2 N/mm^2), similar to the oxygen.

Properties of oxygen. Oxygen is a colourless gas of atomic weight 16, boiling point $-183°C$, with neither taste nor smell. It is slightly soluble in water, and this slight solubility enables fish to breathe the oxygen which has dissolved.

Oxygen itself does not burn, but it very readily supports combustion, as shown by the glowing splinter which is a test for oxygen.

If a piece of red-hot iron is placed in oxygen it burns brilliantly, giving off sparks. This is caused by the iron combining with the oxygen to form an *oxide*, in this case iron oxide (Fe_3O_4).

Oxidation. Most substances combine very readily with oxygen to form oxides, and this process is termed *oxidation*.

Magnesium burns brilliantly, forming a white solid powder, magnesium oxide, i.e.:

$$\text{magnesium} + \text{oxygen} \rightarrow \text{magnesium oxide}$$
$$2Mg + O_2 \rightarrow 2MgO.$$

When copper is heated to redness in contact with oxygen copper oxide is formed:

$$\text{copper} + \text{oxygen} \rightarrow \text{copper oxide}$$
$$2Cu + O_2 \rightarrow 2CuO.$$

Similarly, phosphorus burns with a brilliant flame and forms phosphorus oxide (P_2O_5). Sulphur burns with a blue flame and forms the gas, sulphur dioxide (SO_2).

Silicon, if heated, will combine with oxygen to form silica (SiO_2), which is sand:

$$\text{silicon} + \text{oxygen} \rightarrow \text{silica or oxide of silicon}$$
$$Si + O_2 \rightarrow SiO_2.$$

Burnt dolomite, used as a refractory lining in the basic steel making process, is formed of magnesium and calcium oxides MgO.CaO.

When a chemical action takes place and heat is given out it is termed an exothermic reaction. The combination of iron and sulphur (p. 33), silicon and oxygen, and aluminium and iron oxide (p. 43) are examples. If heat is taken in during a reaction it is said to be endothermic. An example is the reaction which occurs when steam is passed over very hot coke. The oxygen combines with the carbon to form carbon monoxide and hydrogen is liberated, the mixture of the two gases being termed water gas, or:

$$\text{steam} + \text{carbon} \rightarrow \text{carbon monoxide} + \text{hydrogen}$$
$$H_2O + C \rightarrow CO + H_2.$$

The rusting of iron. Moisten the inside of a glass jar so that small iron filings will adhere to the interior surface and invert the jar over a bowl of water, thus entrapping some air inside the jar (Fig. 1.24).

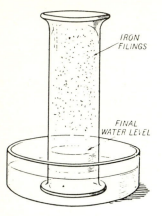

IRON
FILINGS

FINAL
WATER LEVEL

Fig. 1.24

If the surface of the water inside the jar is observed, it is seen that as time passes and the iron filings become rusty, the surface of the water rises and eventually remains stationary at a point roughly $\frac{1}{5}$ of the way up the jar. From the similar experiment performed with the burning candle it can be seen that the oxygen has been used up as the iron rusts and nitrogen remains in the jar. The rusting of iron is, therefore, a process of surface oxidation.

This can further be demonstrated as follows: boil some water for some time in a glass tube (or test tube), in order to expel any dissolved oxygen, and then place a brightly polished nail in the water. Seal the open end of the tube by pouring melted vaseline down on the surface of the water. The nail will now keep bright indefinitely, since it is completely out of contact with oxygen.

Oxidation, from the welder's point of view, is the union of a metal with oxygen to form an oxide, i.e.

metal + oxygen → metallic oxide.

Oxygen reacts with metals in various ways, depending on:

(1) *The character of the metal.* Magnesium burns very completely to form magnesium oxide, while copper, aluminium and chromium form a protective oxide film on their surface at room temperature.

(2) *Temperature*. Zinc at normal temperature only oxidizes slowly on the surface, but if heated to high temperatures it burns with a bright bluish-white flame, forming a white powder, zinc oxide. Nearly all base metals can be converted to their oxide by heating them in oxygen.

(3) *The amount of surface exposed*. The larger the surface area the greater the amount of oxidation.

(4) *The amount of oxygen present*. Oxidation is much more rapid, for example, in a stream of pure oxygen than in air.

(5) *Presence of other substances*. Iron will not rust if no water is present.

Let us now examine the extent to which the more important metals in welding react with oxygen.

Iron and steel. If iron is excessively heated, oxygen is absorbed and oxidation or burning takes place, forming magnetic oxide of iron:

Iron + oxygen → magnetic oxide of iron
$3Fe + 2O_2 \rightarrow Fe_3O_4$.

There are two other oxides of iron, ferric oxide (F_2O_3) or haematite, which is one of the sources of iron from the earth, and ferrous oxide (FeO), which is a black powder which takes fire when heated in air and forms ferric oxide.

Copper is extremely resistant to atmospheric corrosion, since it forms a film of oxide on its surface. This film is very unlike rust on iron, because it protects the metal and offers high resistance to any further attack. In time the oxide becomes changed to compounds having a familiar green colour such as sulphate of copper. When copper is brightly polished and exposed to a clean, dry atmosphere it tarnishes and becomes coated with a thin film of cuprous oxide (Cu_2O). If the temperature of the copper is now raised, the amount of oxidation increases proportionally and at high temperatures the copper begins to scale. The black scale formed is cupric oxide (CuO), while underneath this is another film of cuprous oxide (Cu_2O), which has a characteristic red colour.

Aluminium has a great affinity for oxygen and is similar to copper in that it forms a protective coating (of aluminium oxide, Al_2O_3) on its surface, which protects it against further attack. The depth of the film of oxide formed will depend upon the amount of corrosion, since the film adjusts itself to the amount of corrosive influences.

As the temperature increases little alteration takes place until near the melting point, when the rate of oxidation increases rapidly. It is the formation of this oxide which makes the welding of aluminium almost impossible unless a chemical (termed a flux) is used to dissolve it or an inert gas shield is used to prevent oxidation.

During the welding process, therefore, combination of the metal with oxygen may:

(1) Produce a gaseous oxide of a metal present in the weld and thus produce blow or gas holes.

(2) Produce oxides which, having a melting point higher than that of the surrounding metal, will form solid particles or *slag* in the weld metal.

(3) Produce oxides which will dissolve in the molten metal and make the metal brittle and weak. (The oxide in this case may form along the boundaries of the crystals of the metal.)

Some oxides are heavier than the parent metal and will tend to sink in the molten weld. Others are lighter and will float to the top. These are less troublesome, since they are easier to remove.

Oxides of wrought iron and steel, for example, melt very much below the temperature of the parent metal and, being light, float to the surface as a scale. Thus, if care is taken in the welding process, the oxide is not troublesome.

In the case of cast iron, however, the oxide melts at a temperature above that of the metal; consequently, it would form solid particles in the weld if not removed. For this reason a 'flux' is used which combines with the oxide and floats it to the surface. In welding copper, aluminium, nickel and brass, for example, a flux must be used to remove the oxides formed (see pp. 56–7).

The two most common causes of oxidation in welding are by absorption of oxygen from the atmosphere, and by use of an incorrect flame with excess oxygen in gas welding.

Reduction or deoxidation. Reduction takes places when oxygen is removed from a substance. Evidently it is always accompanied by oxidation, since the substance that removes the oxygen will become oxidized.

The great affinity of aluminium for oxygen is made use of in the thermit process of welding and provides an excellent example of chemical *reduction*.

Suppose we mix some finely divided aluminium and finely divided iron oxide in a crucible or fireclay dish. Upon setting fire to this mixture it burns and great heat is evolved with a temperature as high as 3000 °C. This is due to the fact that the aluminium has a greater affinity for oxygen than the iron has, when they are hot, and as a result the aluminium combines with the oxygen taken from the iron oxide. Thus the pure iron is set free in the molten condition. The action is illustrated as follows:

$$\text{iron oxide} + \text{aluminium} \rightarrow \text{aluminium oxide} + \text{iron}$$
$$Fe_2O_3 + 2Al \rightarrow Al_2O_3 + 2Fe.$$

This is the chemical action which occurs in an incendiary bomb. The detonator ignites the ignition powder which sets fire to the thermit mixture. This is contained in a magnesium–aluminium alloy case (called Elektron) which also burns due to the intense heat set up by the thermit reaction.

Since oxygen has been taken from the iron, the iron has been *reduced* or deoxidized and the aluminium is called the reducing agent. To prevent oxidation taking place in a weld, silicon and manganese are used as deoxidizers. More powerful deoxidizers such as aluminium, titanium and zirconium (triple deoxidized) are added when oxidizing conditions are more severe, as for example in CO_2 welding and also in the flux cored continuous wire feed process without external gas shield, and the deoxidizers control the quality of the weld metal.

Note. Hydrogen is an electro-positive element, while oxygen is an electro-negative element. Therefore, oxidation is often spoken of as an increase in the ratio of the electro-negative portion of a substance, while reduction is an increase in the ratio of the electro-positive portion of a substance.

Examples of:

Oxidizing agents	*Reducing agents*
(1) Oxygen	(1) Hydrogen
(2) Ozone	(2) Carbon
(3) Nitric acid	(3) Carbon monoxide
(4) Chlorine	(4) Sulphur dioxide (at low temperatures)
(5) Potassium chlorate	(5) Sulphuretted hydrogen
(6) Potassium nitrate	(6) Zinc dust
(7) Manganese dioxide	(7) Aluminium.
(8) Hydrogen peroxide	
(9) Potassium permanganate	

Note. Dry SO_2 at welding temperatures behaves as an oxidizing agent and oxidizes carbon to carbon dioxide and many metallic sulphides to sulphates.

Acetylene

Acetylene is prepared by the action of water on calcium carbide (CaC_2). The carbide is made by mixing lime (calcium oxide) and carbon in an electric arc furnace. In the intense heat the calcium of the lime combines with the carbon, forming calcium carbide and, owing to the high temperature at which the combination takes place, the carbide is very hard and brittle. It contains about 63% calcium and 37% carbon by weight and readily absorbs moisture from the air (i.e. it is hygroscopic); hence it is essential to keep it in airtight containers. The reaction is:

calcium oxide
or quicklime + carbon → calcium carbide + carbon monoxide
$$CaO \quad + \quad 3C \quad \rightarrow \quad CaC_2 \quad + \quad CO.$$

The carbon monoxide burns in the furnace, forming carbon dioxide.

When water acts on calcium carbide, the gas acetylene is produced and slaked lime remains:

calcium carbide + water → acetylene + slaked lime
$$CaC_2 \quad + \quad 2H_2O \rightarrow \quad C_2H_2 \quad + \quad Ca(OH)_2.$$

Acetylene is a colourless gas, slightly lighter than air, only very soluble in water, with a pungent smell largely due to impurities. It burns in air with a sooty flame but when burnt in oxygen the flame has a bright blue inner cone. It can be ignited by a spark or even by hot metal and forms explosive compounds with copper and silver so that copper pipes and fittings should never be used with it.

If compressed it is explosive but it is very soluble in acetone which can dissolve 300 times its own volume at a pressure of 1.2 N/mm^2 or 12 bar and this is the method used to store it in the dissolved acetylene cylinders used in oxy-acetylene welding.

Liquid petroleum gas (LPG). Propane C_3H_8, butane C_4H_{10}. Propane is a flammable gas used as a fuel gas with either air or oxygen for heating and cutting operations. Its specific gravity compared with air is 1.4–1.6 (butane 1.9–2.1) so that any escaping gas collects at ground level and an artificial stenchant is added to the gas to warn personnel of its presence since it acts as an asphyxiant. Its boiling point at a pressure of 1 atmosphere is $-42\,°C$ (butane $-7\,°C$) and the air-propane and oxy-propane flames have a greater calorific value than air-natural gas or oxy-natural gas for the same conditions of operating pressures. Flame temperature and hence cutting speeds are lower for oxy-propane than oxy-acetylene but propane is considerably cheaper than acetylene.

Note. The oxy-propane flame cannot be used for welding.

Propane burns in air to form carbon dioxide and water and is supplied in steel cylinders painted red in weights 4.8–47 kg, being sold by weight. It is also supplied by tanker to bulk storage tanks in a similar way to oxygen and nitrogen. Liquid Natural Gas (LNG) is similarly supplied.

Carbon

Carbon is of great importance in welding, since it is present in almost every welding operation. It is a non-metallic element, and is remarkable in that it forms about half a million compounds, the study of which is termed *organic chemistry*.

Carbon can exist in three forms. Two of these forms are crystalline, namely diamond and graphite, but the crystals of a diamond are of a different shape from those of graphite. (Carbon is found in grey cast iron as graphite.) Ordinary carbon is a third form, which is non-crystalline or *amorphous*. Carbon forms with iron the compound ferric carbide, Fe_3C, known as cementite. The addition of carbon to pure iron in the molten state is extremely important, since the character of the iron is greatly changed. Diamond and graphite are allotropes of carbon. Allotropy is the existence of an element in two or more forms.

Carbon is found in organic compounds such as acetylene (C_2H_2), petrol (C_6H_{14}), sugar ($C_{12}H_{22}O_{11}$), ethyl alcohol (C_2H_5OH), propane (C_3H_8), butane (C_4H_{10}), methane (CH_4), natural gas etc.

Graphite used to be considered as a lead compound, but it is now known that it is a crystalline form of carbon. It is greasy to touch and is used as a lubricant and for making pencils.

The oxides of carbon. Carbon dioxide (CO_2) is heavier than air and is easily liquefied. It is formed when carbon is burnt in air; hence is present when any carbon is oxidized in the welding operation.

$$\text{carbon} + \text{oxygen} \rightarrow \text{carbon dioxide}$$
$$C + O_2 \rightarrow CO_2.$$

It will not burn, neither will it support combustion. It turns lime water milky, and this is the usual test for it. When it dissolves in the moisture in the air or rain it forms carbonic acid, which hastens corrosion on steel (see p. 54).

Carbon monoxide is formed when, for example, carbon dioxide is passed through a tube containing red-hot carbon:

$$\text{carbon} + \text{carbon dioxide} \rightarrow \text{carbon monoxide}$$
$$C + CO_2 \rightarrow 2CO.$$

Hence it may be formed from carbon dioxide during the welding process. It is a colourless gas which burns with a blue, non-luminous flame. It is not soluble in water, has no smell and is very poisonous, producing a form of asphyxiation. Exhaust fumes from petrol engines contain a large proportion of carbon monoxide and it is the presence of this that makes them poisonous.

Carbon monoxide readily takes up oxygen to form carbon dioxide. It is thus a reducing agent and it can be made to reduce oxides of metals to the metals themselves.

The following poisonous gases may be formed during welding operations depending upon the process used, the material being welded, its coating and the electrode type:

Gas	*Example of formation*
Carbon monoxide	In CO_2 welding due to dissociation of some of the CO_2 in the heat of the arc.
Oxides of nitrogen (Nitric oxide, NO, Nitrogen dioxide, NO_2)	Oxygen and nitrogen from the air combining, due to the heat and radiation from the arc.
Ozone, O_3	Due to oxygen in the atmosphere being converted to ozone by the ultra-violet radiation from the arc.
Phosgene, $COCl_2$	When trichloroethylene ($CHCl.CCl_2$), used for degreasing, is heated or exposed to ultra-violet radiation from the arc.

Non-poisonous gases such as carbon dioxide and argon act as asphyxiants when the oxygen content of the atmosphere falls below about 18%. Fumes and pollutant gases also occur during welding due for example to the break-up of the electrode coating in metal arc welding and the vaporization of some of the metal used in the welding process. The concentrations of these are governed by a Threshold Limit Value (TLV) and the limits are expressed in parts per million (p.p.m.).

Combustion or burning

The study of combustion is very closely associated with the properties of carbon. When burning takes place, a chemical action occurs. If a flame is formed, the reaction is so vigorous that the gases become luminous. Hydrogen burns in air with a blue, non-luminous flame to form water. In the oxy-hydrogen flame, hydrogen is burnt in a stream of oxygen. This causes intense heat to be developed, with a flame temperature of about 2800 °C.

The oxy-coal-gas flame is very similar, as the coal gas consists of hydrogen, together with other impurities (methane, carbon monoxide and other hydrocarbons). Because of these impurities, the temperature of this flame is much lower than when pure hydrogen is used. The oxy-acetylene flame consists of the burning of acetylene in a stream of oxygen. Acetylene is composed of carbon and hydrogen (C_2H_2), and it is a gas which burns in air with a very smoky flame, the smoke being due, as in the case of the candle, to incomplete combustion of the carbon:

$$\text{Acetylene} + \text{oxygen} \rightarrow \text{carbon} + \text{water}$$
$$2C_2H_2 + O_2 \rightarrow 4C + 2H_2O.$$

By using, however, a special kind of burner, we have almost complete combustion and the acetylene burns with a very brilliant flame, due to the incandescent carbon.

The oxy-acetylene welding flame

When oxygen is mixed with the acetylene in approximately equal proportions a blue, non-luminous flame is produced, the most brilliant part being the blue cone at the centre. The temperature of this flame is given, with others, in the table:

Temperatures of various flames

Oxy-acetylene	3100 °C
Oxy-butane (Calor-gas)	2820 °C
Oxy-propane (liquefied petroleum gas, LPG)	2815 °C
Oxy-methane (natural gas)	2770 °C
Oxy-hydrogen	2825 °C
Air–acetylene	2325 °C
Air–methane	1850 °C
Air–propane	1900 °C
Air–butane	1800 °C

(Metal arc: 6000 °C upwards depending on type of arc)

This process of combustion occurs in two stages: (1) in the innermost blue, luminous cone, (2) in the outer envelope. In (1) the acetylene combines with the oxygen supplied, to form carbon monoxide and hydrogen:

$$\text{acetylene} + \text{oxygen} \rightarrow \text{carbon monoxide} + \text{hydrogen}$$
$$C_2H_2 + O_2 \rightarrow 2CO + H_2.$$

In (2) the carbon monoxide burns and forms carbon dioxide, while the hydrogen which is formed from the above action combines with oxygen to form water:

$$\text{carbon monoxide} + \text{hydrogen} + \text{oxygen} \rightarrow \text{carbon dioxide} + \text{water}$$
$$CO + H_2 + O_2 \rightarrow CO_2 + H_2O.$$

The combustion is therefore complete and carbon dioxide and water (turned to steam) are the chief products of the combustion. This is shown in Fig. 1.25. If insufficient oxygen is supplied, the combustion will be incomplete and carbon will be formed.

OUTER ENVELOPE

INNER CONE

(*HERE ACETYLENE AND OXYGEN FORM CARBON MONOXIDE AND HYDROGEN IN EQUAL VOLUMES*)

(*HERE THE CARBON MONOXIDE AND HYDROGEN COMBINE WITH OXYGEN FROM THE AIR TO FORM CARBON DIOXIDE AND WATER*)

Fig. 1.25. The oxy-acetylene flame.

From this it will be seen that the oxy-acetylene flame is a strong *reducing* agent, since it absorbs oxygen from the air in the outer envelope. Much of its success as a welding flame is due to this, as the tendency to form oxides is greatly decreased. For complete combustion, there is a correct amount of oxygen for a given amount of acetylene. If too little oxygen is supplied, combustion is incomplete and carbon is set free. This is known as a carbonizing or carburizing flame. If too much oxygen is supplied, there is more than is required for complete combustion, and the flame is said to be an oxidizing flame.

For usual welding purposes the neutral flame, that is neither carbonizing nor oxidizing, is required, combustion being just complete with neither excess of carbon nor oxygen. For special work an oxidizing or carbonizing flame may be required, and this is always clearly indicated.

Silicon (Si)

Silicon is an element closely allied to carbon and is found in all parts of the earth in the form of its oxide, silica (SiO_2). In its free state, silica is found as quartz and sand. Silicon is also found combined with certain other oxides of metals in the form of silicates. Silicates of various forms are often used as the flux coverings for arc-welding electrodes and are termed 'siliceous matter'.

Silicon exists either as a brown powder or as yellow-brown crystals. It combines with oxygen, when heated, to form silica, and this takes place during the conversion of iron and steel. Silicon is present, mixed in small proportions with the iron, and, when oxygen is passed through the iron in the molten state, the silicon oxidizes and gives out great heat, an exothermic reaction.

Silicon is important in welding because it is found in cast iron (0.5 to 3.5%) and steel and wrought iron (up to 0.1%). It is found up to 0.3% in steel casings since it makes the steel flow easily in the casting process.

It is particularly important in the welding of cast iron, because silicon aids the formation of graphite and keeps the weld soft and machinable. If the silicon is burnt out during welding the weld becomes very hard and brittle. Because of this, filler rods for oxy-acetylene welding cast iron contain a high percentage of silicon, being known as 'silicon cast iron rods'. This puts back silicon into the weld to replace that which has been lost and thus ensures a sound weld.

By mixing silica (sand) and carbon together and heating them in an electric furnace, silicon carbide or carborundum is formed:

$$\text{silica} + \text{carbon} = \text{silicon carbide} + \text{carbon monoxide}$$
$$SiO_2 + 3C = SiC + 2CO.$$

Carborundum is used for all forms of grinding operations. Silica bricks, owing to their heat-resisting properties, are used for lining furnaces.

Iron (Fe)

Iron has a specific gravity of 7.8, melts at 1530 °C and has a coefficient of expansion of 0.000 012 per degree C.

Pure iron is a fairly soft, malleable metal which can be attracted by a magnet.

All metallic mixtures and alloys containing iron are termed ferrous, while those such as copper, brass and aluminium are termed non-ferrous.

Iron combines directly with many non-metallic elements when heated with them, and of these the following are the most important to the welder:

With sulphur it forms iron sulphide (FeS).
With oxygen it forms magnetic oxide of iron (Fe_3O_4).
With nitrogen it forms iron nitride (Fe_4N).
With carbon it forms iron carbide (Fe_3C, called cementite).
Steel, for example, is a mixture of iron and iron carbide.

Formation of metallic crystals

We have seen that atoms in solid substances take up regular geometrical patterns termed a space lattice. There are many types of space lattice but atoms of pure metals arrange themselves mainly into three of the simpler forms termed: (1) Body-centred cubic, (2) Face-centred cubic, (3) Hexagonal close packed. These are shown in Fig. 1.26. In body-centred cubic, atoms occupy the eight corners of a cube with one atom in the centre of the cube giving a relatively open arrangement. Face-centred cubic has eight atoms at the corners of a cube with six atoms, one in the centre of each face giving a more closely packed arrangement. Hexagonal close packed is formed by six atoms at the corners of a regular hexagon with one in the centre, placed over a similar arrangement and with three atoms in the hollows separating top and bottom layer. The student can very simply obtain the three-dimensional picture of these arrangements by using ping-pong balls to represent atoms. Copper and

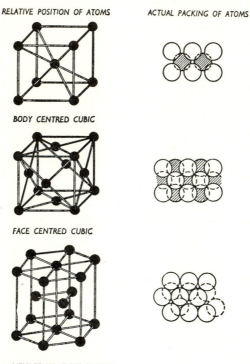

RELATIVE POSITION OF ATOMS   ACTUAL PACKING OF ATOMS

BODY CENTRED CUBIC

FACE CENTRED CUBIC

HEXAGONAL CLOSE PACKED

Fig. 1.26. Types of crystals.

aluminium have a face-centred cubic lattice, magnesium a hexagonal close packed. Iron has a body-centred cubic lattice, below 900 °C (alpha iron, a), this changes to face-centred cubic from 900 °C to 1400 °C (gamma iron, γ), and reverts to body-centred cubic from 1400 °C to its melting point at about 1500 °C (delta iron, δ). These different crystalline forms are allotropic modifications.

When a liquid (or pure molten) metal begins to solidify or freeze, atoms begin to take up their positions in the appropriate lattice at various spots or nuclei in the molten metal, and then more and more atoms add themselves to the first simple lattice, always preserving the ordered arrangement of the lattice, and the crystals thus formed begin to grow like the branches of a tree and, from these arms, other arms grow at right angles, as shown in Fig. 1.27. Eventually these arms meet arms of neighbouring crystals and no further growth outwards can take place.

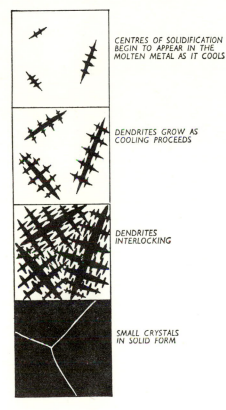

CENTRES OF SOLIDIFICATION BEGIN TO APPEAR IN THE MOLTEN METAL AS IT COOLS

DENDRITES GROW AS COOLING PROCEEDS

DENDRITES INTERLOCKING

SMALL CRYSTALS IN SOLID FORM

Fig. 1.27

The crystal then increases in size, within its boundary, forming a solid crystal, and the junction where it meets the surrounding crystals becomes the crystal or grain boundary. Its shape will now be quite unlike what it would have been if it could have grown without restriction; hence it will have no definite shape.

If we examine a pure metal structure under a microscope we can clearly see these boundary lines separating definite areas (Fig. 1.28), and most pure metals have this kind of appearance, it being very difficult to tell the difference between various pure metals by viewing them in this way. If impurities are present, they tend to remain in the metal which is last to solidify and thus appear between the arms of the dendrites along the grain boundaries.

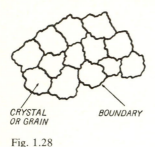

CRYSTAL   BOUNDARY
OR GRAIN

Fig. 1.28

This method of crystallization is termed *dendritic crystallization* and the above crystal is termed a dendrite (Fig. 1.29). It can be observed when frost forms on the window pane, and this gives a good illustration of the method of crystal formation, since the way in which the arms of the dendrite interlock can clearly be seen. (The frost, however, only forms on a flat surface, while the metal crystal forms in three dimensions.) Crystals which are roughly symmetrical in shape are termed equi-axed.

Recrystallization commences at a definite temperature, and if the temperature is increased greatly above this, the grains become much larger in size, some grains absorbing others. Also, the longer that the metal is kept in the heated condition the larger the grains grow. The rate at which cooling occurs in the case of metals also determines the size of the grains, and the slower the cooling the larger the grains (see grain growth, Fig. 2.17). Large, coarse crystals or grains have a bad effect on the mechanical properties of the metal and decrease the strength. If heat conditions are suitable crystals may grow in one direction, then being long and narrow. These are termed columnar crystals. Fig. 2.23a shows a mild steel arc weld run on a steel plate. The upper crystals are small since they cooled quickly or were chilled and had no time to grow. Below these are columnar crystals growing towards the centre of greatest heat. Lower still are equi-axed crystals showing grain growth, while below these is a region of small crystals where recrystallization has occurred.

Cold work distorts the crystals in the direction of the work (see Fig. 2.23a).

(a) Fir tree (dendritic) crystals in the shrinkage cavity of a large carbon steel casting. The crystals have grown mainly in one direction, but the growth of the lateral arms at right angles to the main axes is clearly seen. ($\frac{1}{2}$ actual size.)

Fig. 1.29

(b) Portion of a nickel chrome molybdenum steel ingot showing interlocking of the dendrites. In this case crystallization has started from a series of centres, and the growth of any one dendritic crystal has been restricted by the presence of neighbouring crystals.

The size of the grains, therefore, depends on:

(1) The temperature to which the metal has been raised.
(2) The length of time for which the metal is kept at high temperature.
(3) The rate of cooling.

Crystals of alloys

If one or more metals is added to another in the molten state, they mix together forming a solution, termed an *alloy*.

When this alloy solidifies:

(1) It may remain as a solution, in which case we get a crystal structure similar to that of a pure metal.
(2) The two metals may tend to separate out before solidifying, in which case the crystal structure will be a mixture of the crystals of the two metals, intimately mixed together.

Copper–nickel and chromium–iron are examples of the first kind of crystal formation, while lead–tin and copper–zinc are examples of the second kind.

Welding has a very great effect on the structure and crystal form of metals, and the above brief study will enable the reader to have a clearer understanding of the problem.

Metallic alloys and equilibrium diagrams are dealt with in much greater detail in Chapter 13.

Effects of corrosion on welds in steel

Corrosion is a chemical action on a metal, resulting in the conversion of the metal into a chemical compound.

The rusting of iron, which we have considered, is a good example. In the presence of air and water, the iron eventually changes into oxides and hydroxides of iron and then into hydrated carbonates.

In addition to this type of attack, the matter which is suspended in the atmosphere also assists corrosion. The very small proportion of carbon dioxide in the atmosphere becomes dissolved in the rain, forming very weak carbonic acid, and this attacks the steel, again forming carbonates. In and near large towns the atmosphere contains a very much larger proportion of suspended matter than in the country. Smoke and fumes contain, among other compounds, sulphur dioxide, which again dissolves in rain to form sulphurous acid. This is oxidized into dilute sulphuric acid, which again attacks the steel, the attack being much stronger than in the case of the carbonic acid.

Near the coast, the salt in the atmosphere forms hydrochloric acid and caustic soda, and severe corrosion occurs in these areas.

In addition to this direct form of chemical attack there is a second type of attack which is at first not so apparent. When two different metals are placed in a conducting liquid, such as a dilute acid or alkali, an electric cell or battery is formed, one of the metals becoming electro-positive, while the other becomes electro-negative. The difference of electrical pressure or voltage between these two plates will depend upon the metals chosen.

In the case of a welded joint, if the weld metal is of different structure from the parent metal, we have, if a conducting liquid is present, an electric cell the plates of which are connected together or short-circuited. The currents which flow as a result of this are extremely minute, but nevertheless they greatly accelerate corrosion. This effect is called electrolysis, and its harmful effects are now well known.

The deposited metal in the weld is never of the same composition as the parent metal, although it may have the same properties physically. In the welded region, therefore, dissimilar metals exist, and in the presence of dilute carbonic acid from the atmosphere (or dilute sulphuric acid as the case may be) electrolytic action is set up and the surfaces of the steel become pitted. Now if the weld metal is electro-positive to the parent metal, the weld metal is attacked, since it is the electro-positive plate which suffers most from the corrosive effect. On the other hand, if the parent steel plate is electro-positive to the weld metal, the plate is attacked and since its surface area is much larger than that of the weld, the effect of corrosion will be less than if the weld had been attacked.

Thus, weld metal should be of the same composition throughout its mass to prevent corrosion taking place in the weld itself. It should also be electro-negative to the parent metal to prevent electrolytic action causing pitting of its surface, and it must resist surface oxidation at least as well as the parent metal.

FLUXES

Oxy-acetylene welding

Most metals in their molten condition become oxidized by the absorption of oxygen from the atmosphere. For example, aluminium always has a layer of aluminium oxide over its surface at normal temperature, and has a very great affinity for oxygen. To make certain that the amount of oxidation shall be kept a minimum, that any oxides formed shall be dissolved or floated off, and that welding is made as easy and free from difficulties as possible, fluxes are used. Fluxes, therefore, are chemical compounds used to prevent oxidation and other unwanted chemical reactions. They help to make the welding process easier and ensure the making of a good, sound weld.

The ordinary process of soldering provides a good example. It is well known that it is almost impossible to get the solder to run on to the surface to be soldered unless it is first cleaned. Even then the solder will not adhere uniformly to the surface. If now the surface is lightly coated with zinc chloride or killed spirit (made by adding zinc to hydrochloric acid or spirits of salt until the effervescing action ceases), the solder runs very easily wherever the chloride has been. This 'flux' has removed all the oxides and grease from the surface of the metal by chemical action and presents a clean metal surface to be soldered. This makes the operation much easier and enables a much better bond with the parent metal to be obtained. Fluxes used in oxy-acetylene welding act in the same way. Flux-covered rods are now available for bronze and aluminium welding.

Brass and bronze

A good flux must be used in brass or bronze welding, and it is usual to use one of the borax type, consisting of sodium borate with other additions. (Pure borax may be used.) The flux must remove all oxide from the metal surfaces to be welded and must form a protective coating over the surfaces of the metal, when they have been heated, so as to prevent their oxidation. It must, in addition, float the oxide, and their impurities with which it has combined, to the top of the molten metal.

Aluminium and aluminium alloys

In the welding of aluminium the flux must:

(1) Attack and dissolve the film of aluminium oxide (melting point over 2000 °C) always present on the surface of the metal.
(2) It must prevent further oxidation during welding.
(3) It must melt at a lower temperature than the metal, so that it will dissolve the surface oxide before the metal melts.
(4) It must be lighter when melted than aluminium, so that it will float any impurities to the surface, where they can be easily removed.

Suitable fluxes for aluminium and aluminium alloys contain lithium chloride, potassium chloride, potassium bisulphate and potassium fluorides, while others may contain chlorides of ammonium, sodium and barium. The choice of a suitable high-grade flux is essential to ensure success in aluminium welding. A typical flux is: lithium chloride 0–30%, potassium fluoride 5–15%, potassium chloride 0.6% (all by weight), and remainder sodium chloride.

Most aluminium fluxes are very hydroscopic, that is, they absorb moisture very readily from the atmosphere and become useless, and, as they are very expensive, when not in use they should be kept quite airtight.

Since they work by chemical action, it is necessary to use just sufficient for the purpose. Too much flux is both wasteful and harmful. A good method of ensuring an even supply is by dipping the hot end of the filler rod into the flux and melting the tuft, which clings to the end, along the rod, to form a coating about 100 or 125 mm long. This will ensure approximately the correct supply of flux.

Aluminium flux has a corrosive action if allowed to remain on the aluminium. After welding, therefore, the work should be well scrubbed in hot water to remove all trace of the flux. A 5% nitric acid solution is also an efficient method of removal. Suitable fluxes for welding magnesium alloys such as Elektron are similar to those for welding aluminium and largely consist of lithium and potassium chlorides, and may contain in addition alkaline fluorides (potassium, sodium, magnesium), and magnesium, sodium and barium chlorides. They are for the most part hygroscopic and corrosive and thus particular care must be taken to remove completely all traces of them after welding by scrubbing in hot soapy water and then immersing in a 5% sodium or potassium dichromate solution for an hour or more.

Cast iron

When welding wrought iron and mild steel the oxide which is formed has a lower melting point than the parent metal and, being light, it floats to the surface as a scale which is easily removed after welding. No flux is, therefore, required when welding mild steel or wrought iron.

In the case of cast iron, oxidation is rapid at red heat and the melting point of the oxide is *higher* than that of the parent metal, and it is, therefore, necessary to use a flux which will combine with the oxide and also protect the metal from oxidation during welding. The flux combines with the oxide and forms a slag which floats to the surface and prevents further oxidation. Suitable fluxes contain sodium, potassium or other alkaline borates, carbonates, bicarbonates and slag-forming compounds.

Copper

Copper may be welded without a flux, but many welders prefer to use one, to remove surface oxide and prevent oxidation during welding. Borax is a suitable flux, and its only drawback is that the hard, glass-like scale of copper borate, which is formed on the surface after welding, is hard to remove. Special fluxes, while consisting largely of borax, contain other substances which help to prevent the formation of this hard slag.

To sum up, we may state, therefore, that fluxes are used:

(1) To reduce oxidation.
(2) To remove any oxide formed.
(3) To remove any other impurities.

Because of this, the use of a flux:

(1) Gives a stronger, more ductile weld.
(2) Makes the welding operation easier.

It is important that *too much* flux should never be used, since this has a harmful effect on the weld.

Manual metal arc flux-covered electrodes

If bare wire is used as the electrode in MMA welding many defects are apparent. The arc is difficult to strike and maintain using d.c.; with a.c. it is extremely difficult. The resulting 'weld' lacks good fusion, is porous, contains oxides and nitrides due to absorption of oxygen and nitrogen from the air, and as a result the weld is brittle and has little strength. To remedy these defects electrodes are covered with chemicals or fluxes which:

(1) Enable the arc to be struck and maintained easily on d.c. or a.c. supplies.
(2) Provide a shield of gases such as hydrogen or carbon dioxide to shield the molten metal in its transference across the arc and in the molten pool in the parent plate from reacting with the oxygen and nitrogen of the atmosphere to form oxides and nitrides which are harmful to the mechanical properties of the weld.
(3) Provide a slag which helps to protect the metal in transit across the arc gap when the gas shield is not voluminous, and which when solidified protects the hot metal against oxidation and slows the rate of cooling of the weld; also slag-metal reactions can occur which alter the weld metal analysis.
(4) Alloying elements can be added to the coverings in which case the core wire analysis will not match the weld metal analysis.

We have seen that an oxide is a compound of two elements, one of which is oxygen. Many oxides are used in arc welding fluxes, examples of which are: silicon dioxide SiO_2 (A), manganous oxide MnO (B), magnesium oxide MgO (B), calcium oxide CaO (B), aluminium oxide Al_2O_3 (Am), barium oxide BaO (B), zinc oxide ZnO (Am), ferrous oxide FeO (B). Oxides may be classified thus: acidic, basic and amphoteric, indicated by A, B and Am above (other types are dioxides, peroxides, compound oxides and neutral oxides). An acidic oxide reacts with water to form an acid thus:

$$SO_2 \text{ (sulphur dioxide)} + H_2O = \text{sulphurous acid } (H_2SO_3).$$

In the case of silicon dioxide SiO_2 which is insoluble in water, it reacts similarly with fused sodium hydroxide (NaOH) to form sodium silicate and water thus:

$$SiO_2 + 2 NaOH = Na_2SiO_3 + H_2O.$$

Basic oxides interact with an acid to form a salt and water only thus: calcium oxide (CaO) reacts with hydrochloric acid (HCl) to form calcium chloride ($CaCl_2$) and water:

$$CaO + 2 HCl = CaCl_2 + H_2O.$$

Amphoteric oxides can exhibit either basic or acidic properties. Aluminium oxide (Al_2O_3) reacts with dilute hydrochloric acid as a basic oxide thus:

$$Al_2O + 6 HCl = 2AlCl_2 + 3H_2O,$$

but as an acidic oxide when it reacts with sodium hydroxide thus:

$$Al_2O_3 + 2 NaOH = 2 NaAlO_2 + H_2O.$$

When oxides are mixed to form fluxes the ratio of the basic to acidic oxides is termed the basicity and is important, as for example in the fluxes used for submerged arc welding (q.v.) where the flux must be carefully chosen in conjunction with the electrode wire to give the desired mechanical properties to the weld metal.

To illustrate the action of the flux-covered electrode we may consider the reaction between a basic oxide such as calcium oxide (CaO) and an acidic oxide such as silicon dioxide (SiO_2). With great application of heat these will combine chemically to form calcium silicate which is a slag, thus:

$$\text{calcium oxide} + \text{silicon dioxide} \rightarrow \text{calcium silicate}$$
$$CaO + SiO_2 \rightarrow CaSiO_3.$$

Similarly if we use iron oxide (Fe_2O_3) and silicon dioxide (SiO_2), in the covering of the electrode, in the heat of the arc they will combine chemically to form iron silicate, which floats to the top of the molten pool as a slag, protects the hot metal from further atmospheric oxidation and slows down the cooling rate of the weld.

$$\text{iron oxide} + \text{silicon dioxide} = \text{iron silicate}$$
$$Fe_2O_3 + 3SiO_2 = Fe_2(SiO_3)_3.$$

The most common slag-forming compounds are rutile (TiO_2), limestone ($CaCO_3$), ilmenite ($FeTiO_3$), iron oxide (Fe_2O_3), silica (SiO_2), manganese oxide (MnO_2) and various aluminium silicates such as felspar and kaolin, mica and magnesium silicates.

Deoxidizers such as ferrosilicon, ferromanganese and aluminium are also added to reduce the oxides that would be formed in the weld to a negligible amount.

The chemical composition of the covering also has an effect on the electrical characteristics of the arc. Ionizers such as salts of potassium are added to make striking and maintaining the arc easier, and there is a higher voltage drop across the arc when hydrogen is released by the covering as the shielding gas than with carbon dioxide from arcs of the same length. For a given current this higher voltage drop gives greater energy output from the arc, and hydrogen releasing coatings are usually cellulosic giving a penetrating arc, thin slag cover and quite an amount of spatter. Other coatings are discussed in detail in the section in Chapter 5 under the heading 'Classification of electrodes for MMA welding of carbon and carbon-manganese steels'.

For the arc welding of bronze (also copper and brass) the flux must dissolve the layer of oxide on the surface and, in addition, must prevent the oxidation of the metal by providing the usual sheath. These coatings contain fluorides (cryolite and fluorspar) and borates and the rods are usually operated on the positive pole of direct current supply.

The flux of the aluminium rod is a mixture of chlorides and fluorides as for oxy-acetylene welding of aluminium. It acts chemically on the oxide, freeing it, and this enables it to be floated to the surface of the weld. It is corrosive and also tends to absorb moisture from the air; hence the weld should be well cleaned with hot water on completion, while the electrodes should be stored in a dry place. In fact, all electrodes should be kept very dry, since the coatings tend to absorb moisture and the efficiency of the rod is greatly impaired if the covering is damp.

The manual metal arc welding of aluminium and its alloys has been largely superseded by the inert gas shielded-metal arc processes, TIG and MIG. See Chapters 6 and 7.

Materials used for electrode coatings

(1) Rutile. Rutile is a mineral obtained from rutile-bearing sands by suction dredging. It contains about 88–94% of TiO_2 and is probably the

most widely used material for electrode coatings. Ilmenite is a naturally occurring mineral composed of the oxides of iron and titanium $FeTiO_3$ (FeO, TiO_2) with about 45–55% TiO_2. After separation of impurities it is ground to the required mesh size and varies from grey to brown in colour.

(2) Calcium carbonate, or limestone is the coating for the basic coverings of electrodes. The limestone is purified and ground to required mesh size. The slag is very fluid and fluorspar is added to control fluidity. The deposited metal is very low in hydrogen content.

(3) Fluorspar or fluorite is calcium fluoride and is mined, separated from impurities, crushed, screened, and ground and the ore constituents are separated by a flotation process. Too great an addition of this compound to control slag fluidity affects the stability of the a.c. arc.

(4) Solka floc is cellulose acetate and is prepared from wood pulp. It is the main constituent of class 1 electrodes. Hydrogen is given off when it decomposes under the heat of the arc so that there is a large voltage drop and high power giving deep penetration. Arc control is good but there is hydrogen absorption into the weld metal.

(5) Felspar is an anhydrous silicate of aluminium associated with potash, soda, or calcium, the potash felspar being used for electrode coatings. It is used as a flux and a slag-producing substance. The potash content stabilizes the a.c. arc and it is generally used in association with the rutile and iron oxide–silica coatings. The crystalline ore is quarried and graded according to impurities, ground, and the powder finally air-separated. The binders in general use for the materials composing electrode coatings are silicates of potassium and sodium.

(6) Ball clay. A paste for an electrode coating must flow easily when being extruded and must hold liquid present so that it will not separate out under pressure; also the freshly extruded electrode must be able to resist damage when in the wet or green condition. Because of the way in which its molecules are arranged ball clay gives these properties to a paste and is widely used in those classes of electrodes in which the presence of hydrogen is not excluded. Found in Devon and Cornwall, it is mined, weathered, shredded, pulverized and finally sieved.

(7) Iron powder is added to an electrode coating to increase the rate of metal deposition. In general, to produce the same amount of slag with this powder added to a coating it generally has to be made somewhat thicker. To produce the iron powder, pure magnetic oxide of iron is reduced to cakes of iron in a bed of carbon, coke and limestone. The 'sponge cakes' which are formed are unlike ordinary iron in that they can be pulverized to a fine powder which is then annealed. Electrodes may contain up to 50% of iron powder.

(8) Ferromanganese is employed as a deoxidizer as in steel-making to

remove any oxide that has formed in spite of the arc shield. It reacts with iron oxide to form iron and manganese dioxide which mixes with the slag.

(9) Mica is a mineral found widely dispersed over the world. It is mined and split into sheets. The larger sizes are used for electrical purposes such as commutator insulation and the smaller pieces are ground into powder form. It is used in electrode coatings as a flux and it also assists the extrusion and gives improved touch welding properties with increase in slag volume.

(10) Sodium alginate is extracted from certain types of seaweed. It is used in electrode coatings because when made into a viscous paste, it assists extrusion and is especially useful when the coating contains a large proportion of granules.

The flux coating on electrodes may be applied in one of the following ways:

(1) Solid extrusion.
(2) Extrusion with reinforcement.
(3) Dipping.

(1) **Solid extrusion**. This is the way in which most of the present-day electrodes are produced. The flux, in the form of a paste, is forced under pressure around the wire core. The thickness of the flux covering can be accurately controlled and is of even thickness all round the wire core, the method being suitable for high speed production. This covering, how-ever, will not stand up to very rough handling, nor to bending (as is sometimes required when welding awkwardly placed joints) since the covering flakes off.

(2) **Extrusion with reinforcement**. In this method the reinforcement enables the covering to withstand more severe handling and bending without flaking. The reinforcement may be:

(*a*) An open spiral of yarn wound on the rod, the space between the spiral coils being filled with extruded flux.
(*b*) A close spiral of yarn wound over a solid extrusion of flux applied first to the wire coil, and this flux covering streng-thened by a yarn or by a single or double helical wire winding.

(3) **Dipped**. The dipping process has been largely superseded by the other two methods. Certain rods having special applications are, how-ever, still made by this process. Repeated dippings are used to give thicker coatings.

2

Metallurgy

PRODUCTION AND PROPERTIES OF IRON AND STEEL

Before proceeding to a study of iron and steel it will be well to understand how they are produced.

Iron is found in the natural form as iron ores. These ores are of four main types:

(1) Haematite, red or brown Fe_2O_3 containing 40–60% iron.
(2) Magnetite or magnetic oxide of iron Fe_3O_4 containing up to 70% iron.
(3) Limonite, a hydrated ore, $Fe_2O_3 \cdot 3H_2O$, containing 20–50% iron.
(4) Siderite, a carbonate, $FeCO_3$, with iron content 20–30%.

Limonite and siderite are termed lean ores since they are so low in iron. The ore found in England in Lincolnshire, Northamptonshire, Leicestershire and Oxfordshire is one of low iron content and is generally obtained by opencast working.

Iron ore, as mined, contains appreciable amounts of earthy waste material known as gangue, and if this were fed into the furnace with the ore, more fuel would be consumed to heat it up and it would reduce the furnace capacity. Ores are washed, or magnetically separated in the case of the magnetic ores, to remove much of this waste material. They are roasted or calcined to drive off the moisture and carbon dioxide and to remove some of the sulphur by oxidation to sulphur dioxide, and crushed to bring the lumps to a more uniform size.

Agglomeration of ores

Finer particles of ore (fines) cannot be fed into the furnace because they would either be blown out or would seal up the spaces in the burden (coke, ore and flux) necessary for the passage of the blast. The smaller particles can be made to stick together or agglomerated either by sintering or pelletizing.

Sintering. The materials are chiefly iron ore fines, blast furnace flue dust, limestone and/or dolomite and coke breeze or fine anthracite as fuel. They are mixed, moistened and loaded on to a moving grate consisting of pallets through which air can circulate. The mixture is ignited by gas or oil jets and burns, sucking air through the bed. The sinter is tipped from the end of the moving grate, large lumps being broken up by a breaker.

Pelletizing. The ore feed is usually wet concentrates made into a thick slurry to which a small amount of bentonite is added. This is then balled by feeding it into a slowly rotating drum inclined at 5–10° to the horizontal. The green balls are then fed into a vertical shaft furnace or onto a travelling grate as in sintering where they are dried, fired and cooled.

The blast furnace

The furnace is a vertical steel stack lined with refractories. Charging is done at the top and pig iron and slag are tapped from the bottom (Fig. 2.1). Large volumes of gases (including carbon monoxide) are evolved during operation of the furnace and are burnt in stoves which provide the heat to raise the temperature of the air blast to about 1350°C. This reduces the amount of coke required because combustion speed is increased and thus efficiency is increased and there is a reduction in the sulphur content of the pig iron. Four-fifths of the air in the blast is nitrogen which takes no part in the process yet has to be raised in temperature. By enriching the blast with oxygen (up to 30%) the nitrogen volume is reduced and the efficiency increased.

During the operation of the furnace the burning coke produces carbon monoxide which reduces the ore to metal. This trickles down to the bottom of the furnace where the temperature is highest. The limestone is decomposed into lime (calcium oxide) which combines with the silica in the gangue to form a slag, calcium silicate, and the iron begins to take in carbon. Slag is tapped from the upper notches and pig iron from the lower (Fig. 2.1).

Blast furnaces in Europe, the United States and Japan are becoming increasingly larger with greater productivity, the latest British furnace having a hearth of 14 m diameter and a capacity of 10 000 tonnes of liquid iron a day. Modern furnaces can use oil fuel injection, top pressure, high blast temperatures, oxygen enrichment and pre-reduced burden in the quest for greater economy in energy and increased productivity.

Because of the large volume of these furnaces (4600 m³) it is difficult to distribute the reducing gases evenly throughout the burden, so ore size is carefully graded, strong coke is used, and equalization is done by high-top-pressure nitrogen which reduces the velocity of the gas in the lower regions of the furnace, keeping the gas in longer contact with the burden and allowing it to ascend more uniformly thus achieving more efficient reduction.

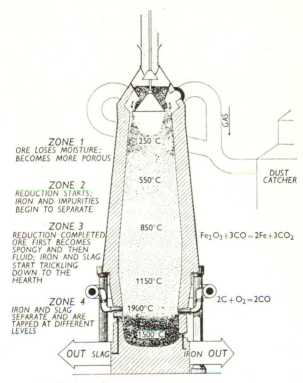

ZONE 1
ORE LOSES MOISTURE;
BECOMES MORE POROUS

ZONE 2
REDUCTION STARTS;
IRON AND IMPURITIES
BEGIN TO SEPARATE

ZONE 3
REDUCTION COMPLETED
ORE FIRST BECOMES
SPONGY AND THEN
FLUID; IRON AND SLAG
START TRICKLING
DOWN TO THE
HEARTH

ZONE 4
IRON AND SLAG
SEPARATE AND ARE
TAPPED AT DIFFERENT
LEVELS

250 C
550°C
850°C
1150°C
1900°C
1500°C

GAS

DUST
CATCHER

$Fe_2O_3 + 3CO = 2Fe + 3CO_2$

$2C + O_2 = 2CO$

OUT SLAG   IRON OUT

Fig. 2.1. The blast furnace.

Furnace construction can be by the stack being welded on to a ring girder which is supported on four columns or there can be a free-standing stack within a structure of four columns. The refractories are carbon and carbon with graphite for tuyères and hearth, and aluminium oxide (alumina) for the stack. Cooling is by forced-draught air or by water for the underhearth and flat copper coolers or staves are used for the stack with open- or closed-circuit cooling water systems.

Typical burdens are 80% sinter, 20% ore or 60/40% sinter, 40/60% pellets; coke and burden are screened, weighed and delivered to the furnace on a charging conveyor, the charging system being either double bell or bell-less with a distribution chute. Furnace charging may be done automatically and can be fully computerized. The gas cleaning plant incorporates dust catcher and water scrubber.

Direct reduction of iron ore

As alternatives to the blast furnace method of producing iron from its ore, other processes can be used, not dependent upon the use of coke. The ore is converted into metallized pellets or sponge iron by removing the oxygen from the iron to leave metallic iron. The amount of metallic iron produced from a given quantity of ore is termed the degree of metallization and is the ratio of the metallic iron produced to the total iron in the ore. The iron left after the removal of the oxygen has a honeycomb structure and is often termed sponge iron.

Direct reduction processes may be classed according to the type of fuel used, either gaseous hydrocarbons using reducing gases produced by reforming from natural gas (methane), or solid fuel such as coal or coke breeze. The gaseous fuel type can be a vertical retort (Hyl), vertical shaft furnace (Midrex) or fluidized bed (HIB) while the solid fuel type uses rotary hearths or kilns (SL/RN, Krupp). A high degree of purity of ore is required for sponge iron because gangue is not removed at the iron-making stage but later in the steel-making process, so that the more gangue present, the less the efficiency of the process. At present pelletized concentrates are used but screened natural lump ore of similar purity can now be used as processing difficulties have been overcome.

Typical of the gaseous type of direct reduction plant installed by British Steel is the Midrex, using natural gas as the reductant. The natural gas is steam-reformed to produce carbon monoxide and hydrogen thus:

$$CH_4 \quad + \quad H_2O \quad + \quad heat \quad \rightarrow \quad CO \quad + \quad 3H_2$$
methane steam   carbon monoxide   hydrogen.

Other hydrocarbons such as naphtha or petroleum can also be used as reductants.

Considering haematite Fe_2O_3 as the ore, the carbon monoxide and hydrogen which are both reducing gases act as follows on the ore, reducing it to metallic iron of spongy appearance, the reduction taking place above 800 °C.

$$Fe_2O_3 + 3CO \rightarrow 2Fe + 3CO_2 + heat \text{ (exothermic).}$$
$$Fe_2O_3 + 3H_2 + heat \rightarrow 2Fe + 3H_2O \text{ (endothermic).}$$

Cold oxide pellets are fed by successive additions into the top of the vertical shaft furnace up which flows a counter-current of heated reducing gases (carbon monoxide and hydrogen). Metallization occurs and the metallized pellets are taken from the bottom of the furnace so that the process is continuous and economical in labour, achieving a metallization of 92–95%, the off-gases being recovered and recycled (Fig. 2.2).

D.R. iron is used chiefly in the electric arc and basic oxygen furnaces and it is evident that the future of this gaseous type of reduction depends upon a continuing supply of natural gas at a price competitive with that of coke.

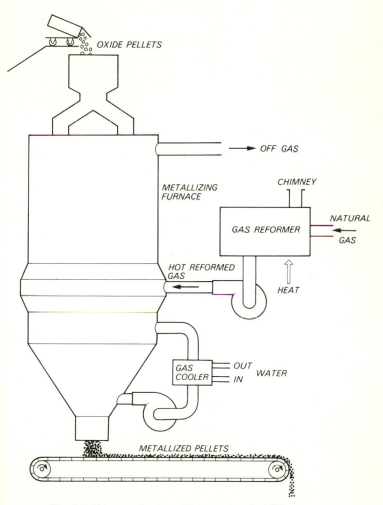

Fig. 2.2. Direct reduction of iron ore by the Midrex process.

Cast iron

Pig iron from the blast furnace is not refined enough for making castings, so in the foundry the iron for casting is prepared as follows: a coke fire is lit at the bottom of a small blast furnace or cupola and then alternate layers of pig iron (broken up into pieces) and scrap and coke together with small quantities of limestone as flux (for purifying and deslagging) are added. When the mass has burnt up, the blast is turned on and the molten iron (melting point 1130°C) flows to the bottom of the furnace from where it is tapped into ladles or moulds direct. Cast iron is relatively cheap to produce and its melt fluidity gives excellent casting properties. It is an alloy of many elements, and an average composition is: iron 94 to 98%, carbon 3 to 4%, silicon below 3%, sulphur below 0.2%, phosphorus below 0.75%, manganese below 1%.

The carbon exists in two forms: chemically combined carbon, and free carbon simply mixed with the iron and known as graphite. The grey look of a fracture of grey cast iron is due to this graphite, which may be from 3 to $3\frac{1}{2}$%, while the chemically combined carbon may range from 0.5 to 1.5%. As the amount of combined carbon increases, so does the hardness and brittleness increase, and if cast iron is cooled or chilled quickly from a very high temperature, the combined carbon is increased, and the free carbon is reduced. As a result this type of cast iron is more brittle and harder than grey iron, and since it has a white appearance at a fractured surface, it is termed white cast iron. This has from 3 to 4% of carbon chemically combined.

Cast iron possesses very low ductility, and for this reason it presents difficulty in welding because of the strains set up by expansion and contraction tending to fracture it.

The properties of cast iron can be modified by the addition of other elements. Nickel gives a fine grain and reduces the tendency of thin sections to crack, while chromium gives a refined grain and greatly increases the resistance to wear. The addition of magnesium enables the graphite normally present in flake form to be obtained in spheroidal form. This SG (spheroidal graphite) cast iron is more ductile than ordinary cast iron (see p. 81).

Wrought iron

Wrought iron is now of historical interest only and is very difficult to obtain. It is manufactured from pig iron by the puddling process which removes impurities such as carbon, sulphur and phosphorus leaving nearly pure iron. A typical analysis is iron 99.5 to 99.8%, carbon 0.01 to 0.03%, silicon 0 to 0.1%, phosphorus 0.04 to 0.2%, sulphur 0.02 to 0.04%, and manganese 0 to 0.25%.

When fractured, wrought iron shows a fibrous or layered stucture but will bend well and is easily worked when hot. It does not harden on cooling rapidly and can be welded in the same way as mild steel.

Steel-making

Pig iron consists of iron together with 3–4% carbon, present either in the combined form as iron carbide or in the free form as graphite, the composition depending upon the type of iron ore used. In addition it contains other elements, the chief of which are manganese, silicon, sulphur and phosphorus. By oxidation of the carbon and these other elements the iron is converted into steel, the composition of which will have the required carbon and manganese percentage with a very small amount of sulphur and phosphorus (e.g. sulphur 0.02% max. and phosphorus 0.03% max.) for a typical welding-quality steel. There are two processes in steel-making, acid and basic, and they differ in the type of slag produced and the refractory furnace lining. In the acid process, low-sulphur and -phosphorus pig iron, rich in silicon, produces an acid slag (silica) and the furnace is lined with silica refractories to prevent reaction of the lining with the slag.

In the basic process, which is largely used nowadays, phosphorus-rich pig irons can be treated with lime (a base) added to the slag to reduce the phosphorus content, and the slag is now basic. The refractory lining of the furnace must now be of dolomite (CaO, MgO) or magnesite (MgO) to prevent reaction of the slag with the furnace lining. This basic process enables widely distributed ores, with high phosphorus and low silica content, to be used.

Much steel was formerly made by the Bessemer process (acid or basic), in which an air blast is blown through the charge of molten iron contained in a steel converter lined with refractories, oxidizing the impurities, with the exothermic reaction providing the necessary heat so that no external heat source is required. The large volume of nitrogen in the air blown through the charge wastes much heat and in addition nitrogen forms nitrides in the steel reducing the deep-drawing prop-erties of the steel, so the process is now little used.

In the *Open Hearth* (acid or basic) process, now rapidly becoming obsolete, the heat required to melt and work the charge is obtained from the burning of a producer gas-and-air mixture over the hearth. Both gas and air are heated to a high temperature (1200 °C) by passing them through chambers of checkered brickwork in which brick and space alternate. In order that the process shall be continuous a regenerative system is used (Fig. 2.3). There are two sets of chambers each, for gas and air. While the gas and air are being heated in their passage through one pair of chambers, the high-temperature waste gases are pre-heating the other pair ready for the change-over. Pig iron and scrap are fed into the furnace by mechanical chargers, the pig iron being fed in the molten condition if the steel-making furnace is near the blast furnace as in integrated plants. The charging doors are on one side and the tapping hole on the other, the hearth capacity being 30–200 tonnes.

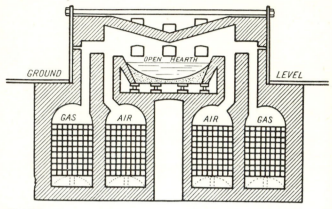

Fig. 2.3

When the charge is molten, iron ore is added and oxidation takes place, carbon monoxide being formed, the carbon content of the melt is reduced and silicon and manganese are also oxidized. Finally deoxidizers (ferromanganese, ferrosilicon, aluminium) are added just before tapping, or as the steel is run into the ladle, to improve the quality of the steel. Ferromanganese (80% Mn, 6% C, remainder iron) reacts with the iron oxide to give iron and manganous oxide which is insoluble in steel, and the excess manganese together with the carbon adjusts the composition of the steel.

The acid process is now declining in use and the basic process produces normal grades of steel. The basic slag, rich in phosphorus, is used as a fertilizer.

Oxygen steel-making

With the introduction of the tonnage oxygen plant situated near the steel furnaces, oxygen is now available at competitive cost in large volumes to greatly speed up steel production. The oxygen plant liquefies air and the oxygen is then fractionally distilled from the nitrogen and argon and is stored in the liquid form as described in the section on liquid oxygen. In the open hearth process an oxygen lance is arranged to blow large volumes of oxygen on to the molten metal in the hearth. With the use of oxygen instead of air there is minimal nitrogen introduced into the steel (below 0.002%), so that its deep-drawing qualities are improved and the time for converting the charge into steel is reduced by as much as 50%. It appears that the basic open hearth process is being replaced by the basic oxygen furnace (BOF) in various forms in Britain, Europe, USA and Japan. The advantages are that iron ores of variable phosphorus content can be used and there are reduced labour and refractory costs. These factors together with the capability of the process to use up to 40% scrap make for reduced costs and higher efficiency.

When the oxygen is blown on to the molten charge it reacts to form iron oxide which combines with lime present as flux to form an oxidizing slag on the melt. Reactions occur between the molten metal and the slag resulting in the removal of phosphorus and lowering of the silicon, manganese and carbon content.

Basic oxygen steel (BOS)

The *basic oxygen steel-making* process together with the electric arc process is responsible for much of present-day steel production. A typical basic oxygen furnace consists of a steel-cased converter lined with dolomite holding up to 400 tonnes of metal. Hot metal is transported in torpedo ladles to the hot-metal pouring station equipped with

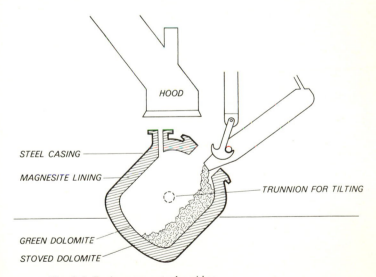

Fig. 2.4. Basic oxygen steel-making.
(*a*) Charging with scrap. The converter is tilted and charged with scrap from a charging box which tips the scrap into the previously heated converter. The scrap represents up to 30% of the total charge.

extraction facilities for fume and kish (graphite which separates from and floats on top of the charge), and selected torpedoes are desulphurized with calcium carbide. The hot pig iron, scrap, flux and any alloys are added and oxygen is injected through a multi-holed, water-cooled lance from a near-by tonnage oxygen plant, on to the surface of the charge. Oxidation is rapid, with the blowing time lasting about 17 minutes, and the converter is then tilted and emptied. The BOF takes between 25 and 35% of scrap metal and efforts are being made to increase this percentage because scrap is cheaper than hot pig iron (Fig. 2.4 *a,b,c,d,e*).

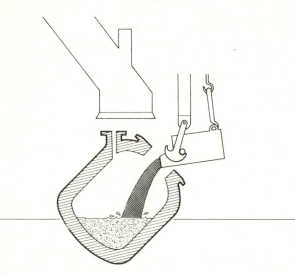

Fig. 2.4. (*b*) Charging with molten iron. Molten iron from the blast furnace is brought to the converter in 'torpedo-cars' and transferred by ladle into the converter.

The Maxhütte bottom-blown furnace (OBM) is a steel-cased converter lined with dolomite with special tuyères in the base of the furnace through which oxygen and powdered lime are introduced. This in turn is surrounded by a protective shield of hydrocarbon fuel gas (natural gas, propane) to protect the refractory lining (Fig. 2.5). The remainder of the process is similar to that already described and the benefits claimed for this process are absence of fuming and splashing and a lower final carbon and sulphur content in the steel, the injection of lime ensuring rapid removal of the phosphorus.

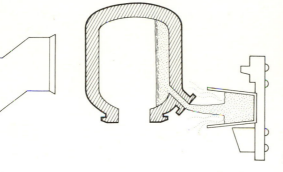

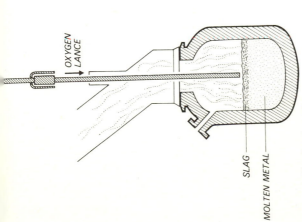

OXYGEN
LANCE

SLAG

MOLTEN METAL

Fig. 2.4 (contd)

(c) The fume-collecting hood is lowered on to the furnace neck. A water-cooled oxygen lance is lowered to within a metre or so of the molten metal surface. Oxygen is blown through the lance and causes turbulence and rise in temperature of the metal. Impurities are oxidized with the 'blow' lasting about fifteen minutes during which time temperatures are carefully controlled and analysis made of the molten metal.

(d) When temperature and metal analysis are satisfactory the hood is lifted, the converter tilted and the steel poured from below the slag which has formed into the teeming ladle from which it passes to the continuous casting plant and is cast into ingots.

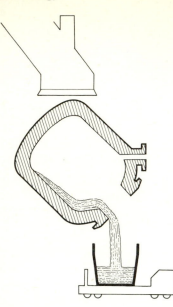

Fig. 2.4. (*e*) Finally the converter is tilted in the opposite direction and the slag which remains is poured into a slag ladle.

Electric arc steel-making

When a welder uses a carbon electrode to produce an arc and give a molten pool on a steel plate he is using the same basic principle as that which is done on a very much larger scale in the electric arc furnace. The heat in this type of furnace comes from the arcs struck between three carbon electrodes connected to a three-phase electric supply and the charge in the furnace hearth, and thus no electrode is required in the furnace hearth.

The furnace shell is mild steel lined with refractories, with two doors (one on smaller furnaces). The three electrodes project through the refractory-lined roof down to the level of the surface of the charge, and roof and electrodes are arranged to swing aside to clear the furnace shell for charging or repair, this type of furnace being batch-charged, though developments are proceeding to feed furnaces continuously either through the roof or side walls. A sealing ring on top of the side walls supports the weight of the roof and the furnace can tip about 15° towards the main door and about 50° forward for tapping. Some of the larger furnaces can also rotate so as to give a variety of melting positions for the electrodes (Fig. 2.6).

Since it is necessary to remove phosphorus and sulphur a basic slag is required, and the furnace must be lined with basic refractories such as dolomite or magnesite for roof, side walls and hearth to give the basic electric arc process. The acid process is only used in cases where melting only is to be done, with little refining.

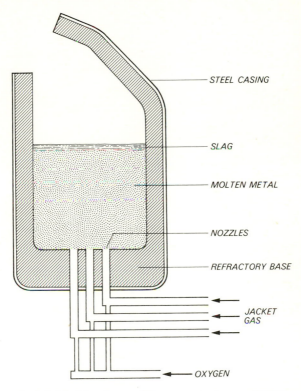

Fig. 2.5. Maxhütte bottom blown oxygen furnace (OBM).

Large transformers of the order of 80–100 MVA capacity feed from the grid supply to that for the arcs. The voltage drop across the arc is a function of the arc length, and the greater the volts drop, the greater the power for a given arc current. As a result the secondary voltage to the arc may be 100–600 V, with currents up to 80 000 A in large furnaces. The three graphite electrodes can vary from 75 to 600 mm in diameter and in length from 1.2 to 2.5 m, and can be raised or lowered either hydrauli-cally or by electric motor, this operation being done automatically so as to keep the arc length correct. Current to the electrodes is taken via water-cooled clamps and bundles of cables to give flexibility.

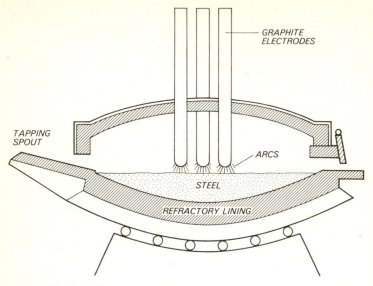

Fig. 2.6. Electric arc furnace.

One of the greatest advantages of the arc furnace is that it can deal with up to 100% scrap charge, and whereas in the past it was used for making high-grade alloy steel it is now used to melt high-percentage scrap charges and even to produce ordinary grades of steel.

The charge is of scrap, iron ore, blast furnace iron, D.R. iron and limestone, depending upon availability. The electrodes are lowered on to the charge, the arcs struck and melting proceeds. The oxygen necessary for the removal of impurities is obtained from the iron ore charge, the furnace atmosphere and in many cases by lance injection of oxygen, the silicon, manganese and phosphorus being removed by oxidation and entering the slag. The carbon is oxidized to carbon monoxide which burns to carbon dioxide.

Melting is done under a basic oxidizing slag, black in colour, and when the desired level of carbon and phosphorus is obtained, the slag is thoroughly removed and the melt deoxidized with ferrosilicon or aluminium. A reducing slag is now made using lime and anthracite or coke dust and the lime reacts with the iron sulphide to form calcium sulphite and iron oxide; the calcium sulphite is insoluble in steel and thus enters the slag, removing the sulphur. This removal requires reducing conditions in the furnace which is not possible with any other furnace in which oxygen is used to burn the fuel so that this sulphur removal is another great advantage of the arc furnace. Alloy additions are made under non-oxidizing conditions which give good mixing, and carbon can be added if required (recarburation) in the form of graphite or coke.

Vacuum refining

Further improvement in steel quality is obtained by vacuum refining. An example is Vacuum Oxygen Decarburization (VOD). A stream of oxygen from a lance is blown on to the surface of the molten steel under partial vacuum in a vacuum chamber. Argon is bubbled through the melt for stirring and alloy additions are made from the top. With this method the carbon content of the steel can be reduced to 0.08% (and lower with higher vacuum), hydrogen and nitrogen contents are reduced and chromium addition recoveries are high.

Induction furnaces

These are melting furnaces generally used for the production of special steels in sizes from 100 to 10 000 kg and giving accurate control over the steel specification (Fig. 2.7).

When discussing the principle of the transformer we will see (p. 270)

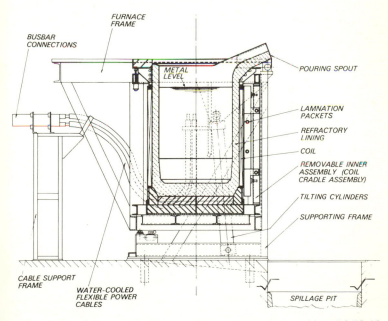

Fig. 2.7. Typical coreless induction furnace in capacities of 0.75–10 tonnes.

that if an alternating current flows in the primary winding, an alternating current is generated in the secondary winding. The alternating magnetic flux due to the primary current generates, or induces, a current in the secondary circuit. In the induction furnace the primary coil is wound around the refractory crucible which contains the metallic charge to be melted. When an alternating current flows in the primary coil eddy currents are induced in the charge and generate the heat required for melting. As the frequency of the alternating current increases, the eddy currents, and thus the heating effect, increase.

Furnaces operate at mains frequency (50 Hz) or at 100, 150, 800, 1600 Hz, etc., and high-frequency furnaces employing static converters to change the frequency operate from 10 to 15 kHz and at voltages up to several kV. The eddy currents produce a stirring action which greatly improves temperature control.

The furnace refractories can be a pre-cast crucible for the smaller furnaces or have a rammed lining of magnesite or, in some cases, silica. The hollow square-shaped copper conductors of the coil are closely wound and water cooled, this being an essential feature to prevent overheating and consequent breakdown. Small furnaces up to 50 kW are very convenient for laboratory and research work and can have capacities as low as a few kilograms. Clean scrap of known analysis can be used in the charge, and as oxidation losses are minimal and there is little slag, there is practically no loss of alloying elements during the melt.

Malleable cast iron

This is made from white cast iron by annealing or graphitizing. The white cast iron is packed in haematite and is heated to about 900°C and kept at this for two or three days, after which the temperature is slowly reduced. In this way, some of the combined carbon of the white cast iron is transformed into free carbon or graphite. Malleable castings are used where strength, ductility and resistance to shock are important, and they can be easily machined.

The 'blackheart' process is similar to the Réaumur or 'whiteheart' process just described except that bone dust, sand and burnt clay are used for packing in place of iron oxide, the temperature being about 850°C. This converts the combined carbon in the cast iron into temper carbon, and after the treatment they contain little or no combined carbon and about $2\frac{1}{2}\%$ graphite. The castings prepared by the former method show a grey fractured surface with a fine grain like mild steel, while those made by the 'blackheart' process have a black fracture with a distinct white rim.

A typical composition of malleable iron is carbon 2 to 3%, silicon 0.6

to 1.2%, manganese under 0.25%, phosphorus under 0.1%, sulphur 0.05 to 0.25%.

THE EFFECT OF THE ADDITION OF CARBON TO PURE IRON

We have seen that the chief difference between iron and mild steel is the amount of carbon present. Steel may contain from 0.03 to 2% carbon, mild or soft steel containing about 0.1% carbon and very hard razor-temper steel 1.7 to 1.9%.

The composition of steel is therefore complicated by these variations of carbon content, and is rendered even more so by the addition of other elements such as nickel, chromium and manganese to produce alloy steels.

Let us consider the structures present in steels of various carbon contents which have been cooled out slowly to room temperature. If we examine a highly polished specimen of wrought iron under a microscope magnifying about 100 times (× 100) Fig. 2.8a, we can see the white crystals of ferrite with the crystal boundaries and also dark elongated bands which are particles of slag entrapped during the rolling process. A specimen with no inclusions is shown in Fig. 2.8b. Now examine a specimen of 0.2% carbon steel under the same magnification. It shows dark areas in with the whiter ferrite, Fig. 2.8c. A 0.4% carbon steel appears with more darker areas, Fig. 2.9c, so that it is evident that an increase in carbon content produces an increase in these dark areas which if observed under first a magnification of 1000, Fig. 2.9a and then of 2500, Fig. 2.9b, are seen to consist of a layered structure, darker areas alternating with lighter ones. The dark areas are iron carbide (Fe$_3$C) or cementite formed by the chemical combination of ferrite and carbon thus: ferrite + carbon → cementite, or iron + carbon → iron carbide. These alternating layers of ferrite and cementite are called pearlite since they have a mother-of-pearl sheen when illuminated. Pearlite contains 0.85% carbon and is known as a eutectoid.[1] When a steel contains 0.85% of carbon the structure is all pearlite and if more than this percentage is present we find that the carbon has combined with more ferrite reducing the area of pearlite and forming cementite in the structure. Pearlite is a ductile structure while cementite is hard and brittle so that as the carbon content increases above 0.85% and more cementite is formed, the steels become very hard and brittle and steels of more than 1.7–1.8% carbon are rarely encountered (Fig. 2.9d and Fig. 2.10). Above 2% the carbon may be present as free carbon termed graphite and when the carbon

[1] See Chapter 13, eutectoid change.

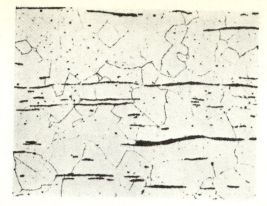

(*a*) Wrought iron, showing grains of ferrite and slag inclusions. × 100.

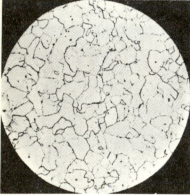

(*b*) Ferrite. × 100.

Fig. 2.8

(*c*) 0.2% carbon steel forging normalized, showing pearlite and ferrite. × 100.

percentage of the iron is between $2\frac{1}{2}\%$ and 4% it is known as cast iron. In grey cast iron which is soft and machinable (but brittle) the carbon is present in the free state as graphite but rapid cooling can cause the carbon to be in the combined form as cementite when we have white cast iron which is hard and not machinable. Hence in a steel the carbon is always in the combined form while in cast iron it may be present either free as graphite or in the combined form as cementite (Fig. 2.11).

SG cast iron

The flakes of graphite present in grey cast iron which reduce its tensile strength can be changed to sphere-shaped particles by adding to the

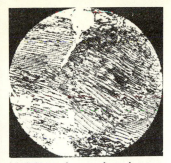

(a) 0.8% carbon tool steel, annealed. × 1000.

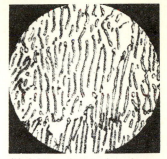

(b) 0.8% carbon tool steel, annealed. × 2500.

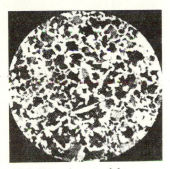

(c) 0.4% carbon steel forging, annealed. × 100.

(d) Cementite structure in 1.2% carbon tool steel, normalized. × 100.

Fig. 2.9

INCREASING PERCENTAGE OF CARBON
INCREASES THE AREAS OF PEARLITE UP TO 0·85% CARBON

ALL FERRITE
PURE IRON

0·2% CARBON

0·4% CARBON

ALL PEARLITE
0·85% CARBON

1·4% CARBON
PEARLITE AND CEMENTITE
WITH MORE THAN 0·85% CARBON EXCESS
CEMENTITE LIES ALONG GRAIN BOUNDARIES

 FERRITE
PEARLITE
CEMENTITE

Fig. 2.10. Structure of steels with varying carbon content.

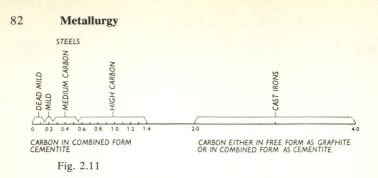

Fig. 2.11

molten iron small amounts of magnesium (or various other substances). This spheroidal graphite (SG) cast iron has greatly increased strength and ductility. In the 'as cast' condition the matrix is normally pearlite, with the carbon in the combined form and is hard, with a tensile strength about twice that of grey cast iron. Normalizing improves its mechanical properties, and stress-relief heat treatment at 550 °C is recommended for more complicated castings. Annealing the SG iron gives a ferritic matrix lowering the tensile strength but improving ductility and elongation compared with pearlitic iron. The iron can be hardened by quenching and tempering and addition of 1–2% nickel increases its hardenability. Nominal composition: C 3.5–3.8%, Si 1.8–2.5%, Mn 0.2–0.6%, P 0.05%, Ni 0–2.0%, Mg 0.04–0.07%.

CARBON AND ALLOY STEELS

An alloy steel may be defined as a steel which owes its properties to the presence of elements other than carbon, manganese up to 1.5% and silicon up to 0.5%.

The purpose of the alloying elements is to give the steel a distinct property which in every case is to increase its toughness, hardness or tensile strength, and to give cleaner and more wear-resistant castings.

Phosphorus has a very injurious influence on steel, making it very brittle, and as a result it is never found in greater percentages than 0.03 to 0.04 for steels which have to stand heavy impacts and shock, while for less important steels, 0.06% may be allowable. Phosphorus produces 'cold shortness', i.e. liability to crack when cold worked.

Sulphur is also injurious because it produces 'hot shortness', that is, liability to crack when hot. For this reason, the maximum sulphur percentage allowed in a steel in approximately the same as that of phosphorus.

Carbon and alloy steels

The chief elements which are alloyed with carbon steel are nickel, chromium, manganese, molybdenum, tungsten, vanadium and silicon. The addition of each element produces a different effect, and evidently there is an enormous variety of steels which can be produced by adding varying percentages of the above elements. We shall only consider very briefly the effect produced on the carbon steel by the alloying elements and not attempt to discuss the variety of steels available under each heading.

Carbon steel

First let us consider the carbon steel before any alloying elements are added. The addition of carbon, as can be seen from the table, greatly affects the properties of the steel and varying the percentage from 0.1% to 1.5% is sufficient to change the steel from very soft and malleable to extremely hard. The melting point of steel depends on the percentage of carbon present, decreasing as the carbon content increases, ranging from 1420 °C for mild steel to 1150 °C for hard steel, of 1.5% carbon. (This is the property on which Lindewelding is based.)

Classification of steel according to carbon content

Approx. carbon content (%)	Examples of use
Below 0.1	Dead soft mild steel for pressing, stamping and flanging, solid drawn tubes.
0.1–0.2	Rivets, nails, tubes, strip, bar and plates.
0.2–0.25	Bars, strips, girders, channel and angle sections, drop forgings and parts to be case hardened.
0.25–0.5	Shafts, tyres, forgings, boiler shells, spades, etc.
0.5–0.7	Springs for automobiles, dies for forging, rails, wire ropes.
0.7–0.8	Setts, engine cylinder liners, hand tools, saws, springs.
0.8–0.9	Cold setts, punches, chisels, shear blades.
0.9–1.1	Cutting tools, milling cutters, screwing dies, wood working tools, punches.
1.1–1.5	Wood working tools, lathe tools, drills, reamers, razors, wire drawing dies.

Steel containing from 0.7–1.4% carbon is often spoken of as cast steel. It was extensively employed in all cases in which a good cutting edge was required, but it has been largely superseded by the present-day range of alloy steels which have a harder cutting edge and are more durable.

Steels for springs generally contain either silicon and manganese (e.g. 0.5% carbon, 1.7% silicon, 0.7% manganese) or chromium (e.g. 0.5% carbon, 0.75% chromium), and they should be correctly heat-treated to develop their maximum tensile strength and elastic properties.

As the carbon content of the steel increases, it becomes increasingly difficult to weld. Steels containing from 0.6 to 1.0% can be *arc welded* using special electrodes. The study of the welding of alloy steels either by arc or oxy-acetylene is still proceeding; hence the welding of the steels now to be studied is possible in certain cases if the correct procedure is adopted, filler rods or electrodes of the correct composition are used, and the correct heat treatment given on completion of the operation if necessary. The weldability of these steels is discussed on pp. 366–70, while a table of typical alloy electrodes available is given in the appendix.

Nickel steels

The addition of nickel to a steel increases the strength and toughness. The nickel lowers the critical cooling rate and tends to decompose carbides present to form graphite so that plain nickel steels usually have a lower carbon content. For the higher carbon content nickel steels, manganese, which stabilizes the carbides, is generally added. Nickel refines the grain and limits grain growth and gives a range of steels varying from those suitable for case hardening to types suitable for highly stressed parts such as crank-shafts, and axle shafts. The addition of 36% nickel gives a steel with a very low coefficient of thermal expansion.

Composition of some steels containing nickel

			Tensile strength		
Ni%	C%	Mn%	N/mm²	hbar	Use
1	0.4	1.5	700	70	Stressed parts
3	0.3	0.6	850	85	Stressed parts
5	0.12	0.4	850	85	Case hardening

Chromium steels and nickel-chromium (stainless) steels

Chromium has the opposite effect from nickel on steel because it raises the upper critical temperature and, due to the formation of carbides, it increases the hardness and strength but reduces the ductility and promotes grain growth so that overheating and maintaining at elevated temperature for too long a period should be avoided. When greater amounts of chromium are added the steel becomes resistant to corrosion, a thin layer of resistant chromium oxide forming on the surface. Stainless steel, which must contain a minimum of 11% chromium to be termed stainless, may be divided into two classes: A, ferritic and martensitic; B, austenitic, according to which phase is predominant.

Group A: Martensitic. These steels are alloyed mainly with chromium, 11–18%, and harden upon cooling from welding temperatures, resulting in embrittlement and a tendency to crack in the weld and heat-affected zone so that in general they are not recommended for welding. These steels are hardenable by heat treatment, are magnetic and have a lower coefficient of expansion and a lower thermal conductivity than mild steel. They are used in engineering plant subject to mildly corrosive conditions, for cutlery, sharp-edged instruments, ball and roller bearings, etc., according to the carbon and chromium content.

If welding is to be carried out the work should be pre-heated to 200–400 °C followed by slow cooling after welding to reduce the hardness and danger of cracking. Post-heating to 650–700 °C is also advised and a basic electrode of the 19% Cr, 9% Ni, 3% Mo type used. Electrodes of similar composition to the parent metal are used only for limited applications such as overlaying.
Examples of this class of steels are:

C%	Cr%	Si%	Mn%	Other elements	Weldability
0.28–0.36	12–14	0.8	1.0	—	Not generally recommended.
0.12	11.5–13.5	0.8	1.0	—	Poor brittle welds but type with 0.06% C max., fair.
0.7–0.9	15.5–17.5	0.8	1.0	0.3–0.7 Mo	Poor. Should not be welded.

Group A: Ferritic. These steels are alloyed with chromium and because of their low carbon content the structure is almost completely ferritic. They are magnetic and though easier to weld than the martensitic group because they are not hardenable to any extent by heat treatment, they suffer from grain growth and embrittlement at temperatures above 900 °C, and from a form of intergranular corrosion in the HAZ.

When welding these steels pre-heating to 200 °C is recommended followed by post-heat at 750 °C which helps to restore ductility. An austenitic stainless steel electrode is recommended for mildly corrosive conditions but if the application is in sulphur-bearing atmospheres these attack the nickel. An example of this ferritic type is: C 0.08%, Cr 16–18%, Si 0.8%, Mn 1.0%. Weldability is fair, but welds tend to be brittle.

Group B: Austenitic. These steels are alloyed with chromium and nickel. The presence of nickel makes the steel austenitic at room temperature, confers high temperature strength, helps corrosion resistance and controls the grain growth associated with the addition of chromium. The chromium tends to form carbides while the nickel tends to decompose them, so that in this group of steels the disadvantages of alloying each element are reduced. The addition of molybdenum to these steels is to improve corrosion resistance and high temperature strength. (Temper brittleness may occur when the steels are tempered in the range 250–400°C.) It is the high-nickel, high-chromium steels which are of great importance in welding since much fabrication is done in these steels.

Those steels containing 17.5–19.5% chromium and 8–10% nickel, known as 18/8 from their nearness to this composition, harden with cold work, are non-magnetic, resistant to corrosion, can be polished, machined and cold-worked. They have a coefficient of thermal expansion 50% greater than mild steel but the thermal conductivity is much less than mild steel so that there is a narrower HAZ when they are welded. They are used in chemical, food, textile and other industrial plant subject to corrosive attack, also for domestic appliances and decorative applications.

Weld decay

When the austenitic steels are heated within the range 600–850 °C, carbon is absorbed by the chromium and chromium carbide is precipitated along the grain boundaries (Fig. 2.12). As a result, the chromium content of the austenite in the adjacent areas is reduced and hence the resistance to corrosive attack is lowered. When welding, a zone of this temperature range exists near the weld and runs parallel to it, and it is in this zone that the corrosion may occur and is known as weld decay, though no corrosive effect occurs in the weld itself. Heat treatment, consisting of heating the part to 1100 °C and water quenching, restores the carbon to solid solution but has the great drawback that much of the fabricated work is too large for heat treatment.

The difficulty is overcome by adding small quantities of titanium, niobium (columbium) or molybdenum to the steel. These elements form carbides very easily and thus no carbon is available for the chromium to form carbides. The austenitic steels with these additions are known as stabilized steels and they contain a very low percentage of carbon (0.03–0.1%). They have good welding properties and need no subsequent heat treatment. The steels with 0.03% carbon may have no stabilizers added but are suitable for welding because the low carbon content precludes the formation of carbides.

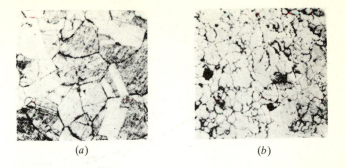

(a)   (b)

(a) Carbide precipitation at the grain boundaries in an 18/8 class stainless steel. The steel is in a state of heat treatment (500–900 °C) in which it is susceptible to intercrystalline corrosion, but it has not yet been subjected to a severe corrosive medium. × 200.

(b) Occurrence of intercrystalline corrosion. The steel is in the same condition as (a), but it has now been subjected in service to a severe corroding medium. × 200.

(c) The effects of intercrystalline corrosion (weld decay) in an 18/8 steel.

Fig. 2.12

Electrodes of 18% Cr–8% Ni; 19% Cr–10% Ni; and 25% Cr–12% Ni, with or without Mo, Nb, Ti and W produce welds which contain residual ferrite and are resistant to hot cracking when the welds are under restraint. Grades with a higher ferrite content under certain conditions in the temperature range 450–900 °C may lose ductility and impact resistance due to the transformation of ferrite into the brittle sigma phase. Most stainless steels however do not encounter these conditions and thus embrittlement does not occur. Manufacturers often indicate the ferrite percentage, e.g. 19% Cr, 9% Ni, 0.05–0.08% C, Nb 10 × carbon content, ferrite 6%.

When welding stainless steel to mild or low-alloy steel dilution (pp. 111–12) occurs and the weld may suffer 20–50% dilution. Root runs in butt joints are greatly affected since the weld metal is in contact with parent metal on both sides. Additional runs are partly in contact with weld metal already laid down and so suffer less dilution.

If a mild or low-alloy steel electrode is used for welding stainless steel to mild or low-alloy steel the weld metal will pick up about 5% Cr and 4% Ni from the stainless steel plate resulting in a hardenable, crack-sensitive weld.

An austenitic steel electrode should be selected such that the weld metal will contain not less than 17% Cr and 7% Ni, otherwise there may not be enough ferrite present to prevent subsequent cracking. Electrodes of 20% Cr, 9% Ni, 3% Mo; or 23% Cr, 12% Ni are the most suitable, since their composition ensures that they accommodate the effects of dilution and there is a sufficiently high ferrite content to give resistance to hot cracking. Electrodes of 20% Cr, 20% Ni are also suitable, except for conditions of high restraint.

If mild steel fittings are to be welded to the exterior of stainless vessels, stainless pads can be first welded to the vessel and the fittings welded to the pads. This reduces the danger of penetration of diluted metal to the face subject to corrosive conditions.

BS 2926 covers the range of chromium–nickel austenitic steels and the chromium steels. The code of composition is : first figure or figures is the chromium content, the second the nickel content and the third the molybdenum content. Nb indicates niobium stabilized, L is the low carbon type and W indicates the presence of tungsten. R is: rutile coating (usually a.c. or d.c.) and B is a basic coating generally d.c. only, electrode +ve.

Example

19.12.3.L.R. is a 19% Cr, 12% Ni, 3% Mo low carbon (0.03% C) rutile-coated electrode.

Identification of stainless and low-alloy steel

If a spot of 30% commercial concentrated nitric acid is placed on grease-free stainless steel, there is no reaction, but on low-alloy or plain carbon steel there is a bubbling reaction.

To determine to which group of stainless steel a specimen belongs, a hand magnet can be used. Ferritic and martensitic steels are strongly magnetic, whilst austenitic steels are generally nonmagnetic. However the austenitic steels become somewhat magnetic when cold worked, but there is considerable difference in magnetic properties between them and the ferritic and martensitic types, which is easily detected. The identification of the various grades within the groups is not easy without laboratory facilities, but an indication is given by the hardness after heat treatment, which varies from 400 HV for a 0.12% C type to 700 HV for a 0.7–0.9% C type.

Percentage composition of some corrosion and heat-resisting steels (the silicon content is 0.2–1.0% and manganese 0.5–2%)

C	Cr	Ni	Mo	Ti	Nb	Weldability
0.1 max.	17–19	8.5–10.5	—	—	—	Good: but for mildly corroding conditions.
0.06 max.	18–19.5	9–11	—	—	—	Good: No heat treatment after welding.
0.03 max.	18–19.5	10–11.5	—	—	—	Good. No subsequent heat treatment.
0.08 max.	16.5–18	12–14	2.25–3	0.6 max. to 0.4C min.	—	Good. May require subsequent heat treatment.
0.1 max.	17–19	8.5–10.5	—	0.7 max to 5C min.	—	Good. No subsequent heat treatment.
0.08 max.	17–19	9–11	—	—	1.0 max. to 10 C min.	Good. No subsequent heat treatment.
0.15 max.	22–24	13–15	—	—	—	Welded with 25–12 and 25–12–Nb electrodes.
0.12	24–26	19–22	—	—	—	Welded with 25–20 and 25–20–Nb electrodes.

Students wishing to study further the metallurgical aspects of stainless steels should consult the Shaeffler's diagram in a text book on metallurgy. In this the various alloying elements are expressed in terms of nickel, which is austenite-forming and chromium, which tends to form ferrite. Using these values in conjunction with the diagram, information can be obtained on the possible behaviour of the steel during welding.

Steel containing molybdenum

Addition of 0.15–0.3% molybdenum to low-nickel and low-chrome steels reduces the tendency to temper brittleness and gives high impact strength. It increases the strength and creep resistance at elevated temperatures and increases the resistance to corrosion of stainless steels.

Examples of steels containing molybdenum

C %	Ni %	Cr %	Mn %	Mo %	
0.25	—	0.5–1.0	0.5	0.5–0.8	Heat treatable. Used for stressed parts up to 500 °C.
0.4	1.5	1.2	0.5	0.3	Heat treatable. Shock and fatigue resistant.

The nickel–chrome–molybdenum steels have high strength combined with good ductility and are used for all applications in engine components such as shafts and gears involving high stress.

Steel containing vanadium

Vanadium reduces grain growth and, due to the formation of its carbide which can be taken into solid solution, gives strength and resistance to fatigue at elevated temperatures. It is usually used in conjunction with chromium and in many ways these steels resemble the nickel-chromium steels.

Examples of steels containing vanadium

C %	Mn %	Si %	Cr %	Mo %	V %	
0.45	0.6	0.25	1.25	—	0.15 min.	Spring steel.
0.13	1.4	0.25	0.6	0.3	0.1	'Supertough'.

Manganese steel

Manganese, in the form of ferromanganese, is used for deoxidation of steel and in most steel there is a manganese content usually less than 1%. Above this value it may be considered an alloying element and lowers the critical temperature. The most widely used manganese steel is that containing 12–14% Mn and 1.2%C. This is austenitic but it hardens greatly with cold work and is widely used for components subject to wear and abrasion since the core retains its toughness while the surface layers harden with cold work to an intense degree. These steels have a wide range of applications in earth-moving equipment, rolls, dredger bucket lips and in all cases where resistance to wear and abrasion is of paramount importance. (See also p. 351.)

Steel containing tungsten

Tungsten reduces the tendency to grain growth, raises the upper critical temperature, forms very hard, stable carbides which remain in solution after oil quenching and renders the steel very hard and suitable for cutting tools and gauges.

High-speed steel usually contains tungsten, chromium, vanadium, molybdenum and cobalt with a carbon content of 0.6–1.5%. Tungsten carbide, made by the sintering process, is used for tool tips for cutting tools, is extremely hard and brittle, and is brazed on to a carbon or alloy steel shank.

Nitralloy steels

These steels contain silicon, manganese, nickel, chromium, molybdenum and aluminium in varying proportion, and their carbon content varies from 0.2 to 0.55%. They are eminently suited for purposes where great resistance to wear is required. After being hardened, by the process of nitriding or nitrogen hardening them, they have an intensely glass hard surface and are suitable in this state for crankshafts, camshafts, pump spindles, shackle bolts, etc. (see Heat Treatment).

THE EFECT OF HEAT ON THE STRUCTURE OF STEEL

Suppose we heat a piece of steel containing a small percentage of carbon and measure its temperature rise. We find that after a certain time, although we continue supplying heat to the steel, the temperature ceases to rise for a short time and then begins to rise again at a uniform rate. Evidently at this arrest point, termed a critical point, the heat which is being absorbed (decalescence), since it has not caused a rise in temperature, has caused a change to occur in the internal structure of the steel.

If the heating is continued, we find that a second arrest or critical point occurs, but the effect is not nearly as marked as the first point. At a higher temperature still, a third critical point occurs, similar in effect to the first.

If the steel is now allowed to cool at a uniform rate, we again have the three critical points corresponding to the three when the steel was heated, but they occur in each case at a slightly lower temperature than the corresponding point in the heating operation. At these points in the cooling operation, the metal gives out heat but the temperature remains steady. The evolution of heat on cooling through the critical range and which is visible in a darkened room is known as recalescence. The second arrest points which occur between upper and lower critical temperatures involve the loss and gain of magnetic properties and need not concern us here.

If the experiment is done with steels of varying carbon content it will be found that the lower critical point is constant at about 720 °C for all steels, but the temperature of the upper critical point decreases with increasing carbon content, until at 0.85% carbon the two critical points occur at the same temperature. Figure 2.13 is part of the iron–carbon equilibrium or constitutional diagram which shows how the structure of any plain carbon steel changes with temperature and shows the limits of temperature and composition in which the various constituents are stable (see Chapter 13 on equilibrium diagrams). We have seen that iron below 900 °C has a body-centred cubic structure (a iron), and in this

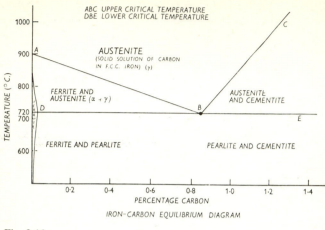

IRON–CARBON EQUILIBRIUM DIAGRAM

Fig. 2.13

state it can only have a trace of carbon in solution (up to 0.03% at 700 °C). When, however, the temperature rises above its upper critical point and the structure changes to face-centred cubic γ iron, this type of iron can have up to 1.7% carbon in solution before it is saturated. Thus when a steel is heated above its upper critical point the carbon, which was present in the combined form as cementite, dissolves in the iron to form a solid solution of iron and carbon, that is, dissolves completely in the iron although the iron is still in the solid state. This solid solution is called austenite (see Fig. 2.15a) and the carbon is diffused uniformly throughout the iron. Now let us lower the temperature of a 0.3% carbon steel slowly from above the upper critical point, about 850 °C. At the upper critical point the structure begins to change. Body-centred cubic crystals of ferrite are precipitated and the carbon content of the remaining austenite begins to increase. This continues as the temperature falls until at the lower critical point the austenite contains 0.85% carbon; it is saturated. The austenite now precipitates cementite (ferrite and carbon chemically combined) at the lower critical point, and it does this in alternate layers with the ferrite that is separating out, forming the areas of pearlite (Fig. 2.14). A 0.5% carbon steel will change in the same way, but begins the change at a lower temperature since the upper critical point is about 800 °C. In the case of a 0.85% carbon steel the transformation begins and ends at about 720 °C, since upper and lower critical points are at the same temperature, and the final structure will be all pearlite. A steel with more than 0.85% carbon will begin to precipitate carbon in the form of cementite at the upper critical point since

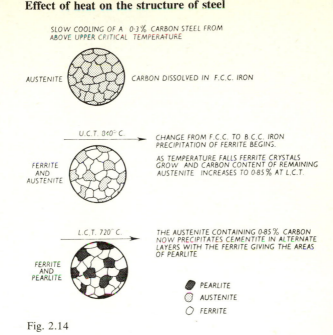

SLOW COOLING OF A 0·3% CARBON STEEL FROM ABOVE UPPER CRITICAL TEMPERATURE

AUSTENITE — CARBON DISSOLVED IN F.C.C. IRON

U.C.T. 840° C.

FERRITE AND AUSTENITE

CHANGE FROM F.C.C. TO B.C.C. IRON PRECIPITATION OF FERRITE BEGINS.

AS TEMPERATURE FALLS FERRITE CRYSTALS GROW AND CARBON CONTENT OF REMAINING AUSTENITE INCREASES TO 0·85% AT L.C.T.

L.C.T. 720° C.

FERRITE AND PEARLITE

THE AUSTENITE CONTAINING 0·85% CARBON NOW PRECIPITATES CEMENTITE IN ALTERNATE LAYERS WITH THE FERRITE GIVING THE AREAS OF PEARLITE

PEARLITE
AUSTENITE
FERRITE

Fig. 2.14

carbon is in excess of 0.85%. At the lower critical point this transforms to pearlite so that the final structure is pearlite and cementite.

Heat treatment of steel

Hardening. In order to harden a carbon steel it must be heated to a temperature of 20–30 °C above its upper critical temperature and kept at this temperature long enough to ensure that the whole mass is at this temperature and the structure is austenitic. Maximum hardness can now be obtained by quenching the steel in water or brine.

Examined under the microscope the structure appears as fine needle-like (acicular) crystals and is known as martensite (Fig. 2.15b); and the steel is hard and brittle. The rapid quench has prevented the normal change from austenite to ferrite and pearlite taking place. Quenching less severely produces bainite, a structure which when viewed under high magnification can be seen as an aggregate of ferrite and carbide particles like finely divided pearlite (Fig. 2.15c). Martensite and bainite are often found together in quenched steels.

The rate of cooling is measured by the fall in temperature per second and can vary from a few to some hundreds of degrees depending upon the method used. The Critical Cooling Rate for a steel is the lowest rate at which a steel can be quenched to give an all-martensitic structure. At lower rates bainite and/or finely divided pearlite will form. If the steel has a large mass the outer layers will cool quickly when quenched, giving maximum hardness, while the core will cool much more slowly and will be softer (mass effect). Although simple shapes quench successfully, more complicated shapes may suffer distortion or cracking or internal stresses may remain. In this case where a rapid quench would lead to complications, an alloy steel (e.g. one containing nickel) can be used instead of a plain carbon steel. The nickel lowers the critical cooling rate by slowing up the rate of transformation of austenite into its decomposition products so that a less drastic quench is required to produce martensite, reducing the mass effect and the risk of distortion, cracking and internal stresses.

(*a*) Austenite in an 18% chromium 8% nickel steel sensitized and etched to show the grain boundaries. × 500.

(*b*) Martensite. × 250.

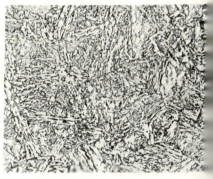

(*c*) Bainite in a low chromium nickel and molybdenum steel transformed over the temperature range 570–430 °C. × 500.

Fig. 2.15

Air-hardening steels usually contain sufficient additions of nickel and chromium to reduce the critical cooling rate so that the steel is hardened by cooling in air, followed by tempering as required.

Quenching media include the following in decreasing order of quenching speed: caustic soda (sodium hydroxide), 5% solution; brine, 5–25% solution; cold water; hot water; mineral, animal, vegetable oils.

Tempering. The hardness and brittleness of a rapidly quenched steel together with the possibility that there may be internal stresses in the steel make the steel unsuitable for use unless the greatest possible hardness is required. The hardness and brittleness can be reduced and internal stresses relieved by tempering, in which process the hard and brittle martensitic structure is transformed to softer and more ductile structures but yet harder and tougher than ferrite and pearlite. To temper a steel it is re-heated to a definite temperature after hardening and then cooled.

(a) Heating to the range 200–250 °C relieves immediate lattice stress but the overall stress pattern persists to fairly high temperatures.

(b) Diffusion of carbon from martensite begins at about 150 °C and is practically complete at 350 °C. Heating to the range 150–350 °C forms a mixture of finely divided ferrite and cementite, not so hard but tougher then martensite.

(c) Coalescence of the carbides (cementite) starts at about 350 °C and is almost complete by about 650 °C, thus tempering in this range produces a structure similar to that in (b) but with larger particles and is softer and more ductile, the particle size depending upon the precise tempering temperature (Fig. 2.16b).

The structures obtained by tempering are stages in the austenite–pearlite transformation due to variations in the size and shape of the carbide particles and the way in which they are found in the ferrite matrix.

Interrupted quenching processes reduce internal stresses and distortion and reduce the possibility of quench cracking. If a steel is heated above its upper critical temperature and is then quenched in a bath of molten metal (lead or tin) or salt, kept at a fixed temperature, the quench is not so drastic and there is less temperature gradient between surface and core, reducing stress and distortion effects. If the steel is held at this temperature for varying periods of time and is then quenched out, time–temperature-transformation curves can be drawn indicating the various structures obtained by varying time and temperature.

A process termed martempering can be used to obtain a martensitic structure without the disadvantages of a drastic quench. The steel is quenched from the austenitic condition into a bath of molten metal kept at a temperature just above that at which martensite can form (260–370 °C) until it has a uniform temperature throughout and is then cooled in air, the structure being fine-grained martensite, and the thermal stresses are minimized. Austempering is an interrupted quenching process in which the steel is quenched from the austenitic condition into a bath of molten metal kept at a temperature below the critical range but above the temperature at which martensite can form. It is held at this temperature until complete transformation has occurred and then cooled to room temperature, the structure being pearlite and bainite.

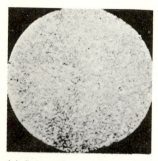

(*a*) Spheroidal pearlite in a 1% carbon steel.

(*b*) Variation in particle size due to different tempering temperatures in a low alloy steel. × 250.

Fig. 2.16

A method often used to obtain a temper on a cutting tool consists of raising the part to bright red heat and then quenching the cutting end of the tool in water. The tool is then removed, any oxide that has formed is quickly polished off and the heat from the part which was not quenched travels by conduction to the quenched end, and the temper colours, formed by light interference on the different thicknesses of layers of surface oxide, begin to appear. The tool is then entirely quenched, when the required colour appears. The colours vary from pale yellow (220 °C), through straw, yellow, purple brown, purple, blue to dark blue (300 °C).

Accurate control of tempering temperature can be obtained by using furnaces with circulating atmospheres; oils for the lower temperature range; liquid salt (potassium nitrate, sodium nitrite, etc.) or liquid metal, e.g. lead.

Annealing. Annealing is the process by which the steel is softened, and internal strains are removed. The process consists of heating the metal to a certain temperature and then allowing it to cool very slowly out of contact with the air to prevent oxidation of the surface. After the first softening is obtained, if the annealing is pro-

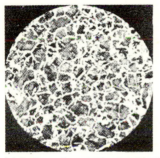

(a) 0.4% carbon steel normalized at 850 °C.

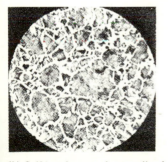

(b) 0.4% carbon steel normalized at 1000 °C.

(c) 0.4% carbon steel normalized at 1200 °C.

Fig. 2.17. Grain growth due to normalizing at increasingly high temperatures. × 100.

longed, large crystals are formed. These, as is usual with all crystals, grow in size and decrease in number as the annealing continues. This is known as crystal growth or grain growth (Fig. 2.17).

As these grow in size, the resistance of the metal to shock and fatigue is greatly lowered; hence over-annealing has the bad effect of promoting grain growth, resulting in reduction in resistance to shock and fatigue.

The annealing temperature should be about 50 °C above the upper critical point and therefore varies with the carbon content of the steel. Low-carbon steel should therefore be heated to about 900 °C, while high-carbon steel should be heated to about 760 °C.

Use is made of the fact that iron (ferrite) recrystallizes at 500–550 °C in the treatment known as process annealing, which is used for mild steel articles which have been cold worked during manufacture. They can be packed in a box with cast iron filings over them, the lid luted on with clay, and then heated to 550–650 °C. Recrystallization of the ferrite takes place (there is little pearlite in the structure) and they are allowed to cool out very slowly in the box, after which considerable softening has taken place (Fig. 2.18).

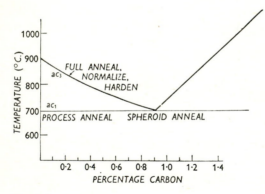

Fig. 2.18. Heat treatment of steel.

High-carbon steels whose structure is mostly pearlite can be annealed by heating to 650–700 °C. At this temperature the cementite forms or balls up into rounded shapes, and the steel is softer and may be drawn and worked. At temperatures above this, pearlite is reformed and the steel becomes hard (Fig. 2.16a).

Normalizing. This process consists in raising the steel only slightly above the upper critical point, keeping it at this temperature for just sufficient time to heat it right through, and then allowing it to cool as rapidly as possible in air. This causes a refining of the structure, since recrystallization takes place, and a coarse structure becomes much finer, since the steel is not held at the high temperature long enough for any grain (or crystal) growth to take place (see Fig. 2.17).

Overheated steel. If steel is exposed to too high a temperature or for too long a time to temperatures above the upper critical point it becomes overheated. This means that a very coarse structure occurs and, on cooling, this gives a similar coarse structure of ferrite and pearlite. This structure results in great reduction in fatigue resistance, impact strength, and a reduced yield point, and is therefore undesirable.

Steel which has been overheated is therefore extremely unsatisfactory. Correct heat treatment will restore the correct structure.

Burnt steel. If steel is heated to too high a temperature, this may result in a condition which cannot be remedied by heat treatment and the steel is said to be 'burnt'. This condition is due to the fact that the boundaries of the crystals become oxidized, due to absorption of oxygen at high temperature, and hence the steel is weakened (Fig. 2.19).

Fig. 2.19. Oxidation along crystal boundaries in mild steel which has been overheated and 'burnt'.

Case-hardening. Case-hardening (and also pack-hardening) is a method by which soft low-carbon steel is hardened on the surface by heating it in contact with carbonaceous material (material containing carbon). Parts to be case-hardened are packed in a box and covered with carbonaceous powder, such as charred leather, powdered bone, animal charcoal, or cyanide of potassium (KCN). The box is then placed in the furnace and heated above the critical temperature (that is, above 900°C, depending on the carbon percentage in the steel). The steel begins to absorb carbon at red heat and continues to do so, the carbon diffusing through the surface. The box is then removed from the furnace and the parts on being taken out can either be directly water or oil quenched. Another favoured method is to allow them to cool out slowly, then heat up to about 800 °C, and quench in oil or water, depending on the hardness required in the case.

In the process known as gas carburizing carbon is introduced into the surface of the part to be hardened by heating in a current of a gas with a high carbon content such as hydrocarbons or hydrocarbons and carbon monoxide. This process is extensively used today and lends itself to automation with accurate control and uniformity of hardness.

The case-hardening furnace is almost always found nowadays as part of the equipment of large engineering shops. Parts such as gudgeon pins, shackle bolts and camshafts and, in fact, all types of components subject to hard wear are case-hardened.

The drawback to the process is that, owing to the quenching process, parts of complicated shape cannot be case-hardened owing to the risk of distortion.

The percentage of carbon in steels suited to case-hardening varies from 0.15 to 0.25%. Above this, the core tends to become hard. The carbon content of the case after hardening may be as high as 1.1%, but is normally of about 0.9%C to a depth of 0.1 mm.

Nitriding or Nitrarding. This process consists of hardening the surface of 'nitralloy steels' (alloy steels containing aluminium and nickel) by heating the steel to approximately 500 °C in an atmosphere of nitrogen. The steel to be nitrarded is placed in the furnace and ammonia gas (NH_3) is passed through it. The ammonia gas splits up, or 'cracks', and the nitrogen is absorbed by the steel, while the hydrogen combines with the oxygen and steam is formed, passing out of the furnace. The parts are left in the furnace for a period depending on the depth of hardening required, because this process produces a hardening effect which decreases gradually from the surface to the core and is not a 'case' or surface hardening. When removed from the furnace, the parts are simply allowed to cool. The nitralloy steel is annealed before being placed in the furnace and the parts can be finished to the finest limits, since the heat of the furnace is so low (500 °C) that distortion is reduced to a minimum, and there is no quenching. Nitralloy steel, after the nitrarding process, is intensely hard, and it does not suffer from the liability of the surface to 'flake' as does very hard case-hardened steel.

It is used extensively today in the automobile and aircraft industries for parts such as crankshafts, pump spindles, etc.

THE EFFECT OF WELDING ON THE STRUCTURE OF STEEL

A typical analysis of all-weld metal-deposit mild steel is: Carbon 0.06–0.08%, manganese 0.43–0.6%, silicon 0.12–0.4%, sulphur 0.02% max, phosphorus 0.03% max, remainder iron. During the welding process, the molten metal is at a temperature of from 2500 to 3000 °C and the weld may be considered as a region of cast steel. Since regions near the weld may be comparatively cool, giving a steep thermal gradient from weld to parent plate, it will be possible to find crystal structures of all types in the vicinity, and great changes may take place as the rate of cooling is altered.

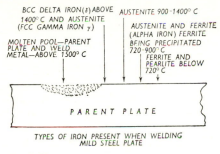

TYPES OF IRON PRESENT WHEN WELDING
MILD STEEL PLATE

Fig. 2.20

A typical cross section from the molten pool to the cold section of the parent plate might reveal the following regions (Fig. 2.20).

(a) The molten pool with parent plate and weld metal mixed at temperatures above melting point 1500 °C.

(b) A region of BCC delta iron and austenite (FCC gamma iron) mixed.

(c) A region of austenite above the upper critical temperature, 900–1400 °C.

(d) A region of austenite and ferrite (BCC alpha iron), where ferrite is being precipitated (between upper and lower critical temperatures).

(e) The parent plate of ferrite and pearlite.

There is a high possibility, in addition, that oxygen or even nitrogen may be absorbed into the weld itself. We have seen that when oxidation occurs on the crystal boundaries, the impact strength and fatigue resistance of the metal are greatly reduced and, hence a weld which has absorbed oxygen will show these symptoms. The formation of iron nitride (Fe_4N) also makes the weld brittle. The nitride is usually present in the form of fine needle-shaped crystals visible under the high-powered microscope (Fig. 2.21). The weld must be safeguarded from these defects as much as possible.

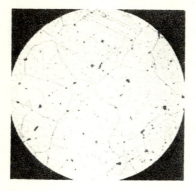

Fig. 2.21. Nitride needles in ferrite.

Evidently, also, the composition of the filler rod or electrode compared with that of the parent metal will be of great importance, since this will naturally alter the properties of the steel at or near the weld. If the mass of the parent metal is small and cooling is very quick, the weld may be tough and strong but brittle due to the presence of martensite and this will particularly be the case if the carbon content is high. If, however, cooling is slow, structures of varying forms of ferrite and pearlite are found, giving a lower strength and decreased hardness, but at the same time a very much increased ductility and impact strength. Evidently, therefore, the welding of a given joint requires consideration as to what properties are required in the finished weld (tenacity, ductility, impact strength, resistance to wear and abrasion, etc.). When this is settled the method of welding and the rate of cooling can be considered, together with the choice of suitable welding rods.

These considerations are of particular importance in the case of the welding of alloy steels, since great care is necessary in the choice of suitable welding rods, which will give the weld the correct properties required. In many cases, heat treatment is advisable after the welding operation, to remove internal stresses and to modify the crystal structure, and this treatment must be given to steels such as high tensile and chrome steels. The study of welding of these steels is, however, still proceeding (pp. 357 and 366).

During the welding process, the part of the weld immediately under the flame or arc is in the molten condition, the section that has just been welded is cooling down from this condition, while the section to be welded is comparatively cold. This, therefore, is virtually a small steel-casting operation, the melting and casting process taking place in a very short time and the weld metal after deposition being 'as cast' steel.

As a result we expect to find most of the various structures (martensite, bainite) that we have considered, and the point of greatest interest to the welder is, what structures will remain on cooling. Evidently the structure that remains will determine the final strength, hardness, ductility and resistance to impact of the weld. Since these structures will be greatly affected by the absorption of any elements that may be present, it will be well to consider these first.

Oxygen

Oxygen may be absorbed into the weld, forming iron oxide (Fe_3O_4) and other oxides such as that of silicon. This iron oxide may also be absorbed into the weld from the steel of the welding rod or electrode. If iron oxide is formed it may react with the carbon in the steel to form carbon monoxide, resulting in blowholes. If this iron oxide is present in any quantity (as is the case when using bare wire electrodes in arc welding, or excess of oxygen in the oxy-acetylene process), oxidation of the weld will occur and this produces a great increase in the grain size, which is easily observed on the microphotograph. Even normalizing will not then produce a fine grain. This oxygen absorption, therefore, has a bad effect on the weld, reducing its tensile strength and ductility and decreasing its resistance to corrosion. Covered arc welding electrodes may contain deoxidizing material to prevent the formation of iron oxide, or sufficient silica to act on the iron oxide to remove it and form iron silicate (slag).

Nitrogen

The percentage of nitrogen in weld metal can vary considerably, and the results of experiments performed have led to the following conclusions:

(a) There is very low absorption by the oxy-acetylene process (maximum 0.02%).

(b) There is much greater absorption in arc welding (0.15 to 0.20%) which is influenced by (1) the current conditions that may cause the nitrogen content to vary from 0.14 to 0.2%, (2) the nature of the atmosphere. By the use of electrodes covered with hydrogen-releasing coatings, e.g. sawdust, the nitrogen content may be brought down to 0.02%.

(c) As regards the thickness of the coatings, use of a very thick covering may reduce the nitrogen from 0.15 to 0.03%.

Nitrogen is found in the weld metal trapped in blowholes (although nitrogen itself does not form the blowholes) and as crystals of iron nitride (Fe_4N), known as nitride needles (Fig. 2.21). Nitrogen, however, may be in solution in the iron, and to cause the iron-nitride needles to appear the weld has to be heated up to about 800–900 °C. The nitrogen tends to increase the tensile strength but *decreases* the ductility of the metal. Low-nitrogen steels are now supplied when required for deep pressing since they are less prone to cracking.

Hydrogen

Hydrogen is absorbed into mild steel weld metal during arc welding with covered electrodes. The hydrogen is present in the composition of many flux coatings and in its moisture content. It begins to diffuse out of the weld metal immediately after the welding process, and continues to do so over a long period. The presence of this hydrogen reduces the tensile strength of the weld. (An experiment to illustrate this is given on p. 301.)

Carbon

If we attempt to introduce carbon into the weld metal from the filler rod, the carbon either oxidizes into carbon monoxide during the melting operation or reacts with the weld metal and produces a porous deposit. Arc welded metal cools more quickly than oxy-acetylene welded metal and hence the former may be expected to give a less ductile weld, but the quantity of carbon introduced in arc welding (pickup) is too small to produce brittle welds in this way.

The effect on the weld metal of the carbon contained in the parent

Fig. 2.22. Crack in weld metal showing structurally altered layers.

metal is, however, important, especially when welding medium or high carbon steels. In this case, the carbon may diffuse from the parent metal, due to its relatively high carbon content, into the weld metal and form, near the line of fusion of weld and parent metal bands of high carbon content sufficient to produce hardness and brittleness if cooled rapidly (Fig. 2.22).

Structural changes

The question of change of structure during welding depends amongst other things on:

(1) The process used, whether arc or oxy-acetylene.
(2) The type and composition of the filler rod and, if arc welding is employed, the composition of the covering of the electrode.
(3) The conditions under which the weld is made, i.e. the amount of oxygen and nitrogen present.
(4) The composition of the parent metal.

The change of structure of the metal is also of great importance, as previously mentioned. This will depend largely upon the amount of carbon and other alloying elements present and upon the rate at which the weld cools.

In arc welding the first run of weld metal flowing on to the cold plate is virtually chill cast. The metal on the top of the weld area freezes quickly as the heat is removed from it and small chill crystals are formed. Below, these crystals grow away from the sides towards the hotter regions of the weld metal and thus growth is faster than the tendency to form new dendrites so that columnar crystals are formed.

Because of the high temperature of the molten metal in arc welding these crystals have enough time to grow. On each side of the weld is a heat-affected zone in which the temperature has been raised above the recrystallization temperature and in which therefore grain growth has occurred. Beyond this zone the plate structure is unaffected (Figs. 2.23a, 2.25).

When a second run is placed over the first there will be region thus:

(1) A refined area in the first run where recrystallization temperature is exceeded and the columnar crystals of the first run are reformed as small equi-axed crystals.
(2) A region between this and the parent plate where grain growth has taken place because the temperature has been well above recrystallization temperature.
(3) The second run will form columnar crystals on its below surface

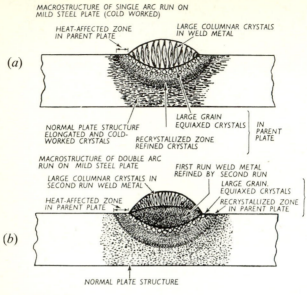

MACROSTRUCTURE OF SINGLE ARC RUN ON
MILD STEEL PLATE (COLD WORKED)

HEAT-AFFECTED ZONE
IN PARENT PLATE

LARGE COLUMNAR CRYSTALS
IN WELD METAL

(a)

NORMAL PLATE STRUCTURE
ELONGATED AND COLD-
WORKED CRYSTALS

RECRYSTALLIZED ZONE
REFINED CRYSTALS

LARGE GRAIN
EQUIAXED CRYSTALS
IN PARENT PLATE

MACROSTRUCTURE OF DOUBLE ARC
RUN ON MILD STEEL PLATE

FIRST RUN WELD METAL
REFINED BY SECOND RUN

LARGE COLUMNAR CRYSTALS IN
SECOND RUN WELD METAL

HEAT-AFFECTED ZONE
IN PARENT PLATE

LARGE GRAIN,
EQUIAXED CRYSTALS

RECRYSTALLIZED ZONE
IN PARENT PLATE

(b)

NORMAL PLATE STRUCTURE

Fig. 2.23

layers because of the quick cooling when in contact with the atmosphere (Figs. 2.23b, 2.26).

In gas welding the heat-affected zone will be wider than in arc welding because, although the flame temperature is below that of the arc, the arc is more localized and the temperature is raised more quickly. When the flame is applied to the plate its temperature is raised so that chill casting does not occur. Grain growth, however, will be more pronounced because the heat is applied for a longer period than in arc welding (Fig. 2.24a, b).

A consideration of this subject makes very evident the reason why austenitic alloy steels present such a problem in welding. These steels owe their properties to their austenitic condition, and immediately they are subjected to the heat of the welding process, they have their structure greatly modified. It is nearly always imperative that after welding steels of this class, they should be heat treated in order to correct as far as possible this change of structure. In addition, owing to the number of alloying elements contained in these steels, it becomes very difficult to obtain a weld whose properties do not differ in a marked degree from those of the parent metal. Hence the welding of alloy steels must be considered for each particular steel and with reference to the particular

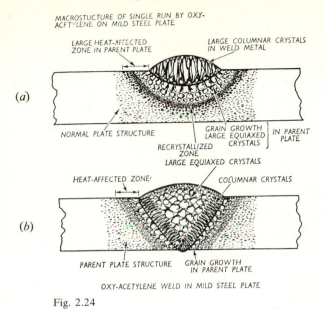

MACROSTUCTURE OF SINGLE RUN BY OXY-ACETYLENE ON MILD STEEL PLATE

LARGE HEAT-AFFECTED ZONE IN PARENT PLATE

LARGE COLUMNAR CRYSTALS IN WELD METAL

(a)

NORMAL PLATE STRUCTURE

RECRYSTALLIZED ZONE

GRAIN GROWTH
LARGE EQUIAXED CRYSTALS } IN PARENT PLATE

LARGE EQUIAXED CRYSTALS

HEAT-AFFECTED ZONE

COLUMNAR CRYSTALS

(b)

PARENT PLATE STRUCTURE

GRAIN GROWTH IN PARENT PLATE

OXY-ACETYLENE WELD IN MILD STEEL PLATE

Fig. 2.24

requirements and service conditions. In addition, great care must be taken in selecting a suitable electrode or filler rod.

The microscope can be used extensively to observe the effect of welding on the structure. When once the observer is trained to recognize the various structures and symptoms, the microscope provides accurate information about the state of the weld. The microscope can indicate the following points, which can hardly be found by any other method and, as previously mentioned, can also indicate fine hair cracks unperceived in X-ray photographs, together with any slag inclusions and blowholes of microscopic proportions. Faults in structure indicated by the microscope using various magnifications are:

(1) True depth of penetration of the weld as indicated by the crystal structure.
(2) The actual extent of the fusion of weld and parent metal (Fig. 2.25).
(3) The actual structure of the weld metal and its condition (Figs. 2.25, 2.26, 2.27, 2.28).
(4) The area over which the disturbance due to the heating effect of the welding operation has occurred.
(5) The amount of nitrogen and oxygen absorption as seen by presence of iron oxide and iron nitride crystals.

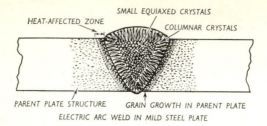

HEAT-AFFECTED ZONE — SMALL EQUIAXED CRYSTALS — COLUMNAR CRYSTALS

PARENT PLATE STRUCTURE — GRAIN GROWTH IN PARENT PLATE

ELECTRIC ARC WELD IN MILD STEEL PLATE

Fig. 2.25

(a)

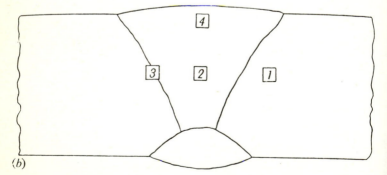

(b)

Fig. 2.26. Section of a flat butt weld in rolled steel plate (×3). Macrographs of the regions labelled 1, 2, 3 and 4 are shown in Figs. 2.27 and 2.28.

(1) Rolled steel plate. ×75

(2) Weld metal showing refined structure. ×75

Fig. 2.27

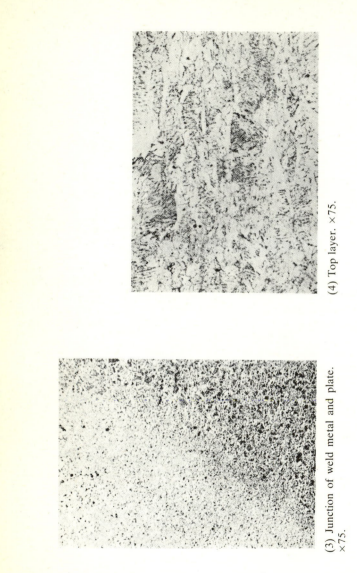

(3) Junction of weld metal and plate. ×75.

(4) Top layer. ×75.

Fig. 2.28

Dilution

When two metals are fusion welded together by metal arc, TIG, MIG or submerged arc processes, the final composition consists of an admixture of parent plate and welding wire. The parent plate has melted in with the filler and has diluted it and this dilution may be expressed as a percentage thus:

$$\text{percentage dilution} = \frac{\text{weight of parent metal in weld}}{\text{total weight of weld.}} \times 100.$$

If there are 15 parts by weight of parent plate in 75 parts by weight of weld metal then the dilution of $15/75 \times 100 = 20\%$.

Average values of dilution for various processes are:

Metal arc	25–40%
Submerged arc	25–40%
MIG (spray transfer)	25–50%
MIG (dip transfer)	15–30%
TIG	25–50%

and it can be seen that minimum dilution is obtained using MIG (dip transfer). Many factors affect dilution. Evidently with a single-run weld there will be greater dilution than with multi-runs and there is always considerable dilution in any root run. The greater the amount of weaving the greater the dilution.

When dissimilar metals are to be welded together the final weld will suffer dilution from each parent plate and for a successful weld it must do so without major defects including liability of cracking. In addition the physical and mechanical properties and resistance to corrosion must be as near as possible to the parent plate. As all these criteria are not generally obtainable a welding wire must be chosen which gives the optimum properties for a given situation.

Example

A plate of 9% Ni steel is welded with an inconel welding wire of composition 80% Ni–20% Cr. What will be the approximate composition of the final weld if there is 40% dilution.

With 40% dilution the plate will contribute 40% and the welding wire 60%.

	Nickel	Iron	Chromium
9% nickel plate	$\frac{40}{100} \times 9 = 3.6$	$\frac{40}{100} \times 91 = 36.4$	—
80% Ni–20% Cr welding wire	$\frac{60}{100} \times 80 = 48$	—	$\frac{60}{100} \times 20 = 12$
	51.6	36.4	12

Therefore the approximate composition of weld metal is:

51.5% Ni   36.5% Fe   12% Cr

A plate of monel (70% Ni–30% Cu) is to be welded to a plate of stainless steel (18% Cr–12% Ni–70% Fe) using an Incoweld A wire (75% Ni–15% Cr–8% Fe). Assuming 30% dilution, what will be the approximate composition of the final weld?

With 30% dilution each plate will contribute 15% and the welding wire 70%.

	Nickel	Chromium	Copper	Iron
Monel plate	$\frac{15}{100} \times 70 = 10.5$	—	$\frac{15}{100} \times 30 = 4.5$	—
Stainless steel plate	$\frac{15}{100} \times 12 = 1.8$	$\frac{15}{100} \times 18 = 2.7$	—	$\frac{15}{100} \times 70 = 10.5$
Welding wire	$\frac{70}{100} \times 75 = 52.5$	$\frac{70}{100} \times 15 = 10.5$	—	$\frac{70}{100} \times 8 = 5.6$
	64.8	13.2	4.5	16.1

Therefore the approximate composition of weld metal is:

65% Ni,   13% Cr,   4.5% Cu,   16% Fe.

Pick-up. This is the term applied to the absorption or transfer of elements from parent plate or non-consumable electrode into the weld metal and is closely associated with dilution. When overlaying plain low-carbon steels with nickel-base alloys there is a tendency for the weld metal to pick up iron from the parent plate, resulting in a lowering of corrosion resistance of the overlay. The pick-up must be kept as low as possible. When welding cast iron with the metal arc processes carbon pick-up can lead to the undesirable excessive precipitation of carbides which are hard and brittle. Tungsten pick-up can occur when welding with the gas-shielded tungsten electrode (TIG) process when excessive currents are used, resulting in the pick-up of tungsten particles from the electrode into the weld metal.

Cracking in steel

A crack is a fissure produced in a metal by tearing action. Hot, or solidification cracking is caused in the weld metal itself by tearing of the grain boundaries before complete solidification has taken place and while the metal is still in the plastic state. The crack may be continuous or discontinuous and often extends from the weld root and may not extend to the face of the weld. Cold cracking occurs in both weld metal and adjacent parent plate (HAZ) and may be due to excessive restraint on the joint, insufficient cross-sectional area of the weld, presence of hydrogen in the weld metal or embrittlement in the HAZ of low alloy steels.

Factors which may promote hot cracking

Current density: a high density tends to promote cracking.

The distribution of heat and hence stress in the weld itself.
Joint restraint and high thermal severity.
Crack sensitivity of the electrode.
Dilution of the weld metal.
Impurities such as sulphur, and high carbon or nickel content.
Pre-heating which increases the liability to cracking.
Weld procedure. High welding speeds and long arc increase sensitivity and crater cracking indicates a crack sensitivity.

Factors which may promote cold cracking

Joint restraint and high thermal severity.
Weld of insufficient sectional area.
Hydrogen in the weld metal.
Presence of impurities.
Embrittlement of the HAZ (low-alloy steels).
High welding speeds and low current density.

THE EFFECT OF DEFORMATION ON THE PROPERTIES OF METALS

Cold working

When a ductile metal is subjected to a stress which exceeds the elastic limit, it deforms plastically by an internal shearing process known as *slip*.

Plastic deformation is permanent so that when the applied stress is removed the metal remains deformed and does not return to its original size and shape as is the case with *elastic* deformation.

Plastic deformation occurs in all shaping processes such as rolling, drawing and pressing, and may occur locally in welded metals owing to the stresses set up during heating and cooling. When plastic deformation is produced by cold working it has several important effects on the structure and properties of metal:

(1) The metal grains are progressively elongated in the direction of deformation.

(2) With heavy reduction by cold work the structure becomes very distorted, broken up, and fibrous in character.

(3) If there is a second constituent present, as in many alloys, this becomes drawn out in threads or strings of particles in the direction of working, thus increasing the fibrous character of the material.

(4) The deformation of the structure is accompanied by progressive hardening, strengthening and loss of ductility and by

an increased resistance to deformation. Cold worked metals and alloys are therefore harder, stronger and less ductile than the same materials in the undeformed state. The properties of cold worked materials may also differ in different directions owing to the fibrous structure produced.

(5) If the deformation process does not act uniformly on all parts of the metal being rolled or drawn, then internal stresses may be set up. These stresses may add to a subsequent working stress to which the metal is subjected, and they also render many metals and alloys subject to a severe form of intercrystalline corrosion, known as stress-corrosion; for example, the well-known season cracking of cold worked brass.

(6) If the deformation process is carried to its limit, the metal loses all of its ductility and breaks in a brittle manner. Brittle fracture of a similar kind can be produced in metals without preliminary deformation if they are subjected to certain complex stress systems which prevent deformation by slipping (three tensile stresses acting perpendicularly to each other). Such conditions can occur in practice in the vicinity of a notch and are sometimes set up in the regions affected by welding.

The effect of heat on cold worked metals: annealing and recrystallization

On heating a cold worked metal, the first important effect produced is the relief of internal stress. This occurs without any visible change in the distorted structure. The temperature at which stress relief occurs varies from 100°C up to 500°C according to the particular metal concerned. In general, the higher the melting point, the higher is the temperature for this effect.

On further heating the cold worked metal, no other changes occur until a critical temperature known as the 'recrystallization temperature' is reached. At the recrystallization temperature, the distorted metal structure is able to re-arrange itself into the normal unstrained arrangement by recrystallizing to form small equi-axed grains. The mechanical properties return again to values similar to those which the metal possessed before the cold working operation, i.e. hardness and strength fall, and ductility increases. However, if the extent of the distortion was very severe (suppose for example there had been a 60% or more reduction by rolling) the recrystallized metal may exhibit different properties in different directions.

If the metal is heated to a higher temperature than the minimum required for recrystallization to occur, the new grains grow progressively larger and the strength and working properties deteriorate.

The resoftening which accompanies recrystallization is made use of during commercial cold working processes to prevent the metal becoming too brittle, as, for example, after a certain percentage of reduction by cold rolling, the metal is annealed to soften it, after which further reduction by cold working may be done.

The control of both working and annealing operations is important because it affects the grain size, which, in turn, controls the properties of the softened material. It is generally advantageous to secure a fine grain size. The main factors determining the grain size are:

(1) The prior amount of cold work – the recrystallized grain size decreases as the amount of prior cold work increases.

(2) The temperature and time of the annealing process. The lowest annealing temperature which will effect recrystallization in the required time produces the finest grain size.

(3) Composition. Certain alloying elements and impurities restrict grain growth.

The temperature at which a cold worked metal or alloy will recrystallize depends on:

(1) Its melting point. The higher the melting point the higher the recrystallization temperature.

(2) Its composition and constitution. Impurities or alloying elements present in solid solution raise the recrystallization temperature. Those present as second constituents have little effect (although these are the type which tend to restrict grain growth once the metal or alloy has recrystallized).

(3) The amount of cold work. As the extent of prior cold work increases, so the recrystallization temperature is lowered.

(4) The annealing time. The shorter the time of annealing the higher the temperature at which recrystallization will occur.

These factors must all be taken into account in determining practical annealing temperatures.

Hot working

Metal and alloys which are not very ductile at normal temperatures, and those which harden very rapidly when cold worked, are generally fabricated by hot working processes, namely forging, rolling, extrusion, etc. Hot working is the general term applied to deformation at temperatures above the recrystallization temperature. Under these conditions the hardening which normally accompanies the deformation is continually offset by recrystallization and softening. Thus a hot worked material retains an unstrained equi-axed grain structure, but the size of the grains and the properties of the structure depend largely on the temperature at which working is discontinued. If this is just above the recrystallization temperature, the grains will be unworked, fine and uniform, and the properties will be equivalent to those obtained by cold working followed by annealing at the recrystallization temperature.

If working is discontinued well above the recrystallization temperature, the grains will grow and develop inferior properties, while a duplex alloy may develop coarse plate structures similar to those present in the cast state and having low shock-strength.

On the other hand, if working is continued until the metal has cooled below its recrystallization temperature, the grains will be fine but will be distorted by cold working and the metal will therefore be somewhat harder and stronger but less ductile than when the working is discontinued at or above the recrystallization temperature.

Insoluble constituents in an alloy become elongated in the direction of work as in cold working and produce similar directional properties.

Hot working is used extensively for the initial 'breaking down' of large ingots or slabs, even of metals which can be cold worked. This is because of the lower power required for a given degree of reduction by hot work. Hot work also welds up clean internal cavities but it tends to give inferior surface properties since some oxide scale is often rolled into the surface. Further, it does not permit such close control of finishing gauges, and cannot be used effectively for finishing sheet or wire products.

Impurities which form low melting point constituents can ruin the hot working properties of metals and alloys – for example, excess sulphur in steel.

Cold working as a major fabrication process is restricted to very ductile materials such as pure and commercial grades of copper, aluminium, tin and lead, and to solid solution alloys, such as manganese – aluminium alloys, brasses containing up to 35% zinc; bronzes, containing up to 8% tin; copper–nickel alloys; aluminium bronzes, containing up to 8% aluminium; nickel silvers (copper–nickel–zinc alloys) and tin–base alloys such as pewter. It is also used in the finishing working stages of many other metals and alloys to give dimensional control, good surface quality and the required degree of work hardening.

Iron, steels, high zinc or high aluminium copper alloys, aluminium alloys such as those with copper, magnesium and zinc, pure zinc and pure magnesium, and the alloys of these metals are all generally fabricated by hot working, though cold working may be used in the late stages, especially for the production of wire or sheet products.

Iron and Steel
Pure iron is very ductile and can be both hot and cold worked.

The carbon content of steel makes it less ductile than iron. Dead mild steel of up to 0.15% carbon is very suitable for cold working and can be flanged and used for solid drawn tubes. Although it is slightly hardened by cold work the modulus of elasticity is unaffected. If steel is hot worked at a temperature well above recrystallization temperature, grain growth takes place and the impact strength and ductility are reduced, hence it should be worked at a temperature just above recrystallization point, which is 900–1200 °C for mild steel and 750–900 °C for high-carbon steel. Thus, in general, hot working increases the tensile and impact strength compared with steel in the cast condition, and as the carbon content increases, the steel becomes less ductile and it must be manipulated by hot working.

Copper

Copper is very suitable for cold working as its crystals are ductile and can suffer considerable distortion without fracture, becoming harder however as the amount of cold work increases. Since welding is performed above the annealing temperature it removes the effect of cold work.

Copper is very suitable for hot working and can be hot rolled, extruded and forged.

Brasses

Brasses of the 85/15 and 70/30 composition are very ductile. Brass can be heavily cold worked without suffering fracture. This cold work modifies the grain structure, increases its strength and hardness and gives it varying degrees of temper.

Since these brasses are so easily cold worked there is little advantage to be gained by hot work, which, however, is quite suitable.

For brass of the 60/40 type ($a + \beta$ structures) see pp. 121–2.

The β structure, which is zinc rich, is harder than the a structure and will stand little cold work without fracture. If the temperature is raised, however, to about 600 °C it becomes more easily worked.

Brasses of this composition therefore should be hot worked above 600–700 °C giving a fine grain and fibrous structure. Below 600 °C for this structure can be considered cold work.

Copper–nickel alloys

80/20 copper–nickel alloy is suitable for hot and cold work but is particularly suitable for the latter due to its extreme ductility.

Nickel–chrome alloys are generally hot worked.

Bronzes. Gunmetal (Cu 88%, Sn 10%, Zn 2%) must be hot

worked above 600 °C and not cold worked. Phosphor bronze with up to 6–7% tin can be cold worked.

Aluminium bronze is similar to brass in being of two types:

(1) α structure containing 5–7% Al, 93% Cu.

(2) $\alpha + \beta$ structure containing 10% Al, 90% Cu.

(1) The α structure, which is the solid solution of aluminium in copper is ductile and this type is easily worked hot or cold.

(2) The $\alpha + \beta$ structure is rendered more brittle by the presence of the β constituent which becomes very hard and brittle at 600 °C due to the formation of another constituent which, in small quantities, increases the tensile strength, hence this alloy must be hot worked.

Aluminium. Pure aluminium can be worked hot or cold but weldable alloys of the work hardening type such as Al–Si, Al–Mn, and Al–Mn–Mg, are hardened by cold work which gives them the required degree of temper and they soften at 350 °C. Duralumin is hot short above 470 °C and too brittle to work below 300 °C so it should be worked in the range 400–470 °C. Y alloy can be hot worked.

NON-FERROUS METALS

Copper

Copper is found in the ore copper pyrites ($CuFeS_2$) and is first smelted in a blast or reverberatory furnace, and is then in the 'blister' or 'Bessemer' stage. In this form it is unsuitable for commercial use, as it contains impurities such as sulphur and oxygen. Further refining may be carried on by the furnace method, in which oxidation of the sulphur and other impurities occurs, or by an electrical method (called electrolytic deposition), resulting in a great reduction of the impurities.

In the refining and melting processes, oxidation of the copper occurs and the excess oxygen is removed by reducing conditions in the furnace. This is done by thrusting green wooden poles or tree trunks under the surface of the molten copper, which is covered with charcoal or coke to exclude the oxygen of the air. The 'poling', as it is called, is continued until the oxygen content of the metal is reduced to the limits suitable for the work for which the copper is required.

The oxygen content of the copper is known as the 'pitch' and poling is ended when the 'tough pitch' condition is reached.

Oxygen in copper. In this condition the oxygen content varies from 0.025 to 0.08%. The oxygen exists in the cast copper as minute particles of cuprous oxide (Cu_2O) (Fig. 2.29a).

The amount of oxygen in the copper is most important from the point of view of welding, since the welding of copper is rendered extremely difficult by the presence of this copper oxide. When molten, copper oxide forms a eutectic with the copper and this collects along the grain boundaries, reducing the ductility and increasing the tendency to crack. Any hydrogen present, as occurs when there are reducing conditions in the flame, reduces the copper oxide to copper and water is also formed. This is present as steam which causes porosity and increases the liability of cracking. For welding purposes, therefore, it is much preferable to use copper almost free from oxygen, and to make this, deoxidizers such as phosphorus, silicon, lithium, magnesium, etc., are added to the molten metal, and they combine with oxygen to form slag and thus *deoxidize* the copper (Fig. 2.29b). The welding of 'tough pitch' copper depends so much on the skill of the welder that it is always advisable to use 'deoxidized copper' for welding, and thus eliminate any uncertainty.

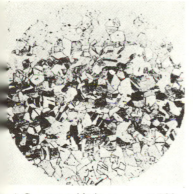 

(a) Cuprous oxide in copper. × 100.   (b) Deoxidized copper. × 75.

Fig. 2.29

Arsenic in copper. When arsenic up to 0.5% is added to copper, the strength and toughness is increased. In addition to this, it increases its resistance to fatigue and raises by about 100 °C the temperature at which softening first occurs and enables it to maintain its strength at higher temperatures. Arsenic is undesirable in copper intended specifically for welding purposes, since it makes welding more difficult. Arsenical copper can be welded by the same method as for ordinary copper, and if care is taken the welds are quite satisfactory. As with ordinary copper, it may be obtained in the deoxidized or the tough pitch form, the former being the more suitable for welding.

Properties of copper. Copper is a red-coloured metal having a melting point of 1083 °C and a density of 8900 kg/m³. The mechanical properties of copper depend greatly on its condition, that is, whether it is in the 'as cast' condition or whether it has been hot or cold worked, hammered, rolled, pressed, or forged.

The tensile strength 'as cast' is about 160 N/mm². Hot rolling and forging, followed by annealing, modifies its structure and increases its strength to about 220 N/mm². Cold working by hammering, rolling, drawing and pressing hardens copper and raises its tensile strength, but it becomes less ductile.

Very heavily cold worked copper may have a tensile strength equal to that of mild steel, but it has very little ductility in this state.

The temper of copper. Copper is tempered by first getting it into a soft or annealed condition, and then the temper required is obtained by cold working it (hammering, rolling, etc.). Thus it is the reverse process from the tempering of steel. Soft-temper copper is that in the annealed condition. It has a Brinell hardness of about 50. After a small amount of cold working, it becomes 'half hard' temper, and further cold working brings it hard temper having a Brinell figure of about 100. Intermediate hardness can of course be obtained by varying the amount of cold working.

Annealing. Copper becomes hard and its structure is deformed when cold worked, and annealing is therefore necessary to soften it again. To anneal the metal, it is usual to heat it up to about 500 °C, that is, dull red heat, and either quench it in water or let it cool out slowly, since the rate of cooling does not affect the properties of the pure metal. Quenching, however, removes dirt and scale and clears the surface. The surface of the copper can be further cleaned or 'pickled' by immersing it in a bath of dilute sulphuric acid containing 1 part of acid to 70 parts of water. If nitric acid is added, it accelerates the cleaning process. If copper has a surface polish, heating to the annealing temperature will cause the surface to scale, which is undesirable; hence annealing is usually carried out in a non-oxidizing atmosphere in this case.

Crystal or grain size. Under the microscope, cold worked copper shows that the grains or crystals of the metal have suffered distortion. During the annealing process, as with steel, recrystallization occurs and new crystals are formed. As before, if the annealing temperature is raised too high or the annealing prolonged too long, the grains tend to grow. With copper, however, unlike steel, the rate of growth is slow, and this makes the annealing operation of copper subject to a great deal of latitude in time and temperature. This explains why it is immaterial whether the metal cools out quickly or slowly after annealing.

The main grades in which copper is available are (1) oxygen-bearing (tough pitch) high conductivity, (2) oxygen-free high conductivity, (3) phosphorus deoxidized. The various British Standards designations for wrought copper and copper alloys and filler-rod alloys are given in the section in Chapter 6 on copper welding by the TIG process.

Alloys of copper

The alloys of copper most frequently encountered in welding are copper–zinc (brasses and nickel silvers); copper–tin (phosphor bronzes and gunmetal) copper–silicon (silicon bronzes); copper–aluminium (aluminium bronzes); copper–nickel (cupro-nickels); and heat-treatable alloys such as copper–chromium and copper–beryllium. These are also discussed in the section on the welding of copper by the TIG process (Chapter 6).

Brasses or copper–zinc alloys. Zinc will dissolve in molten copper in all proportions and give a solution of a uniform character. Uniform solution can be obtained when solidified if the copper content is not less than about 60%. For example, 70% copper–30% zinc consists of a uniform crystal structure known as 'alpha' (a) solid solution and is shown in Fig. 2.30a. If the percentage of zinc is now increased to about 40%, a second constituent structure, rich in zinc, appears, known as 'beta' (β) solid solution, and these crystals appear as reddish in colour, and the brass now has a duplex structure as shown in Fig. 2.30b. These crystals are hard and increase the tensile strength of the brass but lower the ductility. The alpha-type brass, which can be obtained when the copper content has a minimum value of 63%, has good strength and ductility when cold and is used for sheet, strip, wire and tubes. The alpha–beta, e.g. 60% copper–40% zinc, is used for casting purposes, while from 57 to 61% copper types are suitable for hot rolling, extruding and stamping. Hence a great number of alloys of varying copper–zinc content are available. Two groups, however, are of very great importance, as they occur so frequently:

(1) Cartridge brass: 70% copper and 30% zinc, written 70/30 brass.
(2) Yellow or Muntz metal: 60% copper and 40% zinc, written 60/40 brass.

The table illustrates the composition and uses of various copper–zinc alloys.

Composition of copper–zinc and copper–tin alloys

% Composition by weight				
Cu	Zn	Sn	Other elements	Uses
90	10			
85	15			Architectural and decorative work.
80	20			
70	30			Cartridge brass. For deep drawing and where high strength and ductility are required.
66	34			2/1 brass.
62–65	38–35			Common brass.
60	40			Yellow or Muntz metal. Works well when hot. For brass sheets and articles not requiring much cold work during manufacture.
76	22		2 Al	Aluminium brass.
61–63.5	rem	1–1.4		Naval brass.
70–73	rem	1–1.5	0.02–0.06 As	Admiralty brass.
63–66	rem		0.75–1.5 Pb ⎤	Leaded brass, casts well. Easily hot
61–64	rem		1.0–2.0 Pb ⎬	stamped and extruded. Machines
58–60	rem		1.5–2.5 Pb ⎦	well.
58	38		Mn, Fe, Ni or Sn approx 4%	Manganese bronze, high tensile alloys for castings and bearings.
Rem			2.7–3.25 Si 0.75–1.25 Mn	Silicon bronze.
Rem		3–4.5	0.02–0.4 P	3% phosphor bronze.
Rem		4.5–6.0	0.02–0.4 P	5% phosphor bronze.
Rem		6.0–7.5	0.02–0.4 P	7% phosphor bronze.

Note. Material can be supplied as annealed (O); various tempers produced by cold work and partial annealing and indicated by $\frac{1}{4}$H, $\frac{1}{2}$H etc.; spring tempers (SH and ESH); solution treated (W) and precipitation hardened (P).

Properties of brass. Brass is a copper–zinc alloy with a golden colour which can be easily cast, forged, rolled, pressed, drawn and machined. It has a good resistance to atmospheric and sea-water corrosion and therefore is used in the manufacture of parts exposed to these conditions. As the copper content in the brass is decreased, the colour changes from the reddish colour of copper to yellow and then pale yellow.

The density varies from 8200 to 8600 kg/m^3, depending on its com

(a)

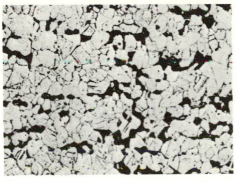

(b)

Fig. 2.30. (a) Rolled and annealed cartridge brass (70/30). This brass has a simple structure, containing only crystals of alpha solid solution, that is zinc dissolved in copper. × 100. (b) Rolled and annealed yellow metal (60/40). This brass is a mixture of alpha crystals (white areas), and beta crystals (black areas), richer in zinc. × 100.

position. The heat and electrical conductivity decrease greatly as the zinc content increases, and the melting point is lowered as the copper content decreases, being about 920 °C for 70/30 brass.

Brass for brazing purposes can vary greatly in composition, depending upon the melting point required; for example, three brazing rod compositions are 54% copper, 46% zinc; 50% copper, 50% zinc (melting at 860 °C) and 85% copper, 15% zinc. The choice of the alloy therefore depends on the work for which it is required. For welding brass, the filler rod usually contains phosphorus or silicon, which act as deoxidizing agents, that is, they remove any oxygen from the weld.

As the copper content is reduced, there is a slight increase in the tensile strength.

Annealing. Examination of cold worked brass under the microscope shows that, as with copper, distortion of the crystals has taken place. When its temperature is raised to about 600 °C, recrystallization takes place and the crystals are very small. The rate of growth depends on the temperature, and the higher the temperature the larger the crystals or grains. The annealing time (as with steel) also affects their size. In over-annealed brass, having large crystals, they may show up on the surface after cold working as an 'orange-peel' effect.

Annealing at too high a temperature may also cause pitting or deterioration of the surface by scaling.

Brass can either be quenched out in water or allowed to cool out slowly after annealing. If quenched, the surface scale is removed, but care must be taken with some brasses lest the ductility suffers.

Temper. Brass is tempered in the same way as copper, that is, by cold working. In the annealed condition it is 'soft temper' (60 to 80 Brinell). A little cold working brings it to 'half-hard temper' and further work gives it a 'hard temper' (150 to 170 Brinell). More cold working still, produces a 'spring-hard temper' with a Brinell number of 170 to 180.

Elasticity. The tensile strength of brass varies with the amount of cold working, and it is sufficiently elastic to allow of its being used as springs when in the spring hard-temper condition.

Bronzes or copper–tin alloys

Gunmetals are zinc-containing bronzes, e.g. 88% Cu, 10% Sn, 2% Zn. Wrought phosphor bronzes contain up to 8% tin and up to 0.4% phosphorus, while cast forms contain at least 10% tin with additions of lead to promote free machining and pressure tightness.

Gunmetal was chiefly used, as its name suggests, for Admiralty and Army Ordnance work, but is now used chiefly where resistance against corrosion together with strength is required. Lead bronze has lead added to improve its properties as a bearing surface and to increase its machinability.

Phosphor bronze has largely replaced the older bearing bronze for bearings owing to its increased resistance to wear. Phosphorus, when added to the copper–tin bronze, helps greatly to remove the impurities, since it is a powerful reducing agent and the molten metal is made much purer.

Bronze welding rods of copper–tin and copper–zinc composition are very much used as filler rods and electrodes in welding. They can be used for the welding of steel, cast iron, brass, bronze and copper, and have several advantages in many cases over autogenous welding since, because of their lower melting point, they introduce less heat during the welding operation. Manganese bronze can be considered as a high tensile brass.

Nickel and nickel alloys

Nickel is a greyish-white metal melting at 1450 °C, has a specific gravity of 8.8, and has a coefficient of linear expansion of 0.000 013 per degree C.

It resists caustic alkalis, ammonia, salt solutions and organic acids, and is used widely in chemical engineering for vats, stills, autoclaves, pumps, etc. When molten, it absorbs (1) carbon, forming nickel carbide (Nl_3C), which forms graphite on cooling; (2) oxygen, forming nickel oxide (NiO), which makes the nickel very brittle, and (3) sulphur, forming nickel sulphide (NiS).

Magnesium and manganese are added to nickel in order to deoxidize it and render it more malleable.

Nickel is widely used as an alloying element in the production of alloy steels, improving the tensile strength and toughness of the steel, and is used in cast iron for the same purpose. In conjunction with chromium it gives the range of stainless steels.

Copper–Nickel alloys

Nickel and copper are soluble in each other in all proportions to give a range of cupro-nickels which are ductile and can be hot and cold worked. The more important alloys are the following:

90% Cu, 10% Ni used for heat exchangers for marine, power, chemical and petrochemical use; feed water heaters, condensers, evaporators, coolers, radiators, etc.

80% Cu, 20% Ni used for heat exchangers, electrical components, deep-drawn pressings and decorative parts.

70% Cu, 30% Ni, which has the best corrosion resistance to sea and other corrosive waters and is used for heat exchangers and other applications given for 90/10% alloy.

The alloys used for resistance to corrosion usually have additions of iron (0.5–2.0%) and manganese (0.5–1.5%), while alloys for electrical uses are free from iron and have only about 0.2% Mn. The actual composition of Monel is 29% Cu, 68% Ni, 1.25% Fe, 1.25% Mn.

Other alloys are, 75% Cu, 25% Ni used for coinage, and nickel silvers

Nickel and nickel alloys

Material	Major constituents	Applications
Nickel	almost pure nickel.	High resistance to corrosion in contact with caustic alkalis, dry halogen gases and organic compounds generally.
Monel	nickel, copper.	Has good corrosion resistance with good mechanical properties. A variation responds to thermal hardening of the precipitation type and has good corrosion resistance, with the mechanical properties of heat-treatable alloy steels.
Inconel	nickel, chromium, iron.	Is oxidation-resistant at high temperatures with good mechanical properties and is resistant to food acids. Widely used for heat treatment and furnace equipment.
	nickel, chromium, iron with additions of molybdenum and niobium.	High level of mechanical properties without the need for heat treatment. Good oxidation resistance and resists corrosive attack by many media. Other variations are age-hardenable with high strength at elevated temperatures, have outstanding weldability and can be welded in the heat-treated condition.

Incoloy	Oxidation-resistant at elevated temperatures with good mechanical properties. Variations with a lower silicon content used for pyrolysis of hydrocarbons as in cracking or reforming operations in the petroleum industry.	nickel, chromium, iron.
	Resistant to hot acid and oxidizing conditions, e.g. nitric–sulphuric–phosphoric acid mixtures.	nickel, chromium, iron with copper and molybdenum additions
Nimonic	Used for gas turbine parts, heat treatment and other purposes where both oxidation resistance and high-temperature mechanical properties are required.	nickel, chromium. nickel, chromium, cobalt. nickel, chromium, iron.
Brightray	Heating elements of electric furnaces, etc.	nickel, chromium. nickel, chromium, iron.
Nilo	These have a controlled low and intermediate coefficient of thermal expansion and are used in machine parts, thermostats and glass-to-metal seals.	nickel, iron.

which contain 10–30% Ni, 55–63% Cu with the balance zinc and are extensively used for cutlery and tableware of all kinds, being easily electro-plated (EPNS). Nickel is added to brass and aluminium bronze to improve corrosion resistance.

Aluminium

Aluminium is prepared by electrolysis from the mineral *bauxite*, which is a mixture of the oxides of aluminium, silicon and iron.

The aluminium oxide, or alumina as it is called, is made to combine with caustic soda to form sodium aluminate, thus:

alumina + caustic soda → sodium aluminate + water

$$Al_2O_3 + \qquad 2NaOH \rightarrow 2NaAlO_2 \qquad + H_2O.$$

This solution of sodium aluminate is diluted and filtered to remove iron oxide, and aluminium hydroxide is precipitated. This is dried and calcined leaving aluminium oxide (Al_2O_3). The aluminium oxide is placed together with cryolite (Na_3AlF_6) and sometimes fluorspar (CaF_2) into a cell lined with carbon, forming the cathode or negative pole of the direct current circuit. Carbon anodes form the positive pole and hang down into the mixture. The p.d. across the cell is about 6 volts and the current which passes through the mixture and fuses it may be up to 100 000 amperes. As the current flows, the cryolite is electrolysed, aluminium is set free and is tapped off from the bottom of the cell, while the fluorine produced reacts with alumina, forming aluminium fluoride again. Because such large currents are involved, aluminium production is carried out near cheap sources of electrical power.

Aluminium prepared in this way is between 99 and 99.9% pure, iron and silicon being the chief impurities. In this state it is used for making sheets for car bodies, cooking utensils, etc., and for alloying with other metals. In the 99.5%-and-over state of purity it is used for electrical conductors and other work of specialized nature.

Properties of aluminium. Aluminium has a whitish colour and is very light. Its density is 2700 kg/m³, i.e. it weighs less, volume for volume, than $\frac{1}{3}$ the weight of copper (8900 kg/m³), and just more than $\frac{1}{3}$ of iron (7900 kg/m³); its melting point is 659 °C and its tensile strength varies from 90 to 120 N/mm² according to purity and the amount of cold working performed on it. It has high ductility, can be hammered and rolled into rod and sheet form, and casts well. In contact with air, a very thin invisible film of oxide (alumina) forms on its surface about 1.25×100^{-5} mm thick. This makes welding difficult and either fluxes are used to remove this oxide, the flux being a chemical compound which readily attacks and dissolves the oxide, or the weld is performed within an inert gas shield as in argon arc welding.

Aluminium and its alloys readily absorb hydrogen when in the molten condition and much absorption leads to porosity.

Annealing. Aluminium can be annealed by heating to 350–400 °C and then cooling in air. A workshop test for the annealing temperature is that a piece of dry stick rubbed on the work becomes charred, or a piece of soap rubbed on it leaves a brown mark, at which point heating should be stopped.

Resistance to corrosion. The property is due to the formation of a film of oxide on the surface, which protects the metal underneath. The thickness of the film depends upon the corrosive condition. In a non-corrosive condition, say indoors, the film is invisible and is extremely thin, but out-of-doors the film may appear on the surface as a grey-coloured coat. This film will protect the aluminium against many corrosive influences such as sea water. An electrochemical treatment known as anodizing may be used to thicken the oxide film and increase its protective value.

Aluminium is a very good conductor of heat and electricity. Owing to its good thermal conductivity, a considerable quantity of heat has to be applied during welding operations.

Aluminium alloys

Although aluminium is very ductile it has a low tensile strength which varies somewhat according to its purity and amount of cold working. By the addition of other elements such as magnesium, manganese, silicon, copper and zinc to form alloys the mechanical properties are improved. Some alloys are used in the 'as cast' or wrought condition whilst others are modified by heat treatment or by cold working.

(1) The aluminium silicon alloys containing 10–13% silicon (a eutectic is formed with 11.7% silicon) are very fluid when molten and have considerable strength, ductility and a high resistance to corrosion with a small contraction on solidification, and are used for castings.

(2) The addition of manganese or magnesium or of both to aluminium gives a range of work-hardening alloys which have good resistance to corrosion and shock.

(3) The addition of copper, magnesium and silicon gives a range of heat-treatable alloys which age-harden due to the presence of intermetallic compounds, and have a high strength/weight ratio.

(4) The aluminium–zinc–magnesium–copper alloys have poor corrosion resistance but include the high strength aircraft alloys which after full heat treatment give the highest strength/weight ratio.

Wrought aluminium alloys are of two types:

(1) heat-treatable (H) – the heat treatment improves their strength,
(2) non-heat-treatable (N) – these are not strengthened by heat treatment but by cold work.

Heat-treatable alloys are generally supplied in the heat-treated condition indicated by TB, TD, etc. (see below). If they are softened by welding, the properties of the original material in the HAZ can be restored by heat treatment but the ductility of the weld may not match that of the parent plate. The heat treatment comprises the following processes:

(1) *Annealing*, which is a softening process, can be performed on both heat-treatable and non-heat-treatable alloys. The alloy is heated to 320–420 °C for a given time (time and temperature depending upon the particular alloy) and is then allowed to cool slowly for heat-treatable alloys and at any rate of cooling for non-heat-treatable alloys. Material thus treated is designated by O.

(2) *Solution treatment*. The heat-treatable alloy is raised in temperature to 450–540 °C, the exact temperature depending upon the alloy, and then is rapidly quenched in a suitable quenching liquid. Immediately after quenching the material is ductile and any forming operation should be done in this period. As time passes the strength and hardness increase at room temperature up to a period of 4–5 days, and it is then in the TB state. Further heat treatment known as precipitation treatment enables greater strength to be acquired.

(3) *Precipitation treatment*. The heat-treatable alloy is heated to a temperature about 100–200 °C lower than that for solution treatment for a period varying from 2 hours to 20 hours, the time and temperature depending upon the alloy and its dimensions. Cooling is done in air sufficiently rapidly to prevent

over-ageing. After solution and precipitation treatment the alloy is in the TF condition.

Work or strain-hardenable alloys harden and increase in tensile strength with cold work such as rolling, pressing and drawing, and heating softens them.

The various conditions in which material is supplied are indicated below (BS 1475).

M	As manufactured. Material which acquires some temper from shaping processes in which there is no special control over thermal treatment or amount of strain hardening.
O	Annealed. Material which is fully annealed to obtain the lowest strength condition.
H1–H8	Strain-hardened. Material subjected to the application of cold work after annealing (or hot forming) or to a combination of cold work and partial annealing/stabilizing in order to secure the specified mechanical properties. The designations are in ascending order of tensile strength.
TB	Solution heat treated and naturally aged. Material which receives no cold work after solution heat treatment except as may be required to flatten or straighten it. Properties of some alloys in this temper are unstable.
TD	Solution heat treated, cold worked and naturally aged.
TE	Cooled from an elevated temperature shaping process and precipitation treated.
TF	Solution heat treated and precipitation treated.
TH	Solution heat treated, cold worked and then precipitation treated.

Wrought alloys have a second prefix letter indicating the form of the material thus:

Sheet and strip	S	Extruded round tube and hollow sections	V
Plate	P	Forgings	F
Wire	G	Rivet stock	R
Drawn tube	T	Bolts and screws stock for forging	B
Bars, rods and sections	E	Clad sheet and strips	C

For example, NS 3 is a work hardening alloy in sheet or strip form whilst NG 3 is the same in wire form.

Composition of aluminium alloy sheet showing essential alloying elements

BS	ISO		Alloying elements			
NS 3		Al.Mn 1	0.8–1.5 Mn			
NS 4	NT 4	Al.Mg 2	1.7–2.4 Mg			
NS 5	NT 5	Al.Mg 3.5	3.1–3.9 Mg			
NS 8	NT 8	Al.Mg 4.5 Mn	4.0–4.9 Mg	0.5–1.0 Mn		
		Al.Mg.Si	0.4–0.9 Mg	0.3–0.7 Si		
HS 15	HT 15	Al.Cu 4 Si.Mg	3.9–5.0 Cu	0.2–0.8 Mg	0.5–1.0 Si	
					0.4–1.2 Mn	
HS 30	HT 30	Al.Si.Mg.Mn	0.5–1.2 Mg	0.7–1.3 Si	0.4–1.0 Mn	
HT 20		Al.Mg 1.S.Cu	0.5–0.4 Cu	0.8–1.2 Mg	0.40	

Composition of aluminium alloy wire showing essential alloys

BS	ISO	Alloying elements			
NG 2	Al.Si 12	10.0–13.0 Si			
NG 21	Al.Si 5	4.5–6.0 Si			
NG 3	Al.Mn 1	0.8–1.5 Mg			
NG 4	Al.Mg 2	1.7–2.4 Mg			
NG 5	Al.Mg 3.5	3.1–3.9 Mg			
NG 6	Al.Mg 5	4.5–5.5 Mg			
NG 61	Al.Mg 5.2 Mn.Cr	5.0–5.5 Mg	0.6–1.0 Mn	0.05–0.20 Cr	
HG 9	Al.Mg.Si	0.4–0.9 Mg	0.3–0.7 Si		
HG 15	Al.Cu 4 Si.Mg	3.9–5.0 Cu	0.2–0.8 Mg	0.5–1.0 Si	
				0.4–1.2 Mn	
HG 20	Al.Mg 1 Si.Cu	0.15–0.4 Cu	0.8–1.2 Mg	0.4–0.8 Si	
				0.2–0.8 Mn	
				0.15–0.35 Cr	

Magnesium

Magnesium is an element of specific gravity 1.8, but although it has a relatively high specific heat capacity (1.1×10^3 joules per kg °C), the volume heat capacity is only $\frac{3}{4}$ of that of aluminium. It melts at 651 °C and its specific latent heat of fusion is lower than that of aluminium, so that for a given section the heat required to melt magnesium is about $\frac{2}{3}$ that required for an equal weight of aluminium. It has a high coefficient of expansion and a high thermal conductivity so that the danger of distortion is always present. It oxidizes rapidly in air above its melting point and though it burns with an intense white flame to form magnesium oxide there is little danger of fire during the welding process (TIG and MIG). The alloying elements added to magnesium are aluminium, zinc, manganese, zirconium, thorium and silver together with the rare earth metals cerium, lanthanum, neodymium and praseodymium. The alloys are heat treated in a similar way to the aluminium alloys. Although zinc refines the cast structure and increases the strength, it promotes weld cracking.

Magnesium alloys. Typical composition – major alloying elements

	Type	Composition % (remainder Mg)
Wrought	ZW3	Zn 3.0, Zr 0.6
	ZW1	Zn 1.3, Zr 0.6
	ZW6	Zn 5.5, Zr 0.6
	ZTY	Zn 0.5, Zr 0.6, Th 0.75
	AM503	Mn 1.5
	AZ31	Al 3.0, Zn 1.0, Mn 0.3
	AZM	Al 6.0, Zn 1.0, Mn 0.3
	AZ855	Al 8.0, Zn 0.4, Mn 0.3
	AL80	Al 0.8, Be 0.005, Zr 0.55
Cast	Z5Z	Zn 4.5, Zr 0.7
	RZ5	Zn 4.0, Zr 0.7, rare earths 1.2
	TZ6	Zn 5.5, Zr 0.7, Th 1.8
	MSRA	Zr 0.6, fractionated rare earths 1.7, Ag 2.5
	MSRB	Zr 0.6, fractionated rare earths 2.5, Ag 2.5
	ZREI	Zn 2.2, Zr 0.6, rare earths 2.7
	ZT1	Zn 2.2, Zr 0.7, Th 3.0
	MTZ	Zr 0.7, Th 3.0
	A8	Al 8.0, Zn 0.5, Mn 0.3

The wrought alloys are not heat treated after welding but the Mg–Al alloys are stress relieved to avoid stress corrosion cracks. No treatment is required with the Mg–Zn or Mg–Mn alloys.

STRESSES AND DISTORTION IN WELDING

Stresses set up in welding

In the welding process, whether electric arc or oxy-acetylene, we have a molten pool of metal which consists partly of the parent metal melted or fused from the side of the joint, and partly of the electrode or filler rod.

As welding proceeds this pool travels along and heat is lost by conduction and radiation, resulting in cooling of the joint. The cooling takes place with varying rapidity, depending on many factors such as size of work, quantity of weld metal being deposited, thermal conductivity of the parent metal and the melting point and specific heat of the weld metal.

As the weld proceeds we have areas surrounding the weld in varying conditions of expansion and contraction, and thus a varying set of forces will be set up in the weld and parent metal. When the weld has cooled, these forces which still remain, due to varying conditions of expansion and contraction, are called *residual stresses* and they are not due to any external load but to internal forces.

The stresses will cause a certain deformation, of the joint. This deformation can be of two kinds: (1) elastic deformation, or (2) plastic deformation.

If the joint recovers its original shape upon removal of the stresses, it has suffered elastic deformation. If, however, it remains permanently distorted, it has suffered plastic deformation. The process of removal of these residual stresses is termed *stress relieving*.

Stresses are set up in plates and bars during manufacture due to rolling and forging. These stresses may be partly reduced during the process of welding, since the metal will be heated and thus cause some of these stresses to disappear. This may consequently reduce the amount of distortion which would otherwise occur.

Stresses, with their accompanying strains, caused in the welding process, are thus of two types:

(1) Those that occur while the weld is being made but which disappear on cooling.
(2) Those that remain after the weld has cooled.

If the welded plates are free to move, the second causes distortion. If the plates are rigid, the stresses remain as residual stresses. Thus we have to consider two problems: how to prevent distortion, and how to relieve the stress.

Distortion is dependent on many factors, and the following experiment will illustrate this.

Take two steel plates about 150 mm × 35 mm × 8 mm. Deposit a straight layer of weld metal with the arc across one face, using a small current and a small electrode. This will give a narrow built-up layer. No distortion takes place when the plate cools (Fig. 2.31). Now deposit a layer on the second plate in the same way, using a larger size electrode and a heavier current. This will give a wider and deeper layer.

On cooling, the plate distorts and bends upwards with the weld on the inside of the bend (Fig. 2.32). In the first operation the quantity of heat given to the plate was small, due to the small mass of weld laid down. Thus contraction forces were small, and no distortion occurred. In the second operation, much more weld metal was laid down, resulting in a much higher temperature of the weld on the upper side of the plate. On cooling, the upper side contracted therefore more than the lower, and due to the pull of the contracting line of weld metal, distortion occurred.

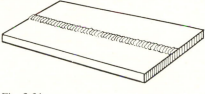

Fig. 2.31

Evidently, if another layer of metal was laid on the second plate in the same way, increased distortion would occur. Thus from experience we find that:

(1) An increase in speed tends to increase distortion because a larger flame (in oxy-acetylene), and a larger diameter electrode and increased current setting (in electric arc) have to be used, increasing the amount of localized heat.

(2) The greater the number of layers of weld metal deposited the greater the distortion.

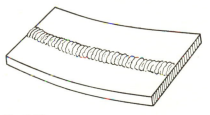

Fig. 2.32

Let us now consider some typical examples of distortion and practical methods of avoiding it.

Two plates, prepared with a V joint, are welded as shown. On cooling, the plates will be found to have distorted by bending upwards (Fig. 2.33).

BEFORE WELDING   AFTER WELDING

Fig. 2.33

Again, suppose one plate is set at right angles to the other and a fillet weld is made as shown. On cooling, it will be found that the plates have pulled over as shown and are no longer at right angles to each other (Fig. 2.34).

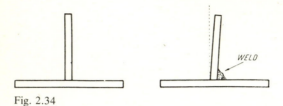

Fig. 2.34

This type of distortion is very common and can be prevented in two ways:

(1) Set the plates at a slight angle to each other, in the opposite direction to that in which distortion will occur so that, when cool, the plates are in correct alignment.
(2) Clamp the plates firmly in a fixture or vice so as to prevent their movement.

Since we have seen that the amount of distortion depends on several factors, such as speed of welding and number of layers, the amount of bias to be given to the plates in the opposite direction will be purely a matter of experience.

If the plates are fixed in a vice or jig, so as to prevent movement, the weld metal or parent metal must stretch or give, instead of the plates distorting. Thus there is more danger in this case that residual stresses will be set up in the joint.

A very familiar case is the building up of a bar or shaft. Here it is essential to keep the shaft as straight as possible during and after welding so as to reduce machining operations. Evidently, also, neither of the two methods of avoiding distortion given above can be employed. In this case distortion can be reduced to a minimum by first welding a deposit on one side of the shaft, and then turning the shaft through 180° and welding a deposit on the opposite diameter. Then weld on two diameters at right angles to these, and so on, as shown in Fig. 2.35. The contraction due to layer 2 will counteract that due to layer 1, layer 4 will counteract layer 3, and so on.

If two flat plates are being butt welded together as shown, after having been set slightly apart to begin with, it is found that the plates will tend to come together as the welding proceeds.

This can be prevented either by tack welding each end before com-

Fig. 2.35

mencing welding operations, clamping the plates in a jig to prevent them moving, or putting a wedge between them to prevent them moving inwards (Fig. 2.36). The disadvantage of tack welding is that it is apt to impair the appearance of the finished weld by producing an irregularity where the weld metal is run over it on finishing the run.

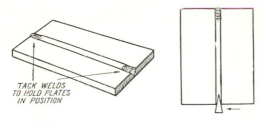

TACK WELDS
TO HOLD PLATES
IN POSITION

Fig. 2.36

Step welding or back stepping is often used to reduce distortion. In this method the line of welded metal is broken up into short lengths, each length ending where the other began. This has the effect of reducing the heat in any one section of the plate, and it will be seen that in this way when the finish of step 2 meets the beginning of step 1 we have an expansion and contraction area next to each other helping to neutralize each other's effect (Fig. 2.37).

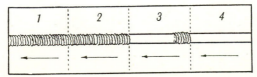

ARROWS AND NUMBERS SHOW DIRECTION AND ORDER OF WELDING

Fig. 2.37

In the arc welding of cast iron without pre-heating, especially where good alignment at the end of the welding process is essential, beware of trying to limit the tendency to distort while welding, by tacking too rigidly, as this will frequently result in cracking at the weakest section, often soon after welding has been commenced. Rather set the parts in position so that after welding they have come naturally into line and thus avoid the setting up of internal stresses. Practical experience will enable the operator to decide how to align the parts to achieve this end.

In *skip welding* a short length of weld metal is deposited in one part of the seam, then the next length is done some distance away, keeping the sections as far away from each other as possible, thus localizing the heat (Fig. 2.38). This method is very successful in the arc welding of cast iron.

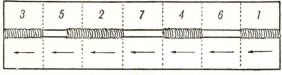

ARROWS AND NUMBERS SHOW DIRECTION AND ORDER OF WELDING

Fig. 2.38

To avoid distortion during fillet welding the welds can be done in short lengths alternately on either side of the leg of the T, as shown in Fig. 2.39, the welds being either opposite each other as in the sketch (*a*) or alternating as in the sketch (*b*). It is evident that a great deal can be done by the operator to minimize the effects of distortion.

In the case of cast iron, however, still greater care is needed, because whereas when welding ductile metals there is the plasticity of the parent metal and weld metal to cause a certain yielding to any stresses set up, cast iron, because of its lack of ductility, will easily fracture before it will

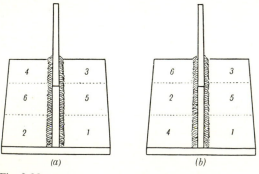

Fig. 2.39

distort, unless the greatest care is taken. This has previously been mentioned in the effects of expansion and contraction.

In the welding of cast iron with the blowpipe, pre-heating is always necessary unless the casting is of the simplest shape. The casting to be welded is placed preferably in a muffle furnace and its temperature raised gradually to red heat.

For smaller jobs pre-heating may be carefully carried out by two or more blowpipes and the casting allowed to cool out in the hot embers of a forge.

Residual stresses and methods of stress relieving

In addition to these precautions, however, the following experiment will show how necessary it is to follow the correct welding procedure to prevent fracture. Three cast-iron plates are placed as shown in Fig. 2.40 and are first welded along the seam $A-B$. No cracking takes place, because they are free to expand and contract.

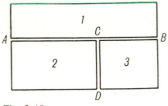

Fig. 2.40

If we now begin at D and weld to C, that is, from the free end to the fixed end, cracking will in all probability occur; whereas, if we weld from C to D, no cracking occurs.

When we weld from D to C the ends D and C are rigid and thus there is no freedom in the joint. Stresses are set up in cooling, giving tendency to fracture. Welding from C to D, that is, from fixed to free end, the plates are able to retain a certain amount of movement regarding each other, as a result of which the stresses set up on cooling are much less and fracture is avoided. Thus, always weld away from a fixed end to a free end in order to reduce residual stress.

Peening. Peening consists of lightly hammering the weld and/or the surrounding parent metal in order to relieve stresses present and to consolidate the structure of the metal. It may be carried out while the weld is still hot or immediately the weld has cooled.

A great deal of controversy exists as to whether peening is advantageous or not. Some engineers advocate it because it reduces the residual stresses, others oppose it because other stresses are set up and the ductility of the weld metal suffers. If done reasonably, however, it undoubtedly is of value in certain instances. For example, in the arc welding of cast iron the risk of fracture is definitely reduced if the short beads of weld metal are lightly peened *immediately* after they have been laid. Care must be taken in peening hot metal that slag particles are not driven under the surface.

Pre- and post-heat treatment

The energy for heat treatment may be provided by oil, propane or natural gas or electricity and the heat may be applied locally, or the welded parts if not too large, may be totally enclosed in a furnace.

The temperature in localized pre-heating, which usually does not exceed 250 °C, should extend for at least 75–80 mm on each side of the welded joint. Post-heat temperatures, usually in the range 590–760 °C, reduce internal stresses and help to soften hardened areas in the HAZ. The work is heated to a given temperature, held at this temperature for a given 'soaking' time, and then allowed to cool, both heating and cooling being subject to a controlled temperature gradient such as 100–200 °C per hour for thicknesses up to 25 mm and slower rates for thicker plate depending upon the code being followed (ASME, BS, CEGB, etc.).

Localized heat

This can be applied using flexible insulated pad and finger electrical heaters Fig. 2.41a. These are available in a variety of shapes with the elements insulated with ceramic beads and supplied from the welding power source or an auxiliary transformer at 60–80 V.

The pads are connected in parallel as required and are covered over with insulating material to conserve heat. Heat can also be applied locally by gas or oil burners as for example on circumferential and longitudinal vessel seams (Fig. 2.41b). There is now available an efficient infra-red heater for pre-heating in which a gas burner heats a rectangular ceramic plaque, protected by an Inconel mesh grill. The plaques each have hundreds of tiny holes and as the gas–air mixture emerges on to the front face of them it is ignited and burns on the plaque surface which becomes intensely hot (1000 °C) and if these are placed 50–75 mm from the seam to be heated heat is transferred mainly by radiation and rapid heating of the work is achieved up to 250 °C. The burners operate in all positions and can be supplied with magnetic feet for easy attachment to the work.

Furnaces

The older furnaces were of firebrick, fired with gas or oil or electrically heated, and were very heavy. The modern 'top hat' furnace has a hearth

Fig. 2.41 (*a*) Flexible ceramic pad heaters strapped round a circular closing seam on a 1.2 m diameter vessel with wall thickness of 25 mm.

with electric heating elements laid in ceramic fibre over which fits a light steel framework insulated inside with mineral wool and ceramic fibre. This furnace is of light thermal mass and can be quickly loaded or unloaded by lifting the unit, consisting of side walls and top, clear of the hearth with its load. The furnace can be heated overnight to take

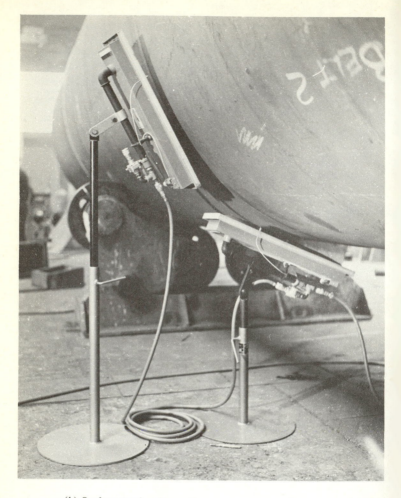

(*b*) Surface combustion units propane or natural gas set up for pre-
heating rotating seams.

advantage of the lower charges for electrical power, the temperature
control of the units being automatic.

These furnaces operate up to 650 °C. If higher temperatures up to
1050 °C are required, additional elements can be added to the furnace
sides with loadings up to 1200 kW. Top hat furnaces are made in a
variety of sizes including multiple hearths in which the top hat can be
moved on to one hearth while the other is being loaded or unloaded. For

furnaces of over 150 m³ volume, with their own gas trains, gas firing is very popular. Burners are chassis-mounted and arranged so that they fire down the side walls. As they are of high-velocity gas they avoid direct impingement on the parts being heated and temperatures are evenly distributed by the high-velocity action of the burners.

Temperatures of pre- or post-heat are generally measured by thermo-couple pyrometers and it is very important that the junction of the ends of the thermo-couple wires should be firmly attached to the place where the temperature is to be measured. Clips welded on in the required position and to which the junction is attached are used but when elements are placed over the junction it is possible for a temperature greater than that which actually exists in the work to be indicated due to the proximity of the heating element to the clip joint. A method to avoid this danger is to weld each of the couple wires to the pipe, spaced a maximum of 6 mm apart, by means of a capacitor discharge weld. The weld is easy to make using an auxiliary capacitor unit and reduces the possibility of erroneous readings. The only consumable is the thermo-couple wire, about 150 mm of which is cut off after each operation.

Summary of foregoing section:

Factors affecting distortion and residual stresses
(1) If the expansion which occurs when a metal is heated is resisted, deformation will occur.
(2) When the contraction which occurs on cooling is resisted, a stress is applied.
(3) If this applied stress causes movement, distortion occurs.
(4) If this applied stress causes no movement, it is left as a residual stress.

Methods of reducing distortion
(1) Decrease the welding speed, using the smallest flame (in oxy-acetylene) and the largest diameter electrode and lowest current setting (in electric arc) consistent with correct penetration and fusion of weld and parent metal.
(2) Line up the work to ensure correct alignment on cooling out of the weld.
(3) Use step-back or skip method of welding.
(4) Use wedges or clamps.

Methods of reducing or relieving stresses
(1) Weld from fixed end to free end.

(2) Peening.
(3) Heat treatment.

Finally, the following is a summary of the chief factors responsible for setting up residual stresses:

(1) Heat present in the welds depending on:
   (*a*) Flame size and speed in oxy-acetylene welding.
   (*b*) Current and electrode size and speed in arc welding.
(2) Qualities of the parent metal and filler rod or electrode.
(3) Shape and size of weld.
(4) Comparative weight of weld metal and parent metal.
(5) Type of joint and method used in making weld (tacking back stepping, etc.).
(6) Type of structure and neighbouring joints.
(7) Expansion and contraction (whether free to expand and contract or controlled).
(8) Rate of cooling.
(9) Stresses already present in the parent metal.

Brittle fracture

When a tensile test is performed on a specimen of mild steel it elongates first elastically and then plastically until fracture occurs at the waist which forms. We have had warning of the final failure because the specimen has elongated considerably before it occurs and has behaved both elastically and plastically. It is possible however, under certain conditions, for mild steel (and certain alloy steels) to behave in a completely brittle manner and for fracture to occur without any previous elongation or deformation even when the applied stress is quite low, the loading being well below the elastic limit of the steel.

This type of fracture which takes place without any warning is termed brittle fracture and although it has been known to engineers for many years it was the failure of so many of the all-welded Liberty ships in World War II that focused attention upon it and brought welding as a method of construction into question.

It is not, however, a phenomenon which occurs only in welded fabrications because it has been observed in riveted constructions; but because welded plates are continuous as opposed to the discontinuity of riveted ones a brittle fracture in a welded fabrication may travel throughout the fabrication with disastrous results, whereas in a riveted fabrication it usually travels only to the edge of the plate in which it commenced.

From investigations into the problem which have been made both in the laboratory and on fractures which have occurred during service, it is apparent that certain significant factors contribute to the occurrence of brittle fracture and we may summarize these as follows:

(1) The ambient temperature when failure occurred was generally low, that is near or below freezing point.

(2) Failure occurred in many cases when the loading on the areas was light.

(3) Failure is generally associated with mild steel but occurs sometimes in alloy steels.

(4) The fracture generally begins at defects such as artificial notches caused by sharp corners, cracks at a rivet hole, a fillet weld corner, poor weld penetration, etc., all behaving generally as a notch.

(5) The age of the structure does not appear to be a significant factor.

(6) Residual stresses may, together with other factors, serve to initiate the fracture.

Once the crack is started it may be propagated as a brittle fracture and can continue at high speeds up to 1200 m per second and at very low loads. It is a trans-granular phenomenon and after failure, tensile tests on other parts of the specimen will show a normal degree of ductility.

We may now consider briefly the conditions which may lead up to the occurrence of brittle fracture.

Notch brittleness

When stress is applied to a bar it may either deform plastically or may break in a brittle manner and the relationship between these types of behaviour determines whether brittle fracture will occur or not.

Consider a specimen of steel in which there is a notch and which is under a tensile stress. The stress on the specimen is below the elastic limit of the main section while at the root of the notch plastic flow may have occurred under a higher stress due to the reduced area.

This localized plastic flow at the notch may cause rapid strain hardening, which may lead to cracking at the root of the notch, and once this crack has begun we have a natural notch of uniform size instead of a variable-size machined notch. Before a crack can be initiated there must therefore be some deformation.

Ductile–brittle transition

The transition from the ductile to the brittle state is affected by temperature, strain rate and the occurrence of notches.

If tests are performed on steel specimens at various temperatures it is found that the yield point stress increases greatly as the temperature is reduced so that the lower the temperature, the greater the brittleness and the liability to brittle fracture. The strain rate has a similar effect on the yield point. As the rate of strain is increased the yield stress rises much more quickly than the fracture stress. A notch may thus localize the stress and increase the strain rate at its root. The size of the ferrite grains also affects the transition temperature and increasing carbon content raises and broadens the transition range. In the case of alloy steels, manganese and nickel lower the transition temperature while carbon, silicon and phosphorus raise it. By keeping a high manganese–carbon ratio, tendency to brittle fracture can be reduced and in general the lowest transition temperatures are obtained with finely dispersed microstructures. We may sum up the preceding by saying that the ductile–brittle transition of steel is affected by both grain size and microstructure, and in general weld metal has similar or even better transition properties than the parent plate. Because of this, brittle fracture is seldom initiated in the weld metal itself but rather in the fusion zone, the heat-affected zone or the parent plate, and the fracture seldom follows the weld. Faults in a weld such as slag inclusions, poros-

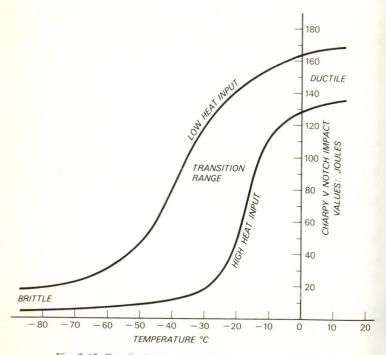

Fig. 2.42. Ductile–brittle transition in a steel weld metal.

ity, lack of fusion, and undercut which occur in the fusion zone may serve as nuclei for a crack from which the brittle fracture may be projected, and since the notch ductility of the weld is usually better than that of the plate the fracture follows the plate.

The notch ductility of a weld can be measured by means of the Charpy V notch test (p. 574) and gives an indication of the resistance to brittle fracture. The impact values in joules are plotted against the temperature, first for a weld with a low heat input and then with a high heat input, and two curves are obtained as in Fig. 2.42, which are for a carbon-manganese weld metal. It can be seen that the lower heat input gives higher impact values, but there is a transition range from ductile to brittle fracture, and as the temperature falls the probability of brittle fracture is greatly increased. The value of 41 J is often taken as the minimum for the weld metal since above this value it is considered that any crack which develops during service would be arrested before it could result in massive brittle fracture. To obtain good low-temperature impact values, low heat input is therefore required, which means that electrodes should be of the basic type and of small diameter, laid down with stringer bead (split weave) technique. Rutile-coated electrodes can be used down to about $-10\,°C$. Basic mild steel electrodes can be used down to about $-50\,°C$, and below that nickel-containing, basic-coated electrodes should be used.

3

Oxy-acetylene welding

PRINCIPLES AND EQUIPMENT

For oxy-acetylene welding, the oxygen is supplied from steel cylinders and the acetylene either from cylinders or from an acetylene generator which can be of the medium-pressure or low-pressure type.

With cylinder gas, the pressure is reduced to 0.06 N/mm² or under, according to the work by means of a pressure-reducing valve and the acetylene is passed to the blowpipe where it is mixed with oxygen in approximately equal proportions, and finally passed into the nozzle or tip to be burnt.

The medium-pressure acetylene generator delivers gas at any desired pressure up to a maximum of 0.06 N/mm² (0.6 bar) in the same ways as cylinder gas.

The low-pressure generator produces gas at a pressure of only a few millimetres water column, necessitating the use of 'injector' blowpipes where the high-pressure oxygen injects or sucks the acetylene into the blowpipe. In some cases, as for example if the supply pipes are small for the volume of gas to be carried or if it is desirable to use the equal pressure type of blowpipe, a booster is fitted to the low-pressure generator and the gas pressure increased to 0.06 N/mm².

Without Home Office approval, the maximum pressure at which acetylene may be used in England is 0.06 N/mm². With approval, the pressure may be increased to 0.15 N/mm², but this is rare and applies only in special cases.

Oxygen

The oxygen for both high- and low-pressure systems is supplied in solid drawn steel cylinders at a pressure of 17.2 N/mm² (172 bar).[1] The cylinders are rated according to the amount of gas they contain, 8500 litres (8.5 m³) being the usual size but very large cylinders containing 800 m³ (2400 cu. ft.) are available.

[1] *Note*. 1 bar 0.1 N/mm² = approximately 15 lbf/in².

The volume of oxygen contained in the cylinder is approximately proportional to the pressure; hence for every 10 litres of oxygen used, the pressure drops about 0.02 N/mm². This enables us to tell how much oxygen remains in a cylinder. The oxygen cylinder is provided with a valve threaded right hand and is painted black. On to this valve, which contains a screw-type tap, the pressure regulator and pressure gauge are screwed. The regulator adjusts the pressure to that required at the blowpipe. Since grease and oil can catch fire spontaneously when in contact with pure oxygen under pressure, they must never be used on any account upon any part of the apparatus. Leakages of oxygen can be detected by the application of a soap solution, when the leak is indicated by the soap bubbles. Never test for leakages with a naked flame.

Liquid oxygen

Liquid oxygen, nitrogen, argon and LPG are available in bulk supply from tankers to vacuum-insulated evaporators in which the liquid is stored at temperatures of -160 to $-200\,°C$ and is very convenient for larger industrial users.

There is no interruption in the supply of gas nor drop in pressure during filling.

The inner vessel, or austenitic stainless steel welded construction, has dished ends and is fitted with safety valve and bursting disc and is available in various sizes with nominal capacities from 844 to 33 900 m³. Nominal capacity is the gaseous equivalent of the amount of liquid that the vessel will hold at atmospheric pressure. The outer vessel is of carbon steel and fitted with pressure relief valve. The inner vessel is vacuum and pearlite insulated from the outer vessel, thus reducing the thermal conductivity to a minimum. The inner vessel A (Fig. 3.1) contains the liquid with gas above, and gas is withdrawn from the vessel through the gaseous withdrawal line B and rises to ambient temperature in the superheater-vaporizer C, from which it passes to the supply pipeline. If the pressure in the supply falls below the required level the pressure control valve D opens and liquid flows under gravity to the pressure-raising vaporizer E, where heat is absorbed from the atmosphere and the liquid vaporizes and passes through the gas withdrawal point H raising the pressure to the required pre-set level which can be up to 1.7 N/mm² (250 lbf/in²), and the valve D then shuts.

In larger units, to allow for heavy gas demand, when the pressure falls on the remote side of the restrictive plate F, liquid flows from the vessel via the withdrawal line G and passes to the superheater-vaporizer where it changes to gaseous form and is heated to ambient temperature, finally passing to the pipeline.

During periods when the VIE is not in use the valve D remains shut.

Heat from the outside atmosphere gradually flows through the insulation between the vessels so that more liquid is vaporized and the pressure of the gas rises. This rate of heat leakage is slow, however, and it usually takes about seven days for the pressure to rise sufficiently to lift the safety valve, so that under normal working conditions there is almost zero loss.

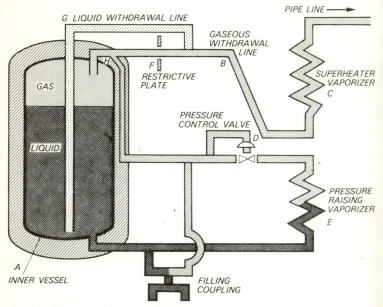

Fig. 3.1. Vacuum insulated evaporator.

Acetylene

In the high-pressure system the acetylene is stored in steel cylinders similar to the oxygen cylinders. Acetylene gas, however, is unstable when compressed to high pressures, and because of this it is contained in the bottles dissolved in a chemical called acetone; hence the name 'dissolved acetylene'.

The acetone is contained in a porous spongy mass of a substance such as charcoal, asbestos, kapok or other such material. Acetone can absorb 25 times its own volume of acetylene at normal temperature and pressure and for every increase of one atmosphere of pressure (0.1 N/mm² or 1 bar) it can absorb an equal amount.

The pressure of the acetylene is usually about 1.5 N/mm² or 15 bar, a typical capacity being 5700 litres (5.7 m³). The gas leaves the cylinder through a valve after passing through a filter pad. The valve has a screw tap fitted and is screwed left-hand. The cylinder is painted maroon and a regulator (also screwed left-hand) is necessary to reduce the pressure to 0.013–0.05 N/mm² or 0.13–0.5 bar as required at the blowpipe.

The amount of dissolved acetylene in a cylinder cannot be determined with any accuracy from the pressure gauge reading since it is in the dissolved condition. The most accurate way to determine the quantity of gas in a cylinder is to weigh it, and subtract this weight from the weight of the full cylinder, which is usually stamped on the label attached to the cylinder. The volume of gas remaining in the cylinder is calculated by remembering that 1 litre of acetylene weighs 1.1 g.

As long as the volume of acetylene drawn from the cylinder is not greater than $\frac{1}{5}$ of its capacity per hour, there is no appreciable amount of acetone contained in gas; hence this rate of supply should not be exceeded. For example, a 5700 litre cylinder can supply up to 1200 litres of gas per hour. The advantages of dissolved acetylene are that no licence is required for storage of the cylinders, there is no fluctuation of pressure in use, and the gas is always dry, clean and chemically pure, resulting in a reliable welding flame, and it avoids the need to charge and maintain a generator and to dispose of the sludge. There is no discernible difference in efficiency between generated and dissolved acetylene when both are used under normal conditions. Acetylene is highly inflammable and no naked lights should be held near a leaking cylinder, valve or tube. Leaks can be detected by smell or by soap bubbles. If any part of the acetylene apparatus catches fire, immediately shut the acetylene valve on the cylinder. The cylinder should be stored and used in an upright position.

The following is a summary of the main safety precautions to be taken when storing and using cylinders of compressed gas.

Storage
(1) Store in a well ventilated, fire-proof room with flame-proof electrical fittings. Do not smoke, wear greasy clothing or have exposed flames in the storage room.
(2) Protect cylinders from snow and ice and from the direct rays of the sun, if stored outside.
(3) Store away from sources of heat and greasy and oily sources. (Heat increases the pressure of the gas and may weaken the cylinder wall. Oil and grease may ignite spontaneously in the presence of pure oxygen.)
(4) Store acetylene cylinders in an upright position and do not store oxygen and combustible gases such as acetylene and propane together.
(5) Keep full and empty cylinders apart from each other.
(6) Avoid dropping and bumping cylinders violently together.

Use

(7) Keep cylinders away from electrical apparatus or live wires where there may be danger of arcing taking place.

(8) Protect them from the sparks and flames of welding and cutting operations.

(9) Always use pressure-reducing regulators to obtain a supply of gas from cylinders.

(10) Make sure that cylinder outlet valves are clear of oil, water and foreign matter, otherwise leakage may occur when the pressure-reducing regulators are fitted.

(11) Do not use lifting magnets on the cylinders. Rope slings may be used on single cylinders taking due precautions against the cylinder slipping from the sling. Otherwise use a cradle with chain suspension.

(12) Deload the diaphragm of the regulator by unscrewing before fitting to a full cylinder, and open the cylinder valve slowly to avoid sudden application of high pressure on to the regulator.

(13) Do not overtighten the valve when shutting off the gas supply; just tighten enough to prevent any leakage.

(14) Always shut off the gas supply when not in use for even a short time, and always shut off when moving cylinders.

(15) Never test for leaks with a naked flame; use soapy water.

(16) Make sure that oxygen cylinders with round bases are fastened when standing vertically, to prevent damage by falling.

(17) Thaw out frozen spindle valves with hot water NOT with a flame.

(18) Use no copper or copper alloy fitting with more than 70% copper because of the explosive compounds which can be formed when in contact with acetylene.

(19) Do not use oil or grease or other lubricant on valves or other apparatus, and do not use any jointing compound.

(20) Blow out the cylinder outlet by quickly opening and closing the valve before fitting the regulator.

(21) Should an acetylene cylinder become heated due to any cause, immediately take it outdoors, immerse in water or spray it with water, open the valve and keep as cool as possible until the cylinder is empty. Then contact the suppliers.

(22) Do not force a regulator on to a damaged outlet thread. Report damage to cylinders to the suppliers.

Note.   Also refer to Form 1704, 'Safety measures required in the use of acetylene gas and in oxy-acetylene processes in factories' (HMSO) and also 'Safety in the use of compressed gas cylinders', a booklet published by the British Oxygen Co. Ltd.

The cylinder outlet union is screwed left-hand for combustible gases and right-hand for non-combustible gases, the thread being $\frac{5}{8}$ in (16 mm) BSP except for CO_2 which is 21.8 mm, 14 TPI male outlet.

The following colour codes are used for cylinders (BS 349, BS 381 C)

Gas or mixture	Colour code
Acetylene, dissolved	Maroon; also with name ACETYLENE.
Argon	Peacock blue.
Argon–CO_2	Peacock blue. Light brunswick green band round middle of cylinder.
Argon–hydrogen	Peacock blue. Signal red band round middle of cylinder.
Argon–nitrogen	Peacock blue. French grey band round middle of cylinder.
Argon–oxygen	Peacock blue. Black band round middle of cylinder. % of oxygen indicated.
CO_2, commercial vapour withdrawal	Black.
CO_2, commercial liquid withdrawal	Black. Narrow white band down length of cylinder (syphon tube).
Helium	Brown.
Hydrogen	Signal red.
Hydrogen – high purity	Signal red. White circle on shoulder.
Hydrogen–nitrogen	Signal red. French grey band round middle of cylinder.
Nitrogen	French grey. Black cylinder shoulder.
Nitrogen (oxygen free)	French grey. Black cylinder shoulder with white circle.
Nitrogen–CO_2	French grey. Light brunswick green band round middle of cylinder and black cylinder shoulder.
Nitrogen–oxygen	French grey. Black band round middle of cylinder and black cylinder shoulder.
Oxygen, commercial	Black.
Propane	Signal red with name PROPANE.
Air	French grey.
Air–CO_2	French grey. Light brunswick green band round middle of cylinder.

Note. Colour bands on shoulder of cylinder denote hazard properties. Cylinders having a red band on the shoulder contain a flammable gas; cylinders having a golden yellow band on the shoulder contain a toxic (poisonous) gas. Cylinders having a red band on the upper shoulder and a yellow band on the lower shoulder contain a flammable and toxic gas. Carbon monoxide has a red cylinder with a yellow shoulder.

Medium-pressure acetylene generators

The medium-pressure generator uses small granulated carbide. 50 kg of good calcium carbide will produce about 14 000 litres of acetylene, an average practical value being 250 litres of acetylene per kilogram of carbide. The most popular sizes are 20/50 mm, 15/25 mm and 10/15 mm. The generator is self-contained (i.e. there is no separate gas holder) and consists of a water tank surmounted by a carbide hopper, in the top of which is a diaphragm. The carbide feed valve is controlled by the diaphragm, which is actuated by the pressure of the gas generated in the tank. When the pressure falls, carbide flows into the tank; as the pressure builds up the flow ceases. The gas pressure at which the generator will work is adjusted by means of a spring fitted to the opposite side of the diaphragm, ensuring close control of pressure with generation strictly in accordance with the demand. The carbide hopper is either made of glass or is fitted with windows so that the quantity of carbide remaining in the hopper can be ascertained at a glance.

Owing to the relatively large volume of water into which the small grains of carbide fall, there is no possibility of overheating and the carbide is completely slaked. The sludge, which collects at the bottom of the tank, and is emptied each time the generator is charged, consists of a thin milky fluid.

The impurities in crude acetylene consist chiefly of ammonia, hydrides of phosphorus, sulphur and nitrogen, and there are also present water vapour and particles of lime.

These impurities must be removed before the gas is suitable for welding use; the gas is filtered and washed and chemically purified by passing it through salts of ferric iron.

The normal method of testing acetylene to ascertain whether it is being efficiently purified is to hold a silver nitrate test paper (a piece of filter paper soaked in a solution of silver nitrate) in the stream of gas for about 10 seconds. If the acetylene is being properly purified, there will be no trace of stain on the silver nitrate paper.

The reducing valve or pressure regulator

In order to reduce the pressure of either oxygen or dissolved acetylene from the high pressure of the storage cylinder to that required at the blowpipe, a regulator or reducing valve is necessary. Good regulators are essential to ensure the even flow of gas to the blowpipe. A reference to Fig. 3.2 will make the principle of operation of the regulator clear. The gas enters the regulator at the base and the cylinder pressure is indicated on the first gauge. The gas then enters the body of the regulator R through the aperture A, which is controlled by the valve V. The pressure inside the regulator rises until it is sufficient to overcome the pressure of the spring S, which loads the diaphragm D. The diaphragm is therefore pushed back and the valve V, to which it is attached, closes the aperture A and prevents any more gas from entering the regulator.

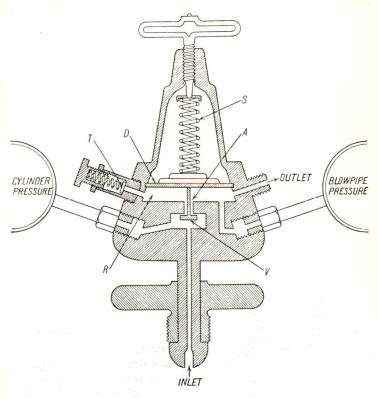

Fig. 3.2. Reducing valve or pressure regulator.

The outlet side is also fitted with a pressure gauge (although in some cases this may be dispensed with) which indicates the working pressure on the blowpipe. Upon gas being drawn off from the outlet side the pressure inside the regulator body falls, the diaphragm is pushed back by the spring, and the valve opens, letting more gas in from the cylinder. The pressure in the body R therefore depends on the pressure of the springs and this can be adjusted by means of a screw as shown.

Regulator bodies are made from brass forgings and single stage regulators are fitted with one safety valve set to relieve pressures of 16–20 bar, and should it be rendered inoperative as for example by misuse, it ruptures at pressures of 70–80 bar and vents to atmosphere through a vent in the bonnet. Single-stage regulators are suitable for general welding with maximum outlet pressure of 2.1 bar and for scrap cutting and heavy-duty cutting, thermic lancing and boring with outlet pressures 8.3–14 bar.

Figure 3.3a and b shows a two-stage regulator. This reduces the pressure in two stages and gives a much more stable output pressure than the single stage regulator.

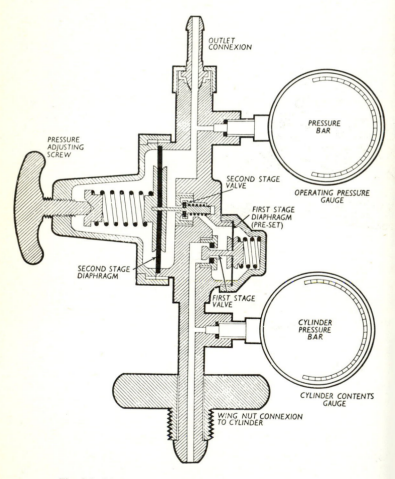

OUTLET
CONNEXION

PRESSURE
BAR

PRESSURE
ADJUSTING
SCREW

OPERATING PRESSURE
GAUGE

SECOND STAGE
VALVE

FIRST STAGE
DIAPHRAGM
(PRE-SET)

SECOND STAGE
DIAPHRAGM

FIRST STAGE
VALVE

CYLINDER
PRESSURE
BAR

CYLINDER CONTENTS
GAUGE

WING NUT CONNEXION
TO CYLINDER

Fig. 3.3. (a) Two-stage regulator.

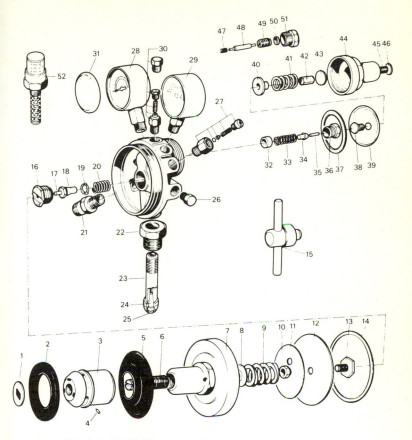

Fig. 3.3. (*b*) Multi-stage regulator.

Key:

1. Disc monogram.
2. Ring cover.
3. Knob.
4. Set screw.
5. Name plate
6. Screw P. A.
7. Bonnet.
8. Spring centre.
9. Spring.
10. Nut.
11. Packing plate.
12. Diaphragm.
13. Diaphragm carrier.
14. Washer.
15. Screw P. A.
16. Nozzle.
17. Valve pin.
18. Valve.
19. Washer.
20. Spring.
21. Outlet adaptor LH.
    Outlet adaptor RH.
22. Inlet nut LH.
    Inlet nut RH.
23. Inlet nipple.
24. Filter.
25. Retaining ring.
26. Plug.
27. Safety valve LP.
28. Gauge (0–30).
29. Gauge (0–4000).
30. Relief valve HP.
31. Gauge glass.
32. Sleeve.
33. Spring.
34. Valve.
35. Plunger.
36. Nozzle and seat.
37. Sealing ring.
38. Diaphragm carrier.
39. Diaphragm.
40. Disc.
41. Spring.
42. Damper plug.
43. Pivot.
44. Bonnet.
45. Screw.
46. Disc anti-tamper.
47. Spring.
48. Valve.
49. Seat retainer.
50. Seat.
51. Valve holder.
52. Indicator assembly.

It really consists of two single stages in series within one body forging. The first stage, which is pre-set, reduces pressure from that of the cylinder to 13–16 bar, and gas at the pressure passes into the second stage, from which it emerges at a pressure set by the pressure-adjusting control screw attached to the diaphragm. High-pressure regulators are designed for inlet pressures of 200 bar and tested to four times working pressure. Two-stage regulators have two safety valves, the first relieves at pressures of 60–75 bar and the second at 16–20 bar, so that if there is any excess pressure there will be no explosion. If a safety valve is blowing, the main valve is not seating and the regulator should immediately be taken out of service and sent for overhaul. Needle-type control valves can be fitted to regulator outlets.

The correct blowpipe pressure is obtained by adjusting the pressure of the spring with the control knob and noting the pressure in bar on the blowpipe pressure gauge. On changing a tip the regulator is set with its finger on the tip number on the scale, and final accurate adjustment of the flame is made with the blowpipe regulating valves. This is a simple and convenient method. The regulators are supplied with a table indicating the suitable pressures for various nozzles which are stamped with their consumption of gas in litres per minute or suitable numbers. With practice the welder soon recognizes the correct pressures for various tips without reference to the table.

Regulators can be obtained for a wide range of gases; oxygen, acetylene, argon, nitrogen, propane, hydrogen, CO_2, etc., with outlet pressures to suit.

To enable two, three or more cylinders of gas to be connected together as may be required when heavy cutting work is to be done and the oxygen consumption is very great, special adaptors are available, and these feed the bottles into one gauge. In this way a much steadier supply of oxygen is obtained.

Two operators may also be fed from one cylinder of oxygen or acetylene by using a branched gauge with two regulators. The type of branched gauge which has only one regulator feeding two outlet pipes is not recommended, since any alteration of the blowpipe pressure by one operator will affect the flame of the other operator.

Owing to the rapid expansion of the oxygen in cases where large quantities are being used, the regulator may become blocked with particles of ice, causing stoppage. This happens most frequently in cold weather, and can be prevented by use of an electric regulator heater. The heater screws into the cylinder and the regulator screws into the heater. The heater is plugged into a source of electric supply, the connexion being by flexible cable.

Hoses

Hoses are usually of a seamless flexible tube reinforced with plies of closely woven fabric impregnated with rubber and covered overall with a tough flexible abrasion-resistant sheath giving a light-weight hose. They are coloured blue for oxygen, red for fuel gases, black for non-combustible gases and orange for LPG, Mapp and natural gas. Available lengths are from 5 to 20 m, with bore diameters 4.5 mm for maximum working pressure of 7 bar, 8 mm for a maximum of 12 bar and 10 mm for a maximum working pressure of 15 bar. Nipple- and nut-type connexions and couplers are available for 4.5 mm ($\frac{3}{8}$ in.), 8 and 10 mm hoses with 6.4 mm ($\frac{1}{4}$ in. BSP) and 10 mm ($\frac{3}{8}$ in.BSP) nuts. A hose check valve is used to prevent feeding back of gases from higher to lower pressures and reduces the danger of a flashback due to a blocked nozzle, leaking valve, etc. It is connected in the hose at the blowpipe end or to economizer or regulator, and consists of a self-aligning spring-loaded valve which seals off the line in the event of backflow. BS 924 J and 796 J apply to hoses.

The welding blowpipe or torch

There are two types of blowpipes: (1) high pressure, (2) low pressure and each type consists of a variety of designs depending on the duty for which the pipe is required. Special designs are available for rightward and leftward methods of welding (the angle of the head is different in these designs), thin gauge or thick plate, etc., in addition to blowpipes designed for general purposes.

The high-pressure blowpipe is simply a mixing device to supply approximately equal volumes of oxygen and acetylene to the nozzle, and is fitted with regulating valves to vary the pressure of the gases as required (Fig. 3.4a, b). A selection of shanks is supplied with each blowpipe, having orifices of varying sizes, each stamped with a number or with the consumption in litres per hour (l/h). Various sizes of pipes are available, from a small light variety, suitable for thin gauge sheet, to a heavy duty pipe. A high-pressure pipe cannot be used on a low-pressure system.

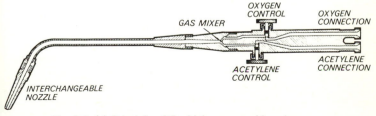

Fig. 3.4. (*a*) Principle of the high-pressure blowpipe.

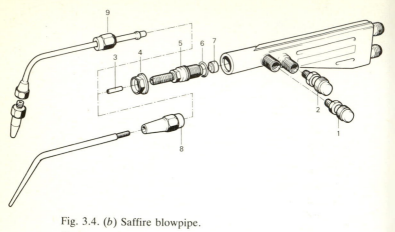

Fig. 3.4. (*b*) Saffire blowpipe.

Key:

1. Spindle assembly RH.
2. Spindle assembly LH.
3. Insert.
4. Nut–locking.
5. Mixer.

6. 'O' ring.
7. Mixer spool.
8. Adaptor nut.
9. Neck assembly.

The low-pressure blowpipe has an injector nozzle inside its body through which the high-pressure oxygen streams (Fig. 3.5). This oxygen draws the low-pressure acetylene into the mixing chamber and gives it the necessary velocity to preserve a steady flame, and the injector also helps to prevent backfiring. The velocity of a 1/1 mixture of oxygen/acetylene may be 200 m per minute, while the maximum gas velocity occurs for a 30% acetylene mixture and may be up to 460 m per min. (These figures are approximate only.)

It is usual for the whole head to be interchangeable in this type of pipe, the head containing both nozzle and injector. This is necessary, since there is a corresponding injector size for each nozzle. Regulating valves, as on the high-pressure pipes, enable the gas to be adjusted as required.

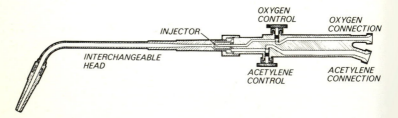

Fig. 3.5. Principle of the low-pressure blowpipe.

The low-pressure pipe is more expensive than the high-pressure pipe; and it can be used on a high-pressure system if required, but it is now used on a very small scale.

A very useful type of combined welding blowpipe and metal-cutting torch is shown in Fig. 3.6. The shank is arranged so that a full range of nozzles, or a cutting head, can be fitted. The design is cheaper than for a corresponding separate set for welding and cutting, and the cutter is sufficient for most work.

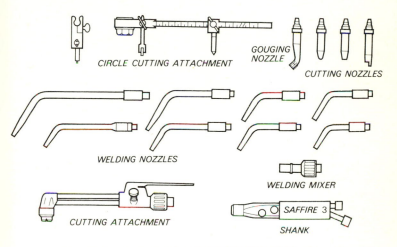

CIRCLE CUTTING ATTACHMENT

GOUGING NOZZLE

CUTTING NOZZLES

WELDING NOZZLES

WELDING MIXER

CUTTING ATTACHMENT

SAFFIRE 3

SHANK

Fig. 3.6. Combined welding and cutting pipes. Will weld sections from 1.6 mm to 32 mm thick, and cut steel up to 150 mm thick with acetylene and 75 mm with propane.

The oxy-acetylene flame

The chemical actions which occur in the flame have been discussed on p. 47, and we will now consider the control and regulation of the flame to a condition suitable for welding.

Adjustment of the flame.[1] To adjust the flame to the neutral condition the acetylene is first turned on and lit. The flame is yellow and smoky. The acetylene pressure is then increased by means of the valve on the pipe until the smokiness has just disappeared and the flame is quite bright. The condition gives approximately the correct amount of acetylene for the particular jet in use. The oxygen is then turned on as

[1] The pressure on oxygen and acetylene gauges is approximately that given in the table on p. 168.

quickly as possible, and as its pressure is increased the flame ceases to be luminous. It will now be noticed that around the inner blue luminous cone, which has appeared on the tip of the jet, there is a feathery white plume which denotes excess acetylene (Fig. 3.7a). As more oxygen is supplied this plume decreases in size until there is a clear-cut blue cone with no white fringe (Fig. 3.7b). This is the neutral flame used for most welding operations. If the oxygen supply is further increased, the inner blue cone becomes smaller and thinner and the outer envelope becomes streaky; the flame is now oxidizing (Fig. 3.7c). Since the oxidizing flame is more difficult to distinguish than the carbonizing or carburizing (excess acetylene) flame, it is always best to start with excess acetylene and increase the oxygen supply until the neutral condition is reached, than to try to obtain the neutral flame from the oxidizing condition.

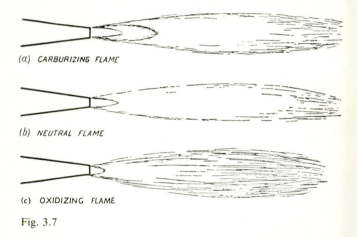

(a) CARBURIZING FLAME

(b) NEUTRAL FLAME

(c) OXIDIZING FLAME

Fig. 3.7

Some welders prefer to regulate the oxygen pressure at the regulator itself. The acetylene is lighted as before, and with the oxygen valve on the blowpipe turned full on, the pressure is adjusted correctly at the regulator until the flame is neutral. In this way the welder is certain that the regulator is supplying the correct pressure to the blowpipe for the particular nozzle being used.

Selection of correct nozzle. As the thickness of the work to be welded increases, the flame will have to supply more heat, and this is made possible by increasing the size of the nozzle. The nozzle selected may cover one or two thicknesses of plate; for example, a nozzle suitable for welding 6.4 mm plate will weld both 4.8 mm and 8 mm plate by suitable regulation of the pressure valves. This is because the blowpipe continues to mix the gases in the correct proportion over a range of pressures. If, however, one is tempted to weld a thickness of plate with a nozzle which is too large, by cutting down the supply of gas at the valves instead of changing the nozzle for one smaller, it will be noticed that explosions occur at the nozzle when welding, these making the operation impossible. These explosions indicate too low a pressure for the nozzle being used.

If, on the other hand, one attempts to weld too great a thickness of metal with a certain nozzle, it will be noticed that as one attempts to increase the pressure of the gases beyond a certain point to obtain a sufficiently powerful flame, the flame leaves the end of the nozzle. This indicates too high a pressure and results in a *hard* noisy flame. It is always better to work with a soft flame which is obtained by using the correct nozzle and pressure. Thus, although there is considerable elasticity as to the thickness weldable with a given nozzle, care should be taken not to overtax it.

Use and care of blowpipe

Oil or grease should upon no account be used on any part of the blowpipe, but a non-oily graphite may be used and is useful for preventing wear and any small leaks.

A backfire is the appearance of the flame in the neck or body of the blowpipe and which rapidly extinguishes itself

A flashback is the appearance of the flame beyond the blowpipe body into the hose and even the gauge, with subsequent explosion. It can be prevented by fitting a flashback arrestor in each pipe at the blowpipe end.

Backfiring may occur at the pipe through several causes:

(1) Insufficient pressure for the nozzle being used. This can be cured by increasing the pressure on the gauge.

(2) Metal particles adhering to the nozzle. The nozzle can be freed of particles by rubbing it on a leather or wooden surface. (The gases should be first shut off and then relit.)

(3) The welder touching the plate or weld metal with the nozzle. In this case the gases should be shut off and then relit.

(4) Overheating of the blowpipe. A can of water should be kept near so as to cool the nozzle from time to time, especially when using a large flame. Oxygen should be allowed to pass slowly through the nozzle, when immersed in the water, to prevent the water entering the inside of the blowpipe.

(5) Should the flame backfire into the mixing chamber with a squealing sound, and a thin plume of black smoke be emitted

from the nozzle, serious damage will be done to the blowpipe unless the valves are immediately turned off. This fault may be cause by particles having lodged inside the pipe, or even under the regulating valves. The pipe should be thoroughly inspected for defects before being relit.

In the event of a backfire, therefore, immediately shut off the acetylene cylinder valve, and then the oxygen, before investigating the cause.

Blowpipe nozzles should be cleaned by using a soft copper or brass pin. They should be taken off the shank and cleaned from the inside, as this prevents enlarging the hole. A clean nozzle is essential, since a dirty one gives an uneven-shaped flame with which good welds are impossible to make. Special sets of reamers can be obtained for this work.

Flashback arrestor. A flashback is potentially dangerous and can cause damage to the regulator. The automatic flashback arrestor, of which the Witt is an example, is connected to the regulator outlet connexion and prevents backfeeding of the gas and damage to the regulator. It quenches the flashback by means of a flame-trap so that there is no leakage of unburnt gas after the flashback and no burning at the flame-trap since the incoming gas is cut off by the automatic stop device. When the pressure in the system builds up beyond the safety limit, a regulating membrane is pushed back, operating a shut-off valve and at the same time a lever fitted with a warning sign to inform the operator that the valve has functioned. No more gas can now enter and the valve remains closed even if the pressure continues to build up and the 'cut-off valve', which operates in both directions, is activated by the slightest back pressure. To re-establish the gas flow the signal lever must be pushed back, and this cannot be done until the pressure has fallen to a safe value. The excess pressure and backed-up gases escape through a by-pass valve when the pressure has risen to a value above that for which the regulator has been set.

Technique of welding

Before attempting any actual welding operations, the beginner should acquire a sense of fusion and a knowledge of blowpipe control. This can be obtained by running lines of fusion on thin-gauge steel plate.

The flame is regulated to the neutral condition and strips of 1.6 or 2 mm steel plate are placed on firebricks on the welding bench. Holding the blowpipe at approximately 60° to the plate, with the inner blue cone near the metal surface, and beginning a little from the right-hand edge of the sheet, the metal is brought to the melting point and a puddle formed with a rotational movement of the blowpipe. Before the sheet has time to melt through into a hole, the pipe is moved steadily forward, still keeping the steady rotating motion, and the line of fusion is made in a straight line. This exercise should be continued on various thicknesses of thin-gauge plate until an even line is obtained, and the underside shows a regular continuous bead, indicating good penetration, the student thus acquiring a sense of fusion and of blowpipe control.

In the following pages various methods of welding techniques are considered, and it would be well to state at this point what constitutes a good weld, and what features are present in a bad weld.

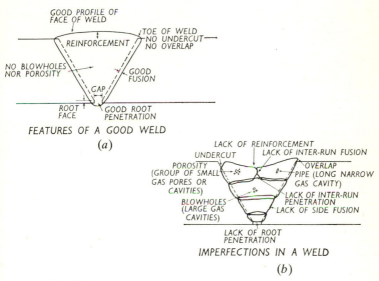

Fig. 3.8

Fig. 3.8*a* indicates the main features of a good fusion weld, with the following features:

    (*a*) Good fusion over the whole side surface of the V.
    (*b*) Penetration of the weld metal to the underside of the parent plate.
    (*c*) Slight reinforcement of the weld above the parent plate.
    (*d*) No entrapped slag, oxide, or blowholes.

Fig. 3.8*b* indicates the following faults in a weld:

    (*a*) Bad flame manipulation and too large a flame has caused molten metal to flow on to unfused plate, giving no fusion (i.e. adhesion).

(*b*) Wrong position of work, incorrect temperature of molten metal, and bad flame manipulation has caused slag and oxide to be entrapped and channels may be formed on each line of fusion, causing undercutting.

(*c*) The blowpipe flame may have been too small, or the speed of welding too rapid, and this with lack of skill in manipulation has caused bad penetration.

(N.B. Reinforcement on the face of a weld *will not* make up for lack of penetration.)

METHODS OF WELDING

The following British Standards apply to this section: BS 1845, *Filler alloys for brazing*; BS 1724, *Bronze welding by gas*; BS 1453, *Filler rods and wires for gas welding; group A, steel filler rods and wires; group B cast iron filler rods; group C, copper and copper alloy filler rods and wires; group D, magnesium alloy filler rods and wires*; BS 1821, *Oxy-acetylene welding of pipe lines, Class 1*; BS 2640, *Oxy-acetylene welding of pipe lines, Class 2.*

Leftward or forward welding

This method is used nowadays for welding steel plate under 6.5 mm thick and for welding non-ferrous metals. The welding rod precedes the blowpipe along the seam, and the weld travels from right to left when the pipe is held in the right hand. The inner cone of the flame, which is adjusted to the neutral condition, is held near the metal, the blowpipe making an angle of 60 to 70° with the plate, while the filler rod is held at an angle of 30 to 40°. This gives an angle of approximately 90° between the rod and the blowpipe. The flame is given a rotational, circular, or side to side motion, to obtain even fusion on each side of the weld. The flame is first played on the joint until a molten pool is obtained and the weld then proceeds, the rod being fed into the molten pool and not melted off by the flame itself. If the flame is used to melt the rod itself into the pool, it becomes easy to melt off too much and thus reduce the temperature of the molten pool in the parent metal to such an extent that good fusion cannot be obtained. Fig. 3.9. will make this clear.

The first exercise in welding with the filler rod is done with the technique just described and consists of running lines of weld on 1.6 or 2 mm plate, using the filler rod. Butt welds of thin plate up to 2.4 mm can be made by flanging the edges and melting the edges down. When a uniform weld is obtained, with good penetration, the exercises can be

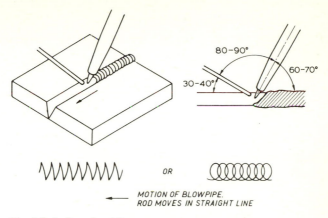

MOTION OF BLOWPIPE.
ROD MOVES IN STRAIGHT LINE

Fig. 3.9. Leftward welding.

repeated on plate up to 3.2 mm thick, and butt welds on this thickness attempted. Above 3.2 mm thick, the plates are bevelled, chamfered, or V'd to an angle of 80 to 90° (Fig. 3.10). The large area of this V means that a large quantity of weld metal is required to fill it. If, however, the V is reduced to less than 80°, it is found that as the V becomes narrower the blowpipe flame tends to push the molten metal from the pool, forward along on to the unmelted sides of the V, resulting in poor fusion or adhesion. This gives an unsound weld, and the narrower the V the greater this effect.

As the plate to be welded increases in thickness, a larger nozzle is required on the blowpipe, and the control of the molten pool becomes more difficult; the volume of metal required to fill the V becomes increasingly greater, and the size of nozzle which can be used does not increase in proportion to the thickness of the plate, and thus welding speed decreases. Also with thicker plates the side to side motion of the blowpipe over a wide V makes it difficult to obtain even fusion on the sides and penetration to the bottom, while the large volume of molten metal present causes considerable expansion. As a result it is necessary to weld thicker plate with two or more layers if this method is used. From these considerations it can be seen that above 6.4 mm plate the leftward method suffers from several drawbacks. It is essential, however, that the beginner should become efficient in this method before proceeding to the other methods, since for general work, including the non-ferrous metals (see later), it is the most used.

The preparation of various thicknesses of plate for butt joints by the leftward method is given in the table accompanying Fig. 3.10.

Edge preparation (Letters refer to consumption Fig. 3.10)	Thickness of plate (mm)	Nozzle size (mm)	Oxygen and acetylene pressure (bar)	Oxygen and acetylene gas consumption (l/h)
(a)	0.9–1.6	0.9–1	0.14	28
		1.2–2	0.14	57
(b)	2.4–3.0	2–3	0.14	86
		2.6–5	0.14	140
(c)	3.0–4.0	3.2–7	0.14	200
		4.0–10	0.21	280

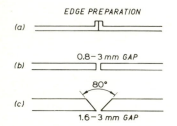

Fig. 3.10. Leftward welding: edge preparations.

Rightward welding

This method was introduced some years ago to compete with electric arc welding in the welding of plate over 4.8 mm thick, since the leftward method has the disadvantage just mentioned on welding thick plate. This method has definite advantages over the leftward method on thick plate, but the student should be quite aware of its limitations and use it only where it has a definite advantage.

In this method the weld progresses along the seam from left to right, the rod following the blowpipe. The rod is given a rotational or circular motion, while the blowpipe moves in practically a straight line, as illustrated in Fig. 3.11. The angle between blowpipe and rod is greater than that used in the leftward method.

When using this method good fusion can be obtained without a V up to 8 mm plate. Above 8 mm the plates are prepared with a 60° V, and since the blowpipe has no side to side motion the heat is all concentrated in the narrow V, giving good fusion. The blowpipe is pointing backward towards the part that has been welded and thus there is no likelihood of the molten metal being pushed over any of the unheated surface giving poor fusion.

A larger blowpipe nozzle is required for a given size plate than in leftward welding, because the molten pool is controlled by the pipe and rod but the pipe has no side to side motion. This larger flame gives greater welding speed, and less filler rod is used in the narrower V. The metal is under good control and plates up to 16 mm thick can be welded in one pass. Because the blowpipe does not move except in a straight line, the molten metal is agitated very little and excess oxidation is prevented. The flame playing on the metal just deposited helps to anneal it, while the smaller volume of molten metal in the V reduces the amount of expansion. In addition, a better view is obtained of the molten pool, resulting in better penetration.

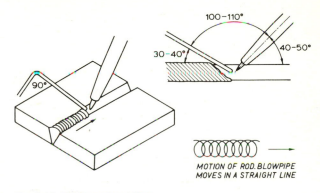

Fig. 3.11. Rightward welding

It is essential, however, in order to ensure good welds by this method, that blowpipe and rod should be held at the correct angle, the correct size nozzle and filler rod should be used, and the edges prepared properly (Fig. 3.12). The rod diameter is about half the thickness of the plate being welded up to 8 mm plate, and half the thickness + 0.8 mm when

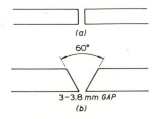

Fig. 3.12. Rightward welding: edge preparations.

welding V'd plate. The blowpipe nozzle is increased in size from one using about 300 litres per hour with the leftward method, to one using about 350 litres per hour, when welding 3.2 mm plate. If too large diameter filler rods are used, they melt too slowly causing poor penetration, and poor fusion. Small rods melt too quickly and reinforcement of the weld is difficult. Rightward welding has no advantage on plates below 6.4 mm thick and is rarely used below this thickness, the leftward method being preferred.

The advantages of the rightward method on thicker plate are:

(1) Less cost per foot run due to less filler rod being used and increased speed.
(2) Less expansion and contraction.
(3) Annealing action of the flame on the weld metal.
(4) Better view of the molten pool, giving better control of the weld.

See p. 176 for 'all-position rightward welding'.

Edge preparation (Letters refer to Fig. 3.12)	Thickness of plate (mm)	Nozzle size (mm)	Oxygen and acetylene pressure (bar)	Oxygen and acetylene gas consumption (l/h)
(a)	4.8–8.2	5–13	0.28	370
		6.5–18	0.28	520
		8.2–25	0.42	710
(b)	8.2–15	10–35	0.63	1000
		13–45	0.35 (heavy duty mixer)	1300

Vertical welding

The preparation of the plate for welding greatly affects the cost of the weld, since it takes time to prepare the edges, and the preparation given affects the amount of filler rod and gas used. Square edges need no preparation and require a minimum of filler rod. In leftward welding square edges are limited to 3.2 mm thickness and less. In vertical welding, up to 4.8 mm plate can be welded with no V'ing with the single operator method while up to 16 mm plate can be welded with no V'ing with the two-operator method, the welders working simultaneously on the weld from each side of the plate. The single-operator method is the most economical up to 4.8 mm plate.

The single-operator method (up to 4.8 mm plate) requires more skill in the control of the molten metal than in downhand welding. Welding is performed either from the bottom upwards, and the rod precedes the flame as in the leftward method, or from the top downwards, in which case the metal is held in place by the blowpipe flame. This may be regarded as the rightward method of vertical welding, since the flame precedes the rod down the seam. In the upward method the aim of the welder is to use the weld metal which has just solidified as a 'step' on which to place the molten pool. A hole is first blown right through the seam, and this hole is maintained as the weld proceeds up the seam, thus ensuring correct penetration and giving an even back bead.

In the vertical welding of thin plate where the edges are close together, such as for example in a cracked automobile chassis, little filler rod is needed and the molten pool can be worked upwards using the metal from the sides of the weld. Little blowpipe movement is necessary when the edges are close together, the rod being fed into the molten pool as required. When the edges are farther apart, the blowpipe can be given the usual semicircular movement to ensure even fusion of the sides.

From Fig. 3.13 it will be noted that as the thickness of the plate increases, the angle of the blowpipe becomes much steeper.

When welding downwards much practice is required (together with the correct size flame and rod), in order to prevent the molten metal

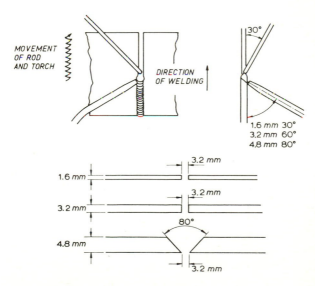

Fig. 3.13. Single-operator vertical welding.

from falling downwards. This method is excellent practice to obtain perfect control of the molten pool.

Double-operator vertical welding

The flames of each welder are adjusted to the neutral condition, both flames being of equal size. To ensure even supply of gas to each pipe the blowpipes can be supplied from the same gas supply. It is possible to use much smaller jets with this method, the combined consumption of which is less than that of a single blowpipe on the same thickness plate. Blowpipes and rods are held at the same angles by each operator, and it is well that a third person should check this when practice runs are being done. To avoid fatigue a sitting position is desirable, while, as for all types of vertical welding, the pipes and tubes should be as light as possible. Angles of blowpipes and rods are shown in Fig. 3.14.

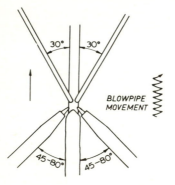

Fig. 3.14. Double-operator vertical welding.

This method has the advantage that plate up to 16 mm thick can be welded without preparation, reducing the gas consumption and filler rod used, and cutting out the time required for preparation. When two operators are welding 12.5 mm plate, the gas used by both is less than 50% of the total consumption of the blowpipe when welding the same thickness by the downhand rightward method. Owing to the increased speed of welding and the reduced volume of molten metal, there is a reduction in the heating effect, which reduces the effects of expansion and contraction.

Overhead welding

Overhead welding is usually performed by holding the blowpipe at a very steep angle to the plate being welded. The molten pool is entirely controlled by the flame and by surface tension, and holding the flame almost at right angles to the plate enables the pool to be kept in position.

Difficulty is most frequently found in obtaining the correct amount of penetration. This is due to the fact that as sufficient heat to obtain the required penetration is applied, the molten pool becomes more fluid and tends to become uncontrollable. With correct size of flame and rod, however, and practice, this difficulty can be entirely overcome and sound welds made. Care should also be taken that there is no under-cutting along the edges of the weld.

A comfortable position and light blowpipe and tubes are essential if the weld is to be made to any fair length, as fatigue of the operator rapidly occurs in this position and precludes the making of a good weld.

The positions of blowpipe and rod for the leftward technique are shown in Fig. 3.15a, while Fig. 3.15b shows their relative positions when the rightward method is used. The rightward method is generally favoured, but it must again be stressed that considerable practice is required for a welder to become skilled in overhead welding.

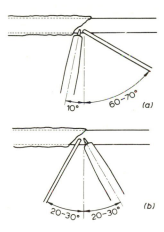

Fig. 3.15. (a) Leftward overhead welding. The flame is used to position the molten metal. (b) Rightward overhead welding. Blowpipe has little motion. Rod moves criss-cross from side to side.

Lindewelding

This method was devised by the Linde Co. of the United States for the welding of pipe lines (gas and oil), and for this type of work it is very suitable. Its operation is based on the following facts:

(1) When steel is heated in the presence of carbon, the carbon will

reduce any iron oxide present, by combining with the oxygen and leaving pure iron. The heated surface of the steel then readily absorbs any carbon present.

(2) The absorption of carbon by the steel lowers the melting point of the steel (e.g. pure iron melts at 1500°C, while cast iron with $3\frac{1}{2}\%$ carbon melts at 1130 °C).

In Lindewelding the carbon for the above action is supplied by using a carburizing flame. This deoxidizes any iron oxide present and then the carbon is absorbed by the surface layers, lowering their melting point. By using a special rod, a good sound weld can be made in this way at increased speed. The method is almost exclusively used on pipe work and is performed in the downhand position only.

Block welding

Block welding is a method especially applicable to steel pipes and thick walled tubes in which the weld is carried out to the full depth of the joint in steps. This can more easily be understood by reference to the figure. The first run is laid giving good penetration for as great a length as is convenient, say AB (Fig. 3.16).

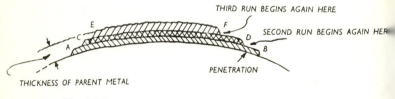

THIRD RUN BEGINS AGAIN HERE

SECOND RUN BEGINS AGAIN HERE

PENETRATION

THICKNESS OF PARENT METAL

Fig. 3.16

The second run is now started a little way short of A and finished short of B, as CD. The third run is laid in a similar manner from E to F, if required, and so on for the full depth of the weld required. We thus have a series of ledges or platforms at the beginning and end of the weld. The welding is now continued with the first run from B with full penetration. The second run starts at D and has the ledge BD deposited before it gets to where the first run started. Similarly, with the third run which starts at F. Upon completing the weld in the case of a pipe, first run finishes at A, second one at C and third one at E, giving a good anchorage on to the previous run.[1]

[1] Refer also to BS 1821 and 2640, *Class I and class II oxy-acetylene welding of steel pipelines and assemblies for carrying fluids.*

Fillet welding

In making fillet welds (Fig. 3.17), care must be taken that, in addition to the precautions taken regarding fusion and penetration, the vertical plate is not undercut as in Fig. 3.18b, and the weld is not of a weakened section. A lap joint may be regarded as a fillet. No difficulty will be experienced with undercutting, since there is no vertical leg, but care should be taken not to melt the edge of the lapped plate.

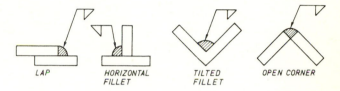

Fig. 3.17. Types of fillet joints.

The blowpipe and rod must be held at the correct angles. Holding the flame too high produces undercutting, and the nozzle of the cone should be held rather more towards the lower plate, since there is a greater mass of metal to be heated in this than in the vertical plate (Fig. 3.18a and b).

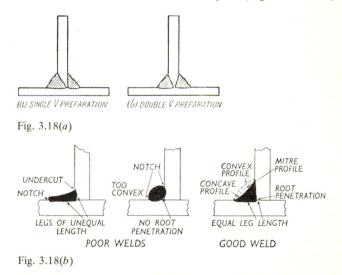

Fig. 3.18(a)

Fig. 3.18(b)

Figures 3.19 and 3.20 show the angles of the blowpipe and rod, the latter being held at a steeper angle than the blowpipe. Fillet welding requires a larger size nozzle than when butt welding the same section plate, owing to the greater amount of metal adjacent to the weld. Because of this, multi-jet blowpipes can be used to great advantage for fillet welding. The single V (Fig. 3.18*a*) preparation is used for joints which are subjected to severe loading, while the double V preparation is used for thick section plate when the welding can be done from both sides. The type of preparation therefore depends entirely on the service conditions, the unprepared joint being quite suitable for most normal work.

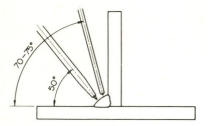

Fig. 3.19. Fillet weld.

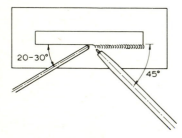

Fig. 3.20. Fillet weld.

All-position rightward welding

This method can be used for vertical, overhead and horizontal–vertical positions, the blowpipe preceding the rod as for downhand welding. For vertical welding the blowpipe is held 10° below the horizontal line (welding upwards) while the rod is held alongside the pipe at 45–60° to the vertical plate. Overhead welding is done similarly. The advantages are similar to those of the downhand position but considerable practice is required to become proficient. See Fig. 3.21.

For the preparation and welding of steel pipes the student should refer to BS 1821, *Class I steel pipelines* and BS 2640, *Class II steel pipelines*.

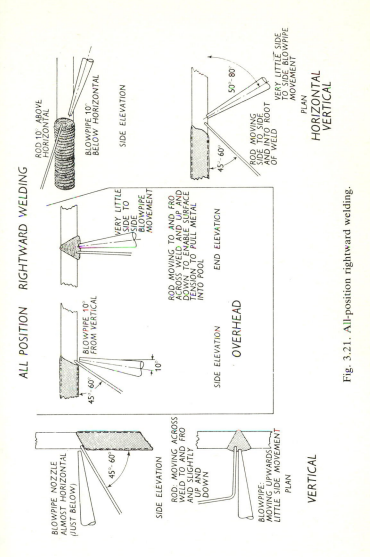

Fig. 3.21. All-position rightward welding.

Steel filler rods for oxy-acetylene welding

Rod type	Composition %					Applications
	C	Si	Mn	Ni	Cr	
Mild steel (copper coated)	0.1	—	0.6	0.25	—	General utility low-carbon-steel rod for low and mild carbon steel and wrought iron. UTS 386 N/mm² hardness 120 Brinell, melting point 1490°C.
Medium carbon steel (copper coated)	0.25–0.3	0.3–0.5	1.3–1.6	0.25	0.25	For medium carbon steels, high strength with toughness, UTS 552 N/mm², hardness 150 Brinell, melting point 1400°C.
Pipe-welding steel	0.1–0.2	0.1–0.35	1.0–1.6	—	—	Low carbon steel for high-strength welds in steel pipes. UTS 492 N/mm², hardness 145 Brinell, melting point 1450°
	Plus Al, Ti, Zr as deoxidizers up to 0.15% max.					

Note. Phosphorus and sulphur 0.04% max. for all types.

CAST IRON WELDING[1]

Cast iron, because of its brittleness, presents a different problem in welding from steel. We may consider three types – grey, white and malleable.

The grey cast iron is softer and tougher than the white, which is hard and brittle. The good mechanical properties of grey cast iron are due to the presence of particles of free carbon or graphite, which separate out during slow cooling. When the cooling is rapid, it is impossible for the cementite (iron carbide) to decompose into ferrite and graphite; hence the structure consists of masses of cementite embedded in pearlite, this giving the white variety of cast iron with its hardness and brittleness.

The other constituents of cast iron are silicon, sulphur, manganese and phosphorus. Silicon is very important, because it helps to increase the formation of graphite, and this helps to soften the cast iron. Manganese makes the casting harder and stronger. It has a great affinity for sulphur and, by combining with it, prevents the formation of iron sulphide, which makes the metal hard and brittle. Phosphorus reduces the melting point and increases the fluidity. If present in a greater proportion than 1% it tends to increase the brittleness. Sulphur tends to prevent the formation of graphite and should not exceed 1%. It is added to enable the outer layers to have a hard surface (chill casting), while the body of the casting is still kept in the grey state.

The aim of the welder should be always, therefore, to form grey cast iron in the weld (Fig. 3.22).

[1] See also pp. 68 and 183.

Fig. 3.22. Oxy-acetylene fusion weld in cast iron. × 100.

Preparation. The edges are V'd out to 90° on one side only up to 10 mm thickness and from both sides for greater thickness. Pre-heating is essential, because not only does it prevent cracking due to expansion and contraction, but by enabling the weld to cool down slowly, it causes grey cast iron to be formed instead of the hard white unmachinable deposit which would result if the weld cooled off rapidly. Pre-heating may be done by blowpipe, forge or furnace, according to the size of the casting.

Blowpipe, flame, flux and rod. The neutral flame is used, care being taken that there is not the slightest trace of excess oxygen which would cause a weak weld through oxidation, and that the metal is not overheated. The inner cone should be about 3 to 4 mm away from the molten metal. If it touches the molten metal, hard spots will result. The flux (of alkaline borates) should be of good quality to dissolve the oxide and prevent oxidation, and it will cause a coating of slag to form on the surface of the weld, preventing atmospheric oxidation. Ferro-silicon and super-silicon rods, containing a high percentage of silicon, are the most suitable rods to use.

Technique. The welding operation should be performed on the dull red-hot casting with the rod and blowpipe at the angles shown in Fig. 3.23 and done in the leftward manner. The rod is dipped into the flux at intervals, using only enough flux to remove the oxide. It will be noticed that as the flux is added, the metal flows more easily and looks brighter, and this gives a good indication as to when flux is required. The excessive use of flux causes blowholes and a weak weld. The rod should be pressed down well into the weld, removing a good percentage of the slag and oxide. Very little motion of the blowpipe is required and the rod should not be stirred round continuously, as this means the formation of more slag with more danger of entrapping it in the weld. Cast-iron welds made by beginners often have brittle parts along the edges, due to their not getting under the oxide and floating it to the top.

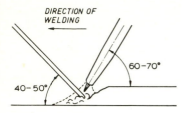

Fig. 3.23. Cast iron welding.

After treatment. The slag and oxide on the surface of the finished weld can be removed by scraping and brushing with a wire brush, but the weld should not be hammered. The casting is then allowed to cool off very slowly, either in the furnace or fire, or if it has been pre-heated with the flame, it can be put in a heap of lime, ashes or coke, to cool. Rapid cooling will result in a hard weld with possibly cracking or distortion.

Malleable cast iron welding

This is unsuitable for welding with cast-iron rods because of its structure. If attempted it invariably results in hard, brittle welds having no strength. The best method of welding is with the use of bronze rods, and is described below.

BRAZE WELDING AND BRONZE WELDING

Braze welding is the general term applied to the process in which the metal filler wire or strip has a lower melting point than that of the parent metal, using a technique similar to that used in fusion welding, except that the parent metal is not melted nor is capillary action involved as in brazing. When the filler metal is made of a copper-rich alloy the process is referred to as bronze welding.

By the use of bronze filler wires such as those given in the accompanying table, a sound weld can be made in steel, wrought iron, cast iron and copper and between dissimilar metals such as steel and copper, at a lower temperature than that usually employed in fusion welding. The bronze filler metal, which has a lower melting point than the parent metal, is melted at the joint so as to flow and 'wet' the surfaces of the joint to form a sound bond without fusion of the parent metal. There is less thermal disturbance than with fusion welding due to the lower temperatures involved and the process is simple and relatively cheap to perform. Unlike capillary brazing (see pp. 209–13) the strength of the joint is not solely dependent upon the areas of the surfaces involved in the joint but rather upon the tensile strength of the filler metal. A bronze weld has relatively great strength in shear, and joints are often designed to make use of this (Fig. 3.24a, b and c).

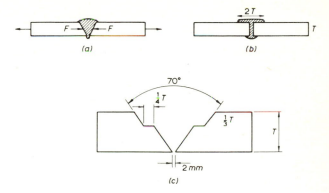

Fig. 3.24. When specimen (a) is subjected to a tensile pull as shown, the faces F of the joint are partly in shear. (c) 'Shear V' preparation.

The final strength of the welded joint depends upon the bond between filler metal and parent metal so that thorough cleanliness of the joints and immediate surroundings is essential to ensure that the molten filler metal should flow over the areas and 'wet' them completely without excessive penetration of the parent metal and there should be freedom from porosity. Care, therefore, should be taken to avoid excessive overheating. Figures 3.25 and 3.26 show various joint designs for plate and tube.

If bronze filler wires are used on alloys such as brasses and bronzes the melting points of wire and parent metal are so nearly equal that the result is a fusion weld.

British Standards designations, compositions and recommended usage of filler alloys for bronze welding

Alloy designation		Composition % (by weight) Deleterious impurities, e.g. Al and Pb, are each restricted to 0.03% max.		Recommended for usage on	Approx. melting point (°C)	Applications
BS 1453	BS 1845 (Group CZ)					
C2	CZ6	Cu 57.00 Si 0.20 Zn balance Sn optional	to 63.00 to 0.50 to 0.50 max.	Copper Mild steel	875–895	A silicon–bronze used for copper sheet and tube mild steel, deep drawing steel and line production applications.
C4	CZ7	Cu 57.00 Si 0.15 Mn 0.05 Fe 0.10 Zn balance Sn optional	to 63.00 to 0.30 to 0.25 to 0.50 to 0.50 max.	Copper Cast iron Wrought iron	870–900	Similar to C2. (CZ6).
C5	CZ8	Cu 45.00 Si 0.15 Ni 8.00 Zn balance Sn optional Mn optional Fe optional	to 53.00 to 0.50 to 11.00 to 0.50 max. to 0.50 max. to 0.50 max.	Mild steel Cast iron Wrought iron	970–980	A nickel–bronze for bronze welding steel and malleable iron, building up worn surfaces and welding Cu–Zn–Ni alloys of similar composition.
C6	——	Cu 41.00 Si 0.20 Ni 14.00 Zn balance Sn optional Mn optional Fe optional	to 45.00 to 0.50 to 16.00 to 1.00 max. to 0.20 max. to 0.30 max.	Cast iron Wrought iron	—— ——	Similar to C5. (CZ8).

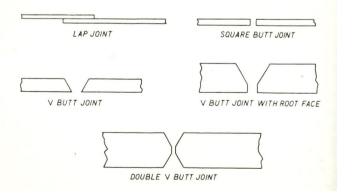

LAP JOINT   *SQUARE BUTT JOINT*

V BUTT JOINT   *V BUTT JOINT WITH ROOT FACE*

DOUBLE V BUTT JOINT

Fig. 3.25. Typical joint designs for bronze welding (sheet and plate).

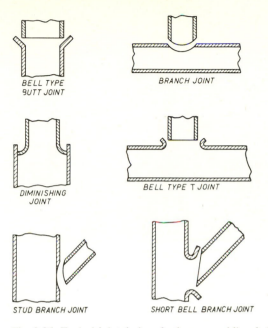

Fig. 3.26. Typical joint designs for bronze welding (tube).

General method of preparation for bronze welding. All impurities such as scale, oxide, grease, etc., should be removed, as these would prevent the bronze wetting the parent metal. The metal should be well cleaned on both upper and lower faces for at least 6 mm on each side of the joint, so that the bronze can overlap the sides of the joint, running through and under on the lower face.

Bronze welding is unlike brazing in that the heat must be kept as local as possible by using a small flame and welding quickly. The bronze must flow in front of the flame for a short distance only, wetting the surface, and by having sufficient control over the molten bronze, welding may be done in the overhead position. Too much heat prevents satisfactory wetting. We will now consider the bronze welding of special metals using a flux of a mixture of alkaline fluoride and borax, but bronze filler rods are also supplied flux-coated.

Cast iron

Bevel the edges to a 90° V, round off the sharp edges of the V, and clean the casting well. Pre-heating may be dispensed with unless the casting is of complicated shape, and the welding may often be done without dismantling the work. If pre-heating is necessary it should be heated to 450°C, and on completion cooling should be as slow as possible as in the fusion welding of cast iron (Fig. 3.27).

Fig. 3.27. Bronze weld in cast iron.

Blowpipe, flame and rod. The blowpipe nozzle may be about two sizes smaller than for the same thickness steel plate, and the flame is adjusted so as to give a slight excess of oxygen. If a second deposit is to be run over the first, the flame is adjusted to a more oxidizing condition still for the subsequent runs, the inner cone being usually only about $\frac{3}{4}$ of its neutral length. The best flame condition can easily be found by trial. Suitable filler alloys are given in the table, those containing nickel giving greater strength, the bronze flux being of the borax type.

Technique. The leftward method is used with the rod and blowpipe held as in Fig. 3.28 the inner cone being held well away from the molten metal. The rod is wiped on the edges of the cast iron and the bronze wets the surface. It is sometimes advisable to tilt the work so that the welding is done uphill, as shown in Fig. 3.28. This gives better control. Do not get the work too hot.

Vertical bronze welding of cast iron can be done by the two-operator method, and often results in saving of time, gas and rods and reduces the risk of cracking and distortion.

The edges are prepared with a double 90° V and thoroughly cleaned for 12 mm on each side of the edges. The blowpipes are held at the angle shown in Fig. 3.29a. Blowpipe and rod are given a side to side motion, as indicated in Fig. 3.29b as the weld proceeds upwards, so as to tin the surfaces.

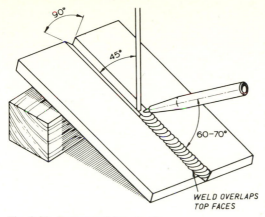

Fig. 3.28. Bronze welding cast iron.

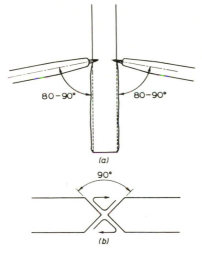

Fig. 3.29. (*a*) Two-operator vertical bronze welding of cast iron. (*b*) Showing motion of pipe and rod.

Malleable cast iron

The bronze welding of malleable casting may be stated to be the only way to ensure any degree of success in welding them. Both types (black-heart and whiteheart) can be welded satisfactorily in this way, since the heat of the process does not materially alter the structure. The method is the same as for cast iron, using nickel bronze rods (C5) and a borax-type flux.

Steel

In cases where excessive distortion must be avoided, or where thin sections are to be joined to thick ones, the bronze welding of steel is often used, the technique being similar to that for cast iron, except, of course, that no pre-heating is necessary.

Galvanized iron

This can be easily bronze welded, and will result in a strong corrosion-resisting joint, with no damage to the zinc coating. If fusion welding is used, the heat of the process would of course burn the zinc (or galvanizing) off the joint and the joint would then not resist corrosion.

Preparation, flame and rod. For galvanized sheet welding, the edges of the joint are tack welded or held in a jig and smeared with a silver–copper flux. Thicker plates and galvanized pipes are bevelled 60 to 80° and tacked to position them. The smallest possible nozzle should be used (for the sheet thickness) and the flame adjusted to be slightly oxidizing. Suitable filler alloys are given in the table as for steel.

Technique. No side to side motion of the blowpipes is given, the flame being directed on to the rod, so as to avoid overheating the parent sheet. The rod is stroked on the edges of the joint so as to wet them. Excessive flux *must* be washed off with hot water.

Copper

Tough pitch copper can be readily bronze welded owing to the much lower temperature of the process compared with fusion welding (Fig 3.30).

Preparation, flame and rod. Preparation is similar to that for cast iron. Copper tubes can be bell mouthed (Fig. 3.31). Special joints are available for multiple branches. The blowpipe nozzle should be small and will depend on the size of the work or the diameter of the pipe to be welded, and should be chosen so that the bronze flows freely but no overheating occurs. The flame should be slightly oxidizing, and if a second run is made, it should be adjusted slightly to be more oxidizing still (inner cone about $\frac{3}{4}$ of its normal neutral length). Suitable filler alloys are given in the table and are used with a bronze-base flux.

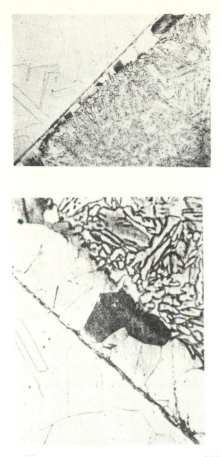

Fig. 3.30. (*b*) Bronze weld in copper. × 250.

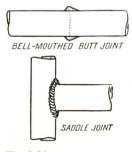

BELL-MOUTHED BUTT JOINT

SADDLE JOINT

Fig. 3.31

Technique. The method is similar to that for cast iron, and the final difference between the bronze-welded and brazed joint is that the former has the usual wavy appearance of the oxy-acetylene weld, while the latter has a smooth appearance, due to the larger area over which the heat was applied. The bronze joint is, of course, much stronger than the brazed one.

Brasses and zinc-containing bronzes

Since the filler rod now melts at approximately the same temperature as the parent metal, this may now be called fusion welding. When these alloys are heated to melting point, the zinc is oxidized, with copious evolution of fumes of zinc oxide, and if this continued, the weld would be full of bubble holes and weak (Fig. 3.32). This can be prevented by using an oxidizing flame, so as to form a layer of zinc oxide over the molten metal, and thus prevent further formation of zinc oxide, and vaporization.

Preparation. The edges and faces of the joint are cleaned and prepared as usual, sheets above 3.2 mm thickness being V'd to 90°. Flux can be applied by making it into a paste, or by dipping the rod into it in the usual manner, or a flux-coated rod can be used.

Flame and rod. Suitable filler alloys are given in the table, while a brass rod is used for brass welding, the colour of the weld then being similar to that of the parent metal. Owing to the greater heat conductivity, a larger size jet is required than for the same thickness of steel plate. The flame is adjusted to be oxidizing, as for bronze welding cast iron, and the exact flame condition is best found by trial as follows. A small test piece of the brass or bronze to be welded is heated with a neutral flame and gives off copious fumes of zinc oxide when molten. The acetylene is now cut down until no more fumes are given off. If any blowholes are seen in the metal on solidifying, the acetylene should be reduced slightly further. The inner cone will now be about half its normal neutral length. Too much oxygen should be avoided, as it will form a thick layer of zinc oxide over the metal and make the filler rod less fluid.

The weld is formed in the 'as cast' condition, and hammering improves its strength. Where 60/40 brass rods have been used the weld should be hammered while hot, while if 70/30 rods have been used the weld should be hammered cold and finally annealed from dull red heat.

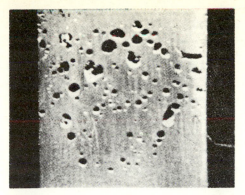

(*a*) Unsatisfactory brass weld made with neutral flame. Unetched. ×
2.5.

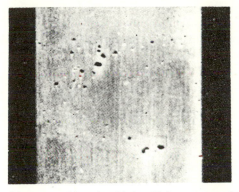

(*b*) Brass weld made with insufficient excess of oxygen. Unetched.

(*c*) Correct brass weld made with adequate excess of oxygen.
Unetched. × 2.5.

Fig. 3.32

Tin bronze

Tin bronze cannot be welded using an oxidizing flame. Special rods and fluxes are available, however, with which good welds can be made. Urgent repairs may be safely carried out using a *neutral* flame and silicon–bronze rod with borax type flux.

Gilding metal

For the weld to be satisfactory on completion, the weld metal must have the same colour as the parent metal. Special rods of various compositions are available, so that the colours will 'match'.

Aluminium bronze

Aluminium bronze can be welded using a filler rod of approximately 90% copper, 10% aluminium (C13, BS 2901 Pt 3), melting point 1040 °C, a rod also suitable for welding copper, manganese bronze and alloy steels where resistance to shock, fatigue and sea-water corrosion is required. The aluminium bronze flux (melting point 940 °C) can be mixed with water to form a paste if required.

Preparation. The edges of the joint should be thoroughly cleaned by filing or wire brushing to remove the oxide film which is difficult to dissolve.

Up to 4.8 mm thickness no preparation is required – just a butt joint with gap. Above 4.8 mm the usual V preparation is required and a double V above 16 mm thick. Sheets should not be clamped for welding, as this tends to cause the weld to crack, they must be allowed to contract freely on cooling, and it is advisable to weld a seam continuously and not make starts in various places.

Flame and rod. A neutral flame is usually used – any excess of acetylene tends to produce hydrogen with porosity of the weld, while excess oxygen causes oxidation. Flame size should be carefully chosen according to the thickness of the plate – too small a flame causes the weld metal to solidify too quickly while there is a danger of burning through with too large a flame.

The filler rod should be a little thicker (0.8 mm) than the sheet to be welded to avoid overheating, and it should be added quickly to give complete penetration without deep fusion.

Technique. The leftward method is used with a steep blowpipe angle (80°) to start the weld, this being reduced to 60–70 ° as welding progresses. The parent metal should be well pre-heated prior to starting welding and during welding a large area should be kept hot to avoid cracks. The rod should be used with a scraping motion to clean the molten pool and remove any entrapped gas.

In welding the single-constituent (or a phase) aluminium bronze, i.e. 5–7% Al, 93% Cu, the weld metal should be deposited in a single run or at most two runs to avoid intergranular cracks. Since the metal is hot short in the range 500–700 °C it should cool quickly through the range, and should not be peened. Cold peening is sometimes an advantage. The two constituent (a and β) or duplex aluminium bronzes contain 10% aluminium. They have a wide application, are not as prone to porosity, and are easier to weld than the 7% Al type, and also their hot short range is smaller.

After treatment. Stresses can be relieved by heat treating at low temperature, and any required heat treatment can be carried out as required after welding.

COPPER WELDING

Tough pitch copper (that containing copper oxide), is difficult to weld, and so much depends on the operator's skill that it is advisable to specify deoxidized copper for all work in which welding is to be used as the method of jointing. Welds made on tough pitch copper often crack along the edge of the weld if they are bent (Fig. 3.33a), showing that the weld is

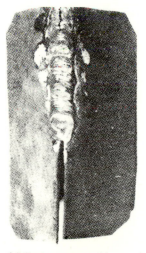

(a) Poor copper weld. Crack developed when bent. Fig. 3.33

(b) Oxy-acetylene weld in deoxidized copper. × 100.

unsound due to the presence of oxide, often along the lines of fusion. A *good* copper weld (Fig. 3.33*b*), on the other hand, can be bent through 180° without cracking and can be hammered and twisted without breaking. This type of copper weld is strong and sound, free from corrosion effects, and is eminently satisfactory as a method of jointing.

Preparation. The surfaces are thoroughly cleaned and the edges are prepared according to the thickness, as shown in Fig. 3.34. In flanging thin sheet the height of the flange is about twice the plate thickness and the flanges are bent at right angles. Copper has a high co-efficient of expansion, and it is necessary therefore to set the plates diverging at the rate of 3–4 mm per 100 mm run, because they come together so much on being welded. Since copper is weak at high temperatures, the weld should be well supported if possible and an asbestos sheet between the weld and the backing strip of steel prevents loss of heat.

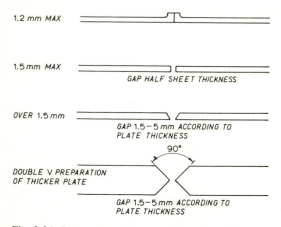

Fig. 3.34. Preparation of copper plates for welding.

Tacking to preserve alignment is not advised owing to the weakness of the copper tacks when hot. When welding long seams, tapered spacing clamps or jigs should be used to ensure correct spacing of the joint, care being taken that these do not put sufficient pressure on the edges to indent them when hot. A very satisfactory method of procedure is to place a clamp *C* at the centre of the seam and commence welding at a point say about one-third along the seam.

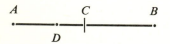

Welding is performed from *D* to *B* and then from *D* to *A*.

Because of the high conductivity of copper it is essential to pre-heat the surface, so as to avoid the heat being taken from the weld too rapidly. If the surface is large or the metal thick, two blowpipes must be used, one being used for pre-heating. When welding pipes they may be flanged or plain butt welded, while T joints can be made as saddles.

Blowpipe, flame and rod. A larger nozzle than for the same thickness of steel should be used and the flame adjusted to be neutral or very slightly carbonizing. Too much oxygen will cause the formation of copper oxide and the weld will be brittle. Too much acetylene will cause steam to form, giving a porous weld, therefore close the acetylene valve until the white feathery plume has almost disappeared. The welding rod should be of the deoxidized type, and many alloy rods, containing deoxidizers and other elements such as silver to increase the fluidity, are now available and give excellent results.

The weld may be made without flux, or a flux of the borax type used. Proprietary fluxes containing additional chemicals greatly help the welding operation and make it easier.

Technique. The blowpipe is held at a fairly steep angle, as shown in Fig. 3.35, to conserve the heat as much as possible. Great care must be taken to keep the tip of the inner cone 6–9 mm away from the molten metal, since the weld is then in an envelope of reducing gases, which prevent oxidation. The weld proceeds in the leftward manner, with a slight sideways motion of the blowpipe. Avoid agitating the molten metal, and do not remove the rod from the flame but keep it in the molten pool. Copper may also be welded by the rightward method, which may be used when the filler rod is not particularly fluid. The technique is similar to that for rightward welding of mild steel, with the flame adjusted as for leftward welding of copper.

Welding can also be performed in the vertical position by either single- or double-operator method, the latter giving increased welding speed.

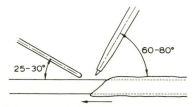

Fig. 3.35. Copper welding.

After treatment. Light peening, performed while the weld is still hot, increases the strength of the weld. The effect of cold hammering is to consolidate the metal, but whether or not it should be done depends on the type of weld and in general it is not advised. Annealing, if required on small articles, can be carried out by heating to 600–650 °C.

ALUMINIUM WELDING

The welding of aluminium, either pure or alloyed, presents no difficulty (Fig. 3.36) provided the operator understands the problems which must be overcome and the technique employed.

Fig. 3.36. Oxy-acetylene weld in aluminium. × 45.

The oxide of aluminium (alumina Al_2O_3), which is always present as a surface film and which is formed when aluminium is heated, has a very high melting point, much higher than that of aluminium, and if it is not removed it would become distributed throughout the weld, resulting in weakness and brittleness. A good flux, melting point 570°C, is necessary to dissolve this oxide and to prevent its formation.

Pure aluminium

Preparation. The work should be cleaned of grease and brushed with a wire brush. Sheets below 1 mm thickness can be turned up at right angles (as for mild steel) and the weld made without a filler rod. Over 3.2 mm thick the edges should have a 90° V and over 6 mm thick a double 90°V. Tubes may be bevelled if thick or simply butted with a gap between them. It is advisable always to support the work with backing strips of asbestos or other material, to prevent collapse when welding. Aluminium, when near its melting point, is extremely weak, and much trouble can be avoided by seeing that no collapsing can occur during the welding operation.

Blowpipe, flame, flux and rod. The flame is adjusted to have a very slight excess of acetylene and then adjusted to neutral, and the rod of pure aluminium or 5% silicon–aluminium alloy should be a little thicker than the section to be welded. A good aluminium flux must be used and should be applied to the rod as a varnish coat, by heating the end of the rod, dipping it in the flux and letting the tuft, which adheres to the rod, run over the surface for about 150 mm of its length. This ensures an even supply. Too much flux is detrimental to the weld.

Technique. The angles of the blowpipe and rod are shown in Fig. 3.37 (a slightly larger angle between the blowpipe and rod than for mild steel), and the welding proceeds in the leftward manner, keeping the inner cone well above the molten pool. The work may be tacked at about 150 mm intervals to preserve alignment, or else due allowance made for the joint coming in as welded, like mild steel. As the weld progresses and the metal becomes hotter, the rate of welding increases, and it is usual to reduce the angle between blowpipe and weld somewhat to prevent melting a hole in the weld. Learners are afraid of applying sufficient heat to the joint as a rule, because they find it difficult to tell exactly when the metal is molten, since it does not change colour and is not very fluid. When they do apply enough heat, owing to the above difficulty, the blowpipe is played on one spot for too long a period and a hole is the result.

If the rightward technique is used the blowpipe angle is 45° and the rod angle 30–40°. Distortion may be reduced when welding sheets, and the flame anneals the deposited metal.

The two-operator vertical method may be employed (as for cast iron) on sheets above 6 mm thickness, the angle of the blowpipes being 50–60° and the rods 70–80°. This method gives a great increase in welding speed (Fig. 3.37).

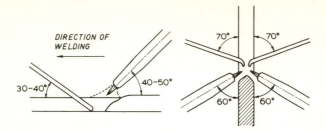

Fig. 3.37. (*a*) Aluminium flat welding. Rightward technique may also be employed using approximately the same angles of blowpipe and rod. (*b*) Aluminium welding by the double-operator method.

After treatment. All the corrosive flux must be removed first by washing and scrubbing in hot soapy water. This can be followed by dipping the article in a 5% nitric acid solution followed by a washing again in hot water.

Where it is not possible to get at the parts for scrubbing, such as in tanks, etc., the following method of removal is suitable. Great care, however, should be taken when using the hydrofluoric acid as it is dangerous, and rubber gloves should be worn, together with a face mask.

A solution is made up as follows in a heavy duty polythene container.

Nitric acid – 100 g to 1 litre water.

Hydrofluoric acid – 6 g to 1 litre water.

The nitric acid is added to the water first, followed by the hydrofluoric acid.

Articles immersed in this solution for about ten minutes will have all the flux removed from them, and will have a uniformly etched surface. They should then be rinsed in cold water followed by a hot rinse, the time of the latter not exceeding three minutes, otherwise staining may occur.

Hammering of the completed weld greatly improves and consolidates the structure of the weld metal, and increases its strength, since the deposited metal is originally in the 'as cast' condition and is coarse grained and weak. Annealing may also be performed if required.

Aluminium alloy castings and sheets

The process for the welding of castings is very similar to that for the welding of sheet aluminium. See p. 132 for explanation of alloy coding letters.

Preparation. The work is prepared by V'ing if the section is thicker than 3 mm, and the joint is thoroughly cleaned of grease and impurities. Castings such as aluminium crank-cases are usually greasy and oily (if they have been in service), the oil saturating into any crack or break which may have occurred. If the work is not to be pre-heated, this oil *must* be removed. It may be washed first in petrol, then in a 10% caustic soda solution, and this followed by a 10% nitric acid or sulphuric acid solution. A final washing in hot water should result in a clean casting. In normal cases in which pre-heating is to be done, filing and a wire brush will produce a clean enough joint, since the pre-heating will burn off the remainder.

Aluminium alloys. Recommended filler rods

Casting alloys (BS 1490)	Composition % (remainder Al)	Filler rod
LM2	2–2.5 Cu, 9–11.5 Si	NG2 (10–12 Si)
LM4	2–4 Cu, 4–6 Si	NG2 or NG21
LM5	3–6 Mg	NG6 (4.6–5.5 Mg)
LM6	10–13 Si	NG21 (4.5–6 Si)
LM8	3.5–6 Si, 0.3–0.8 Mg	NG21 or NG2
LM9	10–13 Si	NG21 or NG2
LM18	4.5–6 Si	NG21
LM20	10–13 Si, 0.4 Cu	NG2
Wrought alloys (BS 1470–1477)		
1, 1A, 1B, 1C	99.99, 99.8, 99.5, 99.0 Al	G1A, G1B, G1C
N3	1–1.5 Mn	NG3
H9	0.4–0.9 Mg, 0.3–0.7 Mn	NG21 or NG6
H20	0.15–0.4 Cu, 0.8–1.2 Mg, 0.4–0.8 Si, 0.2–0.8 Mn, 0.15–0.35 Cr	NG21 or NG6
H30	0.5–1.2 Mg, 0.7–1.3 Si, 0.4–1.0 Mn	NG21 or NG6

If the casting is large or complicated, pre-heating should be done as for cast iron.[1] In any case it is advantageous to heat the work well with the blowpipe flame before commencing the weld.

---

[1] Large complicated castings can be pre-heated to about 400°C, smaller castings to 300–350°C and small castings to 250–300°C. No visible change in the appearance of the aluminium occurs at these temperatures.

Blowpipe, flame, rod and flux. The blowpipe is adjusted as for pure aluminium and a similar flame used. The welding rod should preferably be of the same composition as the alloy being welded (see table on p. 197) but for general use a 5% silicon–aluminium rod is very satisfactory. This type of rod has strength, ductility, low shrinkage, and is reasonably fluid. A 10% silicon–aluminium rod is used for high silicon castings, while 5% copper–aluminium rods are used for the alloys containing copper such as Y alloy and are very useful in automobile and aircraft industries. The deposit from this type of rod is harder than from the other types.

When welding the Al–Mg alloys the oxide film consists of both aluminium and magnesium oxide making the fluxing more difficult so that as the magnesium proportion increases welding may become more difficult. Alloys containing more than $2\frac{1}{2}$% Mg, e.g. N5 and N8 are difficult to weld and require considerable experience as to the high strength alloys H15 and H30. The inert gas arc processes are to be preferred for welding these alloys.

Since there is also a loss of Mg in the welding process note that the filler rod recommended has a greater Mg content than the parent plate. The flux used is similar to that for pure aluminium and its removal must be carried out in the same way.

Technique. The welding is carried out as for aluminium sheet, and the cooling of the casting after welding must be gradual.[1]

After-treatment. After welding, the metal is in the 'as cast' condition and is weaker than the surrounding areas of parent metal, and the structure of the deposited metal may be improved by hammering. The area near the welded zone, however, is annealed during the welding process and failure thus often tends to occur in the area alongside the weld, and not in the weld itself. In the case of heat-treatable alloys, the welded zone can be given back much of its strength by first lightly hammering the weld itself and then heat-treating the whole of the work.

For this to be quite successful it is essential that the weld should be of the same composition as the parent metal. If oxidation has occurred, however, this will result in a weld metal whose structure will differ from that of the parent metal and the weld will not respond to heat treatment. Since many of this type of alloy are 'hot short', cracking may occur as a result of the welding process. Duralumin, Y alloy, hiduminium, etc., demand great care in welding because of these factors.

---

[1] When repairing cracked castings, any impurities which appear in the molten pool should be floated to the top, using excess flux if necessary.

If sheets are anodized, the welding disturbs the area and changes its appearance. Avoidance of overheating, localizing the heat as much as possible, and hammering, will reduce this disturbance to a minimum, but heat treatment will make the weld most inconspicuous.

WELDING OF NICKEL AND NICKEL ALLOYS

The alloys include Monel (nickel 69.4%, copper 29.1%, iron 1.2%, manganese 1.2%, carbon 0.12%) and Inconel (nickel 80%, chromium 20%) and modifications of these compositions to give variations in properties.

Oxy-acetylene welding is used only for welding nickel 200, monel alloys (90/10, 80/20, 70/30), Brightray alloys, Inconel 600, Incoloy DC and 800, and Nimonic 75. The welding of NiLO alloys is not usually performed.

Preparation. Sheets thinner than 1 mm can be bent up through an angle of about 75°, as shown in Fig. 3.38, and the edges melted together. The ridge formed by welding can then be hammered flat. Sheets thicker than 1 mm are bevelled with the usual 90° V and butted together. For corner welds on sheet less than 1 mm the corners are flanged as shown, while for thicker plate the weld is treated as an open corner joint (see Fig. 3.38). Castings should be treated as for cast iron.

In tube welding, preparation should be an 80° V with no root gap.

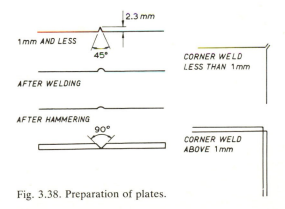

Fig. 3.38. Preparation of plates.

Blowpipe, flame, flux and rod

High-purity acetylene as supplied from DA cylinders is required, and in general the blowpipe nozzle size is the same as that for the same thickness mild steel. For nickel 200 a size larger can be used. No flux is required for nickel 200 and for Incoloy DS the use of flux is optional and special fluxes are available for the other alloys. Flux containing boron should not be used with alloys containing chromium as it tends to cause hot cracking in the weld. The joint is tacked in position without flux and flux is made into a thin paste and painted on to both sides of the joint and the filler rod and allowed to dry before welding. Fused flux remaining should be removed by wire brushing and unfused flux with hot water. Flux remaining (after welding the nickel chromium alloys) can be removed with a solution of equal parts of nitric acid and water for 15–30 minutes followed by washing with water. Flux removal is important when high-temperature applications are involved, as the flux may react with the metal.

Technique

Leftward technique is used. Weaving and puddling of the molten pool should be avoided as agitation of the pool causes porosity due to the absorption of gases by the high-nickel alloys, and the filler rod should be kept within the protective envelope of the flame to prevent oxidation. Keep the flame tip above the pool but for Monel K 500 let it just touch the pool. Nickel 200 melts sluggishly, inconel 600, Nimonic 75 and the Brightray alloys are more fluid while Monel 400 and Incoloy DS flow easily. A slightly carburizing soft flame (excess acetylene) should be used for the nickel and nickel-copper alloys, whilst for the chromium-containing alloys the flame should be a little more carburizing.

There are no pronounced ripples on the weld surfaces. They should be smooth without roughness, burning or signs of porosity.

Filler metals and fluxes

Material	Filler metal	Flux
Monel 400	Monel 40	Monel.
Monel K 500	Monel 64	Monel K 500.
Inconel 600		
Incoloy 800		Inconel.
Incoloy DS	NC 80/20	Boron-free.
Nimonic 75		
Brightray alloys		

Note. Use of a flux is optional with Incoloy DS.

When welding tube with 80° feather edge preparation and no root gap, the first run is made with no filler rod, the edges being well fused together to give an even penetration bead, followed by filler runs. If the work is rotatable, welding in the two o'clock position gives good metal control. If the joint is fixed, vertical runs should be made first downwards followed by a run upwards.

Nickel clad steel[1]

This steel is produced by the hot rolling of pure nickel sheet on to steel plate, the two surfaces uniting to form a permanent bond. It gives a material having the advantages of nickel, but at much less cost than the solid nickel plate. It can be successfully welded by the oxy-acetylene process.

Preparation. The bevelled 90° butt joint is the best type. For fillet welds it is usual to remove (by grinding) the nickel cladding on the one side of the joint, as shown in Fig. 3.39, so as to ensure a good bond of steel to steel.

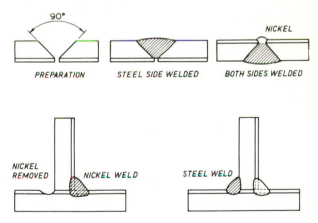

Fig. 3.39. Double fillet weld. Nickel clad steel.

Technique. The weld is made on the steel side first, using a mild steel rod and the same technique as for mild steel. The nickel side is then welded, using a nickel rod and a slightly reducing flame as for welding monel and nickel. The penetration should be such that the nickel penetrates and welds itself into the steel weld.

[1] Also stainless clad steel.

Stainless steels[1]

Stainless steels of the martensitic and austenitic class can be welded by the oxy-acetylene process.

Preparation. Thin gauges may be flanged at 90° as for mild steel sheet and the edges fused together. Thicker sections are prepared with the usual 90° V and the surfaces cleaned of all impurities. The coefficient of expansion of the 18/8 austenitic steels is about 50% greater than that of mild steel, and consequently the sheets should be set diverging much more than for mild steel to allow for them coming together during welding. An alternative method is to tack the weld at both ends first, and then for sheets thinner than 1 mm tack at 25 mm intervals, while for thicker sheets the tacks can be at 50 mm intervals. Cooling clamps are advised for the austenitic steels. The thermal conductivity is less than for mild steel, and thus the heat remains more localized. Unless care is taken, therefore, the beginner tends to penetrate the sheet when welding thin gauges.

Flame. The flame should be very slightly carburizing (excess acetylene), there being the smallest trace of a white plume around the inner cone. The flame should be checked from time to time to make sure that it is in this condition, since any excess oxygen is fatal to a good weld, producing porosity, while too great an excess of acetylene causes a brittle weld.

Rod and flux. The welding rod must be of the same composition as the steel being welded, the two types normally used being the 18/8 class, and one having a higher nickel and chromium content and suitable for welding steels up to a 25/12 chrome–nickel content. Since there is considerable variety in the amounts of the various elements added to stainless steels to improve their physical properties, such as molybdenum and tungsten in addition to titanium or niobium, it is essential that the analysis of the parent metal be known so that a suitable rod can be chosen. The steel makers and electrode makers will always co-operate in this matter. The rod should be of the same gauge or slightly thicker than the sheet being welded. No flux is necessary, but if difficulty is experienced with some steels in penetration, a special flux prepared for these steels should be used, and can be applied as a paste mixed with water.

Technique. Welding is performed in the leftward manner, and the flame is played over a larger area than usual because of the low

[1] See also p. 84 for description of this class of steel.

thermal conductivity. This lessens the risk of melting a hole in the sheet. The tip of the inner cone is kept very close to the surface of the molten pool and the welding performed exactly as for the leftward welding of mild steel. No puddling should be done and the least possible amount of blowpipe motion used.

After treatment. Martensitic stainless steels should be heated to 700 to 800 °C and then left to air cool. The non-stabilized steels require heat treatment after welding if they are to encounter corrosive conditions, otherwise no heat treatment is necessary. The weld decay proof stabilized steels and those with a very low carbon content require no heat treatment after welding.

The original silver-like surface can be restored to stainless steels which have scaled or oxidized due to heat by immersing in a bath made up as follows: sulphuric acid 8%, hydrochloric acid 6%, nitric acid 1%, water 85%. The temperature should be 65°C and the steel should be left in for 10 to 15 minutes, and then immersed for 5 minutes or until clean in a bath of 10% nitric acid, 90% water, at 25 °C.

Other baths made up with special proprietary chemicals with trade names will give a brighter surface. These chemicals are obtainable from the makers of the steel.

Stainless iron

The welding of stainless iron results in a brittle region adjacent to the weld. This brittleness cannot be completely removed by heat treatment, and thus the welding of stainless iron cannot be regarded as completely satisfactory. When, however, it has to be welded the welding should be done as for stainless steel, using a rod specially produced for this type of iron.

HARD SURFACING AND STELLITING

Surfaces of intense hardness can be applied to steel, steel alloys, cast iron or monel metal, by means of the oxy-acetylene flame. Hard surfaces may be deposited either on new parts so that they will have increased resistance to wear at reduced cost, or on old parts which may be worn, thus renewing their usefulness. In addition, non-ferrous surfaces, such as bronze or stellite, may be deposited and are described under their respective headings.

The surfaces deposited may be hard and/or wear and corrosion resisting. It is usual in the case of hard surfaces for the metal to be machinable as deposited, but to be capable of being hardened by quenching or by work hardening, e.g. 12/14% Mn (see table of rods available in the appendix).

We may divide the methods as follows:

(1) Building up worn parts with a deposit similar to that of the parent metal.

This process is very widely used, being cheap and economical, and is used for building up, for example, gear wheels, shafts, keys splines, etc.

The technique employed is similar to that for mild steel, using a neutral flame and the leftward method. The deposit should be laid a little at a time and the usual precautions taken for expansion and distortion. Large parts should be pre-heated and allowed to cool out very slowly, and there are no difficulties in the application.

(2) Building up surfaces with rods containing, for example, carbon, manganese, chromium or silicon to give surfaces which have the required degree of hardness or resistance to wear and corrosion.

These surfaces differ in composition from that of the parent metal, and as a result the fusion method of depositing cannot be used, since the surface deposit would become alloyed with the base metal and its hardness or wear-resisting properties would be thus greatly reduced.

(3) Hard facing with tungsten carbide. The rods consist of a steel tube containing fused granules of tungsten carbide HV 1800 in a matrix of chromium iron HV 850. Varions grades are available with granules of different mesh size. Large granules embedded in the steel surface do not round off as wear takes place, but chip because they are brittle and maintain good serrated cutting edges. Finer mesh granules give a more regular cutting edge so that the mesh of the granules is determined by the working conditions.

Technique. The flame is adjusted to have excess acetylene with the white plume from 2 to 2½ times the length of the inner cone. As explained on p. 174 the heated surface absorbs carbon, its melting point is reduced and the surface sweats. The rod is melted on to this sweating surface and a sound bond is made between deposit and parent metal with the minimum amount of alloying taking place.

This type of deposit, used for its wear-resisting properties, is usually very tough and is practically unmachinable. The surface is therefore usually ground to shape, but in many applications, such as in reinforcing tramway and rail crossings, it is convenient and suitable to hammer the deposit to shape while hot, and thus the deposit requires a minimum amount of grinding. The hammering, in addition, improves the structure.

In the case of the high-carbon deposits, they can be machined or ground to shape and afterwards heat treated to the requisite degree of hardness.

When using tungsten carbide rods the cone of the flame should be played on the rod and pool to allow gases to escape, preventing porosity. Weaving is necessary to give an even distribution of carbide granules in the matrix. Second runs should be applied with the same technique as the first, with no puddling of the first run since this would give dilution of the hard surface with the parent plate.

Another method frequently used to build up a hard surface is to deposit a surface of cast iron in the normal way, using a silicon cast iron rod and flux. Immediately the required depth of deposit has been built up, the part is quenched in oil or water depending on the hardness required.

This results in a hard deposit of white cast iron which can be ground to shape, and which possesses excellent wear-resisting properties. This method is suitable only for parts of relatively simple shape, that will not distort on quenching, such as camshafts, shackle pins, pump parts, etc.

The use of carbon and copper fences in building up and resurfacing results in the deposit being built much nearer to the required shape, reducing time in welding and finishing and also saving material.

When hard surfacing cast iron, the surface will not sweat. In this case the deposit is first laid as a fusion deposit, with neutral flame, and a second layer is then 'sweated' on to this first layer. In this way the second layer is obtained practically free from any contamination of the base metal.

It will be seen, therefore, that the actual composition of the deposit will depend entirely upon the conditions under which it is required to operate. A table of alloy steel rods and uses is given in the appendix.

One of the best known and successful alloys which can be deposited is stellite, an alloy of cobalt, chromium and tungsten with carbon. It has intense resistance to wear and corrosion and preserves these properties at high temperatures, but it is very brittle. By depositing a surface of stellite on a more ductile metal we have an excellent combination.

Tips of stellite can be brazed on to lathe and cutting tools of all types, giving an excellent cutting edge on a less brittle shank. Stellite, however, can be welded directly on to surfaces, and in this form it is used for all types of duty, such as surfaces on shafts which have to stand up to great wear and corrosion, lathe centres, drill tips, etc.

Grade 1 (black tip) HV 610–650 is recommended where the surfaces are subjected to abrasive wear, hardness being an essential feature. Grade 6 (red tip) HV 390–440 gives a stronger and tougher deposit than grade 1. It can be used for valves, tappet heads, and surfaces subject to heavy shock and impact and also for large areas where grade 1 would be liable to crack. Grade 12 (green tip) HV 500–550 gives a tougher deposit than grade 1 and is recommended for building up surfaces subjected not only to abrasion, but shock as well.

Stelliting steel

Preparation. Scale, dirt and impurities are thoroughly removed and the parent metal is pre-heated. Pre-heating and slow cooling are essential to avoid cracking.

Small jets of water, playing on each side of the weld, can be used to limit the flow of heat and reduce distortion, when building up deposits on hardened parts such as camshafts.

Flame and rod. A flame with an excess of acetylene is used, the white plume being about 2 to 2½ times the length of the inner blue cone. Too little acetylene will cause the stellite to foam and bubble, giving rise to blowholes, while too much acetylene will cause carbon to be deposited around the molten metal. The tip should be one size larger than that for the same thickness steel plate, but the pressure should be reduced, giving a softer flame. For small parts 5 mm diameter stellite rods can be used, while thicker rods are used for larger surfaces. Too much heat prevents a sound deposit being obtained, because some of the base metal may melt and mix with the molten stellite, thus modifying its structure.

Technique. The flame is directed on to the part to be surfaced, but the inner cone should not touch the work, both blowpipe and rod making an angle of 25–35° with the plate. When the steel begins to sweat the stelliting rod is brought into flame and a drop of stellite melted on to the sweating surface of the base metal, and it will spread evenly and make an excellent bond with the base metal. The surfacing is continued in this way.

After treatment. The part should be allowed to cool out very slowly to prevent cracks developing.

Stelliting cast iron

Preparation. Clean the casting thoroughly of oil and grease and pre-heat to a dull red heat.

Technique. Using the same type rod and flame as for stelliting steel, it is advisable first to lay down a thin layer and then build up a second layer on this. The reason for this is that more of the surface of the cast iron is melted than in the case of steel, and as a result the first layer of stellite will be diffused with impurities from the cast iron.

The flame is then played on the cast iron and the rod used to push away any scale. A drop of stellite is then flowed on to the surface, and the flame kept a little ahead of the molten pool so as to heat the cast iron to the right temperature before the stellite is run on. Cast iron flux may be used to flux the oxide and produce a better bond.

Heat treatment for depositing stellite

(a) Small components of mild steel and steel up to 0.4% carbon. Pre-heat with the torch the area to be faced; face the area and cool away from draughts.

(b) Large components of mild steel and steel up to 0.4% carbon; small components of high-carbon and low-alloy steels. Pre-heat to 250–350 °C. Hard face whilst keeping at this heat with auxiliary heating flame. Cover and bring to an even heat with flame and cool slowly in dry kieselguhr, mica, slaked lime, ashes or sand.

(c) Large components of high-carbon or low-alloy steel; cast iron components; bulky components of mild steel with large areas of stellite facing. Pre-heat to 400–500 °C. Hard face whilst at this temperature. Bring to a dull red even heat and cool slowly as for (b).

(d) Air hardening steel (not stainless). When a large area of deposit is required these steels should be avoided. Otherwise pre-heat to 600–650 °C and deposit hard surface whilst at this heat. Then place in a furnace at 650 °C for 30 minutes and cool large components in the furnace and small ones as (b).

(e) 18/8 austenitic stainless steel non-hardening welding type. Pre-heat to 600–650°C. Hard surface whilst at this temperature. Bring to an even temperature and cool out as (b).

(f) 12–14% manganese austenitic steel. Use arc process.

Spray fuse process for depositing stellite

In this process a stellite powder (HV 425–750) or nickel powder (HV 375–750) is sprayed on to the part to be hardened and this layer is then fused on with an oxy-acetylene flame. In this way thin layers up to 2 mm thick can be deposited having little dilution with parent metal. The alloys are self-fluxing, the deposit has a fine structure, and any inclusions are well distributed. Surfaces having sharp corners and sudden change of section should be avoided and the areas should be rough turned and then coarse shot-blasted to obtain a rough surface. The nickel base or cobalt powder is applied with a gas standard powder spray gun using compressed air or combustion gas which projects the powder through an oxy-acetylene flame on to the work. The powder is fed under pressure from a hopper to the gun, and the correct technique is to apply the powder with correct adjustment of the flame temperature and particle velocity so that the particles are in the plastic condition and deform on impact with the work. Large jobs should be pre-heated and the gun held about 150 mm from the surface, and after deposition the deposit is porous like other cold-sprayed deposits. The next operation is to fuse the deposit into a sound, wear-resistant coating securely bonded to the parent metal. Because of stresses remaining in the deposit, fusing should be carried out immediately after spraying. It is done with an oxy-acetylene torch with a multi-jet nozzle, the part being first raised to about 350 °C. One area is selected to begin fusing and the temperature raised to 700–800 °C over a small area and part of this is then raised to 1100 °C when glazing of the surface begins, indicating that fusing is taking place. The torch is moved over the area until it is all fused and the whole part finally brought up to an even temperature, cooling being carried out by covering it with heat-insulating substance or heat treatment being given, if required for the parent plate.

With this method of application localized heat is reduced, so that distortion is minimized and thin deposits can be applied with little dilution. Shrinkage amounting to 25% takes place during fusing so this must be allowed for, together with grinding tolerance if required, when calculating the thickness of the sprayed coat. The latest method of application combines both gun and torch in one unit (termed a powder weld torch) and with this spraying and welding are performed in one operation, greatly reducing the time taken. The powder welding attachment shown in Fig. 3.40 fits on to a standard welding or cutting shank.

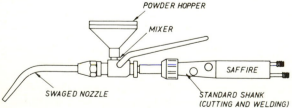

Fig. 3.40. Powder surfacing torch.

BRAZING

Brazing may be defined (BS 499, Part 1, 1965) as 'a process of joining metals in which, during or after heating, molten filler metal is drawn by capillary attraction into the space between closely adjacent surfaces of the parts to be joined'. In general the melting point of the filler metal is above 500 °C but always below the melting temperature of the parent metal.

Since capillarity and hence surface tension are involved in the process it may be convenient to give a brief explanation of some of the principles involved.

The student should refer to BS 1845 which lists the chemical compositions and approximate melting ranges of filler metals grouped under the following headings: Aluminium brazing alloys, silver brazing alloys, copper–phosphorus brazing alloys, copper brazing alloys, brazing brasses, nickel-base brazing alloys, palladium bearing brazing alloys, and gold bearing brazing alloys.

Surface tension

If drops of mercury rest on a level plate it will be noticed that the smaller the drop the more nearly spherical it is in shape, and if any drop is deformed it always returns to its original shape. If the only force acting on any drop were that due to its own weight, the mercury would spread out over the plate to bring its centre of gravity (the point at which the whole weight of the drop may be conceived to be concentrated) to the lowest point so that to keep the shape of the drop other forces must be present. As the drop gets smaller the force due to its own weight decreases and these other forces act so as to make the drop more spherical, that is to take up a shape which has the smallest surface area for a given volume. Other examples of these forces, termed surface tension are the floating of a dry needle on the surface of water, soap bubbles and water dripping from a tap. In the first example the small dry needle must be laid carefully horizontally on the surface. If it is pushed slightly below the surface it will sink because of its greater density. Evidently the surface of the water exhibits a force (surface tension) which will sustain the weight of the needle. If a wire framework $ABCD$, with CD, length x, able to slide along BC and AD, holds a soap film, the film tends to contract, and to prevent this a force F must be applied at right angles to CD (Fig. 3.41). The surface tension is defined as the force per unit length S on a line drawn in the film surface, and since there are two surfaces to the film $F = 2Sx$.

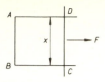

Fig. 3.41

Angle of contact (θ). The angle of contact between a liquid and a solid may be defined as the angle between the tangent to the liquid surface at the point of contact and the solid surface. For mercury on glass the contact angle is about 140°, while for other liquids the angle is acute and may approach zero (Fig. 3.42).

Wetting. If the contact angle approaches zero the liquid spreads and wets the surface and may do so in an upward direction. If the solid and liquid are such that the forces of attraction experienced by the molecules towards the interior of the liquid are less than the forces of attraction towards the solid, the area of contact will increase and the liquid spreads.

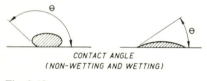

CONTACT ANGLE
(NON-WETTING AND WETTING)

Fig. 3.42

Capillarity. If a narrow bore (capillary) tube with open ends is placed vertically in a liquid which will wet the surface of the tube, the liquid rises in the tube and the narrower the bore of the tube the greater the rise. The wall thickness of the tube does not affect the rise and a similar rise takes place if the tube is replaced by two plates mounted vertically and held close together. If the tube or the plates are held out of the vertical the effect is similar and the vertical rise the same. If the liquid does not wet the tube (e.g. mercury) a depression occurs, and the shape of the liquid surfaces (the meniscus) is shown. The rise is due to the spreading or wetting action already considered – the liquid rises until the vertical upward force due to surface tension acting all round the contact surface with the tube is equal to the vertical downward force due to the weight of the column of liquid (Fig. 3.43).

This wetting action and capillary attraction are involved in the brazing process. The flux which melts at a lower temperature than the brazing

alloys wets the surfaces to be brazed. removes the oxide film and gives
clean surfaces which promote wetting by a reduction of the contact angle
between the molten filler alloy and the parent plate at the joint. The
molten filler alloy flows into the narrow space or joint between the
surfaces by capillary attraction and the narrower the joint the further
will be the capillary flow. Similarly solder flows into the narrow space
between tube and union when 'capillary fittings' are used in copper pipe
work.

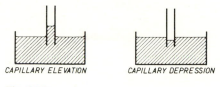

CAPILLARY ELEVATION   *CAPILLARY DEPRESSION*

Fig. 3.43

Brazing can be performed on many metals including copper, steel and
aluminium, and in all cases cleanliness and freedom from grease is
essential. The filler alloy used for aluminium brazing has already been
mentioned in the section on oxy-acetylene welding. In the case of the
nickel alloys, time and temperature are important. For example copper
alloys mix readily with nickel 200 or monel 400 and can pick up sufficient nickel to raise the melting point and hinder the flow of the filler
metal. Also chromium and aluminium form refractory oxides which
make brazing difficult, so that the use of a flux is necessary. A wide
range of brazing alloys are available having a variety of melting
points.

Aluminium brazing

The fusion welding of fillet and lap joints in thin aluminium sections
presents considerable difficulty owing to the way in which the edges melt
back. In addition the corrosive fluoride flux is very liable to be entrapped between contracting metal surfaces so it is advantageous to
modify the design wherever possible so as to include butt joints instead
of fillet and lap.

In many cases, however, corner joints are unavoidable and in these
cases flame brazing overcomes the difficulty. It can be done more
quickly and cheaply than fusion welding, less skill is required and the
finished joint is neat and strong.

Aluminium brazing is suitable for pure aluminium and for alloys
such as LM4, LM18 and for the aluminium–manganese and
aluminium–manganese–magnesium alloys, as long as the magnesium
content is not greater than 2%.

Preparation. It is always advisable to allow clearance at the joints since the weld metal is less fluid as it diffuses between adjacent surfaces. Clearance joints enable full penetration of both flux and filler rod to be obtained.

Surface oxide should be removed by wire brush or file and grease impurities by cleaning or degreasing. Burrs such as result from sawing or shearing and other irregularities should be removed so that the filler metal will run easily across the surface. Socket joints should have a 45° belling or chamfer at the mouth to allow a lead in for the flux and metal, and to prevent possibility of cracking on cooling, the sections of the surfaces to be jointed should be reduced to approximately the same thickness where possible.

Blowpipe, flame, flux and rod. The blowpipe and rod are held at the normal angle for leftward welding, and a nozzle giving a consumption of about 700 litres of both oxygen and acetylene each per hour is used. The flame is adjusted to the excess acetylene condition with the white acetylene plume approximately $1\frac{1}{2}$ to 2 times the length of the inner blue cone. The rod can be of 10–13% silicon–aluminium alloy melting in the range 565–595°C (compared with 659°C for pure aluminium). This is suitable for welding and brazing the high silicon aluminium alloys and for general aluminium brazing. Another type of rod containing 10–13% Si and 2–5%Cu is also used for pure aluminium and aluminium alloys except those with 5% silicon, or with more than 2% magnesium, but is not so suitable due to its copper content if corrosive conditions are to be encountered, but it has the advantage of being heat-treatable after brazing, giving greater mechanical strength.

In general, ordinary finely divided aluminium welding flux prepared commercially is quite suitable for brazing. but there is is also available a brazing flux of similar composition which has a lower melting point.

BS 1845, Filler metals for aluminium brazing (Group AL)

Type	Major alloying elements only (%)				Melting range (°C)	
	Silicon	Copper	Iron	Aluminium	Solidus	Liquidus
AL1	10–13	2–5	0.6	Rem.	535	595
AL2	10–13	0–0.1	0.6	Rem.	565	595
AL3	7–8	0–0.1	—	Rem.	565	610
AL4	4.5–6.0	0–0.1	—	Rem.	565	630

Technique. The flame is held well away from the joint to be brazed and pre-heating is done with the outer envelope of the flame for about three minutes – this procedure ensures an even temperature, which is essential so that first the flux flows evenly into the joint followed by the filler metal which must displace *all* the flux, otherwise if islands of flux are entrapped corrosion of the joint will occur.

The rod is warmed, dipped in the flux and the tuft which adheres to the rod is touched down on the heated joint. When the correct temperature has been reached the flux will melt and flow over and between the surfaces smoothly and easily. The blowpipe is now lowered to the normal welding position, some rod is melted on to the joint and the blow-pipe moved forward along the seam running the filler rod into the joint. The blowpipe is then raised and brought back a little and lowered again, the above operation being repeated – the blowpipe thus describes an elliptical motion – each operator modifying the technique according to his individual style (Fig. 3.44).

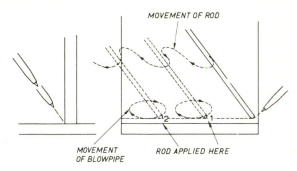

Fig. 3.44. Aluminium brazing.

Flux should only be added when the filler rod does not appear to be running freely. Too much flux is detrimental to the finished joint and great care should be taken that the filler rod flows freely into the joint, so as to displace all the flux.

After treatment. The corrosive flux should be removed by the treatment as given for aluminium welding p. 196.

GENERAL PRECAUTIONS

The following general precautions should be taken in welding:

(1) Always use goggles of proved design when welding or cutting. The intense light of the flame is harmful to the eyes and in addition small particles of metal are continually flying about and may cause serious damage if they lodge in the eyes. Welding filters or glasses are graded according to BS 679 by numbers followed by the letters GW or GWF. The former are for welding operations without a flux and the latter with flux because there is an additional amount of glare. The grades range from 3/GW and 3/GWF to 6/GW and 6/GWF, the lightest shade having the lowest grade number. For aluminium welding and light oxy-acetylene cutting, 3/GW or 3/GWF is recommended, and for general welding of steel and heavier welding in copper, nickel and bronze, 5/GW or 5/GWF is recommended. A full list of recommendations is given in the BS.

(2) When welding galvanized articles the operator should be in a well-ventilated position and if welding is to be performed for any length of time a respirator should be used. (In case of sickness caused by zinc fumes, as in welding galvanized articles or brass, milk should be drunk.)

(3) In heavy duty welding or cutting and in overhead welding, asbestos or leather gauntlet gloves, ankle and feet spats and protective clothing should be worn to prevent burns. When working inside closed vessels such as boilers or tanks, take every precaution to ensure a good supply of fresh air.

(4) In welding or cutting tanks which have contained inflammable liquids or gases, precautions must be taken to prevent danger of explosion. One method for tanks which have contained volatile liquids and gases is to pass steam through the tank for some hours according to its size. Any liquid remaining will be vaporized by the heat of the steam and the vapours removed by displacement.

Tanks should never be merely swilled out with water and then welded many fatal explosions have occurred as a result of this method of preparation. Carbon dioxide in the compressed form can be used to displace the vapours and thus fill the tank, and is quite satisfactory but is not always available. Tanks which have contained heavier types of oil such as fuel oil, tar, etc., present a more difficult problem since air and steam will not vaporize them. One method is to fill the tank with water, letting the water overflow for some time. The tank should then be closed and turned until the fracture is on top. The water level should be adjusted (by letting a little water out if necessary) until it is just below the fracture. Welding can then be done without fear as long as the level of the water does not drop much more than a fraction of an inch below the level of the fracture.

The welder should study the Department of Employment memorandum on *Safety measures for the use of oxy-acetylene equipment in factories* (Form 1704). Toxic gases and fine airborne particles can provide a hazard to a welder's health. The Threshold Limit Value (TLV) is a system by which concentrations of these are classified, and is explained in the Department of Employment Technical data notes No. 2, *Threshold limit values.* (See p. 323 for publications on safety and health in welding and repair of drums, tanks, etc.)

4

Basic electrical principles

ELECTRICAL TECHNOLOGY

Sources of electrical power

The principal sources of electrical power of interest to the welder are (1) batteries and accumulators, (2) generators.

Batteries generate electrical energy by chemical action. Primary batteries, such as the Leclanché (used for flash lamps and transistor radios), continue giving out an electric current until the chemicals in them have undergone a change, and then no further current can be given out.

Secondary batteries or accumulators are of two types: (i) the lead–acid, and (ii) the nickel–iron alkaline. In the former, for example, there are two sets of plates, one set of lead peroxide and the other set of lead, immersed in dilute sulphuric acid (specific gravity 1.250, i.e. 4 parts of distilled water to 1 part of sulphuric acid). Chemical action enables this combination to supply an electric current, and when a current flows from the battery both the lead peroxide plates and the lead plates are changed into lead sulphate, and when this change is complete the battery can give out no more current. By connecting the battery to a source of electric power, however, and passing a current through the battery in the opposite direction from that in which the cell gives out a current, the lead sulphate is changed back to lead peroxide on one set of plates and to lead on the other set. The battery is then said to be 'charged' and is ready to supply current once again.

It may also be noted here that when two different metals are connected together and a conducting liquid such as a weak acid is present currents will flow, since this is now a small primary cell. This effect is called 'electrolysis' and will lead to corrosion at the junction of the metals (see p. 54).

Generators can be made to supply direct current or alternating current as required and are described later. For welding purposes generators of special design are necessary. When welding with alternating current, 'transformers' are used to transform or change the pressure of the supply to a pressure suitable for welding purposes.

The electric circuit

The electric circuit can most easily be understood by comparing it to a water circuit. Such a water circuit is shown in Fig. 4.1*a*. A pump drives or forces the water from the high-pressure side of the pump through pipes to a water meter *M* which measures the flow of water in litres per hour. From the meter the water flows to a control valve the opening in which can be varied, thus regulating the flow of water in the circuit. The water is led back through pipes to the low-pressure side of the pump. We assume that no water is lost in the circuit. Figure 4.1*b* shows an electric circuit corresponding to this water circuit. The generator, which supplies direct current (d.c.), that is, current flowing in one direction only, requires energy to be consumed in driving it. The electrical pressure

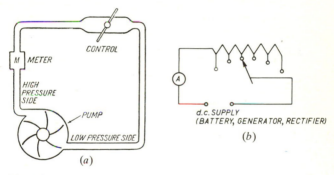

(*a*)

d.c. SUPPLY
(BATTERY, GENERATOR, RECTIFIER)

(*b*)

Fig. 4.1

available at the terminals of the generator when no current is flowing in the circuit is known as the electro-motive force (e.m.f.) or in welding as the open circuit voltage. The current is carried by copper wires or cables, which offer very little obstruction or resistance to the flow of current through them, and the current flows from the high-pressure side of the generator (called the positive or +ve pole) through a meter which corresponds to the water meter. This meter, known as an ampere meter or ammeter, measures the flow of current through it in amperes, A (amps. for short), just as the water meter measures the flow of water in litres per hour. This ammeter may be connected at any point in the circuit so that the current flows *through* it, since the current is the same at all points in the circuit. From the ammeter the current flows through a copper wire to a piece of apparatus called a 'resistor'. This consists of wire usually made of an alloy, such as manganin, nichrome or eureka, which offers considerable obstruction or resistance to the passage of a current.

The number of turns of this coil in the circuit can be varied by means of a switch, as shown in the figure. This resistor corresponds to the water valve by which the flow of water in the circuit is varied. The more resistance wire which we include in the circuit, the greater is the obstruction to the flow of the current and the less will be the current which will flow, so that as we increase the number of turns or length of resistance wire in the circuit, the reading of the ammeter, indicating the flow of current in the circuit, becomes less. The current finally flows through a further length of copper wire to the low-pressure (negative or −ve) side of the generator.

In the water circuit we can measure the pressure of water in N/mm² by means of a pressure gauge. We measure the difference of pressure or potential (p.d.) between any two points in an electric circuit by means of a voltmeter, which indicates the difference of pressure between the two points in volts. Figure 4.2 shows the method of connexion of an ammeter and a voltmeter in a circuit.

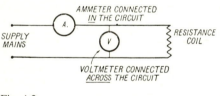

Fig. 4.2

Fall in potential – voltage drop. Let us consider the circuit in Fig. 4.3 in which three coils of resistance wire are connected to each other and to the generator by copper wires as shown, so that the current will flow through each coil in turn. (This is termed connecting them in *series*.) Suppose that the current flows from the +ve terminal *A* through the coils and back to the −ve terminal *H*. Throughout the circuit from *A* to *H* there will be a gradual fall in pressure or potential from the high-pressure side *A* to the low-pressure side *H*. Let us place a voltmeter across each section of the circuit in turn and find out where this fall in pressure or *voltage drop* occurs. If the voltmeter is first placed across *A* and *B*, we find that no difference of pressure or voltage drop is registered. This is because the copper wire connecting *A* and *B* offers very little obstruction indeed to the passage of the current, and hence, since there is no resistance to be overcome, there is no drop in pressure.

If the voltmeter is placed across B and C, however, we find that it will register a definite amount. This is the amount by which the pressure has dropped in forcing the current against the obstruction or resistance of BC. Similarly, by connecting the meter across DE and FG we find that a voltage drop is indicated in each case, whereas if connected across CD, EF, or GH, practically no drop will be recorded, because of the low resistance of the copper wires.

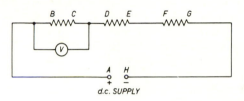

Fig. 4.3

If we add up the voltage drops across BC, DE and FG, we find that it is the same as the reading that will be obtained by placing the voltmeter across A and H, that is, the sum of the voltage drops in various parts of a circuit equals the pressure applied.

The question of voltage drop in various parts of an electric circuit is important in welding. Figure 4.4 shows a circuit composed of an ammeter, a resistance coil, and two pieces of carbon rod called electrodes. This circuit is connected to a generator or large supply battery, as shown. When the carbons are touched together, the circuit is completed and a current flows and is indicated on the ammeter. The amount of current flowing will evidently depend on the amount of resistance in circuit.

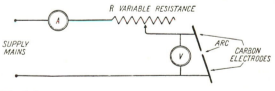

Fig. 4.4

If now the carbons are drawn apart about 4 mm, the current still flows across the gap between the carbons in the form of an arc. This is the 'carbon arc', as it is termed. We can control the current flowing across the arc by varying the amount of resistance R in the circuit, while if a voltmeter is placed across the arc, as shown, it will register the drop in pressure which occurs, due to the current having to be forced across the resistance of the gap between the electrodes. We also notice that the greater the distance between the electrodes the greater the voltage drop. The metallic arc used in arc welding is very similar to the carbon arc and is discussed fully later.

It has been mentioned that copper wire offers little obstruction or resistance to the passage of a current. All substances offer some resistance to the passage of a current, but some offer more than others. Metals, such as silver, copper and aluminium, offer but little resistance, and when in the form of a bar or wire the resistance that they offer increases with the length of the wire and decreases with the area of cross-section of the wire. Therefore the greater the length of a wire or cable, the greater its resistance; and the smaller the cross-sectional area, the greater its resistance. Thus if we require to keep the voltage drop in a cable down to the lowest value possible as we do in welding, the longer that the cable is, the greater must we make its cross-sectional area. Unfortunately, increasing the cross-sectional area makes the cable much more expensive and increases its weight, so evidently there is a limit to the size of cable which can be economically used for a given purpose.

Series and parallel groupings. We have seen that if resistors are connected together so that the current will flow through each one in turn, they are said to be connected in series (Fig. 4.5*a*). If they are connected so that the current has an alternative path through them, they are said to be connected in parallel or shunt (Fig. 4.5*b*).

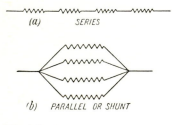

(a)   SERIES

(b)   PARALLEL OR SHUNT

Fig. 4.5

An example of the use of a parallel circuit is that of a shunt for an ammeter. The coil in an ammeter has a low resistance and will pass only a small current so that when large currents as used in welding are involved, it is usual to place the ammeter in parallel or shunt with a resistor (termed the ammeter shunt) which is arranged to have a resistance of such value that it carries the bulk of the current, for example 999/1000 of the total current. In this case, if the welding current were

100 A, 99.9 A would pass through the shunt and 0.1 A through the instrument coil, but the ammeter is calibrated with the particular shunt used and would read 100 A (Fig. 4.6).

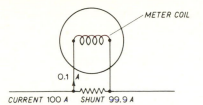

Fig. 4.6. Ammeter shunt.

Conductors, insulators and semi-conductors
Substances may be divided into two classes from an electrical point of view: (1) conductors, (2) insulators.

Conductors. These may be further divided into (*a*) good conductors, such as silver, copper and aluminium, which offer very little obstruction to the passage of a current, and (*b*) poor conductors or resistors, which offer quite a considerable obstruction to the passage of a current, the actual amount depending on the particular substance. Iron, for example, offers six times as much obstruction to the passage of a current as copper and is said to have six times the resistance. Alloys, such as manganin, eureka, constantin, german silver, no-mag and nichrome, etc., offer much greater obstruction than iron and have been developed for this purpose, being used to control the current in an electric circuit. No-mag, for example, is used for making resistance banks for controlling the current in motor and arc welding circuits, etc., while nichrome is familiar, since it is used as the heating element in electric fires and heating appliances, the resistance offered by it being sufficient to render it red hot when a current flows. Certain rare metals, such as tantalum, osmium and tungsten, offer extremely high obstruction, and if a current passes through even a short length of them in the form of wire they are rendered white hot. These metals are used as filaments in electric-light bulbs, being contained in a bulb exhausted of air, so as to prevent them oxidizing and burning away.

Insulators. Many substances offer such a great obstruction to the passage of a current that no current can pass even when high pressures are applied. These substances are called insulators, but it should be remembered that there is no such thing as a perfect insulator, since all substances will allow a current to pass if a sufficiently high pressure is applied. In welding, however, we are concerned with low voltages. Amongst the best and most familiar insulators are glass, porcelain, rubber, shellac, mica, oiled silk, empire cloth, oils, resins, bitumen, paper, etc. In addition, there is a series of compounds termed synthetic resins (made from phenol and formalin), of which 'bakelite' is a well-known example. These compounds are easily moulded into any desired shape and have excellent insulating properties. Plastics such as polyvinyl chloride (PVC) and chloro-sulphonated-polyethylene (CSP) are now used for cable insulation in place of rubber. PVC is resistant to oil and grease and if ignited does not cause flame spreading. At temperatures near freezing it becomes stiff and more brittle and is liable to crack, while at high temperature it becomes soft. As a result PVC insulation should not be used where it is near a heat source such as an electric fire or soldering iron.

The insulating properties of a substance are greatly dependent on the presence of any moisture (since water will conduct a current at fairly low pressures) and the pressure or voltage applied. If a person is standing on dry boards and touches the +ve terminal of a supply of about 200 volts, the −ve of which is earthed, very little effect is felt. If, however, he is standing on a wet floor, the insulation of his body from the earth is very much reduced and a severe shock will be felt, due to the much larger current which now passes through his body. As the voltage across an insulator is increased, it is put in a greater state of strain to prevent the current passing, and the danger of breakdown increases.

All electrical apparatus should be kept as dry as possible at all times Much damage may result from wet or dampness in electrical machines.

Semi-conductors. Most materials are either conductors or insulators of an electric current, but a small group termed semi-conductors fall in between the above types. To compare the resistance of various substances a cube of 1 metre edge is taken as the standard and the resistance between any pair of opposite faces of this cube is termed the resistivity, and is measured in ohm-metre.

Conductors have a low resistivity, copper for example being about 1. $\times 10^{-8}$ ohm-metre. Insulators have a high resistivity varying between 10^{10} and 10^{18} ohm-metre. Silicon and germanium are semi-conductor and their resistivity depends upon their purity. If a crystal of silicon has very small amount of antimony or indium added as an impurity it resistivity is lowered, and the greater the amount of impurity the lower the resistivity. If a small amount of antimony is added to one half of silicon crystal and a small amount of indium to the other half, the junction between these types (termed n type and p type) acts as a barrier layer, so that the crystal acts as a conductor in one direction and offers a high resistance in the other, in other words it can act as a rectifier. This is the principle of the solid state silicon diode (because it has two elements or connexions) used as a rectifier for d.c. welding supplies.

Welding cables

A cable to conduct an electric current consists of an inner core of copper or aluminium covered with an insulating sheath. Welding cables have to carry quite large currents and must be very flexible and as light as possible and to give this flexibility the conductor has very many strands of small diameter. The conductor is covered by a thick sheath of tough rubber (TRS) or synthetic rubber (CSP) to give the necessary insulation at the relatively low voltages used in welding. As great flexibility is often not required in the return lead, a cable of the same sectional area but with fewer conductors of larger diameter can be used with a saving in cost.

CSP is a tough flexible synthetic rubber with very good resistance to heat, oils, acids, alkalis, etc., and is a flame retardant. It can be used at higher current densities than a TRS cable of the same sectional area, and its single-sheath construction gives good mechanical strength.

Copper is usually used as the conductor, but aluminium conductors are now used and are lighter than the same sectional area copper, are economical, less liable to pilfering but are larger in diameter and less flexible. In cases where considerable lengths of cable are involved a short length of copper conductor cable can be plugged in between the aluminium cable and electrode holder to give increased flexibility and less liability of conductor fracture at the holder.

Copper-clad aluminium conductors are available with 10% or 20% cladding, increasing the current-carrying capacity for a given cable size and reducing corrosion liability. Clamped or soldered joints can be used and in general TRS cables with copper conductors are probably the best and most economical choice in cases where high-current rating and resistance to corrosive conditions are not of prime importance.

In order to select the correct size of cable for a particular power unit it is customary to indicate for a given cable its current-carrying capacity in amperes (allowing for a permissible rise in temperature) and the voltage drop which will occur in 10 metres length when carrying a current of 100 A as shown in the table. For greater lengths and currents the voltage drop increases proportionally.

Duty cycle. Cables for welding range from those on automatic machines in which the current is carried almost continuously, to very intermittent manual use in which the cable has time to cool in between load times. To obtain current rating for intermittently loaded cables the term duty cycle is used. The duty cycle is the ratio of the time for which the cable is carrying the current to the total time, expressed as a percentage. If a cable is used for 6 minutes followed by an off load period of 4 minutes the duty cycle is $6/10 \times 100 = 60\%$. Average duty cycles for various processes are: automatic welding, up to 100%; semi-automatic, $30–85\%$; manual, $30–60\%$. Welding cables (BS 638) have many conductors of very small diameter to increase flexibility and may be divided into the following classes:

(1) Single core high conductivity tinned copper (HCC) conductors, paper taped and covered with tough rubber.
(2) Single core HC tinned copper conductors, paper taped and covered with chlorosulphonated polyethylene (CSP).
(3) Single core aluminium conductors (99.5% pure), paper taped and covered with CSP. The CSP cables have a 25% increase in sheath thickness without additional weight.

TRS and CSP insulated cables, copper conductors

Cross sectional area of conductor in mm²	Number and diameter of wires. No./mm	Max. overall diameter, mm	Current rating A max. duty cycle						d.c. volts drop 100A/m of cable at	
			100%		60%		30%		20°C	60°C
			CSP	TRS	CSP	TRS	CSP	TRS		
16	513/0.20	11.5	135	105	175	135	245	190	1.19	1.38
25	783/0.20	13.0	180	135	230	175	330	245	0.78	0.90
35	1107/0.20	14.5	225	170	290	220	410	310	0.55	0.64
50	1566/0.20	17.0	285	220	370	285	520	400	0.39	0.45
70	2214/0.20	19.5	355	270	460	350	650	495	0.28	0.32
95	2997/0.20	22.0	430	330	560	425	790	600	0.20	0.24
120	608/0.50	24.0	500	380	650	490	910	690	0.16	0.18
185	925/0.50	29.0	660	500	850	650	1200	910	0.10	0.12

Ohm's law

Let us now arrange a conductor so that it can be connected to various pressures or voltages from a battery, say 2, 4, 6 and 8 volts, as shown in Fig. 4.7. A voltmeter V is connected so as to read the difference of pressures between the ends of the conductor, and an ammeter is connected so as to read the current flowing in the circuit. Connect the switch first to terminal 1 and read the voltage drop on the voltmeter and the current flowing on the ammeter, and suppose just for example that the readings are 2 volts and 1 ampere. Then place the switch on terminals 2, 3 and 4 in turn and read current and voltage and enter them in a table, as shown below. The last column in the table represents the ratio of the voltage applied to the current flowing, and it will be noted that the ratio is constant, that is, it is the same in each case, any small variations being due to experimental error.

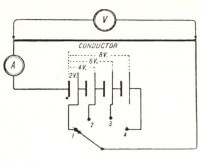

Fig. 4.7

In other words, when the voltage across the conductor was doubled, the current flowing was doubled; when the voltage was trebled, the current was trebled, and so on. This result led the scientist Ohm to formulate his law, thus: The ratio of the steady pressure (or voltage) across the ends of a conductor, to the steady current (or amperes) flowing in the conductor, is constant (provided the temperature remains steady throughout the experiment, and the conductor does not get hot. If it does, the results vary somewhat. Ohm called this constant the *resistance* of the conductor. In other words, the resistance of a conductor is the ratio of the pressure applied to its ends, to the current flowing in it.

Voltage or pressure drop	Current flowing (amperes)	Ratio: $\dfrac{\text{voltage}}{\text{current}}$
2	1	$\frac{2}{1} = 2$
4	2	$\frac{4}{2} = 2$
6	3	$\frac{6}{3} = 2$
8	4	$\frac{8}{4} = 2$

If now a difference of pressure of 1 volt applied to a conductor causes a current of 1 ampere to flow, the resistance of the conductor is said to be 1 ohm, that is, the ohm is the unit of resistance, just as the volt is the unit of pressure and the ampere the unit of current.

That is
$$\frac{1 \text{ volt}}{1 \text{ ampere}} = \text{ohm}$$

or, expressed in general terms,

$$\frac{\text{voltage}}{\text{current}} = \text{resistance}.$$

Another way of expressing this is

$$\text{voltage drop} = \text{current} \times \text{resistance}.$$

A useful way of remembering this is to write down the letters thus: $\frac{V}{I \mid R}$ (*V* being the voltage drop, *I* the current, and *R* the resistance). By placing the finger over the unit required, its value in terms of the others is given. For example, if we require the resistance, place the finger over *R* and we find that it equals V/I, while if we require the voltage *V*, by placing the finger over *V* we have that it equals $I \times R$. The following typical examples show how Ohm's law is applied to some simple useful calculations.

Example
A pressure of 20 volts is applied across ends of a wire, and a current of 5 amperes flows through it. Find the resistance of the wire in ohms.

By Ohm's law,
$$V = I \times R,$$

i.e. $20 = 5 \times R$ or $R = 4$ ohms.

Example
A welding resistance has a resistance of 0.1 ohm. Find the voltage drop across it when a current of 150 amperes is flowing through it.

By Ohm's law,
$$V = I \times R,$$

i.e. $V = 150 \times 0.1 = 15$ volts drop.

Power is the rate of doing work, and the work done per second in a circuit where there is a difference of pressure of 1 volt, and a current of 1 ampere is flowing, is 1 *watt*, that is,

$$\text{power in watts} = \text{volts} \times \text{amperes}.$$

The unit of work, energy and quantity of heat is the joule (J) which is the work done when a force of 1 newton (N) moves through a distance of 1 metre (m). 1 watt (W) = 1 joule per second (J/s). A Newton is the force which, acting on a mass of 1 kilogram (kg), will give it an acceleration of 1 metre per second per second (1 m/s²).

Example
A welding generator has an output of 80 volts, 250 amperes. Find the output in kilowatts and joules per second.

$$80 \times 250 = 20\ 000\ \text{W}$$
$$= 20\ \text{kW}$$
$$= 20\ 000\ \text{J/s}.$$

This is the actual *output* of the machine. If this generator is to be driven by an engine or electric motor, the power required to drive it would have to be much greater than this, due to frictional and other losses in the machine. A rough estimate of the power required to drive a generator can be obtained by adding on one-half of the output of the generator. For example, in the above,

estimate of power required to drive the generator
= 20 + 10 = 30 kW.

It is always advisable to fit an engine which is sufficiently powerful for the work required, and this approximation indicates an engine which would be sufficient for the work, including overloads.

Energy is expended when work is done and it is measured by the product of the power in a circuit and the time for which this power is developed. If the power in a circuit is 1 watt for a period of 1 hour, the energy expended in the circuit is 1 watt hour.

The practical unit of energy is 1000 watt hours, or 1 kilowatt hour (kWh), usually termed 1 Unit. This is the unit of electrical energy for consumption purposes, and is the unit on which supply companies base their charge.

Example
An electric motor driving a welding generator is rated at 25 kW. Find the cost of running this on full load, per day of 6 hours, with electrical energy at 2.5p per unit.

Energy consumed in 6 hours = 25 × 6 = 150 kWh or Units.
Cost per day = 150 × 2.5 = £3.75p.

Resistance of a conductor
The resistance of R ohms of a conductor is proportional to its length l and inversely proportional to its cross sectional area a, that is $R \propto l/a$. Thus the longer a cable the greater its resistance, and the smaller its cross-sectional area the greater its resistance. To reduce the voltage drop in any cable it should be as short as possible and of as large a cross-sectional area as possible.

Measurement of resistance. Ohmmeter and Megger
The Ohmmeter is basically a sensitive current-measuring instrument having its own battery source of supply. When an unknown resistance is connected to the instrument a certain current will flow, causing the pointer to move over the scale. The lower the resistance in circuit, the greater will be the current which flows and consequently the scale, which is graduated in ohms, is in the reverse direction to that of normal instruments, that is zero ohms on the right-hand side of the scale and

maximum ohms on the left-hand side. Before taking a reading, the terminals (or the prods attached to the terminals) should be short-circuited and the pointer set to zero ohms with the adjustment provided. If a zero ohms reading cannot be obtained the internal battery should be replaced.

Insulation resistance. The insulation resistance of a cable decreases as its length increases and for lighting and power installations the resistance to earth (that is between any of the conductors and earth) should be not less than 1.5 megohms (one megohm = 10^6 ohms). Because of the low voltage of the battery in the Ohmmeter it is unsuitable for measuring the insulation resistance to earth of any installation or appliance since this has normally to be done at twice the installation's normal working voltage, namely 500 volts for a 240 volt installation and 1000 volts for a 450–500 volt installation. This is achieved by using a hand-driven generator incorporated in the instrument. A typical instrument of this type is the 'Megger' which can be used for the measurement of resistance and insulation resistance up to infinity. It has two terminals to which the testing prods are attached and a mechanism ensures that the voltage is limited if the handle is turned too quickly. To perform a test the prods are held one on the conductor and the other on an earth connexion and the handle is rotated. The reading is read directly from the scale graduated similarly to that of the Ohmmeter but ranging from zero ohms to infinity. To test a domestic supply the main switch is placed in the off position and the two outgoing conductors (live and neutral) may be connected together. With all switches in the off position and one prod on the conductor strap and the other on an earth connexion, the handle is rotated and the reading taken.

Heating effect of a current

When a current flows through a conductor, heat, which is a form of energy, is generated because the conductor has some resistance. The heat generated is proportional to the power in the circuit and the duration for which this power is developed. The power is the product of the volts drop and the current so that;

heat developed $\propto$ power $\times$ time $\propto$ volts drop $\times$ current $\times$ time.

If the current is I amperes flowing for t seconds with V volts drop in a circuit of resistance R ohms, then

heat in joules $= V \times I \times t = I^2 \times R \times t$ (since by Ohm's law $V = I \times R$),

so that the heating effect $\propto$ (current)^2.

Thus if the current in any cable is doubled, four times as much heat is generated in a circuit; if the current is trebled, nine times as much heat is generated. This loss due to the heating effect is known as the I^2R loss.

The following definitions are useful.

An ampere is that steady current which, passing through two parallel straight, infinitely long conductors of negligible cross-sectional area, one metre apart in a vacuum, produces a force of 2×10^{-7} newtons per metre length on each conductor.

The ohm is that resistance in which a current of 1 ampere flowing for 1 second generates 1 joule of heat energy.

The volt is the potential difference across a resistor having a resistance of 1 ohm and carrying a current of 1 ampere.

Overload protection. One use of the heating effect of a current is to protect electrical apparatus from excessive currents which would cause damage (Fig. 4.8). A heating coil of nickel chrome wire carries the main current to the apparatus, for example it may be the imput supply to a welding transformer. Near the coil is a bi-metal strip made of two thin metal strips which can be of brass and a nickel alloy rolled together to form a single laminated strip. The brass, which has the greater co-efficient of expansion, is placed on the side nearer the heater. When excessive currents flow the position of the strip is such that it curls away with the brass strip on the outer circumference and the contact points break, interrupting the supply to the coil which holds the main contacts in, thus disconnecting the apparatus. The fixed contact can be adjusted to be more or less in contact with the moving contact, thus varying the value of overload current required to break the circuit.

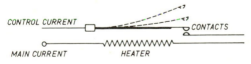

Fig. 4.8. Thermal overload trip.

The simple electric circuit of the welding arc

If a metal arc is to be operated from a source of constant pressure, a resistance must be connected in series with it in order to obtain the correct voltage drop across the arc and to control the current flowing in the circuit. This series resistance can be of the variable type, so that the current can be regulated as required. The ammeter A in Fig. 4.9 indicates the current flowing in the circuit, while the voltmeter V_1 reads the supply voltage, and the voltmeter V_2 indicates the voltage drop across the arc. By placing the switch S on various studs, the resistance is varied, and it will be noted that one section of the resistance marked X cannot be cut out of circuit. This is to prevent the arc being connected directly across the supply mains. If this happened, since the resistance of the arc is fairly low, an excessive current would flow and the supply mains would be 'short-circuited', and furthermore the arc would not be stable.

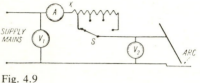

Fig. 4.9

The loss of energy in this series resistance is considerable, since a voltage of about 50 to 60 V *is required to strike the arc, and then a voltage of about 25 to 30* V is required to maintain it. If then, as in Fig. 4.10, the supply is 60 V and 100 A are flowing in the arc circuit, with 25 volts drop across the arc, this means that there is a voltage drop of 35 V across the resistance. The loss of power in the resistance is therefore (35×100) W $= 3.5$ kW, whereas the power consumed in the arc is (25×100) W $= 2.5$ kW.

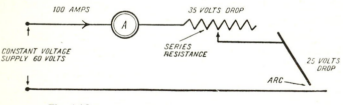

Fig. 4.10

In other words, more power is being lost in the series resistance than is being used in the welding arc. Evidently, therefore, since the 60 V is required to strike the arc, some other more economical means must be found for the supply than one of constant voltage.

Modern welding generators are designed so that there is a high voltage of 50 to 60 V for striking the arc, but once the arc is struck, this voltage falls to that required to maintain the arc, and as a result only a small series resistance is required to control the current, and thus the efficiency of the operation is greatly increased. This type of generator is said to have a 'drooping characteristic'.

This can be illustrated thus: suppose the voltage of the supply is 60 V when no current is flowing, that is, 60 V is available for striking the arc; and suppose that the voltage falls to 37 V when the arc is struck, the voltage drop across the arc again being 25 V and a current of 100 A is flowing (Fig. 4.11). The power lost in the resistance is now only (12 × 100) W or $1\frac{1}{5}$ kW, which is just less than one-half the loss in the previous example, when a constant voltage source was used.

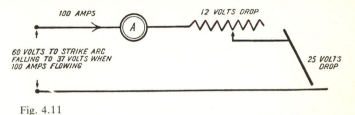

100 AMPS

A

12 VOLTS DROP

*60 VOLTS TO STRIKE ARC
FALLING TO 37 VOLTS WHEN
100 AMPS FLOWING*

*25 VOLTS
DROP*

Fig. 4.11

Contact resistance. Whenever poor electrical contact is made between two points the electrical resistance is increased, and there will be a drop in voltage at this point, resulting in heat being developed. If bad contact occurs in a welding circuit, it often results in insufficient voltage being available at the arc. Good contact should always be made between cable lugs and the generator and the work (or bench on which the work rests). The metal plate on the welding bench to which one of the cables from the generator is fixed is often a source of poor contact, especially if it becomes coated with rust or scale. When attaching the return cable to any point on the work being welded, the point should always be scraped clean before connecting the cable lug to it, and in this respect, especially for repair work, a small hand vice, bolted to the cable lug, will enable good contact to be made with the work when the jaws are lightly clamped on any desired point, and this is especially useful when no holes are available in the article to be welded.

Capacitors and capacitance

Principle of the capacitor. Let two metal plates A and B facing each other a few millimeters apart, be connected through a switch and centre zero milliammeter to a d.c. source of supply (Fig. 4.12*a*). When the switch is closed, electrons flow from A to B through the circuit, and A has a positive and B a negative charge. The needle of the meter moves in one direction as the electrons flow and registers zero again when the p.d. between A and B equals that of the supply. There is no further flow of current and the plates act as a very high resistance in the circuit. The number of electrons transferred is termed the charge, unit charge being the coulomb, which is the charge passing when a current of 1 ampere flows for 1 second.

Between the plates in the air, which is termed the dielectric, there exists a state of electrical stress. If the two plates are now brought quickly together (with the switch still closed) the needle flicks again in the same direction as previously, showing that more electrons have flowed from A to B and the plates now have a greater charge. When the plates are brought nearer together the positive charge on A has a greater neutralizing effect on the charge on B so that the p.d. between the plates is lowered and a further flow of electrons takes place. This arrangement of plates separated by a space of dielectric is termed a capacitor and its function is to store a charge of electricity.

Now disconnect the plates from the supply by opening the switch and short-circuit the plates through a resistor R (Fig. 4.12b). Electrons flow from B to A, the needle of the meter flicks in the opposite direction, and the charge of electricity which is transferred represents the quantity of electricity which the capacitor will hold and is termed its capacitance. The capacitor is now discharged and in practice the quantity of charge passing is determined by discharging it through a ballistic galvanometer. The moving portion of this instrument has considerable mass and therefore inertia and the angle through which the movement turns is proportional to the quantity of electricity which passes.

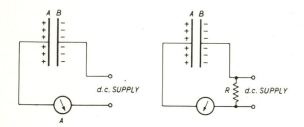

Fig. 4.12. (a) Charging current, (b) discharging current.

Using the same plates as before and about 2 mm apart, charge them through a ballistic galvanometer and note the angle of deflection. Now slide a piece of glass, bakelite, mica or other insulating medium between the plates and after discharging the capacitor repeat the experiment. It will be noted that the angle of deflexion of the meter has increased showing that the capacitance of the capacitor has increased due to the presence of the different dielectric. Similarly if the plates are made larger the capacitance is increased, so that the capacitance depends upon:

(1) The area of the plates. The greater the area, the greater the capacitance.

(2) The distance apart of the plates. The nearer together the plates the greater the capacitance.

(3) The type of dielectric between the plates. Glass, mica and paper give a greater capacitance than air.

Dielectric strength. If the p.d. across the plates of a capacitor is continuously increased, a spark discharge will eventually occur between the plates puncturing the dielectric (if it is a solid). If the dielectric is, say, mica or paper, the hole made by the spark discharge means that at this point there is an air dielectric between the plates, and now it will not stand as high a p.d. across the plates as it did before breakdown.

When capacitors are connected in series (Fig. 4.13b) the sum of the voltage drop across the individual capacitors equals the total volts drop across the circuit. When capacitors are in parallel (Fig. 4.13a) the volts drop across each is the same as that of the supply but the total capacitance is equal to the sum of their individual capacitance.

Capacitance is measured in farads (F). A capacitor has a capacitance of one farad if a charge of one coulomb (C) produces a potential difference of 1 volt between the plates. This is a very large unit and the sub-multiple is the micro-farad (μF). $10^6 \mu$F = 1 farad.

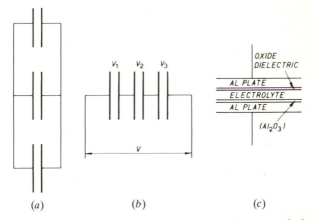

Fig. 4.13. (a) Capacitors in parallel. Total capacitance equals the sum of the individual capacitances. (b) Capacitors in series. Sum of volts across each equals total drop across the circuit. (c) a.c. electrolytic capacitor.

Types of capacitors. The types of capacitors usually met with in welding engineering are the Mansbridge and the electrolytic.

In the Mansbridge type two sheets of thin aluminium foil, usually long and narrow, are separated from each other by a layer of impregnated paper to form the dielectric. Connexions are made to each sheet and the whole is rolled up tightly and placed in an outer container with connexions from the two sheets, one to each terminal. The working voltage and capacitance (in μF) are stamped on the case. These capacitors are rather bulky for their capacitance.

Electrolytic capacitors, on the other hand, have very thin dielectrics and thus can be made in high capacitances but small bulk. If a sheet of aluminium foil is placed in a solution of ammonium borate and glycerine and a current is passed from the sheet to the solution, a thin film of aluminium oxide is formed on the surface of the aluminium. The microscopically thin film is an insulator and insulates the foil from the liquid which is the electrolyte, so that the current quickly falls to zero as the film of oxide forms. The current which continues to flow is the leakage current and is extremely small.

In the wet type of electrolytic, the solution is hermetically sealed in the aluminium canister which contains it, but in the dry type the solution is soaked up in gauze, and the foil, which must be absolutely clean and free from contamination, is rolled up in the gauze, one terminal being connected to the foil and the other to the gauze, and the whole hermetically sealed in an outer case. When connected to a d.c. supply, the initial current which passes forms the dielectric film and the capacitor functions. The thinner the dielectric, the greater the capacitance, so that for a safe working voltage of say 25 V, the capacitor will be of small bulk even for a capacitance of 1000 μF. The foil terminal is positive so that the capacitor is polarized, and wrong polarity connexions will ruin the unit.

The capacitor just described cannot be used on a.c. with its continuously changing polarity but an a.c. electrolytic capacitor has been developed which, although not continuously rated, is very suitable for circuits where intermittent use is required, as for example the series capacitor which is used to suppress the d.c. component of current when a.c. TIG welding aluminium and its alloys. This type consists of two aluminium electrodes, in foil form separated by gauze soaked in the electrolyte (Fig. 4.13c). When a current flows in either direction, a molecularly thin film of aluminium oxide is formed on each sheet providing, the dielectric. This film is not a perfect insulator and a leakage current flows, which increases with increasing voltage. This leakage current quickly makes good any imperfections in the oxide film but the voltage rating of the unit is critical since excess voltage will produce excess leakage current and lead to breakdown. Capacitors of this type are of small size for capacitances of 1000 μF, and when connected in parallel, are suitable for a.c. circuits in which large currents are flowing, as in welding.

WELDING GENERATORS

Magnetic field

Pieces of a mineral called lodestone or magnetic oxide of iron possess the power of attracting pieces of iron or steel and were first discovered centuries ago in Asia Minor. If a piece of lodestone is suspended by a thread, it will always come to rest with its ends pointing in a certain direction (north and south); and if it is rubbed on a knitting-needle (hard steel), the needle then acquires the same properties. The needle is then said to be a magnet, and it has been magnetized by the lodestone. Modern magnets of tungsten and cobalt steel are similar, except that they are magnetized by a method which makes them very powerful magnets.

Suppose a magnetized knitting-needle is dipped into some iron filings. It is seen that the filings adhere to the magnet in large tufts near its ends. These places are termed the poles of the magnet. If the magnet is now suspended by a thread so that it can swing freely horizontally, we find that the needle will come to rest with one particular end always pointing northwards. This end is termed the north pole of the magnet, while the other end which points south is termed the south pole.

Let us now suspend two magnets and mark clearly their north and south poles, then bring two north poles or two south poles near each other. We find that they repel each other. If, however, a north pole is brought near a south pole, we find that they attract each other, and from this experiment we have the law: *like poles repel, unlike poles attract* (Fig. 4.14).

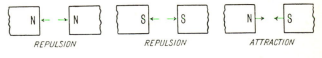

Fig. 4.14

If we attempt to magnetize a piece of soft iron (such as a nail), by rubbing it with a magnet, it is found that it will not retain any magnetic properties. For this reason hard steel is used for permanent magnets.

Iron filings provide an excellent means of observing the area over which a magnet exerts its influence. A sheet of paper is placed over a bar magnet and iron filings are sprinkled over the paper, which is then gently tapped. The filings set themselves along definite lines and form a pattern. This pattern is shown in Fig. 4.15.

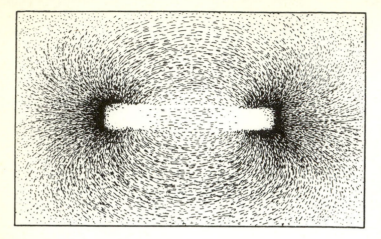

Fig. 4.15. Magnetic field of a bar magnet.

In the three-dimensional space around the magnet there exists a magnetic field and the iron filings set themselves in line with the direction of action of the force in the field, that is in the direction of the magnetic flux.

It should be noted that the iron filings map represents the field in one plane only, whereas the flux exists in all directions around the magnet. Figure 4.16 shows the flux due to two like poles opposite each other and clearly indicates the repulsion effect, while Fig. 4.17 shows the attraction between two unlike poles. The normal flux per unit area (B) is termed the flux density.

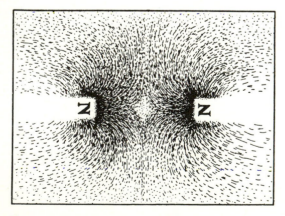

Fig. 4.16. Repulsion between like poles.

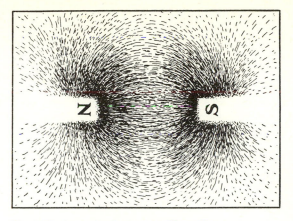

Fig. 4.17. Attraction between unlike poles.

Magnetic field due to a current

If a magnetic needle (or compass needle) is brought near a wire in which a current is flowing, it is noticed that the needle is deflected, indicating that there is a magnetic field around the wire. If the wire carrying the current is passed through the centre of a horizontal piece of paper and an iron filing map made, it can be seen that the magnetic lines of force are in concentric circles around the wire (Fig. 4.18).

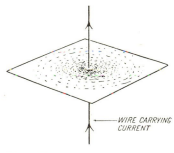

WIRE CARRYING CURRENT

Fig. 4.18

Two wires carrying currents in the same direction will attract each other, due to the attraction of the fields, while if the currents are flowing in opposite directions, there is repulsion between the wires.

The magnetic flux round a wire carrying a current is used to magnetize pieces of soft iron to a very high degree, and these are then termed electro-magnets.

Many turns of insulated wires are wrapped round an iron core and a current passed around the coil thus formed. The iron core becomes strongly magnetized, and we find that the greater the number of turns and the larger the current, the more strongly is the core magnetized. This is true until a point termed 'saturation point' is reached, after which neither increase in the number of turns nor in the current will produced any increase in intensity of the magnetism.

That end of the core around which the current is passing clockwise, when we look at it endways, exhibits south polarity, while the end around which the current passes anti-clockwise exhibits north polarity (Figs. 4.19, 4.20).

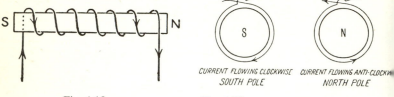

CURRENT FLOWING CLOCKWISE
SOUTH POLE

CURRENT FLOWING ANTI-CLOCKW
NORTH POLE

Fig. 4.19   Fig. 4.20

Relays and contactors

A solenoid is a multi-turn coil of insulated wire wound uniformly on a cylindrical former and the magnetic flux due to a current flowing in a single turn coil and in a solenoid is shown in Fig. 4.21a and b. The flux

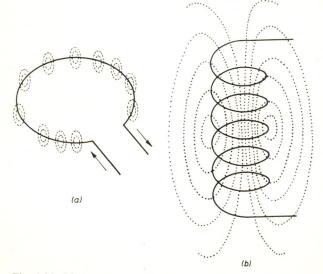

(a)

(b)

Fig. 4.21. (a) Magnetic flux due to a single turn of wire carrying a current. (b) Magnetic flux due to a coil of wire carrying a current.

has greatest intensity within the centre of the coil and if a piece of soft iron is placed with one end just inside the coil it becomes magnetized by induction and drawn within the coil when a current flows (Fig. 4.22). In semi-automatic processes such as TIG and MIG, it is important that the various services required, namely, welding current, gas and water, can be controlled from a switch on the welding gun or welding table. This remote-control operation is performed by the use of relays by which small currents in the control circuit, often at lower voltages than the mains (110 V a.c., 50 V d.c.), operate contactors which make and break the main circuit current.

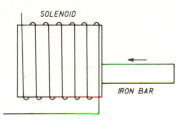

Fig. 4.22

The control wires to the switch are light and flexible and they control the main welding current, which may be several hundred amperes. Figure 4.23 shows a simple layout for the control of a welding current circuit. When *M* is pressed, the control current passes through *M* and the contactor coil *C* which is energized, the iron core *I* moves up, the main

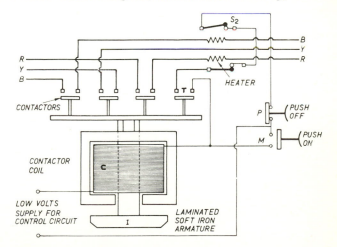

Fig. 4.23. Relay control with overload protection on two phases.

contacts are bridged and the main current flows. Also the contacts T are made and because they are in parallel with those of M, when M is released, the control current passes through P, S_1, S_2 and T, keeping the coil energized so that the contactors are held in. Upon pressing P, the coil circuit is broken and the contactors break the main circuit when the iron armature falls. Overload heaters are often fitted on two of the phases instead of three because the current in the phases is usually balanced.

The valves are similarly operated by solenoids. When a current flows through the solenoid an armature attached to the valve moves and operates the valve.

Magnetic field of a generator

The magnetic fields of generators and motors are made in this way. The outside casing of the generator, termed the yoke, has bolted to it on its inside the iron cores called *pole pieces*, over which the magnetizing coils, consisting of hundreds of turns of insulated copper wire, fit. To extend the area of influence of the flux, pole shoes are fixed to the pole pieces (or made in one with them), and these help to keep the coils in position. Generators may have 2, 4 or more poles, and the arrangement of a 2- and a 4-pole machine is sketched in Fig. 4.24.

It will be noticed that a north and south pole always come alternately, thus producing a strong flux density where the conductors on the rotating portion of the machine are fixed. The magnetic circuit is completed through the yoke. The coils are connected so that the current passes through them alternately clockwise and anti-clockwise when looked at from the inside of the generator, so as to give the correct polarity (this can be tested by using a compass needle), and the current which flows through the coils is termed the magnetizing or *excitation* current (Fig. 4.25).

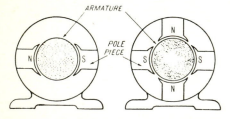

Fig. 4.24

Fig. 4.25

   The larger the air gap between the poles of an electro-magnet the stronger must the magnetizing force be to produce a given flux in the gap. This means that the greater the gap between the pole pieces of a machine and the rotating iron core, called the armature, the greater must the magnetizing current be to produce a field of given strength. For this reason the gap between pole pieces and rotating armature must be kept as small as possible, yet without any danger of slight wear on the bearings causing the armature to foul the pole pieces (the machine is then said to be pole bound). In addition, this gap should be even at each pole piece all round the armature. Excessive air gaps result in an inefficient machine.

Generation of a current by electrical machines

The following explanation of the principle of operation of a generator is an outline only and will serve to give the operator an idea of the function of the various parts of the machine.

   For a current to be generated we require (1) a magnetic flux, (2) a conductor, (3) motion (producing change of magnetic flux).

   The magnetic field causes a magnetic flux to be set up, and the conductor is surrounded by this flux. Any change of flux caused by the change in position of the conductor or by change in value of the field will cause a current to be generated in the conductor.

Generation of alternating current

Let us consider the first case. N and S (Fig. 4.26) are the poles of a magnet and AB is a copper wire whose ends are connected to a milliammeter. (This is an instrument that will measure currents of the order of $\frac{1}{1000}$ amp.) If the conductor AB is moved upwards, we find that the needle of A swings in one direction, while if AB is moved downwards, it swings in the opposite direction. By moving AB up and down, we generate a current that flows first in one direction, B to A, and then in the other direction, A to B. This is termed an alternating current, and the current is said to be induced in the conductor.

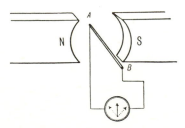

Fig. 4.26

Note. The rule by which the direction of the current in a conductor is found, when we know the direction of the field and the motion, is termed *Fleming's right-hand rule.* This can be stated thus. 'Place the thumb, first finger and second finger of the right hand all at right angles to each other. Point the first finger in the direction of the flux from N to S and turn the hand so that the thumb points in the direction of motion of the conductor. Then the second finger points in the direction in which the current will flow in the conductor.' Figure 4.27 makes this clear.

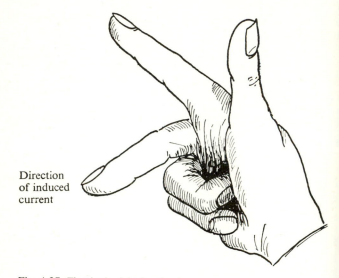

Direction of induced current

Fig. 4.27. Fleming's right-hand rule.

Instead of moving the conductor up and down in this way, the method used for generation is to make it in the form of a coil of several turns and rotate the conductor, and if the coil is wound on an iron core, the field is greatly strengthened and much larger currents are generated.

The ends of the coil are connected to two copper rings mounted on the shaft, but insulated from it, and spring-loaded contacts called brushes bear on these rings, leading the current away from the rotating system (see Fig. 4.28).

From the brushes X and Y, wires lead to the external circuit, which has been shown as a coil of wire, OP, for simplicity.

When the coil is rotated clockwise, AB moves up and CD down. By applying Fleming's rule we find that the current flows from B to A in one conductor and from C to D in the other, as shown by the arrows (Fig. 4.29a). The current will then leave the machine by slip ring Y and flow through the external circuit from O to P and return via slip ring X.

When the coil has been turned through half a turn, as shown in Fig. 4.29b, AB is now moving down and CD up, and by Fleming's rules the currents will now be from D to C in one conductor and from A to B in the other. This causes the current to leave by slip ring X and flow through the external circuit from P to O, returning via slip ring Y. If a milliammeter with centre zero is placed in the circuit in place of the coil OP, the needle of the instrument flicks to one side during the first half turn of the coil and to the other side during the second half turn.

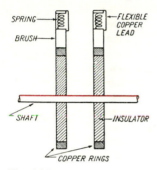

Fig. 4.28

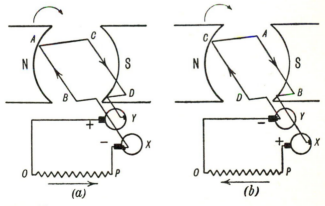

Fig. 4.29

Evidently, therefore, in one revolution of the coil, the direction of the current generated by the coil has been reversed. No current is generated when the coil is passing the position perpendicular to the flux, while maximum current is generated when the coil is passing the position in the plane of the flux. This is illustrated in Fig. 4.30.

One complete rotation of the coil has resulted in the current starting at zero, rising to a maximum, falling to zero, reversing in direction and rising to a maximum and then falling again to zero. This is termed a complete *cycle*, and the number of times this occurs per second (that is, the number of revolutions which the above coil makes per second) is termed the frequency of the alternating current. 1 cycle per second is known as 1 hertz (Hz), named after the German physicist who discovered electro-magnetic waves.

Alternating currents in this country are usually supplied at 50 Hz. In America 60 Hz is largely used. Evidently a.c. has no definite polarity, that is, first one side and then the other becomes +ve or −ve.

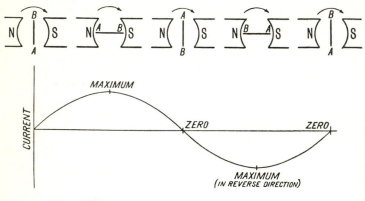

Fig. 4.30

Sinusoidal wave form

The current and voltage generated by a coil rotating in a magnetic field follow a curved path from zero to maximum positions. This curve is known as a sine curve and the voltage and current waves are termed sinusoidal waves.

The e.m.f. generated in a conductor is proportional to the rate at which the conductor cuts the magnetic flux. If the conductor moves across the lines of force so that the flux linkage changes, an e.m.f. is generated. If the conductor moves along a line of force there is no change of flux linkage and no e.m.f. is generated. This can be illustrated in the following way.

The coil AB (Fig. 4.31) is rotating anti-clockwise between the poles N and S of a magnet and the flux is shown in dotted lines. Let the coil turn from AB to A_1B_1 through an angle θ. From A_1 drop a perpendicular A_1X on to AB and join AA_1. In moving from A to A_1 it can be considered that the conductor has moved from A to X across the flux and from X to A

along the flux path. The e.m.f. generated is thus proportional to AX, no e.m.f. being generated in the movement from X to A_1.

But $\dfrac{AX}{AA_1} = \sin \theta$.

$\therefore AX = AA_1 \sin \theta$

and since angle $AA_1X =$ angle AOA_1, the e.m.f. is proportional to the sine of the angle through which the coil is rotated and hence the generated e.m.f. will be a sine curve.

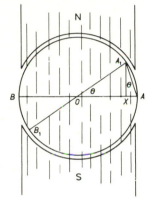

Fig. 4.31

To draw this curve the coil AB (Fig. 4.32) is again rotated anticlockwise between the magnetic poles N and S. Divide the circle into 30° sectors as shown and on the horizontal axis (abscissa) of the graph, divide the 360° of one rotation of the coil into 30° equal divisions. The vertical axis or ordinate represents the e.m.f. generated. When the coil is in the position AB the conductors are moving along the lines of force and no e.m.f. is generated. As it rotates towards A_1B_1 it begins to cut the lines immediately after leaving the AB position. A_1B_1 is the position of the coil after rotating through 30° from AB. Project horizontally from A_1 to meet the vertical ordinate through the 30° ordinate at e_1. This ordinate represents the voltage generated at this point. Let the coil rotate a further 30° to A_2B_2 and again project horizontally to meet the 60° ordinate at e_2. At A_3B_3 the conductors are moving at right angles to the field and generating maximum e.m.f. shown by the e_3 ordinate. Further rotation of the coil results in a reduction of the e.m.f. to zero e_6 when the coil has rotated through 180°. After this, further rotation produces a reversal of the e.m.f. and if the points e_1, e_3, . . . e_6 are joined, the resulting curve is termed as sine wave and represents the generated voltage or e.m.f. Since the current is proportional to the voltage the current wave is also sinusoidal.

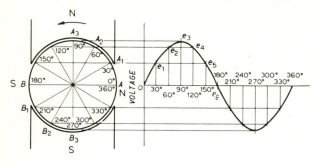

Fig. 4.32

Root Mean Square value of an alternating current

Since an alternating current or voltage is continuously changing from zero to a maximum value some method must be selected to define the true value of an alternating current or voltage. This is done by comparing the direct current required to produce a given heating effect with the corresponding alternating current which produces the same heating effect.

An alternating current of I amperes is that current which will produce the same heating effect as a direct current of I amperes.

If an alternating current equivalent to I amperes d.c. flows through a resistor of $R\Omega$ for t seconds, then the energy generated $= I^2 Rt$ joules (p. 228). Let OXY (Fig. 4.33) be the wave form of this current. Divide it

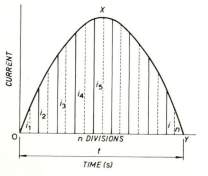

Fig. 4.33

into n areas on equal bases, each base being therefore t/n since OY represents the time in seconds. Draw the mid-ordinates $i_2, i_2, i_3 \ldots i$ for each area. The energy represented by the first area is $i_1^2 R \times t/n$ (mid-ordinate rule for areas).

Energy for second area is $\qquad i_2^2 R \times \dfrac{t}{n}$,

Energy for third area is $\qquad i_3^2 R \times \dfrac{t}{n}$,

Energy for nth area is $\qquad i_n^2 R \times \dfrac{t}{n}$.

Therefore the total energy

$$= i_1^2 R \times \frac{t}{n} + i_2^2 R \times \frac{t}{n} + i_3^2 R \times \frac{t}{n} + \ldots i_n^2 R \times \frac{t}{n} \text{ joules}$$

$$= Rt \left(\frac{i_1^2}{n} + \frac{i_2^2}{n} + \frac{i_3^2}{n} + \ldots \frac{i_n^2}{n} \right) \text{ joules},$$

but the total energy is $I^2 Rt$ joules, therefore

$$I^2 Rt = Rt \left(\frac{i_1^2}{n} + \frac{i_2^2}{n} + \frac{i_3^2}{n} + \ldots \frac{i_n^2}{n} \right)$$

$$\therefore \qquad I^2 = \frac{i_1^2}{n} + \frac{i_2^2}{n} + \frac{i_3^2}{n} + \ldots \frac{i_n^2}{n}$$

$$\therefore \qquad I = \sqrt{\frac{i_1^2 + i_2^2 + i_3^2 + \ldots i_n^2}{n}}$$

That is, the true, effective or virtual value I amperes of the alternating current equals the square root of the mean value of the squares of the current ordinates,

or I = square root of the mean squares, termed the rms value.

If the current wave is sinusoidal this value is 0.707 of the maximum or peak value so that if Im is the maximum value of the current, the true or rms value is

$I = 0.707\ Im$ (Fig. 4.34).

If an alternating current of maximum value Im is flowing in a circuit its effect is the same as that of a d.c. of value 0.707 Im.

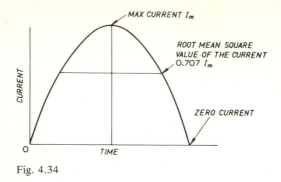

Fig. 4.34

Similarly with an alternating voltage. If the root mean square (rms) value of a supply is 240 V, the maximum value of the voltage is

$$V = 0.707 \ Vm$$
$$240 = 0.707 \ Vm$$
$$Vm = \frac{240}{0.707} = 340 \ \text{V}.$$

This explains why it is possible to get a much greater shock from an a.c. supply of the same rated voltage as a d.c. supply and hence why the earthing of a.c. apparatus is so important. An a.c. supply is always designated by its rms value unless otherwise stated.

Generation of direct current

By an ingenious yet simple device called the commutator (current reverser), this generated alternating current can be changed to direct current, that is, to a current flowing only in one direction. Instead of slip rings, two segments of copper are mounted on the circumference of the shaft, as shown in Fig. 4.35, being separated from each other by a small gap. Brushes bear on these segments as they did on the slip rings previously. As the coil rotates, the segments will first make contact with each brush in turn and thus reverse the connexions to the external circuit.

In Fig. 4.36 a and b, the conductors are lettered as before, but AB is connected to one segment and CD to the other. Brushes X and Y bear on the segments and are connected to the external circuit OP. Upon rotating the coil, the current flows in the coil as previously. It leaves by brush Y (Fig. 4.36a), flows through the external circuit from P to O and back via brush X. In Fig. 4.36b, when the coil has turned through half a turn, the connexions of the coils to the brushes have been reversed by the

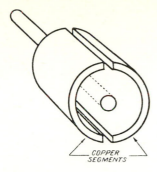

Fig. 4.35

segments of the commutator and the current again leaves via brush Y, through the external circuit in the same way, from P to O, returning via brush X. Thus, though the current in the coil has alternated, the current in the external circuit is uni-directional, or 'direct current' as it is called. Since the current flows from Y to X, through the external circuit, Y is termed the positive pole and X the negative pole.

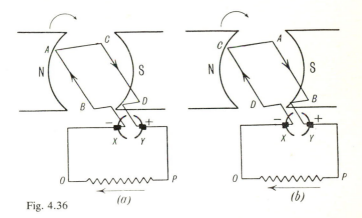

Fig. 4.36   *(a)*   *(b)*

    The brushes pass over the joints or gaps between the segments as the coil passes through the perpendicular position to the field, i.e. when no current is generated; thus there is no spark due to the circuit being broken whilst the current is still flowing (Figs. 4.37, 4.38).

    The current from a direct current generator with a single coil of several turns, as we have just considered, would be a series of pulsations of current, starting at zero, rising to a maximum and decreasing to zero again, but always flowing in the same direction, as shown in Fig. 4.39.

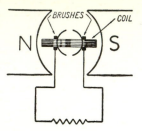
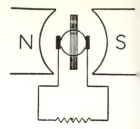

Fig. 4.37. Maximum current position.

Fig. 4.38. Zero current.

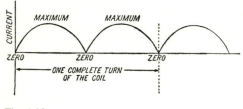

Fig. 4.39

If, now, a second coil is wound and mounted on the shaft at right angles to the first coil and its ends connected to a second pair of commutator segments, the maximum current in one coil will occur when the other coil has zero current; and since there are now four commutator segments, each now only extends round half the length that it did previously. The resulting current from the two coils A and B will now be represented by the thick line; the dotted portion will no longer be collected by the brushes, because of the shortened length of the commutator segment (Fig. 4.40).

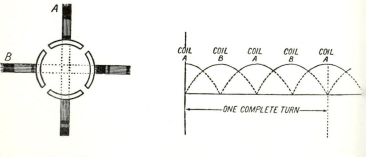

Fig. 4.40

By increasing greatly the number of coils (and consequently the number of commutator segments, since each coil has two segments), the pulsating current can be made less and less, that is, the effect is a steady flow, as shown in Fig. 4.41.

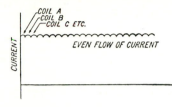

Fig. 4.41

It is not necessary here to enter into details of the various methods of connecting the coils to the segments. Full details of these are given in text-books on electrical engineering. The voltage of a machine is increased by increasing the number of turns of wire in each coil, while the current output of a machine is increased by increasing the total number of coils in parallel on the machine. The output of a machine can also be increased by increasing the speed of the machine and also by increasing the strength of the magnetic flux.

By increasing the number of poles of a machine, its voltage can be increased, while yet keeping its speed the same. Machines of 4 and 6 poles are quite common. In this case there are the same number of sets of brushes as there are poles, i.e. 4 poles, 4 sets of brushes, and so on, and these brushes are connected alternatively, as in Fig. 4.42, so as to give +ve and −ve poles.

Most welding generators are either 2 or 4 poles. For a given output, the greater the number of poles the slower the speed of the machine. As a rule a machine is designed to operate at a given speed, but the output voltage is varied by the resistor known as the field regulator (see later).

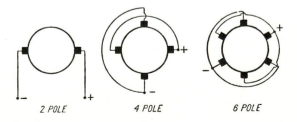

Fig. 4.42. Connexions of brushes.

RECTIFIERS

One method of changing a.c. to d.c. is by the use of an a.c. motor driving a d.c. generator. Static rectifiers perform this operation without the use of moving parts and the types having welding applications are (1) selenium and (2) silicon, and they are used for supplying d.c. for manual metal arc. TIG, MIG, CO_2, etc.

A rectifier should have a low resistance in one direction (forward) so as to allow the current to pass easily, and a high resistance in the opposite direction (reverse) so that very little current will pass. In practice all rectifiers pass some reverse current and this increases with rising temperature so that cooling fins are usually fitted.

Selenium rectifier. This type has largely displaced the copper–copper oxide rectifier for general use and consists of an iron or aluminium base disc coated with selenium, which is a non-metallic element of atomic number 34. A coating of an alloy of lead, cadmium or bismuth is deposited on to the selenium and forms the counter-electrode (Fig. 4.43a).

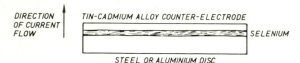

Fig. 4.43. (a) Selenium rectifier.

A current will flow from disc to counter-electrode but not in the reverse direction in which a high resistance is offered, so that the unit acts as a rectifier. To increase the current-carrying capacity the disc is made larger in area and units are connected in parallel. For higher voltages units are connected in series since, if the voltage drop across the unit is too high, the reverse current increases rapidly and the rectifier fails.

Semi-conductor rectifiers. This type is fitted to many of the transformer–rectifier power units for MIG and CO_2 welding. They can supply large output currents and are generally fan cooled. The following greatly simplified explanation will serve to indicate how they operate.

Silicon is a non-metallic element with four valence electrons in its outer shell. These valence electrons form a covalent bond with electrons in the outer shell of neighbouring atoms by completing the stable octet of eight electrons which are shared between the two atoms. One way that atoms combine to form a molecule is by means of this bond. The elements neon and argon, for example, have completed outer shells of eight electrons and are therefore completely inert and form no compounds with other elements (Fig. 4.43*b*).

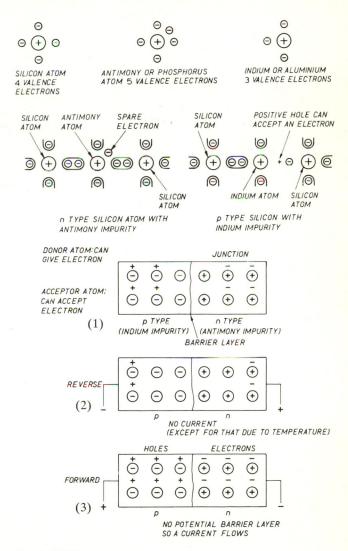

Fig. 4.43 (*b*)

Antimony, phosphorus and arsenic have five electrons in the outer shell, so if a few atoms of antimony are added to a silicon crystal as an impurity, four of the silicon valence electrons from covalent bonds with four of the antimony valence electrons and there is one free electron due to each antimony atom. There are termed donor atoms because they can donate an electron, and silicon with this type of impurity is termed n type (negative); if an electron is donated a positively charged ion remains.

Indium, gallium and aluminium have three valence electrons in the outer shell and if, say, indium is added as an impurity to silicon, these three valence electrons form covalent bonds with three of the four silicon valence electrons, but there is one electron missing so that the remaining silicon electron cannot make the fourth covalent bond. This position, where the electron is missing, is termed a 'positive hole' since it will accept any available electron to form a covalent bond. When the electron enters a positive hole the atom is negatively charged and is a negative ion. These atoms are termed acceptor atoms because they accept an electron, and silicon with this type of impurity is termed p type (positive).

If a silicon crystal is formed, one half being n type and the other half p type, at the junction between the types, electrons will move from n type to p type to fill the holes and the holes will thus move from p type to n type, leaving positive ions in the n type and negative ions in the p type. There is thus a potential barrier set up across the junction in which there are no free holes or electrons so that electrons tending to move across it are repelled by the negative ions, and holes tending to move are repelled by the positive ions (Fig 4.43b(1)).

If a potential difference is placed across the crystal (Fig. 4.43b(2)) so as to make the n type positive and the p type negative, this increases the effect of the potential barrier and no current can flow. This is the reverse direction of the unit. If the p.d. is reversed, making n type negative and p type positive, the applied p.d. now reduces the potential barrier effect, holes and electrons move and a current flows. This is the forward direction of the unit (Fig. 4.43b(3)). Rise in temperature causes liberation of electrons and holes, and an increase in the reverse current. Because there are two connexions the unit is often referred to as a diode.

Both selenium and silicon are used as rectifiers for welding supplies. Silicon has the advantage when used for higher currents, in weight, size and efficiency. It can operate at junction temperatures of $200\,^{\circ}C$ or more and can be more easily cooled. The element silicon however, though it is most plentifully distributed over the earth (sand is silicon dioxide, SiO_2), is very difficult to produce in the pure state suitable for use in rectifiers, but as production difficulties are overcome it will probably be a very strong competitor with other types. In general, at the present time, the higher current rectifiers are silicon and the lower current type, selenium.

With both types, only one half of the a.c. wave flows, the other half being suppressed. This is termed half-wave rectification and the current and voltage consist of uni-directional pulses of 50 per second in the case of a 50 Hz supply. To obtain full wave rectification the rectifier elements can be connected in a 'bridge' system as shown in Fig. 4.44a, the rectifier

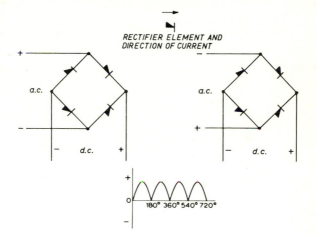

Fig. 4.44. (a) Single-phase full-wave rectification.

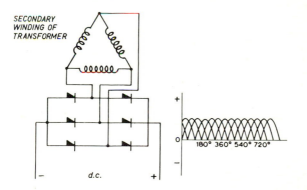

Fig. 4.44 (b)

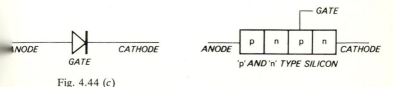

Fig. 4.44 (c)

element being represented by the small triangle pointing in the direction in which the current will flow through it.

To obtain full-wave rectification on a three-phase system the rectifier elements are connected as in Fig. 4.44*b*, the output of the secondary of the transformer being connected in 'delta'. This type is favoured for d.c. welding supply units and gives a balanced load.

Thyristors

A thyristor is a solid-state switch which consumes very little power and is small and compact. Thyristors can be used for a variety of switching operations as for example in motor control, resistance welding circuit control and control of welding power sources. Large thyristors can carry very heavy currents of up to several hundred amperes.

A thyristor has three connexions: an anode, a cathode and a gate electrode. The anode and cathode are the power connexions through which the main current flows, while the gate is the control electrode. There are four alternate layers of p and n type silicon (see section on rectifiers) and anode, cathode and gate are connected as shown in Fig. 4.44*c*.

By varying the voltage (positive or negative) applied across gate and cathode, the thyristor can be made conducting or non-conducting at will. For a fuller explanation of the operation of a thyristor the student should consult the 'Minibook' by Mullard on thyristors.

Ignitron

The ignitron is a rectifier which has three electrodes: an anode, a cathode which is a mercury pool and an igniter, all contained in an evacuated water-cooled steel shroud. The igniter is immersed in the mercury cathode, and when a current is passed through it a 'hot spot' is formed on the surface of the pool. This acts as a source of electrons which stream to the anode if it is kept at a positive potential with respect to the cathode, and a current can now flow from anode to cathode through the ignitron. It will continue doing so as long as the anode remains positive, and large ignitrons can handle currents of several thousands of amperes. The single ignitron behaves as a half-wave rectifier on an a.c. supply, and two connected in reverse parallel can operate as a switch controlling the flow of a.c. in a circuit. The arc within the unit can be struck at any point in any particular half cycle by controlling the current in the igniter circuit.

Description of a typical direct-current generator

A modern direct-current welding generator consists of:
(1) Yoke with pole pieces and terminal box, and end plates.

(2) Magnetizing coils.

(3) Armature and commutator (the rotating portion).

(4) The brush gear.

The yokes of modern machines are now usually made of steel plate rolled to circular form and then butt welded at the joint. The end plates, which contain the bearing housings, bolt on to the yoke, and the feet of the machine are welded on and strengthened with fillets. The pole pieces are of special highly magnetizable iron and are bolted on to the yoke. The coils are usually of double cotton-covered copper wire, insulated and taped over all, and they fit over the pole pieces, being kept in position by the pole shoes. The armature shaft is of nickel steel and the armature core (and often the pole pieces also) is built up of these sheets of laminations of highly magnetizable iron, known by trade names such as Lohys, Hi-mag, etc. Each lamination is coated with insulating varnish, and they are then placed together and keyed on to the armature shaft, being compressed tightly together so that they look like one solid piece.

The insulating of these laminations from each other prevents currents (called eddy currents), which are generated in the iron of the armature when it is rotating from circulating throughout the armature and thus heating it up. This method of construction contributes greatly to the efficiency and cool running of a modern machine. The armature laminations have slots in them into which the armature coils of insulated copper wire are placed (usually in a mica or empire cloth insulation). The coils may be keyed into the slots by fibre wedges and the ends of the coils are securely soldered (or sweated) on to their respective commutator bars (the parts to which they are soldered are known as the commutator risers). A fan for cooling purposes is also keyed on to the armature shaft.

The commutator is of high conductivity, hard drawn copper secured by V rings, and the segments are insulated from the shaft and from each other by highest quality ruby mica. Brushes are of copper carbon, sliding freely in brush holders, and springs keep them in contact with the commutator. The tension of the springs should only be sufficient to prevent sparking. Excessive spring pressure should be avoided, as it tends to wear the commutator unduly. The commutator and brush gear should be kept clean by occasional application of petrol on a rag, which will wash away accumulations of carbon and copper dust from the commutator micas and brush gear. All petrol must evaporate before the machine is started up, to avoid fire risk. The armature usually revolves on dust-proof and watertight ball or roller bearings, which only need packing with grease every few months. Older machines have simple bronze or white metal bearings, lubricated on the ring oil system. These need periodical inspection to see that the oil is up to level and that the oil rings are turning freely and, thus, correctly lubricating the shaft.

Connexions from the coils and brush gear are taken to the terminal box of the machine, and many welding generators have the controlling resistances and meters also mounted on the machine itself.

Connexions of welding generators

In the following sketches, magnetizing coils are shown thus: ϱϱϱϱϱϱϱ, and this represents however many coils the machine possesses, connected so as to form alternate north and south poles, as before explained. The armature, with the brushes bearing on the commutator, is shown in Fig. 4.45.

The current necessary for magnetizing the generator is either taken from the main generator terminals, when the machine is said to be self-excited, or from a separate source, when it is said to be separately excited. Welding generators are manufactured using either of these methods.

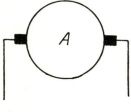

· Fig. 4.45

Separately excited machines

These generally take their excitation or magnetizing current from a small separate generator, mounted on an extension of the main armature shaft, and this little generator is known as the exciter. Current generated by this exciter passes through a variable resistor, with which the operator can control the magnetization current, and then round the magnetizing coils of the generator. This is shown in Figs. 4.46 and 4.51.

By variation of the resistance R, the magnetizing current and hence the strength of the magnetic flux can be varied. This varies the voltage (or pressure) of the machine and thus enables various voltages to be obtained across the arc, varying its controllability and penetration. This control is of great importance to the welder.

This type of machine gives an almost constant output voltage, irrespective of load, and thus, as before explained, would result in large losses in the series resistor, if used for welding. In order to obtain the 'drooping characteristic', so suitable for welding, the output current is

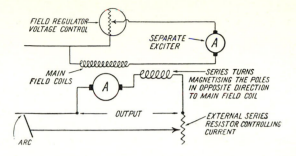

Fig. 4.46. Simple separately excited generator.

carried around some *series* turns of thick copper wire, wound over the magnetizing coils on the pole pieces, and this current passes round these turns so as to magnetize them with the opposite polarity from the normal excitation current. Fig. 4.47 shows how the coils are arranged. Consider then what happens.

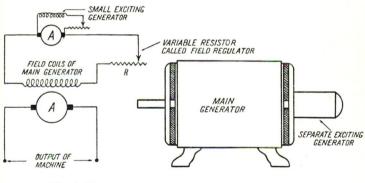

Fig. 4.47

When no load is on the machine, the flux is supplied from the separate exciter and the open circuit voltage of the machine is high, say 60 volts, giving a good voltage for striking. When the arc is struck, current passes through the series winding and magnetizes the poles in the opposite way from the main flux and thus the strength of the flux is reduced and the voltage of the machine drops. The larger the output current the more will the voltage drop, and evidently the voltage drop for any given output current will depend on the number of series turns. This is carefully arranged when the machine is manufactured, so as to be the most suitable for welding purposes.

This type of machine, with control of both current and voltage, is very popular and is reliable, efficient and economical. Because the voltage available at any given instant is only slightly greater than that required to maintain the arc, only a small series resistance is required, this being fitted with the usual variable control.

Self-excited machines

The simplest form of this type of machine is that known as the 'shunt' machine, in which the magnetizing coils take their current direct from the main terminals of the generator through a field-regulating resistor (Fig. 4.48).

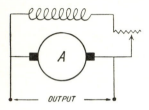

Fig. 4.48. Connexions of a simple shunt machine.

There is always a small amount of 'residual' magnetism remaining in the pole pieces, even when no current is passing around the coils, and, when the armature is rotated, a small voltage is generated and this causes a current to pass around the coils, increasing the strength of the flux and again causing a greater e.m.f. to be generated, until the voltage of the machine quickly rises to normal. Control of voltage is made, as before, by the field regulator. This type of machine is not used for welding because its voltage only drops gradually as the load increases and, as before explained, this would cause a waste of energy in the external series resistor.

Again, this machine is modified for use as a welding generator by passing the output current first round series turns wound on the pole pieces, so as to magnetize them with the opposite polarity from that due to the main flux, and this results, as before, in the voltage dropping to a great extent as the load increases and, thus, the loss of energy in the external resistor is greatly reduced (Fig. 4.49). A machine of this type is termed a differential compound machine and shares with the separately excited machine the distinction of being a reliable, efficient and economical generator for welding purposes. The control of current and voltage are exactly as before.

The rest of the equipment of a direct-current welding generator consists of a main switch and fuses, ammeter and voltmeter. The fuses have an insulating body with copper contacts, across which a piece of copper wire tinned to prevent oxidation is bridged. The size of this wire is chosen so that it will melt or 'fuse' when current over a certain value flows through it. In this way it serves as a protection for the generator against excessive currents, should a fault develop.

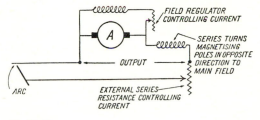

Fig. 4.49

On many machines neither switch nor fuses are fitted. Since there is always some part of the external resistor connected permanently in the circuit of these machines, no damage can result from short circuits, and fuses are therefore unnecessary. The switch is also a matter of convenience and serves to isolate the machine from the electrode holder and work when required.

Interpoles, or commutation poles

Interpoles are small poles situated between the main poles of a generator and serve to prevent sparking at the brushes. The polarity of each interpole must be the same as that of the next main pole in the direction of rotation of the armature, as in Fig. 4.50(a).

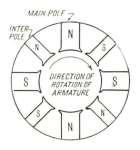

Fig. 4.50 (a)

They carry the main armature current and, therefore, like the series winding on welding generators, are usually of heavy copper wire or strip.
They prevent distortion of the main flux, by the flux caused by the

current flowing in the armature, and thus commutation is greatly assisted.

Most modern machines are fitted with interpoles, as they represent the most convenient and best method of obtaining sparkless commutation.

The following is a summary of the features of a good welding generator:

(1) Fine control of voltage.
(2) Fine control of current.
(3) Excitation must always provide a good welding voltage.
(4) Copper conductors of armature and field of ample size and robust construction, yet the generator must not be of excessive weight.
(5) Well-designed laminated magnetic circuit and accurate armature-pole shoe air gap.
(6) Good ventilation to ensure cool running.
(7) Well-designed ball or roller bearings of ample size and easily filled grease cups.
(8) Well-designed brush gear – no sparking at any load and long-life brushes.
(9) Bearings and brush gear easily accessible.
(10) Large, easily placed terminals enabling polarity to be quickly changed (or fitted with polarity changing switch).
(11) High efficiency, that is, high ratio of output to input energy. (60–65% efficiency is normal for a modern single-operator motor-driven direct current plant.)

Brushless alternators and generators

Brushless alternators and generators have no slip rings or commutators but use a rectifier mounted on the rotating unit (rotor) to supply direct current to excite the rotating field coils, the main current being generated in the stationary (stator) coils.

The rotor has two windings, an exciter winding and a main field winding. The exciter winding rotates in a field provided by exciter coils on the stator and generates a.c., which is passed into silicon diodes (printed circuit connected) mounted on the rotor shaft. The resultant d.c. passes through the rotating field coils of the main generator portion and the rotating field produced generates a.c. in the stationary windings of the main generator portion. If a d.c. output is required, as for welding purposes, this a.c. is fed into a silicon rectifier giving a d.c. output (Fig. 4.50b).

Current for the stator coils on the exciter is obtained from the rectifier supply and variation of the excitation current gives variable voltage control as on a normal generator, and residual magnetism causes the usual build-up.

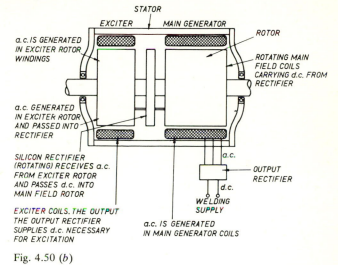

Fig. 4.50 (*b*)

Dual continuous control generator

In the dual continuous control generator, excitation current is supplied by the separate exciting generator shown on the left of Fig. 4.51, and the control of the excitation current is made by the field rheostat which

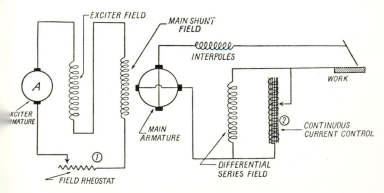

Fig. 4.51. (*a*) Welding generator with separate excitation. Regulation of controls (1) and (2) gives dual continuous control of voltage and current.

therefore controls the output voltage of the machine. Interpoles are fitted to prevent sparking and the continuously variable current control is in parallel with the differential series field, the current control being wound on a laminated iron core so as to give a smoothed output. This generator gives a good arc with excellent control over the whole range and is suitable for all classes of work.

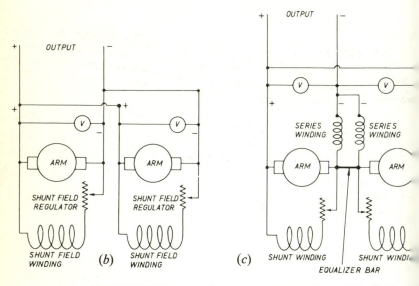

Fig. 4.51. (*b*) Shunt wound generators connected in parallel. (*c*) Compound wound generators connected in parallel.

Generators in parallel

The parallel operation of generators enables the full output of the machines to be fed to a single operator. If two shunt wound generators are run up to speed and their voltages adjusted by means of the shunt field regulators to be equal, they can be connected in parallel by connecting +ve terminal to +ve, and −ve terminal to −ve, and the supply taken from the now common +ve and common −ve terminals. When a load is applied it can be apportioned between either machine by adjustment of the voltage. As the adjustment is made, for example, to increase the voltage of one machine this machine will take an increased share of the load and vice versa (Fig. 4.51*b*).

Welding generators, however, are more often compound wound and if connected in parallel as for shunt machines they would not work satisfactorily, because circulating currents caused by any slight dif ference in voltage between the machines could cause reversal of one of the series fields, and this would lead eventually to one machine only carrying the load. If, however, an equalizer bar is connected from the end of the series field next to the brush connexion on each machine (Fig. 4.51c) the voltage across the series windings of each machine is stabilized and the machines will work satisfactorily. The equalizer connexion should be made when the machines are paralleled as for shunt machines, +ve to +ve and −ve to −ve, and load shared by operation of the shunt field regulators.

Static characteristics of welding power sources

Volt–ampere curves. Variation of the open-circuit voltage greatly affects the characteristics of the arc.

To obtain the volt–ampere curves of a power source:

(1) Set the voltage control to any value.
(2) With arc circuit open, read the open circuit voltage on the voltmeter.
(3) Short-circuit the arc.
(4) Vary the current from the lowest to highest value with the current control and, for each value of current, read the voltage. (Voltage will decrease as current increases.) Fig. 4.52a.

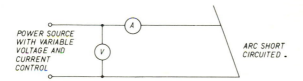

POWER SOURCE WITH VARIABLE VOLTAGE AND CURRENT CONTROL

ARC SHORT CIRCUITED.

Fig. 4.52 (a)

Plot a curve of these readings with voltage and current as axes. This curve is a volt–ampere curve and has a drooping characteristic. Any number of curves may be obtained by taking another value of open-circuit voltage and repeating the experiment. The curves in Fig. 4.52b are the results of typical experiments on a small welding generator.

Suppose Fig. 4.52c is a typical curve. When welding, the arc length is continually undergoing slight changes in length, since it is impossible for the welder to keep the arc length absolutely constant. This change in length results in a change in voltage drop across the arc; the shorter the arc the less the voltage drop. The volt–ampere curve shows us what effect this change of voltage drop across the arc will have on the current flowing. Suppose the arc is shortened and the drop changes from 25–20

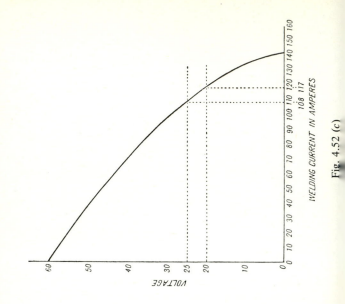

Fig. 4.52 (c)

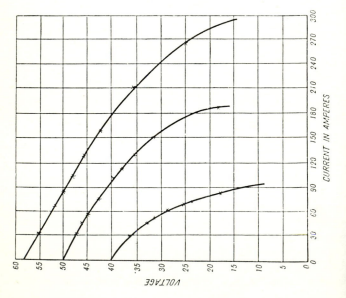

Fig. 4.52 (b)

V. From the curve, we see that the current now increases from 108 to 117 A.

The steeper the curve is where it cuts the arc voltage value, the less variation in current there will be, and, therefore, there will be no current 'surges' and the arc will be steady and the deposit even. Because the slope of the curve controls the variation of voltage with current it is known as 'slope control'. The dynamic characteristic of a power source indicates how quickly the current will rise when the source is short-circuited.

Variation of current and voltage control. Suppose a current of 100 A is suitable for a given welding operation. If the current control is now reduced, the current will fall below 100 A, but it can be brought back to 100 A by increasing the voltage control. The current control may be again reduced and the voltage raised again, bringing the current again to the same value. At each increase of voltage the volts drop across the arc is increased, so we obtain a different arc characteristic, yet with the same current.

This effect of control should be thoroughly grasped by the operator, since by variation of these controls the best arc conditions for any particular work are obtained.

The curves just considered are known as static characteristics. Now let us consider the characteristics of the set under working conditions; these are known as the dynamic characteristics, and they are best observed by means of a cathode-ray oscilloscope. By means of this instrument the instantaneous values of the current and voltage under any desired conditions can be obtained as a wave trace or graph, called an oscillograph.

The curves drawn in Fig. 4.53 are taken from an oscillograph of the current and voltage variation on a welding power source when the external circuit was being short-circuited (as when the arc was struck) and then open-circuited again.

It will be noticed that the short-circuit surge of current (125 A) is about $1\frac{1}{2}$ times that of the normal short-circuit current (80 A). This prevents the electrode sticking to the work by an excessive flow of current when the arc is first struck, yet sufficient current flows initially to make striking easy. In addition, when the circuit is opened, the voltage rises to a maximum and then falls to a 'reserve' voltage value of 45 V and immediately begins to rise to normal. This reserve voltage ensures stability of the arc after the short-circuit which has occurred and makes welding easier, since short-circuits are taking place continually in the arc circuit as the molten drops of metal bridge the gap.

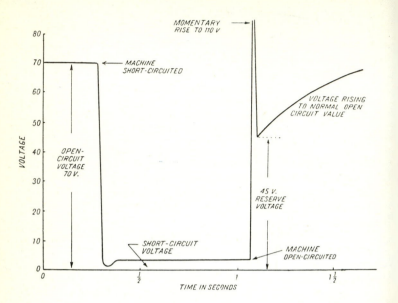

(*a*) Curve showing variations in voltage as machine is short-circuited and open-circuited.

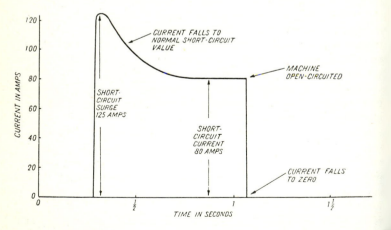

(*b*) Curve showing variation of current due to above variation of voltage.

Fig. 4.53

Motive power for welding generators

Welding generators may be motor- or engine-driven. Sets in semi-permanent positions, such as in workshops, are usually driven by direct current or alternating current motors, and these provide an excellent constant speed drive, since the speed is almost independent of the load. The motor and generator may be build into the same yoke, or may be separate machines. The first method is mostly used in modern machines, as much space is thereby saved.

Motors of either type should be fitted with no-volt and overload tripping gear. The former automatically switches off the supply to the motor in the event of a failure of the supply and thus prevents the motor being started in the 'full on' position when the supply is resumed, while the latter protects the motor against excessive overloading, which might cause damage. This operates by switching the motor off when the current taken by the motor exceeds a certain value, which can be set according to the size of the motor.

Main switch and fuses usually complete the equipment of the motor. The motor-driven set may be mounted on wheels or on a solid bed, depending on whether it is required to be portable or not, and the equipment should be well earthed to prevent shock.

Portable sets for outdoor use are usually engine-driven, and this type of set is extremely useful, since it can be operated independently of any source of electric power. The engines may be of the petrol or diesel type and are usually the four-cylinder, heavy duty type with an adequate system of water cooling and a large fan.

A good reliable governor that will regulate the speed to very close limits is an essential feature of the engine. Many modern sets now have an idling device which cuts down the speed of the machine to a tick-over when the arc is broken for a period (which can be adjusted by the operator), sufficient for him to change electrodes and deslag. This results in a considerable saving in fuel and wear and tear and greatly increases the efficiency of the plant.

Direct drive is mostly favoured for welding generators. Belt drive is not very satisfactory, owing to the rapid application of the load when striking the arc causing slip and putting a great strain on the belt, especially at the fastener. V-Belt drive sets, however, are used in certain circumstances.

ALTERNATING-CURRENT WELDING

Steel fabrication by manual metal arc welding using covered electrodes is now mainly performed using a.c. power sources and this method has certain advantages over the use of d.c. The chief of these are:

(1) The welding transformer (dealt with later) and its controller are very much cheaper than the d.c. set of the same capacity.
(2) There are no rotating parts, and thus no wear and tear and maintenance of plant.
(3) Troublesome magnetic fields causing arc blow are almost eliminated.
(4) The efficiency is slightly greater than for the d.c. welding set.

The following points should be noted concerning a.c. welding:

(1) Covered electrodes must be used. The a.c. arc cannot be used satisfactorily for bare wire or lightly coated rods as with the d.c. arc.
(2) A higher voltage is used than with d.c., consequently the risk of shock is much greater and in some cases, as for example in damp places or when the operator becomes hot and perspires, as in boiler work, a.c. welding can become definitely dangerous, unless care is taken.
(3) Welding of cast iron, bronze and aluminium cannot be done anything like as successfully as with d.c.

The transformer

The supply for arc welding with alternating current is usually from 80 to 100 V, and this may be obtained directly from the supply mains by means of a transformer, which is an instrument that transforms or changes the voltage from that of the main supply to a voltage of 80 to 100 V suitable for welding. Since a transformer has no moving parts, it is termed a 'static' plant.

The action of the transformer can be understood most easily from the following simple experiment, first performed by Faraday.

An iron ring or core (Fig. 4.54) is wrapped with two *insulated* coils of wire: A (called the primary winding) is connected to a source of alternating current, while B (called the secondary winding) is connected to a milliammeter with a centre zero, which will indicate the direction of flow of the current in the circuit. With each revolution of the coil of the a.c. generator, the current flows in the primary first from X to Y and then from Y to X, and a magnetic flux is set up in the iron core which rises and falls very much in the same way as the hair spring of a watch. This rising and falling magnetic flux, producing a change of magnetic flux in the circuit, generates in the secondary coil an alternating current, the current flowing in one direction through the milliammeter when the current in the primary is from X to Y and then in the opposite direction when the current in the primary flows from Y to X. There is no electrical connexion between the two coils, and a current generated in the secondary coil in this way, by a current in the primary, is said to be *induced*. Note that we again have the three factors necessary for generation as stated on p. 241: a conductor, a magnetic flux and motion. In this case, however, it is the change in magnetic flux which takes the place of the motion of the conductor, since this latter is now stationary.

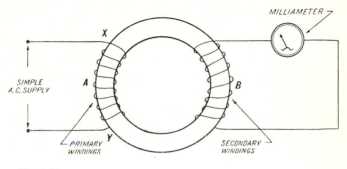

Fig. 4.54

Now let us wind a similar ring (Fig. 4.55) with 400 turns on the primary and 100 turns on the secondary, connect the primary to an alternating supply of 100 V, and connect a voltmeter across each circuit.

It will be found that the voltage across the secondary coil is now 25 V.

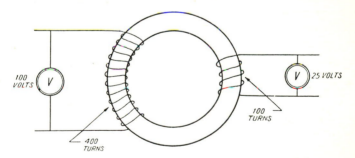

Fig. 4.55

Ratio of $\dfrac{\text{primary turns}}{\text{secondary turns}} = \dfrac{4}{1}$,

ratio of $\dfrac{\text{primary voltage}}{\text{secondary voltage}} = \dfrac{4}{1}$.

Thus we see that the voltage has been changed in the ratio of the number of turns, or

$$\frac{\text{primary turns}}{\text{secondary turns}} = \frac{\text{primary volts}}{\text{secondary volts}}.$$

This is a simple transformer, and since it operates off one pair of a.c. supply conductors, it is called a *single-phase* transformer. The voltage supplied *to* the transformer is termed the input voltage, while that supplied *by* the transformer is termed the output voltage. If the output voltage is greater than the input voltage, it is termed a *step-up* transformer; while if the output voltage is less than the input, it is a *step-down* transformer. Transformers for welding purposes are always step-down, the output voltage being about 85 V. Single-operator transformers have two output tappings of 80 and 100 V, the higher voltage being suitable for light gauge sheet welding (Fig. 4.56). The input voltage to transformers is usually 415 or 240 V, these being the normal mains supply voltages. The alternating magnetic field due to the alternating current in the windings would generate currents in the iron core if it were solid. These currents, known as eddy currents (or Foucault currents), would rapidly heat up the core and overheat the transformer. To prevent this, the core is made up of soft iron laminations varnished with insulating varnish so as not to make electrical contact with each other, and clamped tightly together with bolts passing through insulating bushings to prevent the bolts carrying eddy currents. In this way losses are reduced and temperatures kept lower. Eddy currents are also used for induction heating in the electric induction furnace.

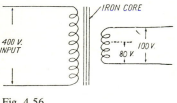

Fig. 4.56

Since the power output cannot be greater than the input (actually it is always less because of losses in the transformer), it is evident that the current will be transformed in the opposite ratio to the voltage. For example, if the supply is 400 V and 50 A are flowing, then if the secondary output is 100 V, the current will be 200 A (Fig. 4.57).

Actually, the output current would be slightly lower than this, since the above assumes a 100% efficient transformer. A transformer on full load has an efficiency of about 97%, so the above may be taken as approximately true.

The highly magnetizable silicon iron core of the transformer is made up of laminations bolted together and the coils fit over these (Fig. 4.58). It will be observed that the magnetic circuit is 'closed', that is, the flux does not have to traverse any air gap.

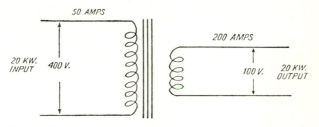

Fig. 4.57

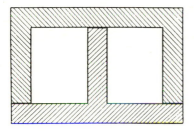

Fig. 4.58

The single-operator welding transformer is made on this principle and is available in sizes up to 450 to 500 A.

The transformer may be of the dry type (air cooled) or it may be immersed in oil, contained in the outer container. Oil-immersed transformers have a lower permissible temperature rise than the dry type and, therefore, their overload capacity (the extent to which they may be used to supply welding currents in excess of those for which they are rated) is much smaller.

Excessive variation in the supply voltage to a transformer welding set may affect the welding operation by causing a variation in welding current and voltage drop across the arc and the open circuit voltage (OCV). For 80 V the ratio is 80/410. If the supply falls to 380 V the OCV of the secondary will now be 80/410 × 380 or approximately 76 V. This fall will reduce arc current by about 4 A if the original current was 100 A: and the arc voltage by 1–1.5 V which will have some effect on welding conditions. Many modern welding units have thyristor regulators by which the arc voltage is kept constant irrespective of any variations of mains voltage, keeping welding conditions constant. Voltage control is by transductor which gives infinite adjustment.

By law the supply authorities must not vary the supply voltage from that specified by more than 6 % so that variation in the case of a specified 410 V supply is from 385–434 V, but this may be exceeded under adverse conditions. Small voltage variations have little effect on welding conditions.

Current control
Current control may be by tapped reactor (choke), flux leakage reactor, saturable reactor or leakage reactance moving coil.

Inductive reactor or choke
An inductive reactor or choke consists, in its simplest form, of a coil of insulated wire wound on a closed laminated iron core (Fig. 4.59). When an e.m.f. is applied to the coil and the current begins to flow, the iron core is magnetized. In establishing itself, this flux cuts the coil which is wound on the core and generates in it an e.m.f. in the opposite direction to the applied e.m.f. and known as the 'back e.m.f.'. It is the result of magnetic induction, and its effect is to slow down the rate of rise of current in the circuit so that it does not rise to its maximum value as given by Ohm's law immediately the e.m.f. is applied; in a very inductive circuit the rise to maximum value may occupy several seconds.

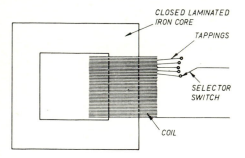

Fig. 4.59. Tapped choke.

The inductance of the circuit is proportional to the square of the number of turns of wire on the coil, so that increasing the number of turns greatly increases the inductive effect. When the current is fully established, there is energy stored in the magnetic circuit by virtue of the magnetic flux in the iron core. When the circuit is broken, the lines of force collapse, and in collapsing cut the coil and generate an e.m.f. in the opposite direction to that when the circuit was made, and which now tends to maintain the current in the original direction of flow. The energy of the magnetic flux is thus dissipated and may cause a spark to occur across the contacts where the circuit is being broken.

The direction of the induced e.m.f. in an inductive circuit is given by Lenz's law, which states: 'The direction of the induced effect in an inductive circuit always opposes the motion producing it.'

If an alternating current is flowing in the coil, the current will reverse before it has time to reach its maximum value in any given direction, and the more inductive the circuit the lower will the value of the current be, so that the effect is to 'choke' the alternating current. If the coil has tappings taken to a selector switch, the inductance of the circuit and hence the amount by which it can control or 'choke' the current can be varied (Fig. 4.59). Another method of varying the inductive effect is to vary the iron circuit so that the flux has a 'leakage path' other than that on which the coils are wound (Fig. 4.60), thus varying the strength of the magnetic flux in the core and hence the inductive effect.

The tapped reactor or choke is used to control the current in metal arc welding a.c. welding units. It can only be used on a.c. supplies and does not generate heat as does a resistor used for control of direct current. Any heat generated is partly due to the iron core, and partly due to the I^2R loss in the windings (Fig. 4.61).

Leakage reactance, moving coil current regulation

Stepless control of the current is achieved in this method by varying the separation of the primary and secondary coils of the transformer. The coils fit on to the iron circuit as shown in Fig. 4.62. The secondary coil supplying the welding current is fixed and the primary coil can be moved up and down by means of a screw thread and nut, operated by a winding handle mounted on top of the unit. As the primary coil is wound so as to approach the secondary coil the inductive reactance is reduced and the current is increased and vice versa, so that the highest current values are when the coils are in the closest proximity to each other. The moving coil carries a pointer which moves over two scales, one high current values and one low values, the different scales being obtained by two tappings on the secondary coil. The inductive reactance does not vary directly with the separation of the coils, and current values get rapidly greater as the coils get near to each other.

The merit of this method of current control is that there is no other item of equipment other than the transformer, thus reducing the initial cost, and there is stepless control operating with the simplest of mechanisms.

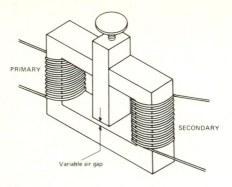

Fig. 4.60. Flux leakage control.

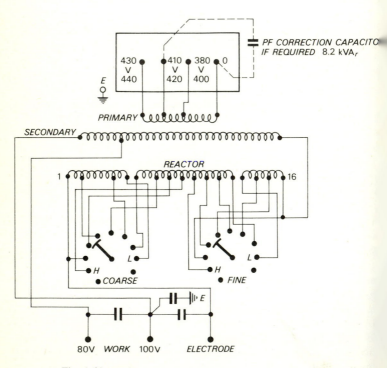

Fig. 4.61

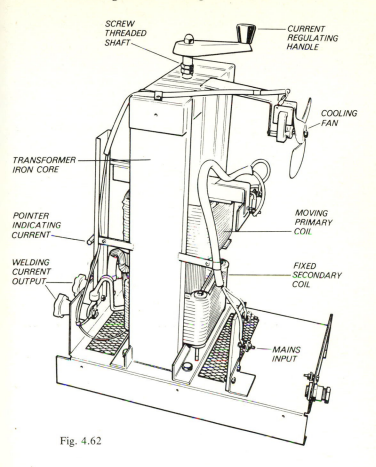

Fig. 4.62

Magnetic induction and saturation

If a coil of insulated wire is wound on an iron (ferro-magnetic) core and a current is passed through the coil, a magnetic flux is set up in the iron core. The strength of this flux depends upon the current in amperes and the number of turns of wire on the coil, that is upon the ampere-turns (AT) so that the magnetizing force (**H**) is proportional to $A \times T$. If a graph is drawn between the magnetizing force and the flux density (**B**) in the core it is known as a **B**/**H** or magnetization curve (Fig. 4.63). It will be noticed that the flux density rises rapidly to X with small increases of **H** and then begins to flatten out until at Y further increases of **H** produce no further increase of flux density **B**. At the point Y the core is said to be magnetically saturated.

Use is made of this to control the current in power units, being known as the saturable reactor method.

A coil A is wound on one limb of a closed laminated iron core and carries d.c. from a bridge connected rectifier X and controlled by a variable resistor Y (Fig. 4.64). A coil carrying the main welding current is wound on the other limb. When there is no current through A, the coil B will have maximum reactance because there is no flux in the core due to A and the welding current will be a minimum. As the current in A is increased by the control Y, magnetic saturation can be reached, at which

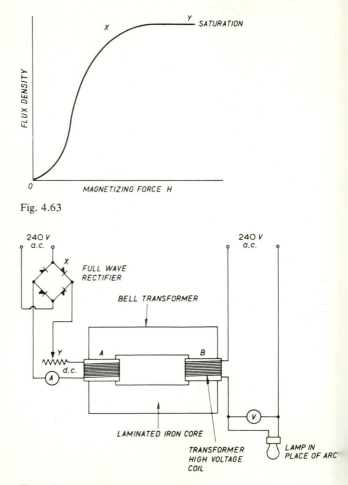

Fig. 4.63

Fig. 4.64. Experiment with saturable reactor control.

point reactance is a minimum and the welding current will be a maximum represented by maximum voltage on V so that between these limits, accurate stepless control of the welding current is achieved.

Behaviour of a capacitor in an a.c. circuit

When a capacitor is connected to a d.c. source a current flows to charge it and no further current flows, the capacitor preventing or blocking the further flow of current.

If the capacitor is now connected to an a.c. source the change of polarity every half-cycle will produce the same change of polarity in the capacitor so that there is a flow of current, first making one plate positive and then half a cycle later making it negative, and the current which is flowing in the circuit, but not through the capacitor, is equal to the charging current. Hence a capacitor behaves as an infinitely high resistor in a d.c. circuit preventing flow of current after the initial charge, while in an a.c. circuit the current flows from plate to plate, not through the capacitor but around the remaining part of the circuit (Fig. 4.65).

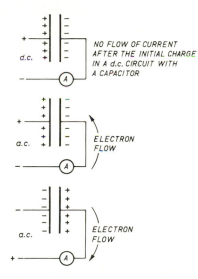

Fig. 4.65. Flow of a.c. in a circuit containing a capacitor.

Phase of current and voltage in an a.c. circuit

When a voltage is applied to a circuit and a current flows, if there is an inductive effect in the circuit the current will fall out of step with the voltage and lag behind it, rising and falling at the same frequency, but lagging a number of degrees behind. If there is capacitance in the circuit, the current will lead the voltage. Zero values of voltage and current do not occur together (Fig. 4.66b) and there is always some energy available, so that in a welding circuit the arc is easier to strike and maintain when a tapped reactor, for example, used for current control is in the circuit, since this produces an inductive effect.

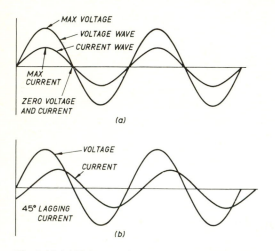

Fig. 4.66. (a) Voltage and current in phase. Both pass through zero and maximum values at the same time. (b) Voltage and current out of phase due to inductance in circuit current lagging 45° behind voltage. Current and voltage now do not pass through zero and maximum values together.

Inductive reactance

In any circuit, the effect of inductance is to increase the apparent resistance of the circuit. This effect is termed inductive reactance and if there is capacitance in the circuit the effect is known as capacitive reactance.

The unit of inductance is the henry (H). A circuit has an inductance of 1 henry if a current, varying at the rate of 1 ampere per second, induces an e.m.f. of 1 volt in the circuit.

If an alternating e.m.f. of V volts at frequency f Hz is applied to a circuit of inductance L henrys and a current of I amperes flows, the volts drop V across the inductor $= 2\pi L f I = IX_L$, where $X_L = 2\pi f L$. Comparing this with the volts drop V across a resistor R ohms carrying a current of I amperes, $V = I \times R$ so that X_L takes the place of R and is known as the inductive reactance.

Impedance

If a circuit contains resistance and inductance in series (Fig. 4.67a) the current I amperes flows through both inductance and resistance and there will be a volts drop IR across the resistor, in phase with the voltage, and a volts drop across the inductor, 90° out of phase with the voltage. This can be represented by a phasor[1] diagram (Fig. 4.65b) using the current common to resistance and inductance as a reference phasor.

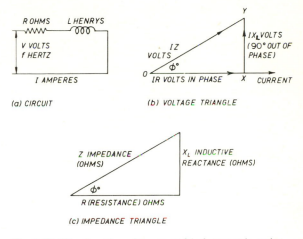

(a) CIRCUIT (b) VOLTAGE TRIANGLE

(c) IMPEDANCE TRIANGLE

Fig. 4.67. Circuit with resistance and inductance in series.

OX represents the volts drop IR in phase with the current.

XY represents the volts drop IX_L across the inductor, 90° out of phase, leading the current.

Then OY is the resultant voltage across the circuit, leading the current by an angle $\varphi°$ so that the current *lags* behind the voltage $\varphi°$, and if $OY = I \times Z$ where Z is the resultant effect of R and X_L, Z is termed the impedance. If the sides of the voltage triangle are all divided by I we have (Fig. 4.67c) the impedance triangle, and from this $Z^2 = R^2 + X_L^2$.

or impedance² = resistance² + inductive reactance².

---

[1] A scalar is a quantity which has magnitude only, e.g. mass or temperature. A vector is a quantity which has magnitude and direction, e.g. velocity or force. Phasors are rotating vectors. They are easier to draw than the more complicated wave form diagrams and are added or subtracted as are vectors.

In the phasor diagram the phasors are drawn to scale to represent the magnitude of the quantity (e.g. volts or amperes) and the angle between the phasors represents the phase displacement, the whole rotating counter-clockwise at an angular velocity measured in radians per second. Spokes on a bicycle wheel provide an analogy. Irrespective of the angular velocity of the wheel the angle between any two given spokes (representing the phasors) remains the same.

The unit of capacitance is the farad (F), see page 233. If an alternating e.m.f. of V volts at frequency f Hz is applied to a circuit of capacitance C farads and a current of I ampere flows:

$$V = \frac{I}{2\pi fC} = IX_C, \text{ where } X_C = \frac{1}{2\pi fC}$$

and is termed the capacitive reactance of the circuit.

Let a current of I amperes flow in a circuit containing a resistor of R ohms and a capacitor of capacitance C farads when an e.m.f. of V volts is applied at frequency f Hz (Fig. 4.68a). The voltage across the resistor, in phase with the current, is IR volts, whilst the volts drop across the capacitor is IX_C volts, the voltage lagging the current by 90°. The phasor diagram in Fig. 4.68b represents this with OX as the reference current phasor.

OX represents the volts drop IR, in phase with the current, XY represents the volts drop across the capacitor, lagging the current by 90°. Then OY represents the resultant voltage across the circuit, lagging the current by an angle $\varphi°$ and $OY = IZ$, where Z is the impedance of the circuit. Fig. 4.68c is the impedance triangle where $Z^2 = R^2 + X_C^2$. Impedance is measured in ohms (apparent).

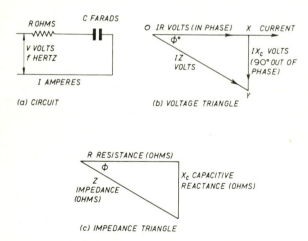

Fig. 4.68. Circuit with resistance and capacitance in series.

Resistance, inductance and capacitance in series
In a circuit such as that used for TIG welding there is inductance in the current controls, capacitance used for blocking the d.c. component, and the resistance of the circuit in series (Fig. 4.69a). The phasor diagram

(Fig. 4.69b) shows that IX_L and IX_C are in opposite directions (anti-phase) and the resultant reactance is $X_L - X_C$, since inductance is usually larger than capacitance. It will be noticed that the impedance is always greater than the ohmic resistance so that, if a d.c. is applied to a circuit designed for a.c., excess currents will flow. Fig. 4.69c is the voltage triangle and Fig. 4.67d the impedance triangle, and from this $Z^2 = R^2 + (X_L - X_C)^2$, or

$$\text{impedance} = \sqrt{[(\text{resistance})^2 + (\text{resultant reactance})^2]} \text{ ohms.}$$

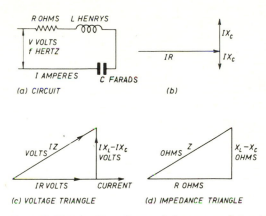

R OHMS   L HENRYS

V VOLTS
f HERTZ

I AMPERES   C FARADS

(a) CIRCUIT

IX_C

IR IX_C

(b)

IZ
VOLTS IX_L-IX_c
           VOLTS

IR VOLTS   CURRENT

(c) VOLTAGE TRIANGLE

Z
OHMS X_L-X_C
          OHMS

R OHMS

(d) IMPEDANCE TRIANGLE

Fig. 4.69. Circuit with resistance, inductance and capacitance in series.

Power factor

Suppose that a current in a circuit is lagging by an angle φ° behind the voltage. To find the true power in the circuit the phasor diagram (Fig. 4.70a) is drawn, in which OV represents the voltage and OB the current lagging the voltage by an angle φ°. The length of OB is drawn to represent the current I in amperes and this is resolved into two components, OA in phase with the voltage and AB 90° out of phase with the voltage.

In the triangle OAB, $OA/OB = \cos\varphi$, therefore $OA = OB\cos\varphi$ and $AB = OC$ and $AB/OB = \sin\varphi$, therefore $AB = OB\sin\varphi$.

The component in phase with the voltage which is the power component is $I\cos\varphi$ amperes, while the component 90° out of phase (the reactive component) which is wattless is $I\sin\varphi$ and produces no useful power. Thus the power in a circuit which is the product of voltage and current is $VI\cos\varphi$. The cosine of any angle cannot be greater than 1, so that the power in a reactive circuit is always less than the product of the volts and amperes. If $\varphi = 0$°, that is the current and voltage are in phase, the power is $VI \cos 0° = VI$, since $\cos 0° = 1$. If $\varphi = 90°$, the power is VI $\cos 90°$, and since $\cos 90° = 0$, the power is zero so that the reactive component produces no power in the circuit.

The factor $\cos \varphi$ is known as the *power factor*. The more inductive the circuit, the more will the current be out of phase with the voltage and the greater will be the angle φ so that $\cos \varphi$ gets smaller and the power becomes less.

If an a.c. welding supply is, for example, 40 V with a current of 120 A at a power factor of 0.7 lagging, the power in the circuit is

$$VI \cos \varphi = 40 \times 120 \times 0.7 = 3360 \text{ W} = 3.36 \text{ kW}.$$

If the power factor were unity or the circuit d.c. the power would be

$$40 \times 120 \times 1 = 4800 \text{ W} = 4.8 \text{ kW}.$$

Thus in any a.c. circuit, the product of the volts and amperes gives the apparent power in volt-amperes or kilovolt-amperes (kVA), while the true power in kilowatts is kVA $\times$ power factor $\cos \varphi$.

If a power triangle is drawn, the kW are in phase with the voltage and the kVA are in phase with the current, and kVA$_r$, represents the reactive component of the power.

In Fig. 4.70b the power triangle is drawn for two different angles of phase difference φ_1 and φ_2, and the reactive components are kVA$_r$ (1) and kVA$_r$ (2) respectively, so that by reducing the angle of lag from φ_1 to φ_2, the reactive kVA is reduced from kVA$_r$ (1) to kVA$_r$ (2) and the total kVA is reduced from kVA (1) to kVA (2) whilst the true power is represented in each case by kW.

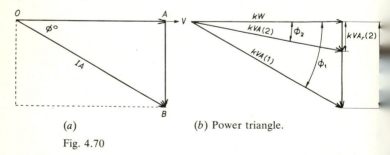

(a)   (b) Power triangle.

Fig. 4.70

Since the kVA is proportional to the current flowing, the supply current can be reduced for a given kW by reducing the angle of lag φ thus reducing the power (I^2R) loss in the supply cables and transformers.

To encourage consumers to have as high a power factor as possible by for example, the installation of banks of capacitors, supply authorities have a tariff based on kVA$_r$ maximum demand which must operate for a given period before registering on a dial and upon which maximum demand the tariff is based, so that a consumer with a low power factor and thus a high kVA$_r$ pays more per unit for energy. Welding equipment, because of the transformers and chokes, tends to give a low power factor.

Fitting of power factor improvement capacitors is an important consideration especially when there is a large transformer load as in the welding industry. Capacitors for power factor improvement are rated in kVA$_r$, the r indicating the reactive (out of phase) component of the power. For example, suppose a typical transformer has a maximum output welding current of 450 A with a input of 440 V, 90 A and a lagging power factor of 0.46 (current lagging) behind the voltage by 62°. If an 8.2 kVA$_r$ bank of capacitors are installed, the input current will now fall to 75 A at 440 V, improving the power factor to 0.57 lagging (55°). In terms of power, the original input was approximately 40 kVA, but with the capacitors improving the power factor the new power input has been reduced to 33 kVA, giving a considerable reduction in power consumed. Thus the saving in energy cost would soon pay for the capital outlay of the power factor improvement capacitors (Fig. 4.71).

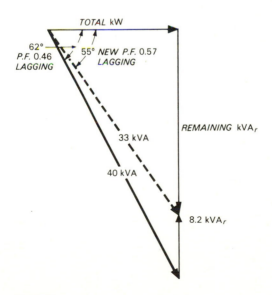

Fig. 4.71

Three-phase welding supply

For convenience in transmission and distribution, alternating current is supplied on the 'three-phase' system. The alternators have three sets of coils set at an angle of 120° to each other, instead of only one coil as on the simple alternator which we considered. These coils can be connected, as shown in Fig. 4.72, and the centre point, termed the star point, is where the beginning of each coil is connected, and the wire from this point is termed the neutral. *A, B* and *C* are the lines. The voltage between *A* and *B, B* and *C, C* and *A* is termed the line voltage and is usually for supply purposes between 440 and 400 V. The voltage between any one of the lines and the neutral wire, termed the phase voltage, is only $1/\sqrt{3}$ of the line voltage, that is, if the line voltage is 400 V the voltage between line and neutral is $400/\sqrt{3} = 230$ V, or if the line voltage is 415 V the phase to neutral voltage is 240 V.

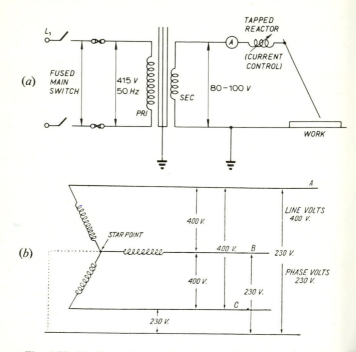

Fig. 4.72. (*a*) Connexions of a simple single-phase single-operator transformer set. (*b*) Three-phase, four-wire system.

Welding supplies for more than one welder are supplied by multi operator sets from the above type of mains supply.

Welding transformers

Welding transformers can be single-phase or three-phase. Single-phase transformers are connected either across two lines with input voltage 380–440 V or across one line and neutral when the voltage is 220–250 V. Evidently the single-phase transformer is an unbalanced load since all three lines are not involved. To balance the load equally on the three lines is not possible in welding using three single-phase transformers (Fig. 4.73a) since the welders are seldom all welding together and using the same current, so that in practice balance is never realized. Three-phase transformers on the other hand give a better balancing of the load even when only one welder is operating (Fig. 4.73b). Single-phase transformers are available for single operator welding with a variety of outputs. As the input voltage is reduced the input current rises for the same power output so that it is usually the smaller output units which are made for line-to-neutral (240 V) connexion. Larger units are connected across two lines to keep the input current down.

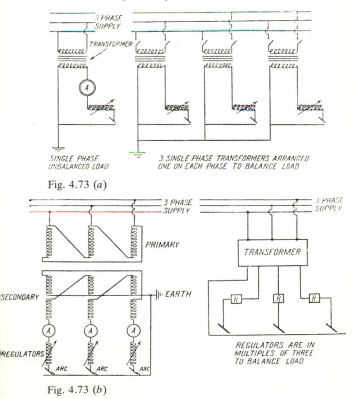

Fig. 4.73 (a)

Fig. 4.73 (b)

The open circuit voltage (OCV) depends upon the particular transformer. Many transformers have 80 to 100 OCV selected as required, for example giving 50–450 A at 80 OCV and 60–375 A at 100 OCV. Other transformers may have a lower OCV of 70 V or even 50 V. The electrode classification indicates the types of electrode coverings suitable for the various OCV, and the striking voltage required is always given with the instructions for use of a particular type of electrode.

Smaller transformers with output currents of about 200 A maximum are air-convection-current cooled: larger units are forced draught (fan assisted) while most of the largest units are oil cooled.

As is the case with most electrical machines the duty cycle is important in order to keep the temperature rise within permissible limits. Although the maximum current specified for a given transformer may be say 200 A, this rating may be only possible for a 25% duty cycle, with for example 180 A at 30% and 100 A at 100% duty cycle (continuous welding). When choosing a unit therefore it is important to estimate the average current settings that will be used and the approximate duty cycle, so that a large enough unit can be selected, i.e. one that will perform the work without excessive temperature rise.

Current control can be by tapped choke, leakage reactance moving coil, etc., and for fine current settings can have 40–50 steps with two selectors, one coarse, one fine, while the leakage reactance moving coil type can have a continuously variable current control operated by hand wheel or lever.

When transformer units are to be used for TIG welding in conjunction with an H.F. unit they are fitted with an H.F. protection circuit because of the high voltages involved. Multi-operator equipment is often used in larger welding establishments at a saving in capital cost, with 6, 8 or 12 welders being supplied from one three-phase transformer, each with this own current regulator. The sizes vary from those for 6 welders each with a maximum welding current of 350 A to the largest units for 12 welders with a maximum current of 450 A each and a rating of 486 kVA (Fig. 4.74).

When using a transformer in which the current is selected by a coarse and fine tapping switch, the current should not be altered whilst welding current is flowing since the arcing which occurs as the selector passes from stud to stud damages the smooth surface of the contact stud. If the transformer is oil cooled, the quenching action of the oil prevents serious arcing, but some oil may be carbonized and this will eventually cause a deterioration of its insulating qualities.

In addition, the three-phase transformer is cheaper to manufacture and install than three single-phase transformers and, because of this, is invariably found wherever many welders have to be supplied, as in shipyards and engineering works. Each welder has his own current regulator, as in the single-phase set. Details of the electrical equipment used in TIG, MIG and CO_2 processes are included under their respective headings.

Note. See also BS 638, *Arc welding plant, equipment and accessories.*

Fig. 4.74. Three-operator set, showing the single three-phase transformer and three welding regulators.

Parallel operation of welding transformers

Welding transformers of similar type can be connected in parallel to give a greater current output than could be provided by either of them used singly. The transformers should have their primaries connected across the same pair of lines and the output welding voltages should be the same in order to prevent circulating currents flowing in the secondary windings before they are connected to a load. There is no problem of phase rotation but the output should be checked for 'polarity'.

This is done by connecting both 'work' terminals together and placing a voltmeter across the 'electrode' terminals as shown in Fig. 4.75. If the voltmeter reads zero the transformers have similar polarity and the electrode terminals can be connected together, and welding performed from the paralleled units. If the voltmeter reads twice the normal output voltage the polarity is reversed and the connexions to one pair of secondary terminals (work and electrode) should be reversed, when the test should show zero voltage and the transformers can now be paralleled.

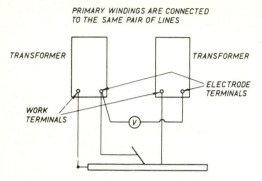

PRIMARY WINDINGS ARE CONNECTED
TO THE SAME PAIR OF LINES

TRANSFORMER

TRANSFORMER

ELECTRODE
TERMINALS

WORK
TERMINALS

V

Fig. 4.75. Paralleling of welding transformers, polarity test.

EARTHING

If a person touches a 'live' or electrified metal conductor, a current will flow from this conductor, through the body to earth, since the conductor is at higher electrical pressure (or potential) than the earth. The shock that will be felt will depend upon how much current passes through the body and this in turn depends upon (1) the voltage of the conductor, (2) the resistance of the human body, (3) the contact resistance between body and earth.

The resistance of the human body varies considerably and may range from 8000 to 100 000 ohms, while the contact resistance between body and earth also has a wide range. Resistance to earth is high if a person is standing on a dry wooden floor and thus a low current would pass through the body if a live conductor is touched, while if a person is standing on a wet concrete floor and touches a live conductor with wet hands the resistance to earth is greatly lowered, a larger current would pass through the body and consequently a greater shock would be felt. It may be stated here that care should be taken to avoid shock when welding in damp situations, especially with a.c. The operator can wear gloves and thus avoid touching the welding terminals with bare hands and he can stand on dry boards.

Most electrical apparatus, such as motors, switch gear, cables, etc., is mounted in, or surrounded by, a metal casing, and if this should come into contact, through any cause whatever, with the live conductors inside, it will then become electrified and a source of danger to any one touching it.

To prevent this danger, *all* metal parts of electrical apparatus *must* be 'earthed', that is, must be connected with the general mass of the earth so that at all times there will be an immediate and safe discharge of energy. Good connexion to earth is essential. If the connexion is poor, its resistance is high and a current may follow an easier alternative path to earth through the human body if the live metal part is touched.

For earthing of electrical installations in houses, the copper pipes of the cold water system are satisfactory since they are sufficient to carry to earth currents likely to be met with in this type of load.

Connexion from the 'earthing system', as it is termed, to earth is made in various ways. Earth plates of cast iron or copper, 1–1.5 m square and buried 1.5–2 m deep, are in general use in this country. They are surrounded by coke and the area round is copiously watered. Tubes, pipes, rods and strips of copper driven deep into the ground are used both in this country and in the USA and the area round them is frequently covered with common salt and again copiously watered.

It is evident that the 'earthing system' must be continuous throughout its length and must connect up and make good contact with every piece of metal likely to come into contact with live conductors. In factories and workshops the cables are carried in steel conduits and this forms the earthing system, the conduit making good contact with all the apparatus which it connects. Any metal part which may become live discharges to earth, through the continuous steel tubing system. To ensure that connexion to earth is well made, extra wires of copper with terminal lugs attached are connected from the conduit to the metal parts of apparatus such as motors and switch gear and ensure a good 'bond' in case of poor connexion developing between the conduit and the metal casing of the apparatus. In the case of portable apparatus such as welding transformers, regulators, welding dynamos (motor driven), drills, hand lamps, etc., an extra earthing wire is run (sometimes included in the flexible tough rubber supply cable) and makes good connexion from the metal parts of the portable apparatus to the main earthing system.

When steel wire or steel tape armoured cable is used, the wire or tape is utilized as the earthing system. In all cases extra wires are run whenever necessary to ensure good continuity with earth, and the whole continuous system is then well connected to the earth plate by copper cables.

In a.c. welding from a transformer it is usual to earth one of the welding supply terminals in addition to the metal parts of the transformer and regulator tanks. This protects the welder in the event of a breakdown in the transformer causing the main supply pressure to come into contact with the welding supply.

Low voltage safety device

As we have seen, if a welder is working in a damp situation or otherwise making good electrical contact through his clothes or boots with the work being welded (as for example inside a boiler or pressure vessel) and he touches a bare portion of the electrode holder or uses bare hands to place an electrode in the holder, his body is making contact across the open circuit voltage of the supply.

If this is d.c. at about 50–60 OCV practically no effect is felt but if the supply is from a transformer at say 80 OCV, this is the rms value and the peak of this is about 113 V so that the welder will feel an electric shock, its severity depending upon how good a contact is being made between electrode and work by the welders body. In some cases the shock can be severe enough to produce a serious effect.

A safety device is available which is attached to the transformer welding unit and consists of step-down transformer and rectifier giving 25 V d.c. with contactors and controls.

When the transformer is switched on, a d.c. voltage of 25 V appears across electrode holder and work terminals. When the electrode is struck on the work and the circuit completed a contactor closes and the 80 OCV of the transformer appears between electrode and work and the arc is struck. Immediately the arc is extinguished the 25 V d.c. reappears across electrode and work terminals thus giving complete safety to the welder. Green and red lights indicate low volts and welding in progress respectively.

5

Manual metal arc welding

Conversion factors

Inch	mm	Preferred mm size	Inch	mm	Preferred mm size
$\frac{1}{64}$	0.4	—	$\frac{5}{16}$	8.0	8
$\frac{1}{32}$	0.8	—	$\frac{3}{8}$	9.5	10
$\frac{1}{16}$	1.6	—	$\frac{7}{16}$	11.1	11
$\frac{3}{32}$	2.4	—	$\frac{1}{2}$	12.7	12
$\frac{1}{8}$	3.2	3	$\frac{9}{16}$	14.4	—
$\frac{5}{32}$	4.0	4	$\frac{5}{8}$	16.0	15
$\frac{2}{16}$	4.8	5	$\frac{1}{16}$	17.6	18
$\frac{1}{4}$	6.4	6	$\frac{3}{4}$	19.0	20
			$\frac{7}{8}$	22.4	22
			1	25.4	25

SWG	mm	SWG	mm
28	0.38	16	1.6
26	0.46	14	2.0
24	0.56	12	2.5
22	0.71	10	3.25
20	0.91	8	4.0
18	1.22	6	5.0
		4	5.9

1 kgF = 9.8 N (10 N approx.).
1 bar = 14.5 lbf/in^2 = 0.1 N/mm^2.
1 tonf/in^2 = 15.4 N/mm^2 = 1.54 hbar.
1000 mb = 1 bar = 14.5 lbf/in^2 = 1 kgf/cm^2.
1 hbar = 100 bar.
1 cu. ft = 0.028 m^3 = 283 litres.

THE ELECTRIC ARC

An electric arc is formed when an electric current passes between two electrodes separated by a short distance from each other. In arc welding (we will first consider direct-current welding) one electrode is the welding rod or wire, while the other is the metal to be welded (we will call this the plate). The electrode and plate are connected to the supply, one to the +ve pole and one to the −ve pole, and we will discuss later the difference which occurs when the electrode is connected to −ve or +ve pole. The arc is started by momentarily touching the electrode on to the plate and then withdrawing it to about 3 to 4 mm from the plate. When the electrode touches the plate, a current flows, and as it is withdrawn from the plate the current continues to flow in the form of a 'spark' across the very small gap first formed. This causes the air gap to become ionized or made conducting, and as a result the current is able to flow across the gap, even when it is quite wide, in the form of an arc. The electrode must always be touched on to the plate before the arc can be started, since the smallest air gap will not conduct a current (at the voltages used in welding) unless the air gap is first ionized or made conducting.

The arc is generated by electrons (small negatively charged particles) flowing from the −ve to the +ve pole and the electrical energy is changed in the arc into heat and light. Approximately two-thirds of the heat is developed near the +ve pole, which burns into the form of a crater, the temperature near the crater being about 6000–7000°C, while the remaining third is developed near to the −ve pole. As a result an electrode connected to the +pole will burn away 50% faster than if connected to the −ve pole. For this reason it is usual to connect medium-coated electrodes and bare rods to the −ve pole, so that they will not burn away too quickly. Heavily coated rods are connected to the +ve pole because, due to the extra heat required to melt the heavy coating, they burn more slowly than the other types of rods when carrying the same current. The thicker the electrode used, the more heat is required to melt it, and thus the more current is required. The welding current may vary from 20 to 600 A, in manual metal arc welding.

When alternating current is used, heat is developed equally at plate and rod, since the electrode and plate are changing polarity at the frequency of the supply.

If a bare wire is used as the electrode it is found that the arc is difficult to control, the arc stream wandering hither and thither over the molten pool. The globules are being exposed to the atmosphere in their travel from the rod to the pool and absorption of oxygen and nitrogen takes place even when a short arc is held. The result is that the weld tends to be porous and brittle.

The arc can be rendered easy to control and the absorption of atmospheric gases reduced to a minimum by 'shielding' the arc. This is done by covering the electrode with one of the various types of covering previously discussed, and as a result gases such as hydrogen and carbon dioxide are released from the covering as it melts and form an envelope around the arc and molten pool excluding the atmosphere with its harmful effects on the weld metal. Under the heat of the arc chemical compounds in the electrode covering also react to form a slag which is liquid and lighter than the molten metal. It rises to the surface, cools and solidifies, forming a protective covering over the hot metal while cooling and protecting it from atmospheric effects, and also slows down the cooling rate of the weld. Some slags are self-removing while others have to be lightly chipped (Fig. 5.1).

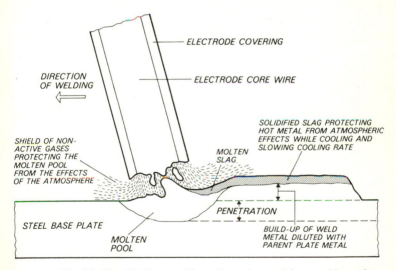

Fig. 5.1. The shielded arc. Manual metal arc weld on steel base plate with a covered electrode.

The electrode covering usually melts at a higher temperature than the wire core so that it extends a little beyond the core, concentrating and directing the arc stream, making the arc stable and easier to control. The difference in controllability when using lightly covered electrodes and various medium and heavily covered electrodes will be quickly noticed by the operator at a very early stage in practical manual metal arc welding.

With bare wire electrodes much metal is lost by volatilization, that is turning into a vapour. The use of covered electrodes reduces this loss.

An arc cannot be maintained with a voltage lower than about 14 V and is not very satisfactory above 45 V. With d.c. sources the voltage can be varied by a switch or regulator, but with a.c. supply by transformer the open circuit voltage (OCV) choice is less, being 80 or 100 V on larger units, down to 50 V on small units.

The greater the volts drop across the arc the greater the energy liberated in heat for a given current.

Arc energy is usually expressed in kilojoules per millimetre length of the weld (kJ/mm) and

$$\text{Arc energy (kJ/mm)} = \frac{\text{arc voltage} \times \text{welding current}}{\text{welding speed (mm/s)} \times 1000}.$$

The volts drop can be varied by altering the type of gas shield liberated by the electrode covering, hydrogen giving a higher volts drop than carbon dioxide for example. As the length of the arc increases so does the voltage drop, but since there is an increased resistance in this long arc the current is decreased. Long arcs are difficult to control and maintain and they lower the efficiency of the gas shield because of the greater length. As a result, absorption of oxygen and nitrogen from the atmosphere can take place, resulting in poor mechanical properties of the weld. It is essential that the welder should keep as short an arc as possible to ensure sound welds.

Transference of metal across the arc gap

When an arc is struck between the electrode and plate, the heat generated forms a molten pool in the plate and the electrode begins to melt away, the metal being transferred from the electrode to the plate. The transference takes place whether the electrode is positive or negative and also when it has a changing polarity as when used on a.c. Similarly it is transferred upwards against the action of gravity as when making an overhead weld. Surface tension plays an important part in overhead welding and a very short arc must be held to weld in the overhead position successfully.

The forces which cause the transfer appear to be due to; (1) its own weight, (2) the electro-magnetic (Lorentz) forces, (3) gas entrainment, (4) magneto-dynamic forces producing movement and (5) surface tension. The globule is finally necked off by the magnetic pinch effect.

If the arc is observed very closely, or better still if photographs are taken of it with a slow-motion cine-camera, it can be seen that the metal is transferred from the electrode to the plate in the form of drops or globules, and these globules vary in size according to the current and type of electrode covering. Larger globules are transferred at longer intervals than smaller globules and the globules form, elongate with a neck connecting them to the electrode, the neck gets reduced in size until it breaks, and the drop is projected into the molten pool, which is agitated by the arc stream, and this helps to ensure a sound bond between weld and parent metal. Drops of water falling from a tap give an excellent idea of the method of transference (see Fig. 5.2). Other methods of transfer known as dip (short circuiting arc) and spray (free flight transfer) are discussed in the section on MIG welding process.

Fig. 5.2. Detachment of molten globule in the metal arc process.

Arc blow

We have seen that whenever a current flows in a conductor a magnetic field is formed around the conductor. Since the arc stream is also a flow of current, it would be expected that a magnetic field would exist around it, and that this is so can be shown by bringing a magnet near the arc. It is seen that the arc is blown to one side by the magnet, due to the interaction of its field with that of the magnet (just as two wires carrying a current will attract each other if the current flows in the same direction in each, or repel if the currents are in opposite directions), and the arc may even be extinguished if the field due to the magnet is strong enough. When welding, particularly with d.c. it is sometimes found that the arc tends to wander and becomes rather uncontrollable, as though it was being blown to and fro. This is known as arc blow and is experienced most when using currents above 200 or below 40 A, though it may be quite troublesome, especially when welding in corners, in between this range. It is due to the interaction of the magnetic field of the arc stream with the magnetic fields set up by the currents in the metal of the work or supply cables. The best methods of correction are:

(1) Weld away from the earth connexion.
(2) Change the position of the earth wire on the work.
(3) Wrap the welding cable a few turns around the work, if possible on such work as girders, etc.
(4) Change the position of the work on the table if working on a bench.

In most cases the blow can be corrected by experimenting on the above lines, but occasionally it can be very troublesome and difficult to eliminate. Alternating-current welding has the advantage that since the magnetic field due to the arc stream is constantly alternating in direction at the frequency of the supply, there is much less trouble with arc blow, and consequently this is very advantageous when heavy currents are being used. Arc blow can be troublesome in the TIG and MIG processes, particularly when welding with d.c.

Spatter

At the conclusion of a weld small particles or globules of metal may sometimes be observed scattered around the vicinity of the weld along its length. This is known as 'spatter' and may occur through:

(1) Arc blow making the arc uncontrollable.
(2) The use of too long an arc or too high an arc voltage.
(3) The use of an excessive current.

The latter is the most frequent cause.

Spatter may also be caused by bubbles of gas becoming entrapped in the molten globules of metal, expanding with great violence and projecting the small drops of metal outside the arc stream, or by the magnetic pinch effect, by the magnetic fields set up, and thus the globules of metal getting projected outside the arc stream.

Spatter can be reduced by controlling the arc correctly, by varying current and voltage, and by preventing arc blow in the manner previously explained. Spatter release sprays ensure easy removal.

Eccentricity of the core wire in an MMA welding electrode

If the core wire of a flux-coated electrode is displaced excessively from the centre of the flux coating because of errors in manufacture, the arc

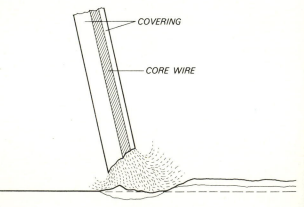

COVERING

CORE WIRE

(*a*)

may not function satisfactorily. The arc tends to be directed towards one side as if influenced by 'arc blow' and accurate placing of the deposited metal is prevented (Fig. 5.3a). A workshop test to establish whether the core wire is displaced outside the manufacturer's tolerance is to clean off the flux covering on one side at varying points down the length of the electrode and measure the distance L (Fig. 5.3b). The difference between the maximum and minimum reading is an approximate indication of the eccentricity.

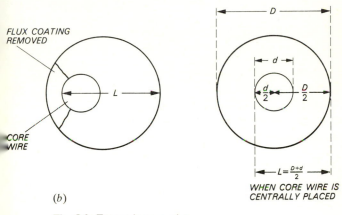

(b)

WHEN CORE WIRE IS
CENTRALLY PLACED

Fig. 5.3. Eccentric core wire.

Electrode efficiency (metal recovery and deposition coefficient)

The efficiency of an electrode is the weight of metal actually deposited compared with the weight of that portion of the electrode consumed. It can be expressed as a percentage thus:

$$\text{Efficiency } \% = \frac{\text{weight of metal deposited}}{\text{weight of metal of the electrode consumed}} \times 100.$$

With ordinary electrodes the efficiency varies from 75 to 95%, but with iron powder electrodes the efficiency can be up to 200%. Efficiencies of 110% and above are indicated in BS 639 (1976) and ISO 2560 by a three-digit figure, giving the efficiency percentage rounded off to the nearest 10 at the beginning of the optional part of the classification.

The efficiency of a particular type of electrode can be obtained by stripping five electrodes of their coverings and weighing them. Take a piece of clean steel plate about 300 × 75 × 12 mm and weigh it and deposit five similar electrodes to the ones stripped, retain the five stub ends and weigh them. Weigh the plate with deposit and subtract the original weight of the plate to give the weight of actual weld metal deposited, say W g. Subtract the weight of the stub ends from the weight of the electrodes stripped and this gives the weight of electrode used up, say W g. Then the percentage efficiency $= \dfrac{W}{w} \times 100$.

Iron powder electrodes

The deposition rate of a given electrode is dependent upon the welding current used, and for maximum deposition rate, maximum current should be used. This maximum current depends upon the diameter of the core wire, and for any given diameter of wire there is a maximum current beyond which increasing current will eventually get the wire red hot and cause overheating and hence deterioration of the covering.

To enable higher currents to be used an electrode of larger diameter core wire must be used, but if sufficient iron powder is added to the covering of the electrode, this covering becomes conducting, and a higher welding current can now be used on an electrode of given core wire diameter. The deposition rate is now increased and in addition the iron powder content is added to the weld metal giving greater efficiency, that is enabling more than the core wire weight of metal to be deposited because of the extra iron powder. Efficiencies of up to 200% are possible, this meaning that twice the core wire weight of weld metal is being deposited. These electrodes can have coverings of rutile or basic type or a mixture of these. The iron powder ionizes easily, giving a smoother arc with little spatter, and the cup which forms as the core wire burns somewhat more quickly than the covering gives the arc directional properties and reduces loss due to metal volatilization. See also Electrode efficiency (metal recovery and deposition coefficient).

Hydrogen-controlled electrodes (basic covered)

If oxygen is present with molten iron or steel a chemical reaction occurs and the iron combines chemically with the oxygen to form a chemical compound, iron oxide. Similarly with nitrogen, iron nitride being formed if the temperature is high enough as in metal arc welding. When hydrogen is present however there is no chemical reaction and the hydrogen simply goes into solution in the steel, its presence being described as x millilitres of hydrogen in y grams of weld metal.

This hydrogen can diffuse out of the iron lattice when in the solid state resulting in a lowering of the mechanical properties of the weld and increasing the tendency to cracking. By the use of basic-hydrogen controlled electrodes, and by keeping the electrodes very dry, the absorption of hydrogen by the weld metal is reduced to a minimum and weld can be produced that have great resistance to cracking even under conditions of very severe restraint.

The coverings of these electrodes are of calcium or other basic carbonates and fluorspar bonded with sodium or potassium silicate. When the basic carbonate is heated carbon dioxide is given off and provides the shield of protective gas thus:

calcium carbonate (limestone) heated → calcium oxide (quicklime) + carbon dioxide.

There is no hydrogen in the chemicals of the covering, so that if they are kept absolutely dry, the deposited weld metal will have a low hydrogen content. Electrodes which will give deposited metal having a maximum of 15 millilitres of hydrogen per 100 grams of deposited metal (15 ml/100 g) are indicated by the letter H in BS 639 1976 classification. The absence of diffusible hydrogen enables free cutting steels to be welded with absence of porosity and cracking and the electrodes are particularly suitable for welding in all conditions of very severe restraint. They can be used on a.c. or d.c. supply according to makers' instructions and are available also in iron powder form and for welding in all positions. Low and medium alloy steels which normally would require considerable pre-heat if welded with rutile-coated electrodes can be welded with very much less pre-heating, the welds resisting cracking under severe restraint conditions and also being very suitable for welding in sub-zero temperature conditions. By correct storage and drying of these electrodes the hydrogen content can be reduced to 5 ml/100 g of weld metal for special applications. Details of these drying methods are given in the section on storage and drying of electrodes (q.v.).

Experiment to illustrate the diffusible hydrogen content in weld metal. Make a run of weld metal about 80 mm long with the metal arc on a small square of steel plate using an ordinary steel welding rod with a cellulose, rutile or iron oxide coating. Deslag, cool out quickly and dry off with a cloth and place the steel plate in a beaker or glass jar of paraffin. It will be noted that minute bubbles of gas stream out of the weld metal and continue to do so even after some considerable time. If this gas is collected as shown in Fig. 5.4 it is found to be hydrogen which has come from the flux covering and the moisture it contains. A steel weld may contain hydrogen dissolved in the weld metal and also in the molecular form in any small voids which may be present. Hydrogen in steel produces embrittlement and a reduction in fatigue strength. If a run of one of these hydrogen-controlled electrodes is made on a test plate and the previous experiment repeated it will be noted that no hydrogen diffuses out of the weld.

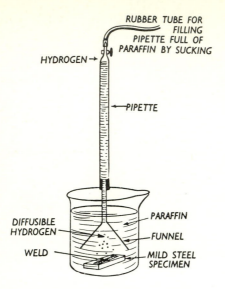

Fig. 5.4. Collecting diffusible hydrogen from a mild steel weld.

Appendix C BS 639, 1976 gives recommended method of determining the quantity of diffusible hydrogen present together with a
drawing of a diffusible hydrogen meter.

Deep penetration electrodes

A deep penetration electrode is defined in BS 499, Part 1 (*Welding
terms and symbols*) as 'A covered electrode in which the covering aids
the production of a penetrating arc to give a deeper than normal fusion
in the root of a joint'. For butt joints with a gap not exceeding 0.25 mm
the penetration should be not less than half the plate thickness, the plate
being twice the electrode core thickness. For fillet welds the gap at the
joint should not exceed 0.25 mm and penetration beyond the root
should be 4 mm minimum when using a 4 mm diameter electrode.

Welding position

Weld slope is the angle between line of the root of the weld and the
horizontal (Fig. 5.5).

Weld rotation. Draw a line from the root of the weld at right
angles to the line welding to bisect the weld profile. The angle that this
line makes with the vertical is the angle of weld rotation.

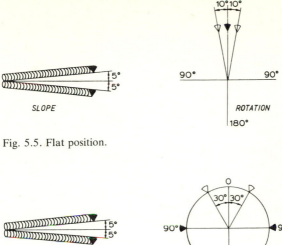

Fig. 5.5. Flat position.

Fig. 5.6. Horizontal–vertical position.

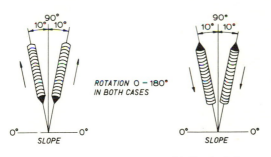

Fig. 5.7. (*a*) Vertical–up position. (*b*) Vertical–down position.

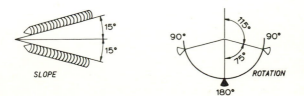

Fig. 5.8. Overhead position.

The table indicates the five welding positions used for electrode classification. Any intermediate position not specified may be referred to as 'inclined'.

Position	Slope	Rotation	Symbol	Fig.
Flat	0–5°	0–10°	F	5.5
Horizontal–vertical	0–5°	30–90°	H	5.6
Vertical–up	80–90°	0–180°	V	5.7a
Vertical–down	80–90°	0–180°	D	5.7b
Overhead	0–15°	115–180°	O	5.8

Storage of electrodes

The flux coverings on modern electrodes are somewhat porous and absorb moisture to a certain extent. (See also pp. 60–62.) The moisture content (or humidity) of the atmosphere is continually varying and hence the moisture content of the covering will be varying. Moisture could be excluded by providing a non-porous covering, but any moisture entrapped would be liable to cause rupture of the coating when the moisture was turned to steam by the heating effect of the passage of the current through the electrode. Cellulosic electrodes absorb quite an appreciable amount of moisture, and it does not affect their properties since they function quite well with a moisture content. They should not be over-dried or the organic compounds of which they are composed tend to char, affecting the voltage and arc properties. The extruded electrodes with rutile, iron oxide and silicate coatings do not pick up so much moisture from the atmosphere and function quite well with a small absorbed content. If they get damp they can be satisfactorily dried out but it should be noted that if they get excessively wet, rusting of the core wire may occur and the coating may break away. In this case the electrodes should be discarded.

Storage temperatures should be about 12 °C above that of external air temperature with 0–60% humidity. Cellulosic electrodes are not so critical: but they should be protected against condensation and stored in a humidity of 0–90%.

Drying of electrodes.

The best drying conditions are when the electrodes are removed from their package and well spaced out in a drying oven which has a good circulation of air. Longer drying times are required if the electrodes are not spaced out. The following table gives an indication of temperatures and times required, but see also the special conditions for drying basic electrodes (Fig. 5.9).

Drying of electrodes: approximate times and temperatures with electrodes spaced apart. Times will vary with air circulation, electrode spacing and oven loading

Electrode type	Diameter mm	Temperature °C	Time in mins air circulation	
			good	poor
Rutile mild steel	1.6–2.5	110	10–30	20–30
	3.2–5.0	110	20–45	30–60
	6.0–10.0	110	30–60	45–120
Cellulose	2.5–6.0	110	10–15	15–20

Fig. 5.9

Hydrogen-controlled (basic) electrodes

The coatings of these electrodes contain no hydrogen-forming compounds, but if moisture is absorbed by the coating it becomes a source of hydrogen and cannot be tolerated. They must therefore be stored in a dry, heated and well-ventilated store on racks above floor level and unused electrodes should be returned to the store rather than left in the colder and moister conditions of the workshop where they could absorb moisture. A temperature of about 12 °C above that of the external air temperature is suitable. Before use they should be removed from their package and spread out in the drying oven, the drying time and temperature depending upon the permissible volume of hydrogen allowable in the weld deposit. Suggested figures are given in the following table.

Hydrogen content in millilitres of hydrogen per 100 grams of weld metal	Temperature °C	Time minutes	Use
10–15 ml H₂/100 g	150	60	To give resistance to H₂ cracking in thick section mild steel, high restrain
5–10 ml H₂/100 g	200	60	High quality welds in p vessel and structural ap
below 5 ml H₂/100 g	450	60	Thick sections to avoid tearing and critical app

In order to obtain high radiographic standards of deposited weld metal the drying periods given above may be extended. The following periods are given as an indication of prolonged drying times such that the electrode coating will not suffer a decrease in coating strength.

Drying temperature	Maximum time
150 °C	72 hours
250 °C	12 hours
450 °C	2 hours

The makers' instructions for drying should be strictly adhered to.

Many electrodes if stored in damp situations get a white fur on their coverings. This is sodium carbonate produced by the action of the carbon dioxide (carbonic acid) of the atmosphere on the sodium silicate of the binder in the flux covering. The fur appears to have little detrimental effect on the weld but shows that the electrodes are being stored in too damp a situation.

ELECTRODE CLASSIFICATION

Classification of covered electrodes for the manual metal arc welding of carbon and carbon–manganese steels. BS 639 1976
There is a compulsory and an optional part of the classification. In the compulsory part a covered electrode for manual metal arc welding is indicated by the letter E. This is followed by a two-digit figure which gives the tensile strength and the yield stress of the weld metal thus:

Table 1

Electrode designation	Tensile strength N/mm² or MPa	Minimum yield stress N/mm² or MPa
E 43	430–550	330
E 51	510–650	360

The following two digits each indicate elongation and impact strength thus:

Table 2. *First digit for elongation and impact strength*

First digit	Minimum elongation % E 43	E 51	Temperature for impact value of 28 J. °C
0	not specified		not specified
1	20	18	+ 20
2	22	18	0
3	24	20	− 20
4	24	20	− 30
5	24	20	− 40

Table 3. *Second digit for elongation and impact strength*

Second digit	Minimum elongation % E 43	E 51	Impact properties Impact value J E 43	E 51	Temperature °C
0	not specified		not specified		
1	22	22	47	47	+ 20
2	22	22	47	47	0
3	22	22	47	47	− 20
4	not relevant	18	not relevant	41	− 30
5	There are no electrodes designated by this digit				− 40
6	not relevant	18	not relevant	47	− 50

Following these digits is a letter (or letters) which indicates the type of covering.

	acid (iron oxide).	O	oxidizing.
R	acid (rutile).	R	rutile (medium coated).
	basic.	RR	rutile (heavy coated).
	cellulosic.	S	other types.

This completes the compulsory portion of the classification which therefore indicates tensile strength, yield strength, elongation and impact values and type of electrode covering.

The optional part of the classification begins with a three-digit number for the nominal electrode efficiency, included only if this is equal to or greater than 110. It is given to the nearest multiple of 10, with values ending in five being rounded off to the next higher ten.

Following this is a digit indicating the welding positions as shown and a digit for the current and voltage conditions as in Table 4.

Welding positions

1	all positions.
2	all positions except vertically down.
3	flat and, for fillet welds, horizontal–vertical.
4	flat.
5	flat, vertical down, and, for fillet welds, horizontal–vertical.
6	any position or combination of positions not classified above.

Table 4. *Welding current and voltage conditions*

Code	Direct current: recommended eletrode polarity	Alternating current: minimum open circuit voltage
0	Polarity as recommended by the manufacturer	Not suitable for use on A.C.
1	+ or −	50
2	−	50
3	+	50
4	+ or −	70
5	−	70
6	+	70
7	+ or −	90
8	−	90
6	+	90

Hydrogen controlled electrodes. The letter H shall be included in the classification for those electrodes which deposit not more than 15 ml of diffusible hydrogen per 100 g of deposited weld metal when stored according to manufacturer's instructions. Recommended drying conditions shall be shown on the packet for hydrogen levels in the following ranges: 10–15 ml, 5–10 ml, 0–5 ml per 100 g of deposited weld metal respectively.

Examples of the use of BS 639 classification.

Example (1)

E .51 .33 .RR .130 .3 .1 .H

E Flux-covered electrode for manual metal arc welding.

51 (Table 1) The all weld metal (AWM) tensile strength lies between 510 and 650 N/mm² and the yield stress is not less than 360 N/mm².

3 (Table 2) The minimum elongation would be 20% with an impact strength of at least 28 J at −20 °C.

3 (Table 3) A second value of elongation, 22% minimum with a minimum impact strength of 47 J at −20 °C.

RR The predominant coating ingredient is rutile and the flux coating diameter to core wire diameter ratio shall be greater than 1.5:1.

130 The electrode is of a type containing iron powder in the flux covering and will deposit weld metal in excess of the core wire weight. The deposit in this case would be from 125 to 134% of the core wire weight.

3 The electrode can be used in the flat position and for horizontal vertical fillet welds.

1 (Table 4) The electrode can be used on a d.c. power supply, with electrode connected to either positive or negative pole or on an a.c. supply having at least 50 OCV.

H The weld metal shall contain not more than 15 ml hydrogen in every 100 g weld metal when the electrode has been stored and used in accordance with the manufacturers instructions.

Note that this is a hydrogen-controlled *rutile* electrode.

Example (2)

E .43 .21 .C .1 .9

E Flux-covered electrode for manual metal arc welding.

43 AWM tensile strength of 430–550 N/mm², yield stress 330 N/mm² minimum.

2 AWM elongation of 22% with an impact strength of 28 J minimum at 0 °C.

1 AWM elongation of 22% with an impact strength of 47 J at 20 °C.

C A cellulosic covered electrode.

1 Can be used in all positions.

9 Can be used on a d.c. supply with electrode positive or on an a.c. supply with an OCV of 90 V.

Note on ISO 2560 – 1973 (E). (Covered electrodes for manual arc welding of mild steel and low alloy steel – Code of symbols for identification.) BS 639 1976 follows closely the ISO 2560 specification, the main differences being that in the ISO standard the tensile strengths are 430–510 N/mm² and 510–610 N/mm², but in each case in view of possible variations in welding and testing, the upper limits of 510 and 610 N/mm² are allowed to be exceeded by 40 N/mm² giving the BS 639 values of 550 and 650 N/mm². The main other difference is that of impact strength and elongation, where the ISO standard has one digit only as from Table 2 whereas BS 639 has two digits, one each from Tables 2 and 3, the second digit covering the amalgamation of the rules of the International Association of Classification Societies and the BS 639 1972 requirements.

The extra Charpy testing at varying temperatures gives the designer a better guide to the selection of electrodes to give resistance to brittle fracture.

American Welding Society (AWS) electrode classification A5. 1–69

Mild steel electrodes. A four-digit number is used, preceded by the letter E indicating electrode. The first two digits indicate the tensile strength of the weld metal in thousands of pounds force per square inch (1000 psi) in the 'as-welded' condition. The third digit indicates the welding position and the fourth digit the current to be used and the type of flux coating. An example of the first and second digits is: E 60xx which indicates that it is a metal arc welding electrode with an 'as-welded' deposit having a UTS of 60 000 lbf/in² minimum or 412 N/mm².

The third and fourth digits are:

E xx10 High cellulose coating, bonded with sodium silicate. Deeply penetrating, forceful, spray type arc, thin friable slag. All positional, d.c. electrode positive only.

E xx11 Similar to E xx10 but bonded with potassium silicate to allow it to be used on a.c. or d.c. positive.

E xx12 High rutile coating, bonded with sodium silicate. Quiet arc medium penetration, all positional, a.c. or d.c. negative.

E xx13 Coating similar to E xx12 but with the addition of easily ionized materials and bonded with potassium silicate to give a steady arc on a low voltage supply. Slag is fluid and easily removed. All positional; a.c. or d.c. negative.

E xx14 Coating similar to E xx12 and E xx13 types with the addition of a medium quantity of iron powder. All positional; a.c. or d.c.

E xx15 Lime-fluoride coating (basic, low hydrogen) type bonded with sodium silicate. All positional. For welding high tensile steels; d.c., positive only.

E xx16 Similar coating to E xx15 but bonded with potassium silicate; a.c. or d.c. positive.

E xx18 Coating similar to E xx15 and E xx16 but with the addition of iron powder. All positional; a.c. or d.c.

E xx20 High iron oxide coating bonded with sodium silicate. For welding in the flat or HV positions. Good X-ray quality; a.c. or d.c.

E xx24 Heavily coated electrode having flux ingredients similar to E xx12 and E xx13 with the addition of a high percentage of iron powder for fast deposition rates. Flat and horizontal positions only; a.c. or d.c.

E xx27 Very heavily coated electrode having flux ingredients similar to E xx20 type, with the addition of a high percentage of iron powder. Flat or horizontal positions. High X-ray quality; a.c. or d.c.

E xx28 Similar to E xx18 but heavier coating and suitable for use in flat or HV positions only; a.c. or d.c.

E xx30 High iron oxide type coating but produces less fluid slag than E xx20. For use in flat position only (primarily narrow groove butt welds). Good radiographic quality; a.c. or d.c.

Example

E 6013. Welding electrode having weld metal of UTS 60 000 lbf/in^2 or 412 N/mm^2, with high rutile coating bonded with potassium silicate. All positional; a.c. or d.c. negative.

Example

E 7018. Welding electrode with weld metal of UTS 70 000 lbf/in^2 or 480 N/mm^2, with basic coating hydrogen controlled, with the addition of iron powder. All positional; a.c. or d.c.

Types of electrode flux coverings

Class	Composition of covering	Characteristics	Uses
Cellulose (C)	Organic material containing cellulose and with some titanium oxide. Hydrogen releasing.	Thin, easily removable slag. Rather high spatter loss. Considerable envelope of shielding gas. Coarse ripple on weld surface, deeply penetrating arc with rapid burn-off rate.	All classes of mild steel welding in all positions; a.c. or d.c. electrode positive.
Acid (A)	Oxides and carbonates of iron and manganese with deoxidizers such as ferro-manganese.	Generally a thick coating which produces a fluid slag of large volume and solidifies in a 'puffed up' manner, is full of holes and easily detached. Smooth weld finish with small ripples. Good penetration. Weld liable to solidification cracking if plate weldability is not good.	Usually in the flat position only but can be used in other positions; a.c. or d.c.
Acid rutile (AR)	Generally a thick coating containing up to 35% rutile. Ilmenite (iron oxide and titanium oxide) is also used.	A fluid slag with other characteristics similar to the acid type of covering.	Similar to class (A).
Rutile (R) (medium coating)	Mixture of titanium oxide and up to 15% organic (cellulose) matter with additions to produce a fluid slag. Coating thickness less than 50% of the core wire diameter.	Heavier coating than the AR type. Smooth weld finish, medium penetration, little spatter; fast freezing, easily detachable slag even from deep grooves.	Widely used for steel welding of all types. All positions; a.c. or d.c. Especially suitable for vertical and overhead positions.
Rutile (RR) (heavy coating)	Coverings of titanium oxide with up to 5% cellulosic matter with calcium fluoride. Coating thickness at least 50% greater than the core wire diameter.	Viscous slag, easy to remove except in deep V. Smooth weld contour.	Mainly in flat position but suitable for all positions; a.c. or d.c.

...ng (O)	Iron oxide with or without manganese oxide and silicates.	Oxidizing slag so that the weld metal has a low carbon and manganese content referred to as 'dead soft'. Reduction of area and impact values are lower than for other types of electrodes. Core wire melts up inside coating forming a cup so that the electrode can be used for 'touch-welding'. Low penetration; solid slag often self-deslagging, with weld of neat appearance.	d.c. or a.c. supply with OCV as low as 45 V.
Basic (B)	Calcium or other basic carbonates and fluorspar bonded with sodium or potassium carbonates. Medium coating. Coating compounds contain no hydrogen. CO_2 releasing.	Brown slag easy to remove. Medium ripple on weld metal, medium penetration. Fillet profile flat or convex. Deposited metal has high resistance to hot and cold cracking because there is a low hydrogen content in the weld. Electrodes must be stored under warm dry conditions and dried before use.	Suitable for d.c. (electrode positive or a.c with OCV of 70 V. Used for mild, low alloy, high tensile and Structural steels. Particularly for conditions of high restraint. For flat, vertical and overhead positions, the latter having a flat deposit.
Any other type (S)	This category is for any electrode coverings not included in the foregoing list. Iron powder electrodes do not come into this category but should be indicated by their efficiency with a three-digit figure.		

WELDERS' ACCESSORIES

Electrode holder

This is an arrangement which enables the welder to hold the electrode when welding. It has an insulated body and head which reduces the danger of electric shock when working in damp situations and also reduces the amount of heat conducted to the hand whilst welding. The electrode is clamped between copper jaws which are usually spring loaded, and a simple movement of a side lever enables the electrode to be changed easily and quickly. In another type the electrode end fits into a socket in the holder head and is held there by a twist on the handle. The welding flexible cable is attached to the holder by clamping pieces or it may be sweated into a terminal lug. The holder should be of light yet robust construction and well insulated, and electrode changing must be a simple operation. Fig. 5.10 shows typical holders.

Fig. 5.10

Head shields and lenses

The rays from the metallic arc are rich in infra-red and ultra-violet radiation, and it is essential that the eyes and face of the welder should be protected from these rays and from the intense brightness of the arc. The welding shield can either fit on to the head (Fig. 5.11a), leaving both hands free, or may be carried in one hand (Fig. 5.11b). It should extend so as to cover the sides of the face, especially when welding is done in the vicinity of other welders, so as to prevent stray flashes reaching the eyes.

The shield must be light and, because of this, preferably made of fibre.

Helmets and hand shields are now available, weighing little more than ordinary shields, that give maximum comfort when welding in restricted conditions. They are similar in appearance to the standard shield but have double-wall construction, the inner face being perforated with small holes through which pure air is supplied to the interior of the helmet. Head and face are kept cool and the operator's eyes and lungs are protected from dust and fumes as well as radiation and spatter as with ordinary shields. The air is supplied from any standard compressor or air line, the supply tube being fitted with a pressure-reducing valve to give a pressure of 1.7 bar.

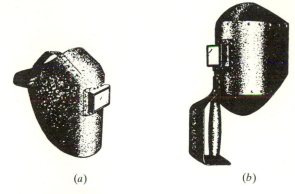

(a)   (b)

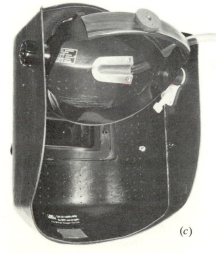

(c)

Fig. 5.11.

The arc emits infra-red and ultra-violet radiation in addition to light in the visible spectrum. The filter or lens is designed to protect the eyes from the ultra-violet and infra-red radiation which would injure the eyes and also to reduce the amount of visible light so that there is no discomfort. The filters are graded by numbers, followed by the letters EW denoting the process, according to BS 679, and increase in shade depth with increasing number for increasing currents. The filters recommended are: up to 100 A, 8 or 9/EW; 100–300 A, 10 or 11/EW; over 300 A, 12, 13, 14/EW. The choice of the correct filter is the safeguard against eye damage. Occasional accidental exposure to direct or reflected rays may result in the condition known as arc flash. This causes no permanent damage but is painful, with a feeling as of sand in the eyes accompanied by watering. Bathing the eyes with eye lotion and wearing dark glasses reduces the discomfort and the condition usually passes with no adverse effects in from 12 to 24 hours. If it persists a doctor should be consulted.

A lens of this type is expensive and is protected from metallic spatter on *both* sides by plain glass, which should be renewed from time to time, as they become opaque and uncleanable, due to spatter and fumes.

The welding area must be adequately screened so that other personnel are not affected by rays from the arc.

Leather or skin aprons are excellent protection for the clothes against sparks and molten metal. Trouser clips are worn to prevent molten metal lodging in the turn-ups, and great care should be taken that no metal can drop inside the shoe, as very bad burns can result before the metal can be removed. Leather spats are worn to prevent this. Gauntlet gloves are worn for the same reason, especially in vertical and overhead welding. In welding in confined spaces, the welder should be fully protected, so that his clothes cannot take fire due to molten metal falling on him, otherwise he may be badly burnt before he can be extracted from the confined space.

The welding bench in the welding shop should have a flat top of sheet metal, about 1.5 m × 0.75 m being a handy size. On one end a vice should be fitted, while a small insulated hook on which to hang the electrode holder when not in use is very handy.

Jigs and fixtures

These are a great aid to the rapid setting up of parts and holding them in position for welding. In the case of repetition work they are essential equipment for economical working. Any device used in this way comes under this heading, and jigs and fixtures of all types can be built easily quickly and economically by arc welding. They are of convenience to the welder, reduce the cost of the operation, and standardize and increase the accuracy of fabrication.

Jigs may be regarded as specialized devices which enable the parts being welded to be easily and rapidly set up, held, and positioned. They should be rigid and strong since they have to stand contractional stresses without deforming unless this is required; simple to operate, yet they must be accurate. Their design must be such that it is not possible to put the work in them the wrong way, and any parts of them which have to stand continual wear should be faced with wear-resistant material. In some cases, as in inert gas welding (q.v.) the jig is used as a means of directing the inert gas on to the underside of the weld (backpurge) and jigs may also incorporate a backing strip.

Fixtures are of a more general character and not so specialized as jigs. They may include rollers, clamps, wedges, etc., used for convenience in manipulation of the work. Universal fixtures are now available, and these greatly reduce the amount of time of handling of the parts to be welded and can be adapted to suit most types of work.

Manipulators, positioners, columns and booms

Positioners are appliances which enable work to be moved easily, quickly, and accurately into any desired position for welding – generally in the downhand position since this speeds up production by making welding easier, and is safer than crane handling. Universal positioners are mounted on trunions or rockers and can vary in size from quite small bench types to very large ones with a capacity of several tons. Manually operated types are generally operated through a worm gearing with safety locks to prevent undesired movement after positioning. The larger types are motor-driven, and on the types fitted with a table for example the work can be swung through any angle, rotated, and moved up and down so that if required it can be positioned under a welding head.

As automatic welding has become more and more important so has the design of positioners and rollers improved. Welding-columns and booms (Fig. 7.16) may be fixed or wheel mounted and have the automatic welding head mounted on a horizontal boom which can slide up and down a vertical column. The column can be swivel mounted to rotate through 360° and can be locked in any position. In the positioning ram-type boom there is horizontal and vertical movement of the boom carrying the welding head and they are used for positioning the head over work which moves beneath them at welding speed. In the manipulating ram-type boom, the boom is provided with a variable speed drive of range of about 150–1500 mm per min enabling the boom to move the welding head over the stationary work. Both types of boom in the larger sizes can be equipped with a platform to carry the operator who can control all movements of head and boom from this position in Fig. 5.12. Various types are shown.

Fig. 5.12. (*a*) Power rotation, manual tilt.

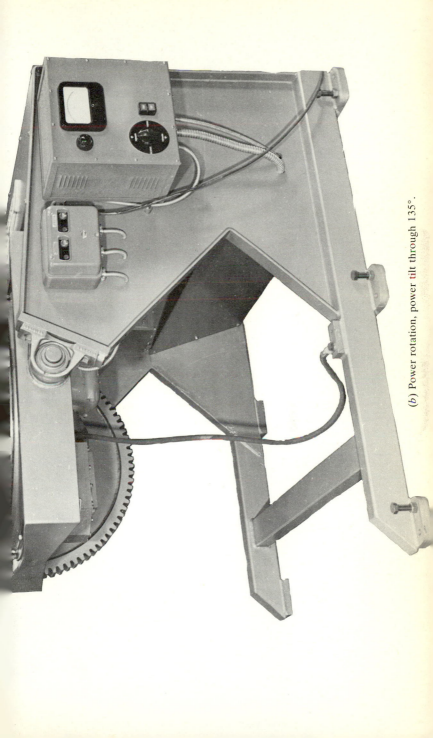

(b) Power rotation, power tilt through 135°.

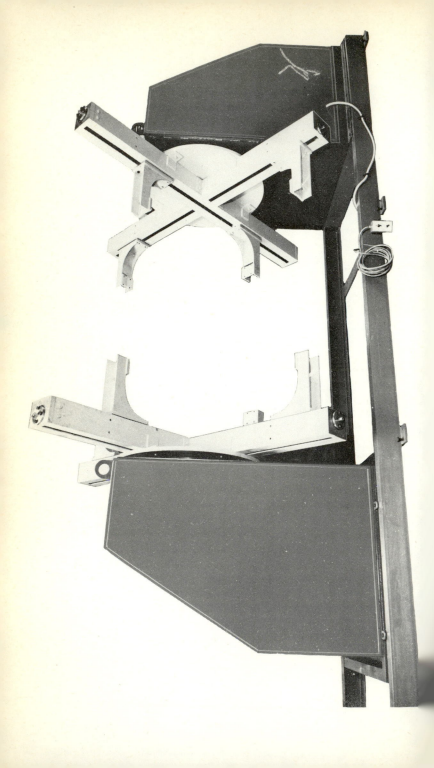

(*d*) Roller bed, travelling type, motorized.

THE PRACTICE OF MANUAL METAL ARC WELDING

Electrode lengths. The actual length must be within ± 2 mm of the nominal value

Diameter, mm	Length
2	200
	250
	300
	350
2.5	250
	300
	350
3.15 or 3.25	350
	450
4 to 10	350
	450
	500
	500
	600
	700
	900

Note. Because the 3.25 mm diameter is widely used this size is being retained until such time as it can be discarded through its lesser usage.

Electrode diameter		Nearest fractional equivalent, in.
mm	inch	
1.6	0.06	—
2.0	0.08	$\frac{5}{64}$
2.5	0.10	$\frac{3}{32}$
3.0	0.12	—
3.25	0.13	$\frac{1}{8}$
4.0	0.16	$\frac{5}{33}$
5.0	0.19	$\frac{3}{16}$
6.0	0.23	—
6.3	0.25	$\frac{1}{4}$
8.0	0.32	$\frac{5}{16}$
10.0	0.37	$\frac{3}{8}$

Safety precautions

Protection of the skin and eyes. Welding gloves should be worn and no part of the body should be exposed to the rays from the arc otherwise burning will result. Filter glasses (p. 316) should be chosen according to BS recommendations. Where there are alternatives the lower shade numbers should be used in bright light or out of doors and the higher shade numbers for use in dark surroundings.

Do not weld in positions where other personnel may receive direct or reflected radiation. Use screens. If possible do not weld in buildings with light coloured walls (white) as this increases the reflected light and introduces greater eye strain.

Do not chip or deslag metal unless glasses are worn.

Do not weld while standing on a damp floor. Use boards.

Switch off apparatus when not in use.

Make sure that welding return leads make good contact, thus improving welding conditions and reducing fire risk.

Avoid having inflammable materials in the welding shop.

Degreasing, using chemical compounds such as trichlorethylene, perchlorethylene carbon tetrachloride or methyl chloride, should be carried out away from welding operations, and chemical allowed to evaporate completely from the surface of the component before beginning welding. The first-named compound gives off a poisonous gas, phosgene, when heated or subjected to ultra-violet radiation, and should be used with care. Special precautions should be taken before entering or commencing any welding operations on tanks which have contained inflammable or poisonous liquids, because of the risk of explosion or suffocation. The student is advised to study the following publications of the Department of Employment (SHW–Safety, health and welfare):

SHW 386. *Explosions in drums and tanks following welding and cutting.* HMSO.

SHW Booklet 32. *Repair of drums and tanks. Explosions and fire risks.* HMSO.

SHW Booklet 38. *Electric arc welding.* HMSO.

Memo No 814. *Explosions – Stills and tanks.* HMSO.

Technical data notes No. 18. *Repair and demolition of large storage tanks.* No. 2. *Threshold limit values.* HMSO.

Health and safety in welding and allied processes. The Welding Institute.

Filters for use during welding. BS 679. British Standards Institution.

Protection for eyes, face and neck. BS 1542. British Standards Institution.

Metal arc welding

In order to assist the operator, tables are given indicating the approximate current values with various types and sizes of electrodes. These tables are approximate only and the actual value of the current employed will depend to a great extent upon the work. In general the higher the current in the range given for one electrode size, the deeper the penetration and the faster the rate of deposit. Too much current leads to undercutting and spatter. Too small a current will result in insufficient penetration and too small a deposit of metal. As a general rule, slightly increase the arc voltage as the electrode size increases.

The angle of the electrode can be varied between 60° and 90° to the line of the weld. As the angle between electrode and plate is reduced, the gas shield becomes less effective, the possibility of adverse effects of the atmosphere on the weld increases, and penetration is reduced.

Full details of currents suitable for the particular electrodes being used are usually found on the electrode packet and should be adhered to; in the following pages it is assumed that, if a.c. is being used, no notice should be taken of the polarity rules, and covered electrodes only should be used.

The technique of welding both mild steel and wrought iron is similar.

Striking and maintaining the arc

Using a medium-coated rod (a.c. or d.c., connected to the +ve or −ve pole according to the type of rod) of say 4 mm diameter and with the correct setting of current, the first exercise for the beginner should be in striking the arc and maintaining it. Mild steel plate 6.4 to 8 mm thick is suitable for the beginning exercises.

Diameter of rod (mm)	Current (amperes)		
	Min.	Max.	Average
1.6	25	45	40
2.5	50	90	90
3.2	60	130	115
4.0	100	180	150
5.0	150	250	200
6.0	200	310	280
6.3	215	350	300
8.0	250	500	350

There are two methods of striking the arc (Fig. 5.13). The first consists of jabbing the tip of the rod on to the plate and then lifting it and drawing the arc to about 3 mm long, while the second method consists of scratching the electrode across the plate with a slight circular motion, so that at the bottom of its travel the arc is struck and further motion of the rod draws the arc to the required length.

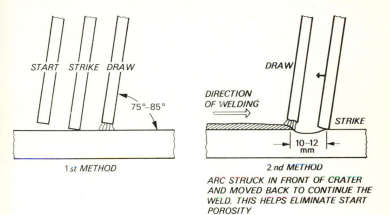

1 *st* METHOD

2 *nd* METHOD

ARC STRUCK IN FRONT OF CRATER AND MOVED BACK TO CONTINUE THE WELD. THIS HELPS ELIMINATE START POROSITY

Fig. 5.13

The second method is generally better for the beginner, who is usually troubled with the electrode 'freezing' or sticking to the plate. This is caused by the heavy current flowing when the plate is touched with the end of the rod, melting the end of the rod and virtually welding it on to the plate unless it is drawn quickly enough. If the electrode sticks to the plate, it should be freed with a sharp twist; if this fails, the electrode should be released from the holder or the supply switched off. At the instant of striking the arc the operator should place the shield over his face and observe the arc through the lens. When the arc has been drawn, the rod should be held at a steep angle (70–80°) to the plate and then moved slowly and evenly along across the plate towards the operator, keeping the arc a constant fairly short length. No side to side or weaving motion should be attempted. The deposited bead must be continuous, free from holes, even, and must penetrate well into the parent metal. The invariable fault of most beginners is that the rate of travel is too fast, resulting in an irregular bead with poor penetration. If the speed of travel is too slow, too much metal is deposited, the crater is too deep and the electrode tends to become red hot. The molten metal can be observed piling up at the rear of the crater, and this is the best indication as to whether the speed of travel is correct (Fig. 5.14*a*). The depth of the crater will indicate the amount of penetration and Fig. 5.14*b* and *c* show incorrect and correct bead sections respectively. Fig. 5.14*d* indicates undercutting caused by too high a current or too great a welding speed.

Upon striking and drawing the arc, especially with electrodes other than mild steel, it may be noticed that there is evidence of porosity at the start where the metal was deposited before the protective gas shield was established. To overcome this, method 2 (Fig. 5.13) can be used or a striking plate can be placed adjacent to the joint and the arc struck on this and then moved over on to the joint line.

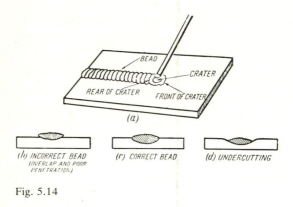

Fig. 5.14

It will be noticed that when welding on small masses of metal, the deposited bead rises high above the level of the metal when it is cold. As welding progresses, however, and the parent metal gets heated up, the penetration becomes deeper and the bead has a lower contour. For this reason the current setting can be higher when welding is commenced and then should be reduced slightly as the metal heats up, resulting in a bead having an even contour throughout its length.

This exercise should be continued until a straight, even, uniform bead about 250 mm long can be run with good penetration.

Next, a heavily coated rod should be tried. This bead has a much heavier slag deposit, because of the thick coating of the rod, and the operator will have difficulty at first in controlling the slag. He should notice that the molten pool has two distinct colours, one dark and the other light. The dark colour is the molten metal and the light colour the molten slag. Upon no account must the slag be allowed to get in front of the molten metal or blowholes will result, and also the dark-coloured portion must be kept continuous, or else this will result in the slag being entrapped in the metal, causing blowholes. The slag can easily be kept at the rear of the pool, and the best way of ensuring this is to progress at an even rate, keeping a constant arc length. A little practice will ensure excellent beads and, when the slag is chipped off, a bright shining layer of weld metal is exposed to view, completely free from oxidation. With some rods the weld practically deslags itself, and care should be taken that pieces do not fly up into the eyes. In any case chipping should be done, and is performed most easily when the weld has cooled down. The operator will have no difficulty with slag of medium-coated rods.

If a bead is to be continued after stopping, as for example to change an electrode, the end of the bead and the crater must be deslagged and brushed clean and the arc struck at the forward end of the crater. The electrode is then quickly moved to the rear end and the bead continued, so that no interruption can be detected. This should be practised until no discontinuity in the finished bead can be observed. Since in welding long runs, the welder may have to change rods many times, the importance of this exercise will be appreciated, otherwise a weakness or irregularity would exist whenever the welding operation was stopped.

Beads can now be laid welding away from the operator and also welding from left to right or right to left. A figure-of-eight about 120 mm long provides good practice in this and in changing direction of the bead when welding. The bead should be laid continuously around the figure. Little difficulty will be experienced in these exercises when using medium- or lightly-coated rods, but control of the slag will be found difficult at first when using heavily-coated rods. These exercises are useful, because in many cases of fabricated or repair work the welding has to be done in difficult positions and the above methods can be used to advantage.

Control of current and voltage

The operator can next familiarize himself with the effect of variation of current and voltage of the arc on the bead. This can be best observed by one operator calling out the reading of the meters while the other operator welds. The machine is set so as to give a voltage of 20 to 24 V across the arc, and using a medium-coated rod the current is set at the lowest value. There is poor penetration with a very shallow crater, and the metal heaps up on the plate, producing overlap. The sound of the arc is a splutter more than a crackle. The current is increased until at, say, 110 A using a 3 mm rod the crater is deeper (about 1.5 mm), giving good penetration, the metal flows well, the arc is very easily controlled, and the sound is a steady crackle. Increasing the current well above this produces an excessively deep crater, giving too much penetration (a hole will be blown through the plate if it is insufficiently thick), the arc is fierce and not so easily controlled, while the deposited bead is flat and the electrode becomes red hot. In addition, there is considerable spatter, and the noise of the arc is a loud crackle with a series of explosions (causing the spatter).

If welding with d.c. and the set has the usual voltage control fitted, now set the current at the correct value, as discovered in the previous exercise (say 110 A for the 3 mm rod), for the 3 mm rod being used, and set the voltage control on its lowest setting, giving say 12 V across the arc. The arc is very difficult, in fact almost impossible, to maintain, and consists of a spluttering in and out, while the metal is deposited in blobs on the plate. The rod tends to stick to the work and the crater formed is very shallow. Increase of voltage improves the arc until at, say, 20 to 24 V for the rod chosen, the weld has every good characteristic and is easily controlled. Increasing the voltage above this to, say, 30 to 35 V across the arc produces a noisy, hissing sound, little penetration and spatter, while the arc is difficult to control and tends to wander. These results are tabulated for convenience of reference.

Condition of welding circuit	Effect
Too low current	Poor penetration; shallow crater; metal heaps up on plate with overlap; arc has unsteady spluttering sound.
Too high current	Deep crater; too deep penetration; flat bead; fierce arc with loud crackle; electrode becomes red hot; much spatter.
Too low voltage	Rod sticks to work; arc difficult to maintain: spluttering sound as arc goes in and out; metal deposited in blobs with no penetration.
Too high voltage	Noisy hissing arc; fierce and wandering arc; bead tends to be porous and flat; spatter.
Correct voltage and current and welding speed	Steady crackle; medium crater giving good penetration; easily controlled stable arc; smooth even bead.

In a.c. welding the remarks regarding the current apply, but since there is no voltage control the remarks on this may be neglected. The good welder will, however, make a note of all these effects, since he may at any time be called upon to weld with a d.c. set.

Weaving

This may be attempted before or after the preceding exercise according to the inclination of the operator. Weaving is a side to side motion of the electrode, as it progresses down the weld, which helps to give better fusion on the sides of the weld, and also enables the metal to be built up or reinforced along any desired line, according to the type of weave used. It increases the dilution of weld metal with parent metal and should be reduced to a minimum when welding alloy steel.

There are many different methods of weaving, and the method adopted depends on the welder and the work being done. The simplest type is shown in Fig. 5.15*a*, and is a simple regular side to side motion, the circular portion helping to pile the metal in the bead in ripples. Fig. 5.15*b* is a circular motion favoured by many welders and has the same effect as Fig. 5.15*a*. Care should be taken with this method when using heavily coated rods that the slag is not entrapped in the weld. Fig. 5.15*c* is a figure-of-eight method and gives increased penetration on the lines of fusion, but care must again be taken that slag is not entrapped in the overlap of the weave on the edges. It is useful when reinforcing and building up deposits of wear-resisting steels. Fig. 5.15*d* is a weave that is useful when running horizontal beads on a vertical plate, since by the hesitating movement at the side of the bead, the metal may be heaped up as required. The longer the period of hesitation at any point in a weave, the more metal will be deposited at this point.

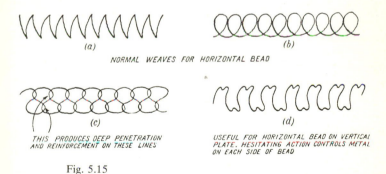

(a)

(b)

NORMAL WEAVES FOR HORIZONTAL BEAD

(c)

THIS PRODUCES DEEP PENETRATION AND REINFORCEMENT ON THESE LINES

(d)

USEFUL FOR HORIZONTAL BEAD ON VERTICAL PLATE. HESITATING ACTION CONTROLS METAL ON EACH SIDE OF BEAD

Fig. 5.15

Weaving should be practised until the bead laid down has an even surface with evenly spaced ripples. The width of the weave can be varied, resulting in a narrow or wide bead as required.

As this stage the d.c. welder may also observe the effect of polarity on the bead. Bare wire and lightly coated rods are difficult to deposit, get overheated, and give an uneven bead when connected to the + ve pole, whereas great improvement is noticed at the same current and voltage setting when connected to the − ve pole. Similarly, heavily-coated rods connected to the + ve pole run well when used with correct current and voltage setting, while if connected to the −ve pole they are more difficult to control and the metal does not flow well.

Since the polarity evidently depends on the type of coating, the welder should employ the polarity recommended for the rods being used. Vertical and overhead welding, together with the welding of cast iron and non-ferrous metals, and special steels, are performed with the rod normally positive.

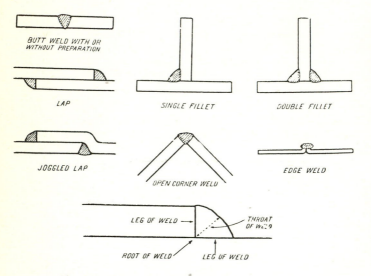

Fig. 5.16. Types of welded joints.

The operator should now be able to proceed to the making of welded joints, and it will be well to consider first of all the method of preparation of plates of various thicknesses for butt welding. Although a U pre-paration may prove to be more expensive to prepare than a V, it requires less weld metal and distortion will be lessened. For thicker sections the stringer bead technique is often used (Fig. 5.17) since the stresses due to contraction are less with this method than with a wide weave.

When a weld is made on a plate inclined at not more than 5° to the horizontal, this is termed the flat position, and wherever possible weld-ing should be done in this position, since it is the easiest from the welder's point of view.

Butt weld data
The following figures may be modified according to conditions, but are approximate for normal work.

Plate thickness (mm)	Method of preparation and gap (mm)	Electrode size (mm)	Number of runs
1.6 and 2.4	Butt, no gap	2.0	1
3.2	Butt, 1.6 gap	2.5 or 3.25	1
4.8	Butt, 3.2 gap	3.25 or 4	1
6.4	Butt, 3.2 gap	4	1
8 and 9.5	60 V 1.6 gap	4	2
12.5	60 V 3.2 gap	3.25, 4 or 5	3
19 and above	60 double V 1.6 gap or U with 4.8 gap	4, 5 or 6	3 or more

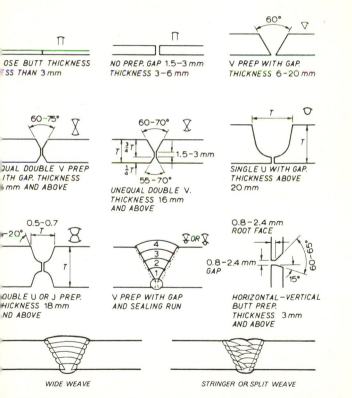

Fig. 5.17. Preparation of welded joints. Butt welds. U preparation reduces the volume of weld metal compared with V preparation. Weld width should not be more than the plate thickness.

Tack welding

Tack welds are essential in welding fabrication to ensure that there is correct line-up of the components to be fabricated. The tacks should be strong enough and of sufficient length and frequency of spacing that they ensure rigidity of the components and yet should be small enough, without need for metal removal, to be welded into the finished weld without any imperfections. Tacks should be made with the smallest electrode using a current some 10–20 A above that for normal welding, thus giving ease of striking. The tacks should be carefully deslagged and if there are any imperfections these should be removed and the weld chipped and remade.

Full data regarding the length of weld per foot of electrode, welding current, and tensile strength, are given with the particular electrodes being used, and therefore are not included here.

Butt welding

It will be seen from the above table that the size of electrode used will depend on the thickness of the plate being welded. Welds should be made first of all on 6.4 and 8 mm plate, preparing them correctly with a 60° V, and should be made both with and without weaving. They can be tested roughly for ductility and absence of blowholes by being bent in the vice, and they can also be tested for tensile strength in the testing machine, as described in Chapter 11.

The operator must get used to breaking open his weld and looking at it, observing any defects, as it is only in this way that good sound welds will be ensured. The effects of distortion can be countered by the methods given on pp. 135–6.

It is essential that penetration should be right through to the bottom of the V and that an 'underbead', as shown in Fig. 5.18, should be visible along the whole length of the weld on the underside of the plate. In many cases especially when making multi-run welds the first run is 'back chipped'. This entails removing some of the deposited metal on the underside of the first run generally by means of a chipping hammer pneumatically or electrically operated so that any defects such as inclusion and porosity are removed and a clean, defect-free surface is thus prepared for a sealing run to be applied.

Open-corner joints can be practised next, a good build-up being aimed at, together with fusion right through to the inside of the corner (Fig. 5.20). This exercise should be done by first tack welding the plates together at the correct angle. As a rough test of the ductility of the weld, absence of blowholes, and strength of the line of fusion, the joint can be flattened by closing the plates on each other. Weaknesses of a major character will then become apparent.

Butt welds on pipes provide good exercise in arc manipulation. The pipes are prepared in the same way as for plates by V'ing. They are then lined up in a clamp, or are lined up and tack welded in position and then placed across two V blocks.

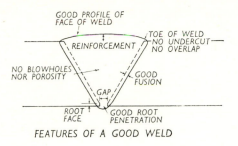

FEATURES OF A GOOD WELD

Fig. 5.18

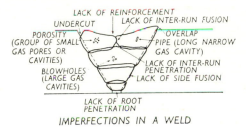

IMPERFECTIONS IN A WELD

Fig. 5.19

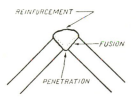

Fig. 5.20. Open corner joint.

Padding and building up shafts

This exercise provides a good test of continued accuracy in laying a bead. A plate of 8 to 9.5 mm mild steel about 150 mm square is chosen and a series of parallel beads are laid side by side across the surface of the plate and so as to *slightly overlap* each other. If the beads are laid side by side with no overlap, slag becomes entrapped in the line where the beads meet, being difficult to remove and causing blowholes (Fig. 5.21). Each bead is deslagged before the next is laid. The result is a built-up layer of weld metal.

After thoroughly cleaning and brushing all slag and impurities from this layer, another layer is deposited on top of this with the beads at right angles to those of the first layer (Fig. 5.22), or they may be laid in the same direction as those of the first layer. This can be continued for several layers, and the finished pad can then be sawn through and the section etched, when defects such as entrapped slag and blowholes can at once be seen.

SLAG TENDS TO GET ENTRAPPED HERE

BY OVERLAPPING THE BEADS THEY ARE EASIER TO DESLAG AND THUS NO BLOW-HOLES ARE FORMED

Fig. 5.21

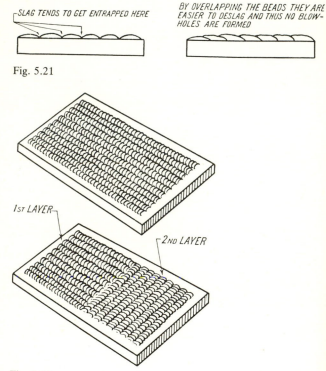

1ST LAYER

2ND LAYER

Fig. 5.22

Odd lengths of steel pipe, about 6 mm or more thick, may be used for the next exercise, which again consists in building up layers as before. The beads should be welded on opposite diameters to prevent distortion (Fig. 5.23). After building up two or more layers, the pipe can be turned down on the lathe and the deposit examined for closeness of texture and absence of slag and blowholes. Let each bead overlap the one next to it as previously mentioned – this greatly reduces the liability of pin-holes in the weld metal after being turned down.

The same method exactly is adopted in building up worn shafts (Fig. 5.24), and a bead may be run around the ends as shown to finish off the deposit. Another method sometimes used consists of mounting the shaft on V blocks and welding spirally. The operator should try both methods.

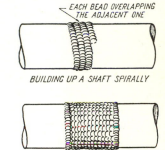

EACH BEAD OVERLAPPING THE ADJACENT ONE

BUILDING UP A SHAFT SPIRALLY

ALTERNATIVE METHODS OF LAYING SEQUENCE OF BEADS. EACH BEAD DESLAGGED BEFORE THE OVERLAPPING LAYER IS RUN

CIRCULAR BEADS TO END
NORMAL METHOD OF BUILDING UP

Fig. 5.23   Fig. 5.24

Lap joints

Preliminary exercises in welding lap joints may be made by tilting the plates so that the weld is flat, as shown in Fig. 5.25. The electrode should point to the centre of the V at 45° to the plate and 60° to the line of weld (Fig. 5.26). The correct penetration can be obtained in the lower plate which has the greater mass by causing the slight weave to hesitate slightly on this plate. Welds should be then made with the plates flat and the metal controlled so as to get a bead of good section. A wedge inserted at W will enable the joint to be broken open for inspection.

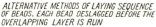

GOOD BUILT-UP SECTION

GOOD FUSION

W

ELECTRODE

60°

45°

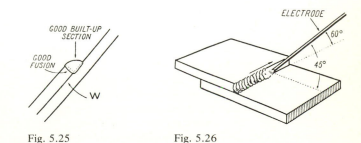

Fig. 5.25   Fig. 5.26

When a uniform regular bead can be obtained, specimens can be prepared and tested for strength in shear on the testing machine. Note that
a transverse weld at T in Fig. 5.27 is 30% stronger than the same length
of weld made longitudinally at L.

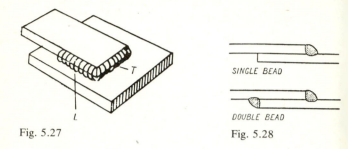

SINGLE BEAD

DOUBLE BEAD

Fig. 5.27   Fig. 5.28

Types of lap joints. No preparation is normally required for lap
joints, and the single joint is used in most cases, since it will stand most
loads. The double lap joint is used in cases where heavier loads will be
encountered (Fig. 5.28).

Fillet welding

Dimensions of fillet welds are indicated specifically by the terms 'leg
length' and 'throat thickness' (Fig. 5.30*a*) (see table at the beginning of
this chapter for the preferred metric sizes).

Welds made on an inside corner joint provide good practice for fillet
welding (Fig. 5.29), since the operator thereby gets used to holding and
controlling the arc in the more confined space between the two plates.

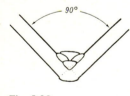

Fig. 5.29

Difficulty is often experienced in making good fillet welds having equal
legs and no undercutting. This is because there is a greater mass of metal
present near the weld than in a butt joint, and in the case of d.c. welding
arc blow may make the arc difficult to control. The weld must penetrate
to the bottom of the corner between the plates (Figs. 5.30*b*, 5.31 and
5.32), and to ensure this a short arc must be held and the speed of travel
must be slow, because of the greater mass of the plates to be heated. To

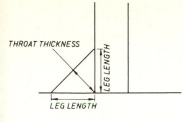

Fig. 5.30 (a)

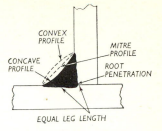

Fig. 5.30 (b)

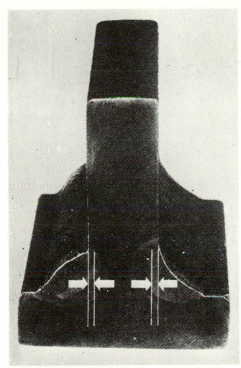

Fig. 5.31. A double fillet weld showing good penetration. Note the convex bead on left fillet and concave bead on right; this depends on the type of electrode used.

long an arc and too high a speed of travel will produce undercutting of the vertical plate (Fig. 5.33). The rod should point at an angle of 34–45° into the corner of the joint, and held at about 60° or steeper to the line of travel. In most cases weaving is not needed (Fig. 5.34a and b).

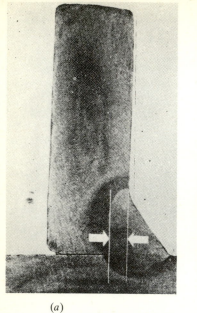
(a)

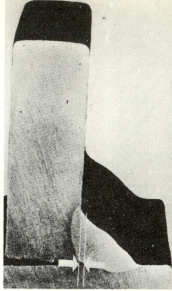
(b)

Fig. 5.32. (a) A fillet weld showing good penetration.
(b) Showing poor penetration at corner of fillet.

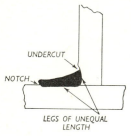

Fig. 5.33

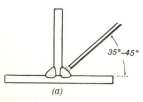
(a)

(b)

Fig. 5.34

Special electrodes for fillet welding help greatly in producing welds having uniform surface, and good penetration with no undercut. Control of the slag often presents difficulties. Keeping the rod inclined at about 60 ° as before stated helps to prevent the slag running ahead of the molten pool. Too fast a rate of travel will result in the slag appearing on the surface in uneven thicknesses, while too slow a rate of travel will cause it to pile up and flow off the bead. Observation of the slag layer will enable the welder to tell whether his speed is correct or not.

The plain fillet or tee joint (Fig. 5.35a) is suitable for all normal purposes and has considerable strength. The single V fillet (Fig. 5.35b) is suitable for heavier loads, and is welded from one side only.

In thick sections a thinner electrode is used for the first bead, followed by final runs with thicker rods, each run being well deslagged before the next is laid.

All practice welds should be either broken open or tested in the machine, as explained in Chapter 11.

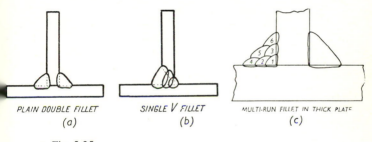

PLAIN DOUBLE FILLET (a)   SINGLE V FILLET (b)   MULTI-RUN FILLET IN THICK PLATE (c)

Fig. 5.35

Vertical welding

All welds inclined at a greater angle than 45 ° to the horizontal can be classed as vertical welds.

Vertical welding may be performed either upwards or downwards, and in both methods a short arc should be held to enable surface tension to pull the drop across into the molten pool. Electrodes 4 mm or smaller are generally used, and special rods having light coatings to reduce difficulties with slag are available. Vertical beads should first be run on mild steel plate, the electrode being held at 60–80 ° to the plate (Fig. 5.36a and b).

Downward welding (Fig. 5.36a) produces a concave bead, and is generally used for lighter runs, since a heavy deposit cannot be laid. If it is, the metal will not freeze immediately it is deposited on the plate, and will drop and run down the plate. This method is, therefore, usually only used as a finishing run over an upward weld because of its neat appearance, or for thin sections.

Upward welding (Fig. 5.36a) produces a convex bead, and is used on sections of above 6 mm thickness. The metal just deposited is used as a step on which to continue the deposit, and the slag flows away from the pool and does not hinder penetration as it does in the downward method.

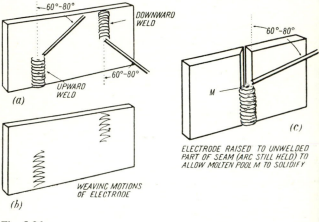

Fig. 5.36

In both methods accurate control of the molten metal by using the correct current setting, and keeping a short arc, is essential. In downward welding the weaving motion is exactly similar to that used in downhand welding, but care must be taken that slag does not flow and remain behind the electrode, or this results in blowholes. If it does lengthen the arc and melt it out, then shorten the arc again and continue. In upward welding, the same side to side motion is used and thus good penetration ensured both at the sides and the bottom of the V. Control of the metal in this method is best obtained by depositing some metal first of all, and then, just as it looks as if the molten metal is going to run out of the pool down the plate, the electrode is raised up the plate, out of the pool, without extinguishing the arc. Figure 5.36c will make this clear. In this way the heat is reduced, the metal is given time to solidify, and the next layer can be deposited. Progression is thus made in a series of layers. In wider joints on thicker plates, solidification takes place on one side of the weld, while the arc is being weaved to the other side, therefore the above-described method may be dispensed with and welding done in the ordinary manner.

Vertical fillet welds are made in exactly the same way as the butt welds just described, and welding is usually performed upwards. Practice welds should be broken open, so that the degree of penetration into the corner can be observed (Fig. 5.37).

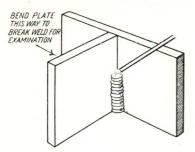

BEND PLATE THIS WAY TO BREAK WELD FOR EXAMINATION

Fig. 5.37. Vertical fillet weld.

Horizontal–vertical welding

This type of weld is extremely useful for many classes of work, as for example in boiler and firebox reinforcement. As with vertical welding, the greatest factor is correct current value, since if too much heat is introduced, the pool becomes too molten and runs down the side of the plate. A weaving motion is used, keeping a short steady arc, and one free from current variations. It will be noticed that the metal can be piled up at will on the top side or bottom side of the bead according to the time that the rod hesitates at the particular side. The student should aim at a deposit of uniform section, as in Fig. 5.38, having good penetration with no overlap.

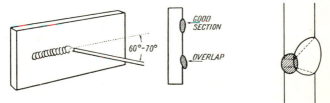

60°–70°

GOOD SECTION

OVERLAP

Fig. 5.38. Horizontal–vertical welding.

Fig. 5.39

In butt welds in this class, the joint may be prepared with only one-half V as shown (Fig. 5.39), leaving a lower edge as a step on which to weld. A small rod is used to ensure penetration through this type of joint with the first bead. The final bead is then a normal horizontal–vertical one.

In reinforcement of vertical surfaces, the horizontal layer may be first laid along the bottom of the surface and succeeding beads built up above this, using the lower bead in each case as a step. The next layer is then deposited with the beads at right angles to those in the first layer as in padding, this second layer consisting of normal vertical beads. An alternative method is to lay all the beads vertically, though this is scarcely so satisfactory.

Overhead welding

This takes a great deal of practice before the operator is able to deposit an even bead. Heavily coated rods must not be used, because of the trouble due to the continual dropping of slag. Medium-coated rods of 4 mm or smaller are generally used, and electrodes recommended for overhead use are most suitable. The most important points about overhead welding are (1) correct control of the current, (2) a very short arc. Correct current control gives a pool that is sufficiently molten to ensure good penetration, but that does not contain enough molten metal to cause it to drip down, while the short arc enables the molten globules to be pulled upwards, against the force of gravity, into the molten pool by surface tension.

Carbon arc welding

The carbon arc (as used in the electric arc furnace) can be used for welding and pre-heating, and is especially useful for the welding of thin sheets without the use of a filler rod.

In carbon arc welding the holder grips the carbon close to the tapered end so as to avoid loss of carbon by vaporization, and since great heat is evolved, gauntlet gloves should be worn. Maximum currents vary from 50 A for 4 mm to 350 A for 8 mm diameter carbons and all types of joints can be welded by this method, the method being similar to oxy-acetylene welding, using the carbon arc in place of the flame, the metal being supplied by a filler rod.

The carbon is connected to the negative pole, since this reduces the amount of carbon introduced into the weld. When welding sheet metal the welding should be done quickly and almost daintily to ensure a neat weld. As with metallic arc welding, practice runs should be made on flat sheet so as to obtain the sense of fusion. Runs may be made first of all without a filler rod, fusion being obtained without melting through the plate. When this is practicable, a bead can be laid using a filler rod.

Thin sheets are prepared for butt welding by flanging the edges. These are then fused together without the use of a filler rod, holding the carbon at right angles to the work. The magnetic field is apt to cause trouble some arc blow when using low currents, and in certain holders for thin gauge work this is prevented by the coil in the head of the holder.

Thicker plates are prepared as usual for metallic arc welding, and the weld is performed with the carbon at right angles to the plate, the filler rod being melted into the joint by holding its end in the molten pool (as in oxy-acetylene welding).

Paste-type fluxes (or autogenizers) can be used as in automatic carbon arc welding, and these help to produce welds having better characteristics.

When the carbon arc is used for pre-heating, it is connected to the −ve pole as before, and the arc moved about over the area to be heated. This heats the whole surface without raising any part of the molten condition.

Cast iron can be welded with the carbon arc. A cast-iron filler rod is used and a flux of the borax type is helpful in producing a sound weld. The welding is done in a similar manner to that explained previously for steel plate, but as usual great care must be taken with the heating and cooling of the casting for fear of cracking.

This method is very like the oxy-acetylene method of welding cast iron, the filler rod being used to float any oxide to the surface. The welds made are fairly machinable, and the deposited metal is stronger than the parent casting.

All grades of tubular granulated tungsten carbide hard surfacing materials as used for oxy-acetylene hard surfacing may be deposited with the carbon arc. The carbon is connected to the +ve pole of a d.c. generator and a current of about 200 A is used with a 6.4 mm diameter carbon. The parent plate should be pre-heated if possible to prevent cracking and the carbon weaved as the electrode is applied, as in oxy-acetylene practice, to obtain an even layer.

Cutting with the carbon arc. The carbon arc, owing to its high temperature, can be used for cutting steel. A high current is required, and the cut must be started in such a spot that the molten metal can flow away easily. The cut should also be wide enough so that the electrode (of carbon or graphite) can be used well down in it, especially when the metal is thick, so as to melt the lower layers. Cast iron is much more difficult to cut, since the changing of the iron into iron oxide is not easily performed owing to the presence of the graphite.

Arc cutting does not produce anywhere near such a neat cut as the flame cutter, and because of this is only used in special circumstances.

Aluminium and aluminium alloys[1]

See p. 132 for explanation of symbols and code letters.

Aluminium can be arc welded using rods coated with fluxes consisting of mixtures of fluorides and chlorides. The flux dissolves the layer of oxide (alumina) on the surface of the metal, and also prevents oxidation during welding. The heat of the arc produces rapid melting and, as a result, beginners find a considerable difficulty in managing the arc, since it is so different from the 'steel' arc. With practice, however, and by following the correct procedure, excellent welds can be made.

Preparation. Sheets thinner than 6.6 mm can be butted together with no preparation, but with a gap between them for penetration. A backing strip of copper is always advisable in welding aluminium, to prevent collapse, since the metal is so weak at high temperatures. Above 6.6 mm thick sheets and castings are prepared with the usual 60° V, as for steel, and the parts should either be tack welded or held in position with clamps or fixtures. The surface must be thoroughly cleaned before attempting to weld (Fig. 5.40). In many cases the welding can be made easier by sprinkling a little of the powdered flux, used for the oxy-acetylene welding of aluminium, along the line of the weld. This helps in the removal of the oxide. Molten aluminium absorbs hydrogen and this results in porosity, so the electrodes should be kept very dry and, if damp, should be heated to 130–160 °C before use.

COPPER BACKING STRIP GROOVED TO ALLOW
REINFORCEMENT

(a) LESS THAN 6 mm THICK

(b) OVER 6 mm THICK

Fig. 5.40

The following types of electrodes are available (BS 1616): pure aluminium (G1C); aluminium–5% silicon (NG21); aluminium–10–12% silicon (NG2); aluminium–7.5–9% silicon; aluminium–1¼% manganese (NG3); aluminium–6% magnesium (NG6). In addition core wires con-

---

[1] *Note*. BS 2901 (1970) gives specifications for *filler rods and wires for inert gas arc welding. Part 1, Ferritic steels; Part 2, Austenitic stainless steels; Part 3, Copper and copper alloys; Part 4, Aluminium and aluminium alloys; Part 5, Nickel and nickel alloys.*

taining copper as the alloying element are suitable for welding the aluminium–copper alloys.

If the alloy to be welded contains less than 2% of copper, silicon, manganese, or magnesium, it should be welded with an aluminium–5% silicon (NG21) electrode. In the case of the aluminium–magnesium alloys, since magnesium is lost in the transference through the arc, they should be welded with electrodes having a higher magnesium content than the parent plate; thus for a weld in N4 ($2\frac{1}{2}$% Mg), NG6 (5% Mg) electrodes should be used.

Aluminium castings containing silicon as the principal alloying element, e.g. LM4, LM6, LM8, LM9, LM18, LM20, can be welded very satisfactorily with the Al–Si rods.

Some castings of different composition from the preceding types may be difficult to weld because of cracking tendencies. Pre-heating and use of a rod of the same composition as the casting often proves successful.

Technique. The rod is connected to the positive pole of a d.c. supply, and the arc struck by scratching action, as explained for mild steel. It will be found that, as a layer of flux generally forms over the end of the rod, it has to be struck very hard to start the arc. The rod is held at right angles to the work and a *short* arc must be held (Fig. 5.41), keeping the end pushed down into the molten pool. This short arc, together with the shielding action of the coating of the rod, reduces oxidation to a minimum. A long arc will result in a weak, brittle weld. No weaving need be performed, and the rate of welding must be uniform. As the metal warms up, the speed of welding must be increased. Pre-heating of wrought aluminium alloys to about 200 °C reduces the possibility of cracking.

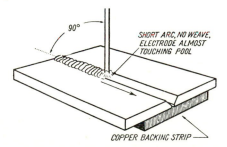

Fig. 5.41. Aluminium welding.

Castings are welded in the same way after preparation, but owing to their larger mass, care must be taken to get good fusion right down into the parent metal, since if the arc is held for too short a time on a given portion of the weld the deposited aluminium is merely 'stuck' on the surface as a bead with no fusion. This is a very common fault. Castings should be pre-heated to 200 °C to reduce the cracking tendency and to make welding easier.

Lap joints and fillet joints should be avoided since they tend to trap corrosive flux, where it cannot be removed by cleaning (Fig. 5.42). Fillet welds are performed with no weave and with the rod bisecting the angle between the plates.

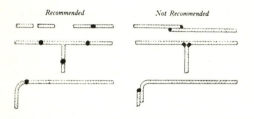

Fig. 5.42. Joints (*a*) recommended and (*b*) not recommended unless special precautions are taken to avoid corrosive fluxes.

After treatment. The flux used is very corrosive and the weld must be thoroughly washed and brushed in hot water after it has cooled out. Immersion in a 5% solution of nitric acid in water is an even better method of removing the flux, this being followed by brushing and washing in hot water.

Cast-iron welding

Cast iron can be welded with the electric arc without the necessity of pre-heating, and this makes the process extremely useful, since much time and expense is thereby saved.

Preliminary considerations. Two types of electrode are generally used for cast-iron welding:

(1) Mild steel, or steel base containing alloying elements.
(2) Nickel, nickel–copper, nickel–iron with high carbon content to assist fluidity and to counter silicon pick-up from the casting.

When steel base weld metal is deposited on cold cast iron, quick cooling results, due to the large mass of cold metal near the weld. This quick cooling results in much of the carbon in an area adjacent to the weld being retained in the combined form (cementite), and thus hardened zone exists near the weld. In addition, the steel weld metal absorbs carbon and the quick cooling causes this to harden also. As result welds made with this type of rod have hard zones and cannot always be machined. If, however, the cooling is made as gradual as possible, a good high-speed steel tool will generally cut quite satisfactorily. In many cases, however, machining is not necessary and therefore the previous drawback is no disadvantage. This type of weld has about three times the strength of the parent metal, and steel base rods in particular give good fusion with the cast iron. Rods up to 3.2 mm diameter are generally used, and a low current ensures the minimum of heat being introduced into the work.

Nickel and nickel alloy electrodes have reduced carbon pick-up and therefore reduce hardening effect. The welds are therefore machinable, though as stated above, a hardened zone still exists near the weld. Where pre-heating and slow cooling are possible the liability to crack is reduced and the hardening effect much less with both types of rods. Pre-heating may be done whenever the casting is of complicated shape and liable to fracture easily, though, with care, even a complicated casting may be welded satisfactorily without pre-heating if the welding is done slowly.

The nickel and nickel alloy electrodes are also used for the welding of SG cast irons, but the heat input will affect the pearlitic and ferritic structures in the heat-affected zone, precipitating eutectic carbide and martensite in a narrow zone at the weld interface even with slow cooling. For increased strength, annealing or normalizing should be carried out after welding. The lower the heat input the less the hardening effect in the HAZ.

Preparation. Cracks in thin casting should be V'd, or better still U'd, as for example with a bull-nose chisel. Thicker castings should be prepared with a single V below 9.5 mm thick and a double V above this. Studding (see p. 348) can be thoroughly recommended for thicker sections. The surrounding metal should be well cleaned. The polarity of the electrode depends on the rod being used and the maker's instructions should be followed, though it is generally +ve. With a.c. an open circuit voltage of 80 V is required and the transformer should be set to give this. Since the heat in the work must be kept to a minimum, a small-gauge electrode, with the lowest current setting that will give sufficient penetration, should be used. A 3.2 mm rod with 70 to 90 A is very suitable for many classes of work. Thick rods with correspondingly heavier currents may be used, but are only advisable in cases where there is no danger of cracking. Full considerations of the effect of expansion and contraction must be given to each particular job.

Technique. The rod is held as for mild steel, and a slight weave can be used as required. Short beads of about 50 to 60 mm should be run. If longer beads are deposited, cracking will occur unless the casting is of the simplest shape. In the case of a long weld the welding can be done by the skip method, since this will reduce the period of waiting for the section welded to cool. It may be found that with steel base rods, welding fairly thin sections, fine cracks often appear down the centre of the weld on cooling. This can often be prevented, and the weld greatly improved, by peening the weld immediately after depositing a run with quick light blows with a ball-paned hammer. If cracks do appear a further light 'stitching' run will seal them. Remember that the cooler the casting is kept, the less will be the risk of cracking, and the better the result. Therefore take time and let each bead cool before laying another. The weld should be cool enough for the hand to be held on it before proceeding with the next bead. In welding a deep V, lay a deposit on the sides of the V first and follow up by filling in the centre of the V. This reduces risk of cracking. If the weld has been prepared by studding (q.v.), take care that the studs are fused well into the parent metal.

In depositing non-ferrous rods, the welding is performed in the same way, holding a short arc and welding *slowly*. Too fast a welding speed results in porosity. In many cases a nickel–copper rod may be deposited first on the cast iron and then a steel base rod used to complete the weld. The nickel–copper rod deposit prevents the absorption of carbon into the weld metal and makes the resulting weld softer. Where a soft deposit is required on the surface of the weld for machining purposes, the weld may be made in the ordinary way with a steel base rod and the final top runs with a non-ferrous rod. The steel base rod often gives a weld which has hard spots in it that can only be ground down, hence this weld can never be completely guaranteed machinable.

Studding

We have seen that whenever either steel base or non-ferrous rods are used for cast iron welding, there is a brittle zone near the line of fusion, and since contraction stresses are set up, a weakness exists along this line. This weakened area can be greatly strengthened in thick section castings by studding. Welds made by this method have proved to be exceptionally strong and durable. Studding consists of preparing the casting for welding with the usual single or double V, and then drilling and tapping holes along the V and screwing steel studs into the holes to a depth slightly deeper than the diameter of the studs. Studs of 4.8 mm diameter and larger are generally used, depending on the thickness of the casting, and they must project about 6 mm above the surface. The number of studs can be such that their area is about $\frac{1}{5}$ to $\frac{1}{4}$ of the area of the weld, though a lesser number can be used in many cases (Fig. 5.43). Welding is performed around the area near each stud, using steel or steel-base electrodes, so as to ensure good fusion between the stud and the parent casting. These areas are then welded together with intermittent beads, as before explained, always doing a little at a time and keeping the casting as cool as possible. This method should always be adopted for the repair, by arc welding, of large castings subjected to severe stress. An alternative method is to weld steel bars across the projecting studs as additional reinforcement.

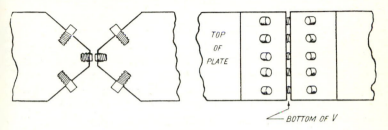

Fig. 5.43. Method of studding.

Copper and bronze welding

Copper can be welded with the d.c. arc, using coated electrodes of copper or bronze. It can also be carbon arc welded, though this is seldom used.

If the weld must have the same characteristics as the parent metal, as for example for electrical conductivity, coated electrodes of deoxidized copper may be used, but great difficulty is encountered owing to the resulting weld being porous. For this reason at present the oxy-acetylene, TIG and MIG processes are much to be preferred for this type of work.

Bronze welding

When the weld on copper need not have the same characteristics as the parent metal, covered copper alloy or bronze electrodes will ensure sound welds, and the process is then termed bronze welding. These rods have a lower melting point than copper. Bronze may be laid on copper, brass or bronze and steel with the same ease, and its application to the welding of cast iron has already been considered. It is also useful for the welding of galvanized sheets, as it has a minimum amount of disturbance to the zinc coating. The following description applies also to the use of a copper rod used for copper welding.

Preparation. Plates below 3 mm thickness need no preparation, but above this they are prepared as for mild steel with a 60° V and the surfaces thoroughly cleaned. The work must be well supported during welding, since copper especially is very weak when red hot.

Asbestos sheet strips can be used as backing bars, this being more essential when using a copper electrode than when using a bronze one. Heavy sections above 5 mm must be pre-heated, because of the high thermal conductivity of the metal, and this can be done either by the carbon arc (connected to the −ve pole), the oxy-acetylene flame or furnace.

Technique. The rod should be connected to the +ve pole except where otherwise stated. The current value is generally the same or slightly less than that for the same gauge mild-steel electrode, with an arc voltage drop of 20 to 25 V. The actual values will depend on the particular job. The electrode should be held steeply to the line of weld, and a short arc held keeping the electrode well down into the molten pool. The weld must be made *slowly*, since a quick rate of welding will produce a porous deposit even when bronze rods are used. As welding proceeds and the part heats up, it is usually advisable to reduce the current, or there is a liability to melt through the weld, especially in thinner sections.

After treatment. Light hammering, while still hot, greatly improves the structure and strength of the weld. The extent to which this should be done depends on the thickness of the metal.

The carbon arc when used to weld copper tends to produce porous welds, and cannot therefore be recommended.

Welding of copper–silicon alloys, such as Everdur

Everdur is an alloy of copper 96% and silicon 3% with additions of Al, Fe, Mn, Sn or Zn, and is typical of this range of alloys. Preparation is similar to that for mild steel.

The rod must be of the same composition as the parent metal and can be coated with a flux made of 90% borax and 10% sodium fluoride. The rod should be connected to the +ve pole and the welding done with the rod held almost vertically to the line of weld. The weld has properties similar to those of the parent metal.

Inconel, 80% nickel, 12–14% chromium, balance mainly iron is welded in exactly the same way, using flux-coated inconel rods.

Aluminium bronze welding

Aluminium bronze containing 5–10% aluminium and the remainder copper can be arc welded easily using aluminium bronze electrodes having an approximate composition 10% Al, 4% Fe and 4% Ni. These rods are suitable for fabrication and repair of high tensile aluminium bronzes. In arc welding the danger of porosity due to absorption of gas and the risk of cracking due to hot shortness is not as great as with gas welding.

The coverings on the electrodes are of fluoride of sodium or potassium; or silicon bonded by sodium or potassium silicate. They should be stored in a dry place as dampness considerably affects them. Drying at 70–100 °C may be carried out if they get damp.

Preparation. Sections up to 5 mm can be butted with a small gap. Over 5 mm the usual 60° V preparation is given, while over 16 mm a double V is required.

The edges of the joints *must* be cleaned of all oxide and scale by filing and brushing, this operation being essential to the production of a good weld. Up to 6 mm thick the electrode diameter should be roughly the same as the plate thickness.

Technique. A d.c. supply is required with the electrode +ve, the rod being held nearly vertically to the weld. It is found advantageous to give, in addition to the usual side to side weave, a small up and down motion of the electrode so that the molten globules are, as it were, helped along into the molten pool before they get too large and are in danger of slight oxidation. Welding rate is faster than on mild steel.

When recommencing welding during a run, the arc should be struck a little way away from where the start is required and then moved over to where the weld must continue. In this way greater heat is obtained at the start and porosity minimized.

Single runs are best if possible, since in multiple layers each must be thoroughly descaled and pre-heated prior to welding, otherwise gas and slag inclusions may result. Arc length is not very important but a short arc is usually preferable with the up and down motion, while current value depends upon the electrodes used and the type of work.

Flux may be cleaned off after welding with a weak sodium or potassium hydroxide (caustic soda) solution.

Hard surfacing

The advantages of hard surfacing are that the surfaces can be deposited on relatively much cheaper base metal to give the wear-resistant or other qualities exactly where required, with a great saving in cost, and in addition built-up parts save time and replacement costs. The chief causes of wear in machine parts are abrasion, impact, corrosion and heat. In order to resist impact the surface must be sufficiently hard to resist deformation yet not hard enough to allow cracks to develop. On the other hand, to resist abrasion a surface must be very hard and if subject to severe impact conditions cracking may occur, so that in general the higher the abrasion resistance the less the impact resistance, and evidently it is not possible to obtain a surface which has the highest values of both impact and abrasion resistance. In choosing an electrode for building up surfaces consideration must be given as to whether high abrasion or high impact resistance is required. High-impact electrodes will give moderate abrasion resistance and vice versa, so that the final choice must be made as to the degree of (1) hardness, (2) toughness, (3) corrosion resistance, (4) temperature of working, (5) type of base metal and whether pre-and post-heating is possible. The main types of wear- and abrasion-resistant surfacing electrodes are:

(1) Fused granules of tungsten carbide in an austenitic matrix. The deposit has highest resistance to wear but is brittle and has medium impact strength. The electrodes are tubular and there-fore moisture-resistant with high deposition rates.

(2) Chromium carbide. These basic-coated electrodes deposit a dense network of chromium carbide in an austenitic matrix and have high resistance to wear with good impact resistance.

(3) Cobalt–chromium–tungsten non-ferrous alloys have a high carbon content, are corrosion-resistant, have a low coefficient of friction and retain their hardness at red heat.

(4) Nickel-base alloys containing chromium, molybdenum, iron and tungsten have good abrasion resistance and metal-to-metal impact resistance. They work-harden and have resistance to high-temperature wear and corrosive conditions.

(5) Air-hardening martensitic steels. These have a high hardness value due to their martensitic structure and there is a variety of electrodes available in this group. Dilution plays an important part in the hardness of the deposit, as with all surfacing applications. A single run on mild steel may be only a little harder than the parent plate, but if deposited on carbon or alloy steel, carbon and alloy pick-up greatly increases the hardness of the deposit.

(6) Austenitic steels: (1) 12/14% manganese deposits develop their hardness by cold work-hardening so that the deposit has the strength of its austenitic core with the hard surface. With approximately 3% Ni there is reduced tendency to cracking and brittleness due to heat compared with plain 12/14% Mn weld metal. (2) Chromium–nickel, chromium molybdenum nickel and chromium–manganese deposits work-harden and give resistance to heavy impact.

Thus, in general, low alloy deposits give medium abrasion with high impact resistance, medium alloy deposits give high abrasion and medium impact properties.

Austenitic, including 13% manganese, deposits give high impact resistance and work-harden as a result of this impact and work to give abrasion-resistant qualities. Of the carbides, chromium deposits give the best abrasion and impact properties while tungsten has the hardest surface and thus the highest resistance to abrasion with medium impact properties.

The table overleaf gives a selection of electrodes available.

Preparation and technique

The surface should be ground all over and loose or frittered metal removed. Because of the danger of cracking, large areas should be divided up and the welding done in skipped sections so that the heat is distributed as evenly as possible over the whole area. Sharp corners should be rounded and thick deposits should be avoided as they tend to splinter or spall. The first runs on any surface are subject to considerable dilution, so it is advisable to lay down a buffer layer using a nickel-based or austenitic stainless steel electrode, especially when welding on high carbon or high alloy steels. The buffer layer should be chosen so as to be of intermediate hardness between the parent metal and the deposit, or two layers should be used with increasing hardness if the parent metal is very soft compared with the deposit.

Mild pre-heating to 150–200 °C is advantageous if the base metal has sufficient carbon or alloying elements to make it hardenable but is not usual on steel below 0.3% C or stainless steel. If the base metal is very hard and brittle, slow pre-heating to 400–600 °C with slow cooling after welding may be necessary to prevent the formation of brittle areas in the HAZ. Below a hardness value of 350 HV, surfaces are machinable generally with carbide tools but above this grinding is usually necessary. In depositing a surface on manganese steel there should be no pre-heating or stress relieving and the electrode should be connected to the +ve pole. Use only sufficient current to ensure fusion, keep a short arc, hold the electrode as in welding mild steel and introduce as little heat as possible into the casting by staggering welding so that the temperature does not rise much above 200 °C. Austenitic stainless steel electrodes should be used for joining broken sections as these electrodes work-harden in the same way as the parent metal and 13% manganese, or 13% manganese, 3.5% nickel electrodes used for building up.

Dilution is about 25–35% in the first layer, and to obtain a dilution of about 5% at least three layers must be laid down with a thickness of 6–10 mm.

Note. The manual metal arc method of surfacing is rather slow compared with semi-automatic and automatic processes such as MIG, submerged arc and hot wire arc plasma so that it is best suited for smaller areas and complex shapes.

Type (R – rutile) (B – basic) (T – tubular)		Hardening	HV	Use	Abrasion	Impact
					H – high M – medium	
pearlitic R	Cr–Mn	air	250	Used also for buffer layers.	M	M
martensitic R	Cr–Mn	air	350	For and with buffer layers.	M	H
martensitic R	Cr–Mn–Mo	air	650	After buffer layers.	H	M
martensitic B	Cr–Mo–V	air	700	After buffer layers for heavy reinforcement.	H	M
martensitic B	Cr–Mo with borides	air	800	Surfacing generally restricted to two layers.	H	M
austenitic B	Cr–Mo–Ni	work	250 500	For joining and depositing on 13% Mn steel and joining this steel to carbon steel.	M	H
austenitic B	Cr–Ni–W–B	work (slight)	450 500	Hard at elevated temperatures, corrosion-resistant.	M	M
austenitic B	13% Mn	work	170 500		H	M
austenitic T	Cr–Mn	work	300 480	For reinforcing 13% Mn steel castings and as buffer layer prior to.	M	H
non-ferrous B	Co–Cr–W		630	Red hard and corrosion-resistant, various grades.	H	M
austenitic T matrix	Chromium-carbide–Mn		560 matrix 1400 carbides	Heat resistant to about 1100 °C.	H	H
non-ferrous T matrix	tungsten-carbide		600 matrix 1800 carbides	Used at all temperatures.	H	M

Corrosion-resistant surfacing

Nickel and its alloys can be laid as surfaces on low carbon and other steel to give corrosion- and heat-resistant surfaces. Electrode diameter should be as large as possible, with minimum current compatible with good fusion to reduce dilution effects. The first bead is laid down at slow speed and subsequent runs overlapped with minimum weaving, reducing dilution to 15–20% compared with 25–35% for the first bead. Subsequent layers should be put down with interpass temperature below 180 °C to minimize dilution and avoid micro-fissures in the deposit. Dilution is reduced to about 5% after three layers with 6–10 mm thickness. Suitable electrodes are in the nickel, Monel, Inconel, Incoloy and Incoweld range, for which the student should refer to manufacturers' instructions.

Tipping tool steel

Cutting tools for lathes, milling machines and high-speed cutting tools of all types can be made by depositing a layer of high-speed steel on to a shank of lower carbon steel. Special electrodes are made for this purpose, and give very good results.

The surface to be tipped is ground so as to receive the deposit. The electrode is connected to the +ve pole when d.c. is used, and held vertically to the line of weld, a narrow deposit being laid as a general rule. More than one layer is generally advisable, since the first layer tends to become alloyed with the parent metal. For this reason the current setting should be as low as possible, giving small penetration. Use a very narrow weave so as to prevent porosity, which is very usual unless great care is taken. Each bead should be allowed to cool out and then be deslagged before depositing the next. The deposited metal when allowed to cool out slowly usually has a Brinell hardness of 500 to 700, depending upon the rod used. The hardness can be increased by heat treatment in the same way as for high-speed tool steel, and the deposit retains its hardness at fairly high temperatures.

These types of rods are very suitable for depositing cutting edges on drills, chisels, shearing blades, dies and tappets, etc. Figure 5.44 shows the method of depositing the surface on a lathe tool.

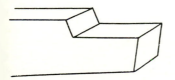

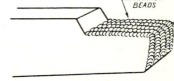

TOOL PRIOR TO DEPOSIT   TOOL TIPPED AND READY FOR GRINDING

Fig. 5.44

Stelliting

Stellite is an alloy of cobalt, chromium, tungsten and carbon, and when deposited on steel, steel alloy or cast iron gives a surface having excellent wear-resisting qualities and one that will stand up well to corrosive action. It preserves its hardness of surface even when red-hot (650°C), and is thus suitable for use in places where heat is likely to be generated.

It may be deposited very satisfactorily with the arc, though the deposit is not as smooth as one deposited with the oxy-acetylene flame. The arc method of application, however, is specially recommended where it is essential not to introduce undue heat into the part, due to danger of warping or cracking. In many cases, especially where the part has large mass, stelliting by the arc saves time and money compared with the flame. Bare rods of stellite may be used when welding with d.c. and are connected to the +ve pole. If a.c. is employed covered rods must be used, and better results are also obtained with covered rods when using d.c., since the deposit is closer grained and the arc more stable.

Covered rods can be obtained or bare rods may be fluxed with a covering of equal parts of calcium carbonate (chalk), silica flour and either borax or sodium carbonate (baking soda), mixed with shellac as a binder.

Preparation. The surface to be stellited must be thoroughly cleaned of all rust and scale and all sharp corners removed. A portable grinder is extremely useful for this purpose. In some cases, where the shape is complicated, pre-heating to prevent cracking is definitely an advantage.

Technique. In d.c. the rod is connected to the +ve pole and approximate currents are 120 to 140 A for 3 mm rod or 200 A for 6 mm rod. Higher currents are used with a.c., namely, 150 to 180 A for 3 mm rod and 280 A for 6 mm rod. A slightly longer arc than usual is held, with the rod nearly perpendicular to the surface, as this helps to spread the stellite more evenly. Care must be taken not to get the penetration too deep, otherwise the stellite will become alloyed with the base metal and a poor deposit will result. Since stellite has no ductility, cooling must be at an even rate throughout to avoid danger of cracking. The surface is finally ground to shape. Lathe centres, valve seats, rock drills and tool tips, cams, bucket lips, dies, punches, shear knives, valve tappet surfaces, thrust washers, stillson teeth, etc., are a few examples of the many applications of hard surfacing by this method.

Stainless steels

Note on BS 2926. This standard includes the chromium–nickel austenitic steels and chromium steels and uses a code by which weld metal content and coating can be identified.

The first figure is the % chromium content, the second figure the % nickel content and the third figure the % molybdenum content. The letter L indicates the low carbon version, Nb indicates stabilization with niobium, W indicates that there is tungsten present. R indicates a rutile coating, usually either d.c. or a.c. and B a basic coating, usually d.c. electrode +ve only. A suffix MP indicates a mild steel core.

For example: 19 .12 .3 .Nb .R is a niobium-stabilized 19% Cr, 12% Ni, 3% Mo rutile-coated electrode.

Stainless steels are welded with very much the same technique as low carbon steels, but because there is such a variety of them care must be exercised in choosing the right type of electrode for the parent plate. The coefficient of expansion is much greater than mild steel so that distortion effects are greater and the thermal conductivity is lower, so that HAZ is narrower and the heat of the arc more localized. They have a high electrical resistivity so that the current for a given electrode diameter must be kept within its current-carrying limits to prevent overheating.

The chief groups (discussed on pp. 85–6) are the martensitic and ferritic plain chromium steels of group A and the austenitic chromium nickel steels of group B.

Martensitic steels contain 12–16% chromium and harden when welded. They should be pre-heated to 200–300 °C, allowed to cool slowly after welding and then post-heated to 700 °C to remove brittleness and prevent cracking. If the weld metal must match the parent metal, basic electrodes of the 13% Cr type should be used for very limited applications only.

By using an austenitic electrode of the 29 .12 .3 .L type the join is more ductile and free from cracking.

Ferritic steels contain 16–30% Cr, do not harden when welded but suffer from grain growth when heated in the 950–1100 °C range, so that they are brittle at ordinary temperatures but may be tougher at red heat. Pre-heating to 200 °C should be carried out and the weld completed, followed by post-heating to 750 °C. For mildly corrosive conditions, electrodes of matching or higher chromium content should be used, e.g. 26.5.1·5, which give a tough deposit. If joint metal properties need not match the parent plate electrodes of 26.20 type can be used. It should be noted that the weld deposit of these electrodes contains nickel which is attacked by sulphurous atmospheres, so they are only suitable for mildly corrosive conditions. Austenitic chromium nickel steels comprise the largest and most important group from the fabrication point of view. They do not harden when welded because of their austenitic structure, the largest group being the 18% Cr 8% Ni type with other groups such as 25% Cr 20% Ni and 18% Cr 12% Ni. Other elements such as molybdenum are added to make them more acid-resistant, and they are available with rutile coatings suitable for d.c. electrode +ve or a.c. minimum OCV 55–80 V depending upon the electrode; or with basic coatings usually for d.c. only, electrode +ve, these being especially suitable for vertical and overhead positions. The table opposite gives a selection of electrodes available.

Stainless steel butt weld preparation

Plate thickness	Preparation		Suggested electrode diameter (mm)	
1.2–1.6	square edge	⎫	1.6	
2.0	square edge	⎬ gap 0–1.6 mm	2.0	
2.5	square edge	⎭	2.5	
3.2	single 60° V	⎫ 1.5–3.2 mm	2.5	
5	single 60° V	⎬ root face	3.2	
6.3	single 60° V	⎭ and gap	3.2 and 4	
10–20	single 60° V or double 70° V		4 and 5	
over 20	double 70° V or double U		4 and 5	

GAP 0–1.6 mm
60°
ROOT FACE AND G...
1.5–3.2 mm
70°
2–3 mm ROOT FACE
0–2 mm ROOT FACE
0–2 mm ROOT GAP
10°–15°
DOUBLE V
DOUBLE ...

Nickel and nickel alloys

The welding of nickel and its alloys is widely practised using similar technique to those used for ferrous metals. The electrodes should be dried before use by heating to 120 °C or, if they are really damp, by heating for 1 to 2 hours at about 260 °C. Direct current from generator or rectifier with electrode positive gives the best results. A short arc should be held with the electrode making an angle of 20–30° to the vertical and when the arc is broken it should first be reduced in length as much as possible and held almost stationary, or the arc can be moved backwards over the weld already laid and gradually lengthened to break it. This reduces the crater effect and reduces the tendency to oxidation. The arc should be restruck by striking at the end of a run and moving quickly back over the crater, afterwards moving forward with a slight weave over the crater area, thus eliminating starting porosity.

Fig. 6.20 shows the most satisfactory joint preparation and it should be remembered that the molten metal of the nickel alloys is not as fluid

Stainless steel manual metal arc welding electrodes

Approximate composition (%)			Coating	Carbon (%)	Ferrite (%)	Applications
Cr	Ni	Mo				
19	9	–	R	0.03	6	For austenitic stainless steels, a few air-hardening steels and manganese steels.
19	9	1.5	R	0.03	8	For austenitic stainless steels of the 18/8/1.5 type and for air-hardening steels such as armour plate; for certain manganese steels and welding stainless steel to unalloyed steel.
19	9	Nb	R	0.04–0.06	6	For plain and Ti and Nb stabilized stainless steels.
19	9	Nb	B	0.05–0.08	6	As preceding electrode. Light slag, flat weld profile. Very suitable for vertical and overhead welding and multi-positional applications such as pipe work.
19	12	3	R	0.04–0.06	6	For welding unstabilized 3% molybdenum bearing stainless steel.
19	12	3 Nb	B	0.05–0.07	6	For in situ pipework and all positional applications for all plain and NB stabilized 3% Mo bearing steels.
19	12	3 Nb	R	0.05–0.06	6	For all plain and Nb and Ti stabilized 3% Mo stainless steels. Particularly for flat butts and fillets giving a smooth neat weld profile.
19	13	4	R	0.03	11	For austenitic stainless steels of the 19.13.4 type except for very severe corrosive conditions. Also for certain low alloy steels and manganese steels.
23	12	–	B	0.05	17	For root runs on the boundary between cladding and backing in clad steels. Also for welding stainless steels to carbon steels.
23	12	2	R	0.04	15	For welding stainless clad plate. Dilution gives a weld metal analysis of 18% Cr 8% Ni with about 6% ferrite. Useful for buttering carbon and low alloy steels.
25	20	–	B	0.03	–	All-positional electrode for welding heat- and corrosion-resistant steels of similar composition and for welding stainless steel to low alloy steel when there is low restraint, otherwise there is danger of hot cracking.
26	5	1.5	R	0.1	–	For heat- and acid-resistant steels of similar analysis and particularly for ferritic chromium steels containing 17% Cr and 24% Cr provided they are not subject to severe sulphurous attack.

as that of steel so that a wider V preparation is required with a smaller root face to obtain satisfactory penetration.

Each run of a multi-run weld should be deslagged by chipping and wire-brushed. Grinding should not be undertaken as it may lead to particles of slag being driven into the weld surface with consequent loss of corrosion resistance. Stray arcing should be avoided and minimum weaving performed because it results in poorer quality of weld deposit due to increased dilution.

Although it is preferable to make welds in the flat position, vertical welds can be made either upwards or downwards holding the electrode at right angles to the line of weld, using reduced current compared with similar flat conditions.

For fillet welds the electrode angle should roughly bisect the angle between the plates and be at right angles to the line of weld. If the plates are of unequal thickness the arc should be held more on to the thicker plate to obtain better fusion, and tilted fillets give equal leg length more easily.

The table gives a list of the alloys suitable for welding by this process and the type of electrode recommended.

Steels containing 9% nickel used for low temperature ($-196\,^\circ$C) are welded with basic coated electrodes using d.c. such as NC 80/20 or Nicrex 9.

| | Electrodes recommended | |
Alloy	Alloy to itself	Alloy to steel
Nickel 200	Nickel 141	Nickel 141
Nickel 201		
Monel 400	Monel 190	Monel 190
Monel K500	Monel 134	Monel 190
Inconel 600	Inconel 132	Inconel 182
	Inconel 182	Incoweld A
Incoloy DS	Incoweld A	Incoweld A
Incoloy 800	Incoweld A	Incoweld A
Incoloy 825	Incoloy 135	Inconel 182
		Incoloy A
Nimonic 75	Inconel 132	Inconel 182
	Inconel 182	Incoweld A
Nilo alloys	Incoweld A	Incoweld A
	Nickel 141	Nickel 141

Clad steels

Stainless clad steel is a mild or low alloy steel backing faced with stainless steel such as 18% Cr 8% Ni or 18% Cr 10% Ni with or without Mo, Ti and Nb, or a martensitic 13% Cr steel, the thickness of the cladding being 10–20% of the total plate thickness.

Preparation and technique. The backing should be welded first and the mild steel root run should not come into contact with the cladding, so that preparation should be either with V preparation close butted with a deep root face, or the cladding should be cut away from the joint at the root (see Fig. 5.45). The clad side is then back grooved and the stainless side welded with an electrode of similar composition.

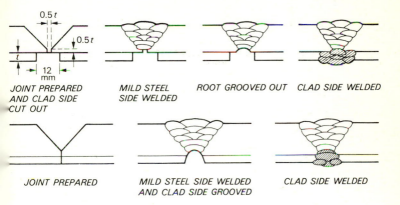

JOINT PREPARED AND CLAD SIDE CUT OUT

MILD STEEL SIDE WELDED

ROOT GROOVED OUT

CLAD SIDE WELDED

JOINT PREPARED

MILD STEEL SIDE WELDED AND CLAD SIDE GROOVED

CLAD SIDE WELDED

Fig. 5.45. Alternative methods of welding clad steel.

Generally an austenitic stainless steel electrode of the 25% Cr 20% Ni should be used for the root run on the clad side because of dilution effects, and at least two layers or more if possible should be laid on the clad side to prevent dilution effects affecting the corrosion-resistant properties. First runs should be made with low current values to reduce dilution effects. For martensitic 13% Cr cladding, pre-heating to 250 °C is advisable followed by post-heating and using an austenitic stainless steel electrode such as 22% Cr 12% Ni, 3% Mo with about 15% ferrite which gives weld metal of approximately 18% Cr 8% Ni with about 6% ferrite. Welding beads not adjacent to the backing plate can be made with 18% Cr 12% Ni 3% Mo electrodes. If the heat input is kept as low as possible welding may be carried out without heat treatment, the HAZ being tempered by the heat from successive runs.

Nickel clad steel

Mild and low alloy steel can be clad with nickel, monel or inconel for corrosion resistance at lower cost compared to solid nickel base material and with an increase in thermal conductivity and greater strength, the thickness of cladding usually being not greater then 6 mm. When welding clad steels it is essential to ensure the continuity of the cladding, and because of this butt joints are favoured. Dilution of the weld metal with iron occurs when welding the clad side, but the electrode alloy can accept this with the exception of monel 60: in this case a buffer layer of monel 61 should be laid down first. A minimum of two runs should be used on the clad side resisting properties, and first runs should be made with low current values to reduce the dilution.

Recommended electrodes are: for nickel 200 use a nickel 141 electrode: for monel 400 use a monel 190 electrode and for inconel 600, an inconel 182 electrode. Preparation of joints is similar to that for stainless clad welding.

WELDING OF PIPE LINES

The following brief account will indicate to the welder the chief methods used in the welding of pipe lines for gas, oil, water, etc. The lengths of pipes to be welded are placed on rollers, so that they can easily be rotated. The lengths are then lined up and held by clamps and tack welded in four places around the circumference, as many lengths as can be handled conveniently, depending on the nature of the country, being tacked together to make a section. The tack welder is followed by the main squad of welders, and the pipes are rolled on the rollers by assistants using chain wrenches, so that the welding of the joint is entirely done in the flat position; hence the name *roll welding*.

After careful inspection, each welded section is lifted off the rollers by tractor-driven derricks and rested on timber baulks, either over, or near the trench in which it is to be laid. The sections are then *bell hole* welded together. The operator welds right round the pipe, the top portion being done down-hand, the sides vertical, and the underside as an overhead weld. Electrodes of 4 and 5 mm diameter are used in this type of weld.

Stove pipe welding

This type of welding has a different technique from conventional positional welding methods and has enabled steel pipelines to be laid across long distances at high rates. The vertically downward technique is used welding from 12 o'clock to 6 o'clock in multiple runs. It often involves welders working in pairs one on each side of the pipe and often one pair of welders is responsible for one type of run only.

Cellulose or cellulose–iron powder coated electrodes are used and give a strongly directional arc with stiff controllable slag. There is a high rate of deposition and complete elimination of back chipping due to the penetrating characteristics of the electrode and the reduced amount of weld metal deposited reduces distortion. The coating provides a gas shield which is less affected by wind than other classes of electrodes, so porosity is eliminated and the multiple runs give grain refinement with good impact strength. Direct current is generally used with the electrode

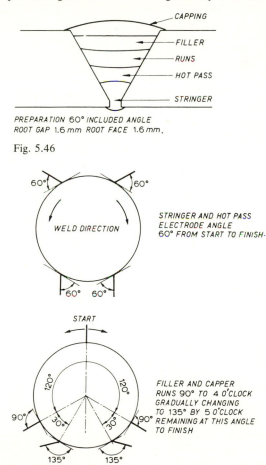

PREPARATION 60° INCLUDED ANGLE
ROOT GAP 1.6 mm ROOT FACE 1.6 mm.

Fig. 5.46

STRINGER AND HOT PASS
ELECTRODE ANGLE
60° FROM START TO FINISH.

FILLER AND CAPPER
RUNS 90° TO 4 O'CLOCK
GRADUALLY CHANGING
TO 135° BY 5 O'CLOCK
REMAINING AT THIS ANGLE
TO FINISH

Fig. 5.47 (a)

positive (reversed polarity) but the capping bead may be made with the
electrode negative (straight polarity) to reduce surface porosity. If a.c. is
used the open circuit voltage should be 95 V minimum.

Preparation is usually with 60° between weld faces, increased sometimes to 70° with a 1.5 mm root face and 1.5 mm root gap; internal
alignment clamps are used. The stringer bead is forced with no weave
into the root, then the hot pass with increased current fuses the sides and
fills up any burn-through which may have occurred. Filler runs, stripper
run and capping run complete the welding, and for clarity the sequence
of operations is shown in Figs. 5.46 and 5.46a. Two types of fabricated
bend and various preparations are shown in Fig. 5.47b.

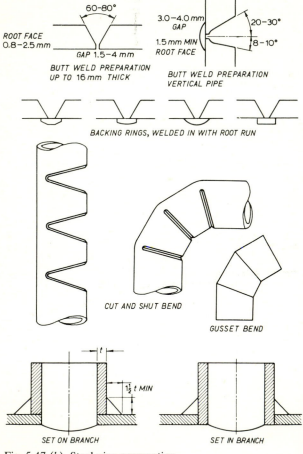

Fig. 5.47 (b). Steel pipe preparation.

Weld pass	Electrode size (mm)	Current (A)	Technique
Stringer bead	4	140–170	Tight drag weld without weaving. Electrode angle 60° to line of travel. Arc should be clearly visible on the inside of the pipe. Some small burn-through areas in this pass are permissible because they are eliminated by the hot pass.
Hot pass	4	160–180	Light drag weld with backward and forward movement of electrode. Electrode angle 60° to line of travel. Hot pass *must* be deposited while stringer bead is still warm.
Filler bead	5	160–180	Normal arc length and weave. Electrode angle 90° except for 4 o'clock to 6 o'clock and 8 o'clock areas where the angle should be increased to 135°. Electrode should be manipulated with a lifting or vertical flicking movement from weld deposit to arc pool.
Stripper bead	5	160–180	Medium long arc and weave. Electrode angle 90°. Used for 2 o'clock to 5 o'clock and 10 o'clock to 7 o'clock areas if shallow at these points.
Capping bead	5	140–170	Medium long arc and rapid side to side weave. Electrode angle 90° except for 4 o'clock to 6 o'clock and 8 o'clock to 6 o'clock areas where the angle is increased to 135°. Manipulate in these areas by a lifting or flicking technique.

The CO_2 semi-automatic process is also used for pipe line welding. The supply unit is usually an engine-driven generator set and the technique used is similar to the stove pipe method, but it is important to obtain good line-up to avoid defects in the penetration and correct manipulation to avoid cold shuts. Fully automatic orbital pipe welders are also available using the TIG process and are used on pipes from 25 to 65 mm outside diameter.

For full details of the various methods of preparation and welding standards for pipes, the student should refer to BS 2633, class 1: *Steel pipework*, and BS 2971, class 2: *Steel pipelines*; BS 2910: *Radiographic examination of pipe joints*, and BS 938: *Metal arc welding of structural steel tubes*.

For river crossings the pipe thickness is increased by 50 to 100% and the lengths are welded to the length required for the crossing. Each joint is further reinforced by a sleeve, and large clamps bolted to the line serve as anchor points. The line is then laid in a trench in the river bed.

For water lines from 1 to $1\frac{1}{2}$ m diameter, bell and spigot joints are often used. On larger diameter pipes the usual joint is the reinforced butt. The pipe is V'd on the inside and welded from the inside. A steel reinforcing band is then slipped over the joint and fillet welded in position.

The types of joint are illustrated in Fig. 5.48.

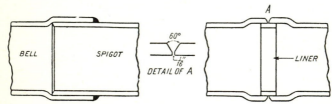

BELL AND SPIGOT JOINT

DETAIL OF A

DOUBLE BELL JOINT WITH LINER
USED FOR OIL 1930–33

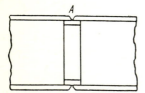

PLAIN JOINT WITH LINER FOR
GAS UP TO 24" DIA. NEEDS
2 WELDING BEADS

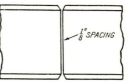

PLAIN JOINT WITHOUT LINER
NOW USED FOR OIL UP TO 12" DIA.
NEEDS 3 WELDING BEADS

Fig. 5.48

WELDING OF CARBON AND CARBON–MANGANESE AND LOW-ALLOY STEELS

Steels with a carbon content up to about 0.25% carbon (mild steels) are easy to weld and fabricate because, due to their low carbon content, they do not harden by heat treatment so that the weld and HAZ does not have hardened zones even though there is quick cooling. As the carbon content of a steel increases it becomes more difficult to weld, because as the weld cools quickly owing to the quenching action of the adjacent cold mass of metal hardened zones are formed in the HAZ, resulting in brittleness and possible cracking if the joint is under restraint. Pre-heat can be used to overcome this tendency to cracking.

One or more alloying elements such as Ni, Cr, Mo, Mn, Si, V and Cu are added to steel to increase tensile, impact and shear strength, resistance to corrosion and heat and in some cases to give these improved properties either at high temperatures or low (cryogenic) temperatures. The alloying elements do not impair the weldability as would an increase in carbon content, but the steels are susceptible to cracking, and it is this problem which has to be considered in more detail.

Cracking may appear as: (1) delayed cold cracking caused by the presence of H_2; (2) high-temperature liquation cracking (hot cracking); (3) solidification cracking; (4) lamellar tearing; hot cracking and cold cracking have been already considered on pp. 112–13. Hot cracks usually appear down the centre of a weld, which is the last part to solidify, while cold cracks occur in the HAZ (Fig. 5.49). These latter may not occur until the weld is subject to stress in service, and since they are often below the metal surface they cannot be seen, so that the first indication of their presence is failure of the joint.

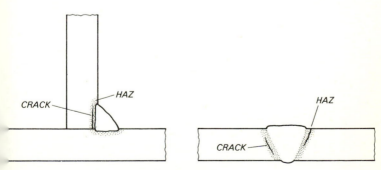

HAZ CRACKING

Fig. 5.49

The factors which lead to cold cracking are:

(1) The composition of steel being welded.
(2) The presence of hydrogen.
(3) The rate of cooling of the welded joint.
(4) The degree of restraint (stress) on the joint.

(1) Composition of the steel and carbon equivalent (CE). The tendency to crack increases as the carbon and alloying element content increases, and since there is a great variety in the types of steel to be welded it is convenient to convert the varying amounts of alloying elements present in a given steel into terms of a simple equivalent carbon steel, thus giving an indication of the tendency to crack. This 'carbon equivalent' can be calculated from a formula such as the following:

$$CE = C\% + \frac{Mn\%}{6} + \frac{Cr\% + Mo\% + V\%}{5} + \frac{Ni\% + Cu\%}{15}$$

(which is a variation of the original formula of Dearden and O'Neill) so that to take a simple example, if steel has the following composition: C 0.13%, Mn 0.3%, Ni 0.5%, Cr 1.0%, application of the formula gives a CE of 0.41%.

(2) The presence of hydrogen in the welded zone greatly increases the tendency to crack, the amount of hydrogen present depending upon the type of electrode used and the moisture content of its coating. A rutile coating may have a high moisture coating giving up to 30 ml of hydrogen in 100 g of weld metal. The hydrogen diffuses into the HAZ, and on cooling quickly a hard martensitic zone exists with a liability of cracks occurring. Even a small amount of hydrogen present can result in cracking in severely restrained joints. Basic (hydrogen controlled) electrodes correctly dried before use (p. 304) result in a very low hydrogen content. Austenitic stainless steel electrodes deposit weld metal in which the hydrogen is retained and does not diffuse into the HAZ. The weld has a relatively low yield strength, and when stressed, yields and reduces the restraint on the joint, so they are used, for example, to weld steels such as armour plate which may crack when welded with basic-coated mild steel electrodes. Gas-shielded processes using CO_2 or argon–CO_2 mixtures give welds of a very low hydrogen content.

(3) Rate of cooling of the welded zone. The rate of cooling depends upon (1) the heat energy put into the joint and (2) the combined thickness of the metal forming the joint. Arc energy is measured in kilojoules per mm length of weld and can be found from the formula

$$\text{Arc energy (kJ/mm)} = \frac{\text{arc voltage} \times \text{welding current}}{\text{welding speed (mm/s)} \times 1000}.$$

The greater the heat input into the joint the slower the rate of cooling so that the use of a large-diameter electrode with high current reduces the quenching effect and thus the cracking tendency. Similarly, smaller-diameter electrodes with lower currents reduce the heat input and give a quicker cooling rate, increasing the tendency to crack due to the formation of hardened zones. Subsequent runs made immediately afterwards are not quenched as is the first run, but if the first or subsequent runs are allowed to cool, conditions then return to those of the first run. For this reason interpass temperature is often stipulated so as to ensure that the weld is not allowed to cool too much before the next run or pass is made. The use of large electrodes with high currents however does not necessarily give good impact properties at low temperatures. For cryogenic work it is essential to obtain the greatest possible refining of each layer of weld metal by using smaller-diameter electrodes with stringer or split-weave technique. As the 'combined thickness', that is, the total thickness of the sections at the joint increases, so the cooling rate increases, since there is increased section through which the heat can be conducted away from the joint. The cooling rate of a fillet joint is greater than that for a butt weld of the same section plate since the combined thickness is greater, Fig. 5.50.

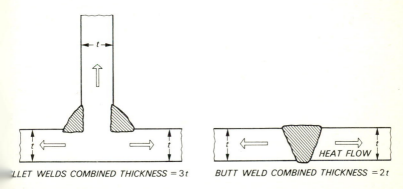

LLET WELDS COMBINED THICKNESS = 3t   BUTT WELD COMBINED THICKNESS = 2t

Fig. 5.50

(4) Restraint. When a joint is being welded the heat causes xpansion which is followed by rapid cooling. If the joint is part of a very igid structure the welded zone has to accommodate the stresses due to these effects and if the weld is not ductile enough cracking may occur. The degree of restraint is a variable factor and is important when estimating the tendency to crack.

Controlled Thermal Severity (C.T.S.) tests in which degrees of restraint are placed upon the joint to be welded and on which pre- and post-heat can be applied are used to establish the liability to crack. (See Reeve test, Chapter 11.)

Hydrogen cracking can be avoided by (1) using basic hydrogen-controlled electrodes, correctly dried and (2) pre-heating.

The temperature of pre-heat depends upon (1) the CE of the steel, (2) the process used and in the MMA process, the type of electrode (rutile or basic), (3) the type of weld, whether butt or fillet and the run out length (x mm electrode giving y mm weld), (4) the combined thickness of the joint, (5) arc energy.

Reference tables are given in BS 5135 (1974) from which the pre-heat temperature can be ascertained from the above variables. Pre-heat temperatures may vary from 0 to 150 °C for carbon and carbon-manganese steels and up to 300 °C for higher carbon low alloy steels containing chromium and molybdenum. The pre-heating temperature is specified as the temperature of the plate immediately before welding begins and measured for a distance of at least 75 mm on each side of the joint, preferably on the opposite face from that which was heated. The combined thickness is the sum of the plate thicknesses up to a distance of 75 mm from the joint. If the thickness increases greatly near the 75 mm zone higher combined thickness values should be used and it should be noted that if the whole unit being welded (or up to twice the distance given above) can be pre-heated, pre-heat temperatures can be reduced by about 50 °C. Austenitic electrodes can generally be used without pre-heat.

Steels used for cryogenic (low temperature) applications can be carbon–manganese types which have good impact properties down to −30 °C and should be welded with electrodes containing nickel; 3% nickel steels are used for temperatues down to −100 °C and are welded with matching electrodes whilst 9% nickel steels, used down to −196 °C (liquid nitrogen) are welded with nickel–chromium–iron electrodes since the 3% nickel electrodes are subject to solidification cracking.

Creep-resistant steels usually contain chromium and molybdenum and occasionally vanadium and are welded with basic-coated low alloy electrodes with similar chromium and molybdenum content. Two types of cracking are encountered: (1) transverse cracks in the weld metal, (2) HAZ cracking in the parent plate. Pre-heating and interpass temperatures of 200–300 °C with post-heat stress relief to about 700 °C is usually advisable.

Lamellar tearing

In large, highly stressed structures cracks may occur in the material of the parent plate or the HAZ of a joint, the cracks usually running parallel to the plate surface (Fig. 5.51). This is known as lamellar tearing, and it is the result of very severe restraint on the joint and poor ductility, due to the presence of non-metallic inclusions running parallel to the plate surface which are difficult to detect by the usual non-destructive tests. Certain types of joint such as T, cruciform and corner are more susceptible than others. Should lamellar tearing occur the joint design should be modified and tests made on the parent plate to indicate its sensitivity to tearing, whilst buttering of the surface may also help.

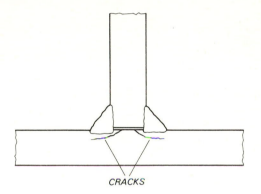

CRACKS

Fig. 5.51. Lamellar tearing.

For structural steels in general, basic-coated electrodes are used and current electrode lists of the electrode makers should be consulted for the most up-to-date information.

When welding these steels the following points should be observed:

(*a*) When tack welds are used to position the work, as is the general practice, they should be well fused into the weld because, due to the rapid cooling and consequent hardening of the area around the tack weld, cracks may develop.

(*b*) Since the chilling effect is most marked on the first run, careful watch should be kept for any cracks which may develop. Any sealing run applied to the back of the joint should preferably be made either while the joint is still hot, or with a large electrode. The least possible number of runs should be used to fill up the V to minimize distortion.

(*c*) When austenitic rods are used to obtain a weld free from cracks, all the runs should be made with this type of rod and ordinary steel rods not used for subsequent runs. Plates thicker than 15 mm are preferably pre-heated to 100–200 °C to avoid cracking. Cold work and interrupted welds should generally be pre-heated.

(*d*) The electrode or holder should not be struck momentarily or 'flashed' by design or accident on to the plate prior to welding, since the rapid cooling of the small crater produced leads to areas of intense hardness that may result in fatigue cracks developing.

(*e*) Hard spots in the parent plate may be softened by post-heat applied locally but this may reduce the endurance value of the joint.

Welding of steel castings

The use of steel castings in welded assemblies generally falls under two methods.

(1) Welding steel castings together to form a complex casting that would be difficult to cast as a whole.
(2) Castings and wrought material welded together to form a 'composite' fabrication.

The same rules regarding weldability apply as when welding wrought material. Castings may be in low-carbon-content mild steel, low-alloy steels, or high-alloy steels (such as austenitic manganese and chrome nickel stainless types) and for the alloy steels the same precautions must be taken against cracking as for the wrought form (q.v.). In general the stringer bead technique is used for thick butt welds since it reduces cracking tendency and horizontal vertical fillets of greater than 8 mm leg length may be multi-run to avoid the tendency towards undercutting of the vertical plate.

6

Tungsten electrode, inert gas shielded welding processes (TIG), and the arc plasma process

TECHNOLOGY AND EQUIPMENT

The welding of aluminium and magnesium alloys by the oxy-acetylene and manual metal arc processes is limited by the necessity to use a corrosive flux. The gas shielded, tungsten arc process (Fig. 6.1) enables these metals and a wide range of ferrous alloys to be welded without the use of a flux. The choice of either a.c. or d.c. depends upon the metal to be welded. For metals having refractory surface oxides such as aluminium and its alloys, magnesium alloys and aluminium bronze, a.c. is used whilst d.c. is used for carbon and alloy steels, heat-resistant and stainless steels, copper and its alloys, nickel and its alloys, titanium, zirconium and silver.

See also BS 3019, *General recommendations for manual inert gas tungsten arc welding.* Part 1, *Wrought aluminium, aluminium alloys and magnesium alloys*; Part 2, *Austenitic stainless and heat resisting steels.*

The arc burns between a tungsten electrode and the workpiece within a shield of the inert gas argon, which excludes the atmosphere and prevents contamination of electrode and molten metal. The hot tungsten arc ionizes argon atoms within the shield to form a gas plasma consisting of almost equal numbers of free electrons and positive ions. Unlike the electrode in the manual metal arc process, the tungsten is not transferred to the work and evaporates very slowly, being classed as 'non-consumable'. Small amounts of other elements are added to the tungsten to improve electron emission.

Gases

Argon in its commercial purity state (99.996%) is used for the metals named above, but for titanium extreme purity is required. Argon with 5% hydrogen gives increased welding speed and/or penetration in the welding of stainless steel and nickel alloys; nitrogen can be used for copper welding on deoxidized coppers only. Helium may be used for aluminium and its alloys and copper, but it is more expensive than argon

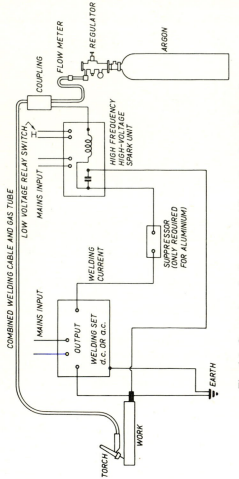

Fig. 6.1. Connexions for inert gas welding using air-cooled torch.

and, due to its lower density, a greater volume is required than with argon to ensure adequate shielding, and small variations in arc length cause greater changes in weld conditions so that manual welding with helium is not as easy as with argon. The mechanized d.c. welding of aluminium with helium gives deep penetration and high speeds.

The characteristics of the arc are changed considerably with change of direction of flow of current, that is with arc polarity.

Electrode positive

The electron stream is from work to electrode while the heavier positive ions travel from electrode to work-piece (Fig. 6.2a). If the work is of aluminium or magnesium alloys there is always a thin layer of refractory oxide of melting point about 2000 °C present over the surface and which has to be dispersed in other processes by means of a corrosive flux to ensure weldability. The positive ions in the TIG arc bombard this oxide and, together with the electron emission from the plate, break up and disperse the oxide film. It is this characteristic which has made the process so successful for the welding of the light alloys. The electrons streaming to the tungsten electrons generate great heat, so its diameter must be relatively large and it forms a bulbous end. It is this overheating with consequent vaporization of the tungsten and possibility of tungsten being transferred to the molten pool (pick-up) and contaminating it that is the drawback to the use of the process with electrode positive. Very much less heat is generated at the molten pool and this is therefore wide and shallow.

Electrode negative

The electron stream is now from electrode to work with the zone of greatest heat concentrated in the workpiece so that penetration is deep and the pool is narrower. The ion flow is from work to electrode so that there is no dispersal of oxide film and this polarity cannot be used for welding the light alloys. The electrode is now near the zone of lesser heat and needs be of reduced diameter compared with that with positive polarity. For a given diameter the electrode, when negative, will carry from four to eight times the current than when it is positive and twice as much as when a.c. is used. (Fig. 6.2b).

Alternating current

When a.c. is used on a 50 Hz supply, voltage and current are reversing direction 100 times a second so there is a state of affairs between that of electrode positive and electrode negative, the heat being fairly evenly distributed between electrode and work (Fig. 6.2c). Depth of penetration is between that of electrode positive and electrode negative and the electrode diameter is between the previous diameters. When the electrode is positive it is termed the positive half-cycle and when negative the negative half-cycle. Oxide removal takes place on the positive half-cycle.

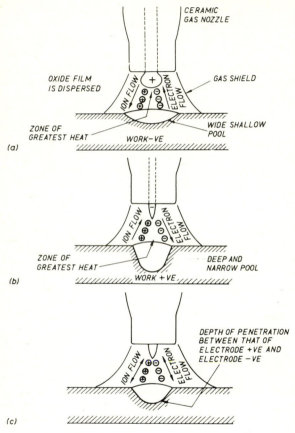

Fig. 6.2. Electron streams between electrode and work: (*a*) d.c., tungsten electrode +ve of large diameter tends to overheat; (*b*) d.c. tungsten electrode −ve of small diameter; (*c*) a.c., electrode diameter between that of electrode +ve and −ve electrode.

Inherent rectification in the a.c. arc

In the a.c. arc the current in the positive half-cycle is less than that in the negative half-cycle (Fig. 6.3*b*). This is known as inherent rectification and is a characteristic of arcs between dissimilar metals such as tungsten.

and aluminium. It is due to the layer of oxide acting as a barrier layer to the current flowing in one direction and to the greater emission of electrons from the tungsten electrode when it is of negative polarity. The result of this imbalance is that an excess pulsating current flows in one direction only and the unbalanced wave can be considered as a balanced a.c. wave, plus an excess pulsating current flowing on one direction only on the negative half-cycle. This latter is known as the d.c. component and can be measured with a d.c. ammeter. (The suppression of this d.c. component is discussed later.) The reduction of current in the positive half-cycle due to the inherent rectification results in a reduction of oxide removal.

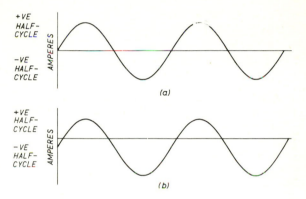

Fig. 6.3. Alternating current: (*a*) balanced wave; (*b*) unbalanced wave, inherent rectification (current in +ve half-cycle less than that in −ve half-cycle).

Partial rectification

A greater voltage is required to strike the arc than to maintain it and re-ignition on the negative half-cycle requires a lower voltage than for the positive half-cycle, partly due to the greater electron emission from the tungsten when it is negative polarity, but actual re-ignition depends upon many factors including the surface condition of weld pool and electrode, the temperature of the pool and the type of shielding gas. There may be a delay in arc re-ignition on the positive half-cycle until sufficient voltage is available and this will result in a short period of zero current (Fig. 6.4*a*) until the arc ignites. This delay reduces the current in the positive half-cycle and this state is known as partial rectification. If the available voltage is not sufficient, ignition of the arc may not occur at all on the positive half-cycle, the arc is extinguished on the one half-pulse and continues burning on the uni-directional pulses of the negative

half-cycle, and we have complete rectification with gradual extinguishing of the arc (Fig. 6.4*b*).

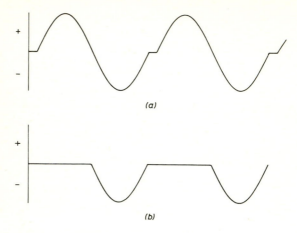

(a)

(b)

Fig. 6.4. (*a*) Partial rectification; (*b*) half-wave rectification (+ve
half-cycle missing).

Re-ignition voltages

To ensure re-ignition of the arc on the positive half-cycle, the available
voltage should be of the order of 150 V, which is greater than that of the
supply transformer. To ensure re-ignition, auxiliary devices are used
which obviate the need for high open-circuit transformer voltages.

Ignition and re-ignition equipment

High frequency, high voltage, spark gap oscillator. This device
enables the arc to be ignited without touching down the electrode on the
work and thus it prevents electrode contamination. It also helps arc
re-ignition at the beginning of the positive half-cycle.

The oscillator consists of an iron-cored transformer with a high voltage secondary winding, a capacitor, a spark gap and an air core transformer or inductive circuit, one coil of which is in the high voltage circuit
and the other in the welding circuit (Fig. 6.5). The capacitor is charged
every half cycle to 3000–5000 V and discharges across the spark gap.
The discharge is oscillatory, that is, it is not a single spark but a series of
sparks oscillating across the spark gap during discharge. This discharge,
occurring on every half-cycle, sets up oscillatory currents in the circuit
and these are induced and superimposed on the welding current through
the inductance of the coils *L* Fig. 6.6.

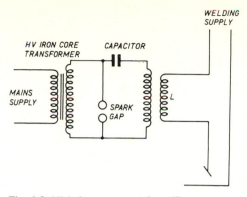

Fig. 6.5. High-frequency spark oscillator.

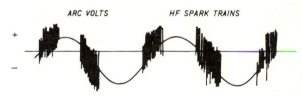

Fig. 6.6. Arc voltage with superimposed HF spark main for re-ignition and stabilization; the HF is injected on both +ve and −ve half-cycles, but is only required at the beginning of the former.

The spark discharge is phased to occur at the beginning of each half-cycle (although for re-ignition purposes it is required only for the positive half-cycle) and is of about 5 milliseconds' or less duration compared with the half-cycle duration of 10 milliseconds (Fig. 6.7). To initiate the arc, the electrode is brought to about 6 mm from the work with the HF unit and welding current switched on. Groups of sparks pass across the gap, ionizing it, and the welding current flows in the form of an arc without contamination of the electrode by touching down. The HF unit can give rise to considerable radio and TV interference and adequate suppression and screening must be provided to eliminate this as far as possible. The use of HF stabilization with the a.c. arc enables this method to be used for aluminium welding, although inherent rectification is still present, but partial rectification can be reduced to a minimum by correct phasing of the spark train.

Surge injection unit. This device supplies a single pulse surge of about 300 V phased to occur at the point when the negative half-cycle changes to the positive half-cycle (Fig. 6.7), so that the pulse occurs at 20 millisecond intervals and it lasts for only a few microseconds. This unit enables transformers of 80 V open-circuit voltage to be used and it does not produce the high frequency radiation of the spark oscillator and thus does not interfere to any extent with radio and TV apparatus.

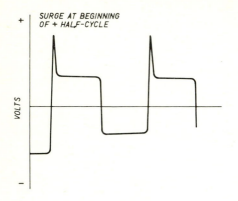

Fig. 6.7. Surge voltage superimposed on arc voltage.

The unit consists of a rectifier which supplies d.c. to a circuit containing resistance and capacitance, a surge valve which supplies the short pulses and a trigger valve which releases the pulses into the welding circuit. The trigger valve which is sensitive to change in arc voltage releases the pulse at the end of the negative half-cycle, just when the positive half-cycle is beginning. As the pulse is usually unable to initiate the arc from cold, a spark oscillator is also included in the unit and is cut out of circuit automatically when the arc is established so that the surge injection unit is an alternative method to the H.F. oscillator for re-igniting the arc on the positive half-cycle.

Suppression of the d.c. component in the a.c. arc.

In spite of the use of the HF unit, the imbalance between the positive and negative half-cycles remains and there is a d.c. component flowing. This direct current flows through the transformer winding and saturates the iron core magnetically, giving rise to high primary currents with such heating effect that the rating of the transformer is lowered, that is, it cannot supply its rated output without overheating. This insertion of banks of electrolytic capacitors in series with the welding circuit has two effects:

(1) They offer low impedance to the a.c. which flows practically uninterrupted, but they offer very high impedance to the d.c. which is therefore suppressed or blocked;

(2) During the negative half cycle the capacitors receive a greater
charge (because of the imbalance) and during the following
positive half-cycle this excess adds to the positive half-cycle
voltage so that it is increased, and if the open-circuit voltage is
greater than about 100 V the arc is re-ignited on each half-
cycle without the aid of high-frequency currents which can
then be used for starting only. The effect of this increase in
voltage on the positive half-cycle is to improve the balance of
the wave (Fig. 6.8), so that the heating effect between electrode and plate is more equal and disposal of the oxide film is
increased.

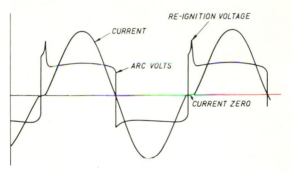

Fig. 6.8. Current and arc voltage wave-forms showing re-ignition.

The value of the capacitor must be chosen so that there is no danger of
electrical resonance in the circuit which contains resistance, inductance
capacitance, which would result in dangerous excessive currents and
voltages irrespective of the transformer output (Fig. 6.9). If external
capacitors are used they may affect the current output of the transformer
by altering the power factor, and this must be taken into account since
the current calibrations on the set will be increased.

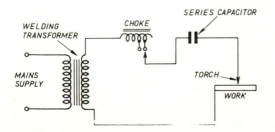

Fig. 6.9. Series capacitor used to block d.c. component.

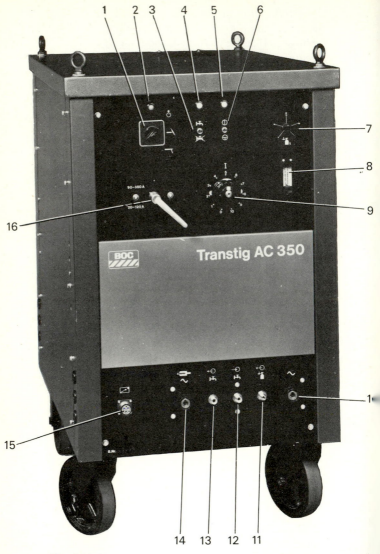

Fig. 6.10 (a)

Symbols used on the unit

○	Off		Pulse time
	TIG welding		Background time
	TIG spot welding		Spot weld time
	Plasma welding		Slope-in time
	Manual arc welding		Slope-out time
	Water on		Gas
	Water off		Water
①	Latching		In
⊖	Non-latching		Out
▲	Current (continuous or pulse)	t	Time
	Background current		Water recirculating unit
	Continuous welding		Tap water
	Pulse welding		Remote control

Fig. 6.10

(*a*) Drooping characteristic a.c. power unit for TIG welding of aluminium and its alloys and for MMA welding if required. Unit includes HF unit for striking and maintaining the arc, d.c. component suppressor, post-weld gas delay to prevent oxidation of electrode, and latching circuit. Fan cooled 28 kVA rating with output of 20 A at 15 V to 350 A at 25 V with 80 OCV.

Key:
1. Process switch.
2. Mains lamp.
3. Water switch.
4. Water warning lamp.
5. Latching lamp.
6. Latching switch.
7. Post-weld gas delay time control.
8. Gas control and flowmeter.
9. Current control.
10. Electrode socket.
11. Gas connector.
12. Water return connector.
13. Water supply connector.
14. Work socket.
15. Remote control socket.
16. Current range switch.

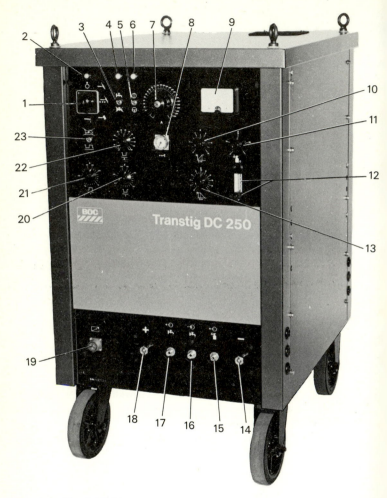

Fig. 6.10 (*b*)

(*b*) Drooping characteristic fan-cooled d.c. power source for TIG (stainless steel), TIG spot and MMA welding with pulse welding facilities. It can also be used for plasma welding in conjunction with a special console. Input; 3-phase, 50 Hz, 220 and 380–440 V. Rating 18 kVA giving an output from 10 A at 10 V to 250 A at 20 V with 68 OCV. Pulse current 10–250 A, pulse time 0.2–3 seconds, background current 5–65 A, background time 0.2–6 seconds.

Key:

1. *Process switch.* TIG, TIG spot, plasma or MMA processes can be selected. When 'off' mains are disconnected from circuits.

2. *Mains indicator lamp.* Indicates that mains and process switch are 'on' or 'off'.

3. *Water switch.* When 'on', remote control switch opens the water valve and water flows to the welding torch.

4. *Water warning lamp.* Lights when water switch is not switched on.

5. *Latching switch.* When 'on' the latching circuit operates and holds the circuit in the 'on' position, taking over from the hand-operated switch on the torch.

6. *Latching lamp.* When this is lit the torch is 'live'.

7. *Main current control.* Sets the level of the welding current for continuous and also the pulse current for pulse welding.

8. *Spot weld timer.* Selects the time for the spot weld and can be adjusted for a variety of timed welds.

9. *Ammeter.* This indicates the welding current but cannot be used for measuring the welding current when short pulse times have been selected because the pointer is relatively slow to respond to variations in current.

10. *Slope-in time control.* Sets the slope-in time, reducing the current for the immediate start, and allows it to rise to the selected welding value in a given period of time, thus reducing the danger of burn-through in thin sections.

11. *Post-weld gas delay time control.* This varies the duration of the post-weld gas flow between 5 and 30 seconds after welding has ceased, protecting the weld from the atmosphere during cooling.

12. *Gas control and flowmeter.* Enables the gas flow to be adjusted and shows the gas flow rate.

13. *Slope-out time control.* Sets the slope-out time by operating with the current control and reducing the current used in the welding operation to the minimum value that the unit can supply over a period of time, thus enabling the crater that would normally be formed at the weld termination to be filled, reducing the tendency to crack.

14. *Negative socket.*

15. *Gas connector.*

16. *Water return connector.*

17. *Water supply connector.*

18. *Positive socket.*

19. *Remote control socket.* (A remote control switch may be attached here that is foot operated, or attached to the torch, enabling welding services to be switched on or off from a remote position.)

20. *Background time switch.* Sets the duration of the background current for pulse welding.

21. *Pulse time switch.* Sets the duration of the pulse current for pulse welding.

22. *Background current control.* Sets the level of the background current for pulse welding.

23. *Pulse switch.* Selects either continuous or pulse welding.

Power sources, a.c. and d.c.

Equipment can be chosen to give a.c., d.c., or both a.c. and d.c. from one set and, power sources are now designed for specific industries: (1) d.c. output for stainless steel fabrication industry; (2) a.c. output for aluminium fabrication industry; (3) a.c./d.c. output for the general fabrication industry.

a.c. power supply. For the light alloys, aluminium and magnesium, a transformer similar to that used for manual metal arc welding is used. It is usually double-wound and oil-immersed and can have primary tappings for either single-phase or three-phase, with a secondary voltage of 80–100 V. Current control is by tapped choke or saturable reactor and the auxiliary devices include HF oscillator, d.c. component suppressor, and in some cases a surge injection unit. These units may be built into the set or fitted externally (Fig. 6.10a).

d.c. power supply. These units consist of a step-down transformer and bridge-connected selenium or silicon rectifier with current outputs from 1 to 400 A according to the range of current required. Stepless current control is obtained by means of a magnetic amplifier or saturable reactor described on p. 278. The smaller units can be used on 200–250 V single-phase supplies and larger ones on a three-phase supply. A spark starter unit is fitted for initiating the arc (Fig. 6.10b).

a.c. and d.c. power supply. These units which supply either a.c. or d.c. as required can be used for manual or automatic tungsten inert gas shielded welding, TIG spot welding, or for manual metal arc process. A timing unit also enables argon spot welding to be performed. A typical unit (Fig. 6.10c) has the following specification. The primary of the main transformer is connected to a three-phase 400–440 V a.c. supply at 50 Hz. The transformer feeds a bridge-connected silicon rectifier which has replaceable diodes; the unit is forced-draught cooled and incorporates an HF oscillator, d.c. component suppressor, reactor and auxiliary devices such as gas purging, crater filling, argon delay and soft start. Output voltages are 75–80 V a.c. and d.c., with current ranges from 20 to 400 A depending upon the rating of the unit. These units enable a.c. or d.c. to be selected as required so that they cover the whole range of welding applications by this process. Manual metal arc supply is obtained from transformer and reactor. Supply for a.c. TIG uses transformer, reactor, HF oscillator, d.c. component suppressor, and the auxiliaries, while d.c. supply for TIG uses transformers, rectifier, reactor, spark starter and the auxiliaries. Power supply for the auxiliaries and the various contactors is usually at 110 V a.c.

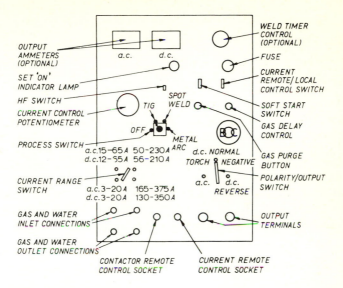

Fig. 6.10. (*c*) TIG welding unit, a.c./d.c., front view.

The rectifier diodes can be checked for faults by first removing the connexion and then measuring the resistance of each diode, first in one direction and then in the other, by means of an ohmmeter. The resistance should be high in one direction and low in the other. If low in both directions, the diode is faulty.

Slope-in (slope-up). When welding thinner sections, the use of the normal welding current at the beginning of the weld often causes burn-through. To prevent this a slope-in control is provided which gives a controlled rise of the current to its normal value over a preselected period of time which can be from 0 to 10 seconds. The soft start control is available on some equipment and performs the same function but it has a fixed time of operation and can be switched in or out as required.

Slope-out (slope-down). The crater which would normally form at a weld termination can be filled to the correct amount with the use of this control. The crater-filling device reduces the current from that used for the welding operation to the minimum that the equipment can supply in a series of steps. The slope-out control performs the same function, but the current is reduced over a period of 20 seconds as a linear function (not steps). The tendency to cracking is reduced by the use of this control (Fig. 6.11).

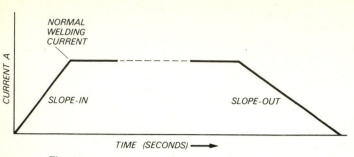

Fig. 6.11

Electrical contactors

The contactors control the various circuits for welding inert gas, water flow and ancillary equipment. The contactor control voltage is of the order of 25 V and if the TIG head is machine mounted, a 110 V supply for the wire feed and tractor is provided. The post-weld argon flow contactor allows the arc to be extinguished without removing the torch with its argon shield from the hot weld area, thus safeguarding the hot electrode end and weld area from contamination whilst cooling. An argon–water purge is also provided and contactors may be foot-switch operated for convenience. Where the ancillary equipment is built into one unit an off-manual metal arc–TIG switch enables the unit to be used for either process.

Gas regulator, flowmeter and economizer

The single-stage gas regulator reduces the pressure in the argon cylinder from 17.5 N/mm² (175 bar)[1] down to 2 bar for supply to the torch (Fig. 6.12a). The flowmeter, which has a manually operated needle valve, controls the argon flow from 0–600 litres/hour to 0–2100 litres/hour according to type (Fig. 6.12b).

The economizer may be fitted in a convenient position near the welder and when the torch is hung from the projecting lever on the unit, argon gas and (if fitted) water supplies are cut off. A micro-switch operated by the lever can also be used to control the HF unit.

Additional equipment

To extend the use of an MMA welding transformer to enable it to be used for the a.c. welding of aluminium and the d.c. welding of stainless and other heat-resistant steels by the TIG process, items of equipment can be added to the transformer to achieve this aim. The operation of these units has already been considered.

[1] Cylinders may now be filled to 200 bar maximum.

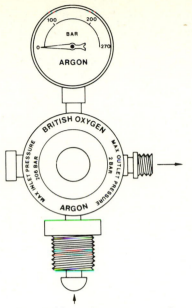

Fig. 6.12. (*a*) Single-stage gas regulator.

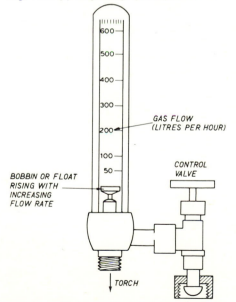

Fig. 6.12. (*b*) Argon flowmeter.

(1) A high-frequency unit which can be used on a.c. or d.c., enabling the arc to be struck without touchdown. This unit also includes a gas–water economizer incorporating a gas purge and gas delay control on the front panel. A latching relay enables the switch on the torch to be pressed for start, after which welding continues until the switch is again pressed for stop.

(2) Contactor unit enabling the arc to be broken instantaneously from a remote control unit, thus avoiding the necessity of withdrawing the arc from the hot workpiece.

(3) Rectifier unit, consisting of single-phase, bridge-connected, silicon diodes, and arc-stabilizing choke enabling the unit to supply d.c. for TIG welding of stainless and heat-resistant steel and for MMA welding with d.c.

(4) Suppressor unit consisting of a bank of electrolytic capacitors enabling a.c. to be used for the welding of aluminium and its alloys.

Torch

There is a variety of torches available varying from lightweight air cooled to heavy duty water cooled types (Fig. 6.13). The main factors to be considered in choosing a torch are:

(1) Current-carrying capacity for the work in hand.

(2) Weight, balance and accessibility of the torch head to the work in hand.

The torch body holds a top-loading compression-type collet assembly which accommodates electrodes of various diameters. They are securely gripped yet the collet is easily slackened for removal or repositioning of the electrode. As the thickness of plate to be welded increases, size of torch and electrode diameter must increase to deal with the larger welding currents required.

Small lightweight air cooled torches rated at 75 A d.c. and 55 A a.c. are ideal for small fittings and welds in awkward places and may be of pencil or swivel head type. Collet sizes on these are generally 0.8 mm, 1.2 mm and 1.6 mm diameter. Larger air cooled torches of 75 A d.c. or a.c. continuous rating or 100 A intermittent usually have collet of 1.6 mm diameter. Air or water cooled torches rated at 300 A intermittent may be used with electrodes from 1.6 to 6.35 mm diameter and can be fitted with water cooled shields while heavy duty water cooled torches with water cooled nozzle of 500 A a.c. or d.c. continuous rating and 600 A intermittent employ larger electrodes. A gas lens can be fitted to the torch to give better gas coverage and to obtain greater accessibility or visibility.

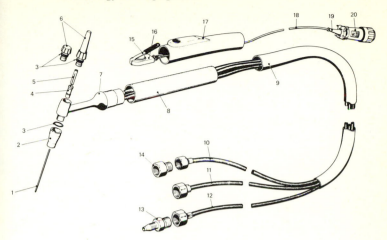

Fig. 6.13. A water-cooled torch.

Key:

1. Thoriated or zirconiated tungsten electrode (0.8, 1.2, 1.6, 2.4, 3.2 mm diameter).
2. Ceramic nozzle.
3. O ring.
4. Collet holder.
5. Collet (sizes as above to take various diameters of electrodes).
6. Electrode cap (long and short).
7. Body assembly.
8. Handle.
9. Sheath.
10. Argon hose assembly.
11. Water hose assembly.
12. Power cable assembly.
13. Adapter power/water; required only in certain cases.
14. Adapter argon; required only in certain cases.
15. Switch actuator.
16. Switch.
17. Switch-retaining sheath.
18. Cable, 2 core.
19. Insulating sleeve.
20. Plug.

Normally, because of turbulence in the flow of gas from the nozzle, the electrode is adjusted to project up to a maximum of 4–6 mm beyond the nozzle. By the use of a lens which contains wire gauzes of coarse and fine mesh, turbulence is prevented and a smooth even gas stream is obtained, enveloping the electrode which, if the gas flow is suitably increased, can be used on a flat surface projecting up to 19 mm from the nozzle orifice, greatly improving accessibility. The lens is screwed on to the torch body in place of the standard nozzle and as the projection of electrode from nozzle is increased the torch must be held more vertically to the work to obtain good gas coverage.

The ceramic nozzles (of alumina or silicon carbide), which direct the flow of gas, screw on to the torch head and are easily removable for cleaning and replacement. Nozzle orifices range from 9.5 to 15.9 mm in diameter and they are available in a variety of patterns for various applications. Ceramic nozzles are generally used up to 200 A a.c. or d.c. but above this water cooled nozzles or shields are recommended because they avoid constant replacement.

Electrodes

The electrode may be of pure tungsten but more generally is of tungsten alloyed with thorium oxide (thoria ThO_2) or zirconium oxide (zirconia ZrO_2). 1% thoriated is used for d.c. welding: 2% thoriated gives good arc striking characteristics on low d.c. values: 1% zirconiated is used for welding aluminium and its alloys.

Tungsten, with a melting point of 3380 °C, has a boiling point of 5950 °C so there is only little vaporization in the welding arc and it retains its hardness when red hot.

The electrodes are supplied with a ground grey finish to ensure good collect contact and standard diameters are 1.2, 1.6, 2.4, 3.2, 4.0, 4.8, 6.4 and 8 mm to a tolerance of 0.075 mm.

Pure tungsten electrodes are generally used for ordinary quality welds. They give a stable a.c. arc when used with balanced wave or HF stabilization and can be used with d.c. and with argon, helium, or argon–helium mixtures. The thoriated tungsten electrodes give easier starting, a more stable arc and less possibility of weld contamination with tungsten particles, and in addition they have a greater current-carrying capacity for a given diameter than pure tungsten. Difficulty is, however, encountered when they are used on a.c. in maintaining a hemispherical end, and as a result zirconiated electrodes are often selected for a.c welding because of the high resistance to tungsten contamination and good arc starting characteristics. They are used therefore for high quality welds in aluminium and magnesium, and like pure tungsten they produce a hemispherical or ball end. Selection of electrode size is usually made by choosing one near the maximum current range for the electrode and work. Too small an electrode will result in overheating and thus contamination of the weld while too large an electrode results in difficult control of the arc. Aim for a shining hemispherical end on the electrode.

Electrode grinding

Usually electrodes need grinding to a point only when very thin materials are to be welded. They should be ground on a fine grit, hard abrasive wheel or abrasive band used only for this purpose to avoid contamination, held at right angles to the wheel face and rotated while grinding.

Fig. 6.14 indicates suitable preparations.

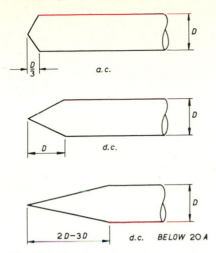

Fig. 6.14. Tungsten electrode preparation.

Electrode current ratings

	d.c.					a.c.						
Electrode diameter (mm)	1.2	1.6	2.4	3.2	4.0	1.2	1.6	2.4	3.2	4.0	4.8	6.0
Max. current (A):						Thoriated may be used but zirconiated is preferable						
Thoriated	70	150	240	380	400	30	60	90	140	195	250	320
Zirconiated	—	—	—	—	—	30	60	90	150	210	275	350

Safety precautions

The safety precautions that should be taken when TIG weldings are similar to those when metal arc welding (p. 323). The EW filters are graded according to welding current thus: up to 50 A, no. 8; 15–75 A, no. 9; 75–100 A, no. 10; 100–200 A, no. 11; 200–250 A, no. 12; 250–300 A, no. 13 or 14.

It will be noted that these filters are darker than those used for similar current range in MMA welding. The TIG (and MIG) arcs are richer in infra-red and ultra-violet radiation, the former requiring some provision for absorbing the extra associated heat, and the latter requiring the use of darker lenses.

It is recommended that the student reads the booklet *Electric arc welding, safety, health and welfare,* new series no. 38, published for the

Department of Employment by HMSO; and also BS 679, *Filters for use during welding*; British Standards Institution.

WELDING TECHNIQUES

Welding of aluminium alloys

Wrought alloys. Weld pure aluminium to N3 with either pure Al or NG3 wire. Welding pure Al or N3 to the Al–Mg alloys is not recommended because of a tendency to weld cracking but NG6 wire gives the best result. When welding pure Al or N3 to the Al–Mg–Si range of alloys special care must be taken to avoid the tendency to crack. Suitable wires are (1) the Al–Si alloy NG21 for easy welding, (2) the Al–Mg alloy wire NG6 for maximum ductility, strength and colour match after anodizing; and (3) the Al–Si alloy NG2 where cracking is particularly troublesome. When welding the Al–Mg alloys N4, N5, N51 and N8 to the Al–Mg–Si alloys H9, H20, H30, the highest strength and corrosion resistance are obtained with NG6 or NG61 wire, but in highly restrained joints, NG21 filler would be more effective in preventing shrinkage cracking. However, this filler wire is not generally recommended for use with Al–Mg alloys because it is liable to form coarse, brittle segregates of Mg_2Si compound at the edge of the fusion zone.

Note. N8 replaces N6 in the metric version of BS 1470 series, N51 is the new supplementary alloy to BS 4300/8/10/12 and NG61 is the new filler alloy in BS 2901.

Casting alloys. The casting alloys with high silicon, e.g. LM2, LM6, LM8 and LM9, can be welded to the Al–Mg–Si alloys with Al–Si NG21 wire. Similarly the Al–Mg casting alloys LM5 and LM10 can be welded to the wrought alloys using NG6 wire. Serious instability of LM10 may occur due to local heating after welding, and this alloy must be solution treated after welding.

Techniques. The shielding gas is pure argon and pre-heating is required for drying only to produce welds of the highest quality. All surfaces and welding wire should be degreased and the area near the joint and the welding wire should be stainless steel wire brushed or scraped to remove oxide and each run brushed before the next is laid.

The angles of torch and filler rod are shown in Fig. 6.15. After switching on the gas, water, welding current and HF unit, the arc is struck by bringing the tungsten electrode near the work (without touching down). The HF sparks jump the gap and the welding current flows.

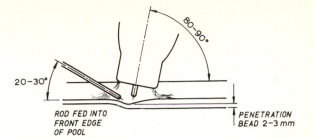

Fig. 6.15. Angles of torch and rod showing typical weld: aluminium alloy plate 11 mm thick 70–80° prep., 1.5 mm root face, no gap, a.c. volts 102, arc volts drop, 17–18 volts: electrode 4.8 mm diameter, current 255 A, arc length approx. 4 mm.

Arc length should be about 3 mm. Practise starting by laying the holder on its side and bringing it to the vertical position, but using the ceramic shield as a fulcrum can lead to damage to the holder and ceramic shield. The arc is held in one position on the plate until a molten pool is obtained and welding is commenced, proceeding from right to left, the rod being fed into the forward edge of the molten pool and always kept within the gas shield. It must not be allowed to touch the electrode or contamination occurs. A black appearance on the weld metal indicates insufficient argon supply. The flow rate should be checked and the line inspected for leaks. A brown film on the weld metal indicates presence of oxygen in the argon while a chalky white appearance of the weld metal accompanied by difficulty in controlling the weld indicates excessive current and overheating. The weld continues with the edge of the portion sinking through, clearly visible, and the amount of sinking which determines the size of the penetration bead is controlled by the welding rate. Preparation for single V butt joints are shown in Fig. 6.16a, while

Thickness of plate (mm)	Single-run Current (A)	
0.9–1.6	50–90	
1.6–2.0	80–120	
3.2	100–140	

Thickness of (plate mm)	Number of runs	Root run current (A)	Filler run current (A)	Filler wire diam. (mm)
		Multi-run		
4.8	2	70–115	120–140	3.2
6.4	2	70–115	120–140	3.2
8.0	3	90–125	140–180	4.8
9.6	3	90–125	140–200	4.8

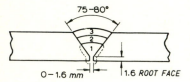

Fig. 6.16. (*a*) Manual TIG process: preparation of aluminium and aluminium alloy plate. Single V butt joints. Flat and vertical.

Thickness of plate (mm)	Number of runs	Root run current (A)	Filler run current (A)	Filler wire diam. (mm)
6.4	2	110–130	120–150	3 down-hand, 5 vert.
8	2	110–130	130–180	5
9.5	3	120–140	140–200	5

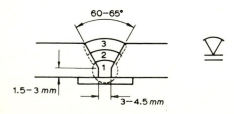

Fig. 6.16. (*b*) Manual TIG process: butt weld preparation, flat and vertical using backing strip, aluminium alloys.

Fig. 6.17*b* shows the backing strip method, and it should be noted that there is a somewhat narrower angle of preparation and heavier current for the backing strip method.

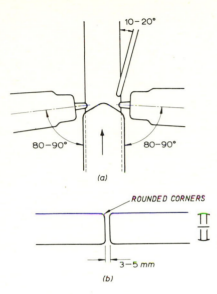

Fig. 6.17. (*a*) Angles of torches and filler wire. (*b*) Double operator vertical welding aluminium alloys.

Aluminium and aluminium alloys: weldability and resistance to atmospheric attack

Alloy BS	ISO	MIG TIG	Resistance and spot	Resistance to atmospheric attack
1	Al 99.99	E	G	E
1A	Al 99.8	E	G	E
1B	Al 99.5	E	V	V
1C	Al 99.0	E	V	V
N3	Al–Mn	E	E	V
N4	Al–Mg2	V	E	V
N51	Al–Mg2.7–Mn	V	E	V
N5	Al–Mg3.5	E	E	V
N8	Al–Mg4.5	E	—	—

Aluminium and aluminium alloys: weldability and resistance to atmospheric attack (contd.)

Alloy BS	ISO	MIG TIG	Resistance and spot	Resistance to atmospheric attack
H9	Al–Mg–Si	V	V	
H15	Al–Cu4–Si–Mg	U	E	
H20	Al–Mg1–Si–Cu	V	V	
H30	Al–Si1–Mg–Mn	V	V	

The vertical double-operator method (Fig. 6.17), using filler wire on one side only, gives sound non-porous welds with accurate penetration bead controlled by one operator 'pulling through' on the reverse side.

Aluminium and aluminium alloys: recommended filler wires for general engineering purposes (BS 1470 *and* 1490)

Alloy BS	ISO	Recommended filler wire
1A	Al 99.8	G1A
1B	Al 99.5	G1A
1C	Al 99.0	G1C or G1b
N3	Al–Mn	NG3
N4	Al–Mg2	NG6 or NG52 or NG5 for optimum corrosion resistance
N51	Al–Mg3	NG6
N5	Al–Mg3.5	NG6
N8	Al–Mg4.5	NG6 or NG61 for optimum strength
H9	Al–Mg–Si	NG21 or NG6
H20	Al–Mg–Si–Cu	NG21 or NG6
H30	Al–Mg–Si–Mn	NG21 or NG6
LM5	Al–Mg	NG6
LM6	Al–Si12	NG21 or NG2
LM8	Al–Si5–Mg	NG21 or NG2

Note. NG2, Al–Si 12; NG6, Al–Mg 5; NG21, Al–Si 5. NG61 is a higher strength version of N6, specially developed for welding N8. It has 5.1–5.5% Mg with Mn and Cr. Butt joints in N8 welded with NG61 have approximately 20 N/mm² greater tensile strength than similar joints welded with N6.

There is excellent argon shielding from both sides but the method is expensive in man-hours of work. Typical figures for a 6 mm plate are: 1 run at 90 to 150 A using 5 mm wire with a 3 mm gap.

Magnesium alloys

The most frequently welded alloys are the wrought alloys ZW1, AZ31, AZM, AM503 and more recently the nuclear alloys Magnox AL80, and Magnox ZR55. Equipment is similar to that for the aluminium alloys and the technique similar, welding from right to left with a short arc and with the same angles of torch and filler rod with pure argon as the shielding gas, and an a.c. supply. Little movement of the filler rod is required but with material over 3 mm thick some weaving may be used. For fillet welding the torch is held so as to bisect the angle between the plates and at sufficient angle from the vertical to obtain a clear view of the molten pool, with enough weave to obtain equal leg fusion. Tilted fillets are used to obtain equal length.

The material is supplied greased or chromated. Degreasing removes the grease and it is usual to remove the chromate by wire brushing from the side to be welded for about 12 mm on each side of the weld and to leave the chromate on the underside where it helps to support the penetration bead. No back purge of argon is required. The surface and edges should be wire brushed and the filler rod cleaned before use. Each run should be brushed before the next is laid.

Backing plates can be used to profile the underbead and can be of mild steel, or of aluminium or copper 6.4–9.5 mm wide, with grooves 1.6–3.2 mm deep for material 1.6–6.4 mm thick (Fig. 6.18).

Jigs together with correct welding sequence can be used to prevent distortion, but if it occurs the parts can be raised to stress relief temperatures of approximately 250 °C (or 330 °C for RZ5 and TZ6 and 350 °C for ZT1) for times varying from 15 minutes for AZ31, AZM, ZW1 and ZW3, 1 hour for A8 and AZ291, 10 hours for ZRE1 and 2 hours for RZ5, TZ6 and ZT1, and pressing or hammering to shape. Ensuring sufficient flow of argon will prevent any oxidized areas, porosity, and entrapped oxides and nitrides.

Welding rods are of similar composition to that of the plate and the table indicates the relative weldability of similar and dissimilar alloys. It is not recommended that the Mg–Zr alloys should be welded to alloys containing Al or Mn. Recommended filler rods are shown in brackets.

For fillet welds the torch is held at about 90 ° to the line of travel and roughly bisecting the angle between the plates, so that the nozzle is clear of either plate and it is often neccessary to have more projection of the wire beyond the nozzle to give good visibility, but it should be kept to a minimum. Tilted fillets give a better weld profile and equal leg length.

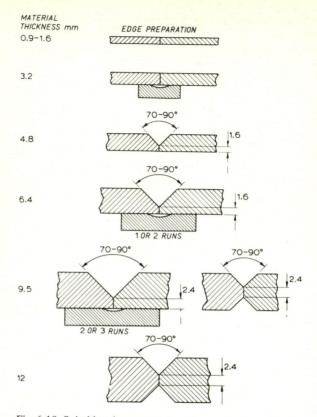

Fig. 6.18. Suitable edge preparations for magnesium alloy butt weld (dimensions in mm).

The tables on pages 401–2 give recommended filler wires for the various alloys and their weldability.

Stainless and heat-resistant steel

The production welding of stainless steel by the process is apt to be rather slow compared with metal arc, MIG and pulse arc.

Areas adjacent to the weld should be thoroughly cleaned with wire brushes and a d.c. supply with torch negative used. The shielding gas is pure argon or argon–hydrogen mixtures (up to 5%) which give a more fluid weld pool, faster rate of deposition (due to the higher temperature of the arc), better 'wetting' and reduction of slag skin by the hydrogen.

The addition of oxygen, CO_2 or nitrogen is not recommended. Joint preparation is given in Fig. 6.19.

Weldability code scheme:

A   Good (a plus sign indicates   B   Fair
     particularly good weldability)   D   Not recommended
C   Possible under certain conditions

Suitability for welding: magnesium wrought alloys

Alloy	AM503	AZM	AZ31	ZW1	ZW3	ZW6	ZTY
AM503	(AM503) A+	(AM503) C	(AM503) B	(ZW1) B	D	D	D
AZM		(AZM) A	(AZM) C	(ZW1) C	D	D	D
AZ31			(AZ31) A	ZW1) C	D	D	D
ZW1				(ZW1) A+	(ZW1) B	D	(ZW1) A
ZW3					(Z2Z2C)[1] C[2]	D	(ZW1) C
ZW6						D	D
ZTY							(ZTY) A+

Note. The nuclear alloys AL80 and ZR55 both have good weldability.

[1] For machine welds only; ZW1 rod can be used for manual welds but with reduced weld efficiency.

[2] ZW3 should be used only when machine welding can be applied or when only simple manual welds are involved.

Suitability for welding: magnesium casting alloys

Alloy	A8	ZRE1	MSR	RZ5	TZ6	ZT1	Z5Z
A8	(A8) A	(ZRE1) C	(MSR) C	(ZRE1) C	(ZT1) C	(ZT1) C	D
ZRE1		(ZRE1) A+	(MSR) A	(ZRE1) A	(ZT1) C	(ZT1) C	D
MSR			(MSR) A	(MSR) A	(ZT1) C	(ZT1) C	D
RZ5				(RZ5) A	(ZT1) C	(ZT1) C	D
TZ6					(TZ6) A	(ZT1) A	D
ZT1						(ZT1) A+	D
Z5Z							D*

* Only simple repair work should be carried out on Z5Z castings.

Suitability for welding: magnesium to wrought alloys

Alloy	AM503	AZM	AZ31	ZW1	ZW3	ZW6	ZTY
A8	(AM503) C	(AZM) A	(AZM) A	D	D	D	D
ZRE1	D	(ZRE1) C	(ZRE1) C	(ZW1) A	(ZW1) B	D	(ZTY) C
MSR	D	(MSR) C	(MSR) C	(MSR) A	(MSR) B	D	(MSR) A
RZ5	D	(ZRE1) C	(ZRE1) C	(ZW1) A	(ZW1) B	D	(ZTY) C
TZ6	D	D	D	(ZW1) A	(ZW1) B	D	(ZTY) A
ZT1	D	D	D	(ZW1) A	(ZW1) B	D	(ZTY) A
Z5Z	D	D	D	D	D	D	D

Technique. The torch is held almost vertically to the line of travel and the filler rod fed into the leading edge of the molten pool, the hot end never being removed from the argon shield. Electrode extension beyond the nozzle should be as short as possible: 3.0–5.0 mm for butt and 6.0–11.0 for fillets. To prevent excessive dilution which occurs with a wide weave, multi-run fillet welds can be made with a series of 'stringer beads'. The arc length should be about 2 mm when no filler wire is used and 3–4 mm with filler. Gas flow should be generous and a gas lens can be used to advantage to give a non-turbulent flow. Any draughts should be avoided.

Nickel alloys

Although this process is very suitable for welding nickel and its alloys it is generally considered rather slow, so that it is mostly done on thinner sections of sheet and tube.

Shielding gas should be commercially pure argon and addition of hydrogen up to 10% helps to reduce porosity and increases welding speeds especially for Monel 400 and Nickel 200. Helium has the same advantages, greatly increasing welding speed. Great care should be

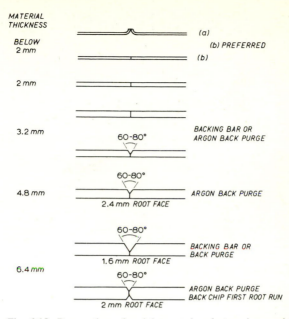

Fig. 6.19. Preparation of stainless steel and corrosion- and heat-resistant steels.

taken to avoid disturbance of the protective inert gas shield by draughts, and the largest nozzle diameter possible should be used, with minimum distance between nozzle and work. The use of a gas lens increases the efficiency of the shield and gives increased gas flow without turbulence. Argon flow is of the order of 17–35 litres per hour for manual operation and higher for automatic welding. Helium flow should be about $1\frac{1}{2}$–3 times this flow rate.

The torch is held as near 90° to the work as possible – the more acute the angle the greater the danger of an aspirating effect causing contamination of the gas shield. Electrodes of pure, thoriated or zirconiated tungsten can be used, ground to a point to give good arc control and should project 3–4.5 mm beyond the nozzle for butt welds and 6–12 mm for fillets.

Stray arcing should be avoided as it contaminates the parent plate, and there should be no weaving or puddling of the pool, the hot filler rod end being kept always within the gas shield, and fed into the pool by touching it on the outer rim ahead of the weld. The arc is kept as short as possible – up to 0.5 mm maximum when no filler rod is used.

Nickel alloys. Joint design for fusion welding.

Type of joint		Material thickness (mm) T	Width of groove (top) (mm) W	Root space (mm) S
Square butt	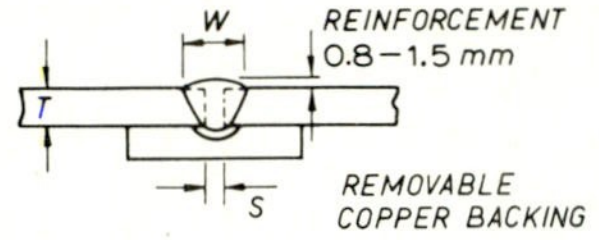 REINFORCEMENT 0.8–1.5 mm, REMOVABLE COPPER BACKING	1.0	3.2	0
		1.2	4.0	0
		1.6	4.8	0
		2.4	4.8–6.4	0–0.8
		3.2	6.4	0–0.8
V groove	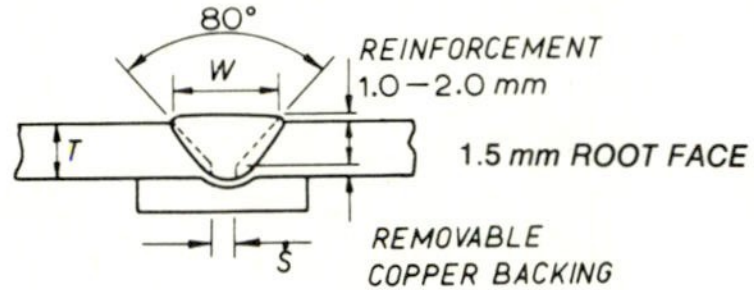 80°, REINFORCEMENT 1.0–2.0 mm, 1.5 mm ROOT FACE, REMOVABLE COPPER BACKING	4.8	8.9	3.2
		6.4	13.0	4.8
		8.0	15.5	4.8
		9.5	18.0	4.8
		12.7	23.0	4.8
		16.0	29.5	4.8
V groove	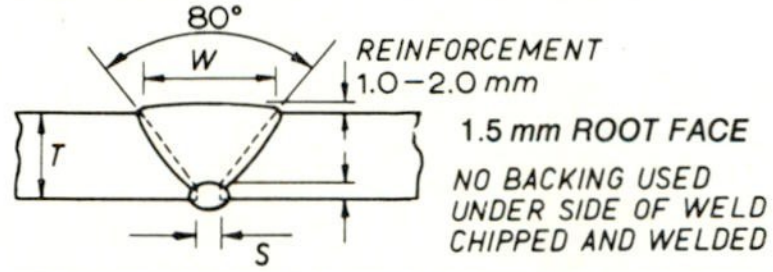 80°, REINFORCEMENT 1.0–2.0 mm, 1.5 mm ROOT FACE, NO BACKING USED UNDER SIDE OF WELD CHIPPED AND WELDED	6.4	10.4	2.4
		8.0	13.0	2.4
		9.5	16.5	3.2
		12.7	21.6	3.2
		16.0	27.0	3.2

Fig. 6.20

Nickel alloys. Joint design for fusion welding.

Type of joint	Material thickness (mm) T	Width of groove (top) (mm) W	Root space (mm) S
Double V groove	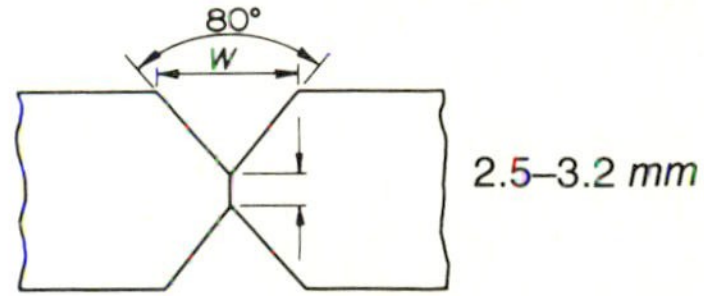 12.7	10.0	3.2
	16	12.5	3.2
	19	15.5	3.2
	25	20.5	3.2
	32	26.0	3.2
U groove	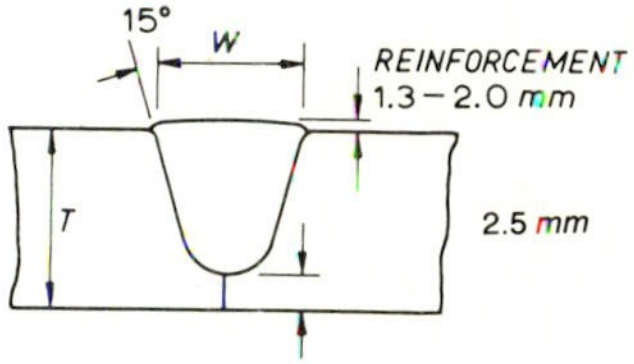 12.7	17.2	3.2
	16	19	3.2
	19	20.5	3.2
	25.4	24.3	3.2
	32	27.2	3.2
	38	30.9	3.2
	44	34.3	3.2
	51	37.7	3.2
Double U groove	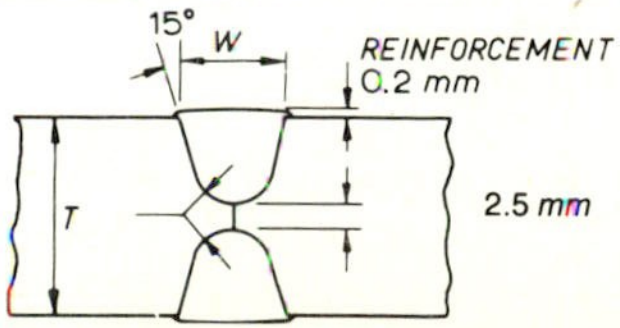 25.4	17.3	3.2
	32	19	3.2
	38	20.6	3.2
	51	24.3	3.2
	64	27.2	3.2

Double V groove — 80°, W, 2.5–3.2 mm

U groove — 15°, W, T, REINFORCEMENT 1.3 – 2.0 mm, 2.5 mm

Double U groove — 15°, W, T, REINFORCEMENT C.2 mm, 2.5 mm

Fig. 6.20

To avoid crater shrinkage at the end of a weld a crater-filling unit (p. 387) can be used or failing this, extension tabs for run-off are left on the work and removed on completion. Fig. 6.20 gives suitable joint preparations.

For vertical and overhead joints the technique is similar to that for steel, but downhand welding gives the best quality welds.

Filler metals for inert-gas shielded metal-arc welding

Material	Filler metal
Nickel 200	Nickel 61
Nickel 201	Nickel 61
Monel 400	Inconel 82
Monel K 500	Inconel 64
Inconel 600	Inconel 82
Inconel 625	Inconel 625
Inconel 718	Inconel 718
Inconel 750	Inconel 69
Incoloy DS	NC 80/20
Incoloy 800	Inconel 82
Incoloy 825	Incoloy 65
Nimonic 75	NC 80/20
Nimonic 80A	Nimonic 90
Nimonic 90	Nimonic 90
Nimonic 263	Nimonic 263
Nimonic PE13	Nimonic PE13
Nimonic PE16	Nimonic PE16
Brightray alloys	NC 80/20
Nilo alloys	Inconel 82 or 92
	Nickel 61

Copper and copper alloys

Direct current, electrode negative is used with argon as the shielding gas for copper and most of its alloys with currents up to 400 A, and angle of torch and rod roughly as for aluminium welding. With aluminium bronze and copper–chromium alloys however, dispersal of the oxide film is difficult when using d.c., and for these alloys a.c. is generally used, but d.c. can be used with helium as the shielding gas. The weld areas should be bronze wire brushed and degreased and each run brushed to remove oxide film. Jigs or tack welds can be used for accurate positioning and to prevent distortion.

Mild steel or stainless steel backing bars, coated with graphite or anti-spatter compound to prevent fusion, can be used to control the penetration bead and also to prevent heat dissipation, or backing may be welded into the joint.

Because of the high coefficient of thermal expansion of copper the root gap has a tendency to close up as welding proceeds, and due to its high thermal conductivity pre-heating from 400–700 °C according to thickness is essential on all but the thinnest sections to obtain a good molten pool. The thermal conductivity of most of the copper alloys is much lower than copper so that pre-heating to about 150 °C is usually sufficient. In the range 400–700 °C a reduction in ductility occurs in most of the alloys so that a cooling period should be allowed between runs. The main grades of copper are (1) tough pitch (oxygen containing) high conductivity, (2) oxygen free, high conductivity, (3) phosphorus deoxidized, and in most cases it is the latter type that is used for pressure vessels, heat exchangers and food processing equipment, etc. If tough pitch copper, in the cast form, is to be welded, a deoxidizing filler rod should be used to give a deoxidized weld and in all cases the weld should be performed as quickly as possible to prevent overheating.

Argon, helium and nitrogen can be used as shielding gases. Argon has a lower heat output than helium and nitrogen because it has a lower arc voltage. Argon can be mixed with helium and the mixture increases the heat output proportionately with helium percentage, so that using the mixture for copper welding a lower pre-heat input is required and penetration is increased. Nitrogen gives the greatest heat output but although the welds are sound they are of rough appearance.

Recommended filler alloys are:

> C7 0.05–0.35% Mn, 0.2–0.35% Si, up to 1.0% Sn, remainder Cu.
> C8 0.1–0.3% Al, 0.1–0.3% Ti, remainder Cu.
> C21 0.02–0.1% B. 99.8% Cu.

Note. Filler alloys for copper and its alloys are to BS 2901, Pt 3, 1970.

Copper–silicon (silicon bronzes). d.c. supply, electrode negative and argon (or helium) shield. These alloys contain about 3% silicon and 1% manganese. They tend to be hot short in the temperature range 800–950 °C so that cooling should be rapid through this range. Too rapid cooling, however, may promote weld cracking.

Filler alloy C9, 96% Cu, 3% Si, 1% Mn. is recommended.

Copper–aluminium (aluminium bronzes). a.c. with argon shield or d.c. with helium shield. The single phase alloys contain about 7% Al and the two phase alloys up to 11% Al with additions of Fe, Mn and Ni. Welding may affect the ductility of both types. Weld root cracking in the 6–8% Al, 2–2.3% Fe alloy used for heat exchangers can be avoided by using a non-matching filler rod. Due to a reduction in ductility at welding temperatures difficulty may be experienced in welding these alloys. The alloy containing 9% Al and 12% Mn with additions of Fe and Ni has good weldability but requires heat treatment after welding to restore corrosion-resistance and other mechanical properties.

Recommended filler alloys are:

C12 6.0–7.0% Al, remainder Cu.
C13 9.0–11.0% Al, remainder Cu.
C20 8.0–10.0% Al, 1.5–3.5% Fe, 4.0–7.0% Ni, remainder Cu.

Copper–nickel (cupro–nickels). d.c. with argon shield and electrode negative. These alloys, used for example for pipe work and heat exchangers, contain from 5–30 % Ni, often with additions of Fe or Mn to increase corrosion-resistance. They are very successfully welded by both TIG and MIG processes but since they are prone to contamination by oxygen and hydrogen, an adequate supply of shielding gas must be used including a back purge in certain cases, if possible.

Recommended filler alloys are:

C17 19–21% Ni, 0.2–0.5% Ti, 0.2–1.0% Mn, remainder Cu.
C19 29–31% Ni, 0.2–0.5% Ti, 0.2–1.0% Mn, remainder Cu.

Copper–tin (phosphor bronzes and gunmetal). d.c. with argon shield. The wrought phosphor bronzes contain up to 8% Sn with phosphorus additions up to 0.4%. Gunmetal is a tin bronze containing zinc and often some lead. Welding of these alloys is most often associated with repair work using a phosphor bronze filler wire but to remove the danger of porosity, non-matching filler wires containing deoxidizer should be used.

Filler alloys recommended are:

C10 4.5–6.0% Sn, 0.4% P, remainder Cu.
C11 6.0–7.5% Sn, 0.4% P, remainder Cu.

Copper–zinc (brass and nickel silvers). a.c. with argon shield, d.c. with helium shield. The copper–zinc alloys most frequently welded are Admiralty brass (70% Cu, 29% Zn, 1% Sn), Naval brass (62% Cu, 36.75% Zn, 1.25% Sn) and aluminium brass (76% Cu, 22% Zn, 2% Al). Porosity, due to the formation of zinc oxide, occurs when matching filler wire is used so that it is often preferable to use a silicon–bronze wire which reduces evolution of fumes. This may, however, induce cracking in the HAZ. Post-weld stress relief in the 250–300 °C range reduces th risk of stress corrosion cracking. The range of nickel silver alloys is often brazed.

Filler alloys recommended are:

C14 70–73% Cu, 1.0–1.5% Sn, 0.02–0.06% As
C15 76–78% Cu, 1.8–2.3% Sn, 0.02–0.06% As.

Work-hardening and precipitation-hardening copper-rich alloys

The work-hardening alloys are not often welded because mechanical properties are lost in the welding operation. Heat-treatable alloys such as copper–chromium and copper–beryllium are welded with matching filler rods using an a.c. supply to disperse the surface oxides. They are usually welded in the solution-treated or in the over-aged condition and then finally heat-treated.

TIG process. Recommended filler wires for copper welding

Type (BS 2870–2875)	Grade	Filler wire	
		Argon or helium shield	Nitrogen shield
C106	Phosphorus deoxidized non-arsenical	C7, C21	C8
C107	Phosphorus deoxidized arsenical		
C101	Electrolytic tough pitch high conductivity	C7, C21	Not recommended
C102	Fire-refined tough pitch high conductivity		
C103	Oxygen-free high conductivity	C7, C21	Not recommended

Edge preparations for butt welds in copper are given in Fig. 7.9*b* p. 452).

AUTOMATIC WELDING

For automatic welding the TIG torch is usually water cooled and may be carried on a tractor moving along a track or mounted on a boom so as to move over the work or for the work to move under the head (Figs. 7.14 and 7.15). The head has a control panel for spark starter, water and gas and current contactor, and the torch has lateral and vertical movement, the arc length being kept constant by a motor-driven movement controlled by circuits operating from the arc length. Filler wire is supplied from the reel to the weld pool by rollers driven by an adjustable speed motor. Heavier currents can be used than with manual operation resulting in greater deposition rates and the accurate control of speed of travel and arc length results in welds of high quality. Plate up to 9 mm thick can be welded in one run, the welding being of course down-hand. Fig. 6.21a illustrates a typical tractor-driven head.

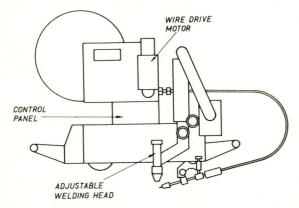

Fig. 6.21. (a) Automatic TIG welding head mounted on electrically driven tractor; 500 A water cooled torch.

Orbital pipe welding unit

This can operate from a standard TIG power source and enables circumferential joints on pipes and tubes to be made automatically. It is of caliper type enabling rapid adjustments to be made, and rotational drive is by a small motor mounted in the handle of the unit, rotational speed being controlled by transistor regulator. A wire feed unit feeds wire to the arc, the feed speed of the wire being controlled by a thyristor regulator. Three sizes are available for pipes with outside diameters of 18–40, 36–80 and 71–160 mm (Fig. 6.21b).

Fig. 6.21. (*b*) Orbital tube welder in action.

Pulsed current TIG welding

When manual welding with the TIG process at a given current setting penetration on thinner sections is achieved by the welder combining judicious manipulation of filler wire with slight variation in the arc length and welding speed and use of a slight weaving motion. When the TIG process is mechanized the above-mentioned variable factors are stabilized, and penetration will depend upon accuracy of fit-up with the correct combination of welding speed and filler wire addition. The pulsed current process enables a close control to be exercised over penetration. A background current provides the ionization of the arc gap and maintains the arc. Regular pulses are provided by a super-imposed pulse current, each pulse (with carefully chosen magnitude) causing a molten pool which just freezes over before the next pulse is applied. The final seam is thus in some ways similar to a series of overlapping spot welds.

TIG power sources are supplied with connexions for the background and pulsed current with the necessary controls. To operate the equip-ment the background current is set between 5 and 12 A and the pulse current is then gradually increased until the required degree of pene-tration is obtained. Reference to Fig. 6.10 shows a power unit with controls for pulse or continuous welding, pulse time, current value and background current. It should be noted that the pulsed current can be varied both in amplitude (current), duration and frequency, all of which affect the weld characteristics.

Fig. 6.22 shows a typical background current and pulse current with rectangular wave form.

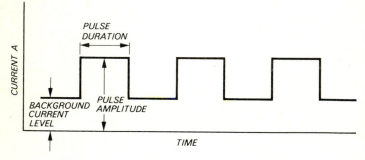

Fig. 6.22. Rectangular wave form pulsed current.

This process can be applied successfully to the welding of sections of differing thickness, to thin sections and edge welds and in cases where build-up of heat during welding causes problems. It is used for the orbital welding of thin pipes as in Fig. 6.21*b*.

Spot welding

Mild steel, stainless and alloy steel and the rarer metals such as titanium can be spot welded using a TIG spot welding gun. The pistol grip water cooled gun carries a collet assembly holding a thoriated tungsten electrode positioned so that the tip of the electrode is approximately 5 mm within the nozzle orifice, thus determining the arc length. Interchangeable nozzles enable various types of joints to be welded flat, or vertically. A variable timing control on the power unit allows the current to be timed to flow to ensure fusion of the joint and a switch on the gun controls gas, water HF unit and current contractor. To make a weld, the nozzle of the gun is pressed on to the work in the required position and when the gun switch is pressed the following sequence occurs: (1) gas and water flow to the gun, (2) the starter ignites the arc and the current flows for a period controlled by the timer, (3) current is automatically switched off and post-weld argon flows to cool electrode tip and weld.

Note. Refer also to BS 3019. *General recommendations for inert gas, tungsten arc welding,* Part 1: *Wrought aluminium, aluminium alloys and magnesium alloys,* Part 2: *Austenitic stainless and heat resisting steels.*

Titanium

Titanium is a silvery coloured metal with atomic weight 47.9, specific gravity 4.5 (cf. Al 2.7 and Fe 7.8) melting point 1800 °C and UTS 310 N/mm² in the pure state. It is used as deoxidizer for steel and sometimes in stainless steel to prevent weld decay and is being used in increasing amounts in the aircraft and atomic energy industries because of its high strength to weight ratio.

Titanium absorbs nitrogen and oxygen rapidly at temperatures above 1000 °C and most commercial grades contain small amounts of these gases; the difficulty in reducing the amounts present makes titanium expensive. Because of this absorption it can be seen that special precautions have to be taken when welding this metal.

Where seams are reasonably straight a TIG torch with tungsten electrode gives reasonable shrouding effect on the upper side of the weld, and jigs may be employed to concentrate the argon shroud even more over the weld. In addition the underbead must be protected and this may be done by welding over a grooved plate, argon being fed into this groove so that a uniform distribution is obtained on the underside. It can be seen therefore that the success of welding in these conditions depends upon the successful shrouding of the seam to be welded.

For more complicated shapes a vacuum chamber may be employed. The chamber into which the part to be welded is placed is fitted with hand holes, and over these are bolted long-sleeve rubber gloves into which the operator places his hands, and operates the TIG torch and filler rod inside the chamber. An inspection window on the sloping top fitted with welding glass enables a clear view to be obtained. All cable entries are sealed tightly and the air is extracted from the chamber as completely as possible by pumps, and filled with argon, this being carried out again if necessary to get the level of oxygen and nitrogen down to the lowest permissible level. The welding may be performed with a continuous flow of argon through the chamber or in a static argon atmosphere. Since the whole success of the operation depends upon keeping oxygen and nitrogen level down to an absolute minimum great care must be taken to avoid leaks as welding progresses and to watch for discoloration of the weld indicating absorption of the gases.

Tantalum

Tantalum is a metallic element of atomic weight 180.88, HV 45, melting point 2910 °C, specific gravity 16.6, it is a good conductor of heat and is used in sheet form for heat exchangers, condensers, small tubes, etc. Like titanium it readily absorbs nitrogen and oxygen at high temperatures and also readily combines with other metals so that it is necessary to weld it with a TIG torch in a vacuum-purged argon chamber as for titanium.

Beryllium

Beryllium is a light steely coloured metallic element, atomic weight 9, melting point 1280–1300 °C, specific gravity 1.8, HV55–60. It is used as an alloy with copper (beryllium bronze) to give high strength with elasticity. As with tantalum and titanium it can be welded in a vacuum-purged argon chamber using a TIG torch. Also it can be pressure welded and resistance welded.

Electron beam welding is now being applied to the welding of the rare metals such as the foregoing, in the vacuum chamber. A beam of electrons is concentrated on the spot where the weld is required. The beam can be focused so as to give a small or large spot, and the power and thus the heating effect controlled. Successful welds in a vacuum-purged argon chamber have been made in beryllium and tantalum.

PLASMA WELDING

Plasma welding is a process which complements and in some cases is a substitute for, the TIG process, offering, for certain applications, greater welding speed, better weld quality and less sensitivity to process variations. The constricted arc allows lower current operation than TIG for similar joints and gives a very stable controllable arc at currents down to 0.1 A, below the range of the TIG arc, for welding thin metal foil sections. Manual plasma welding is operated over a range 0.1–100 A from foil to 3–4 mm thickness in stainless steel, nickel and nickel alloys, copper, titanium and other rare earth metals (not aluminium and magnesium).

Ions and plasma

An ion is an atom (or group of atoms bound into a molecule) which has gained or lost an electron or electrons. In its normal state an atom exhibits no external charge but when transference of electrons takes place an atom will exhibit a positive or negative charge depending upon whether it has lost or gained electrons. The charged atom is called an ion and when a group of atoms is involved in this transference the gas become ionized.

All elements can be ionized by heat to varying degrees (thermal ionization) and each varies in the amount of heat required to produce a given degree of ionization, for example argon is more easily ionized than helium. In a mass of ionized gas there will be electrons, positive ions and neutral atoms of gas, the ratio of these depending upon the degree of ionization.

A plasma is the gas region in which there is practically no resultant charge, that is, where positive ions and electrons are equal in number; the region is an electrical conductor and is affected by electric and magnetic fields. The TIG torch produces a plasma effect due to the shield of argon and the tungsten arc but a plasma jet can be produced by placing a tungsten electrode centrally within a water cooled constricted copper nozzle. The tungsten is connected to the negative pole (cathode) of a d.c. supply and the nozzle to the positive pole (anode). Gas is fed into the nozzle and when an arc is struck between tungsten electrode and nozzle, the gas is ionized in its passage through the arc and due to the restricted shape of the nozzle orifice, ionization is greatly increased and the gas issues from the nozzle orifice, as a high-temperature, high-velocity plasma jet, cylindrical in shape and of very narrow diameter realizing temperatures up to 10 000 °C. This type is known as the non-transferred plasma (Fig. 6.23a). With the transferred arc process used for welding, cutting and surfacing, the restricting orifice is in an inner water-cooled nozzle within which the tungsten electrode is centrally placed. Both work and nozzle are connected to the anode and the tungsten electrode to the cathode of a d.c. supply (in American terms, d.c.s.p., direct current straight polarity). Relatively low plasma gas flow of argon, argon–helium or argon hydrogen) is necessary to prevent turbulence and disturbance of the weld pool, so a further supply of argon is fed to the outer shielding nozzle to protect the weld (Fig. 6.23b). In the lead from work to power unit there is a contactor switch as shown.

A high-frequency unit fed from a separate source from the mains

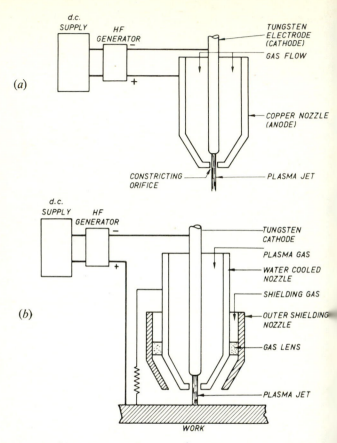

Fig. 6.23. (*a*) Non-transferred arc plasma. (*b*) Transferred arc plasma.

supply initiates the pilot arc and the torch nozzle is positioned exactl
over the work. Upon closing the contactor switch in the work-to-powe
unit connexion, the arc is transferred from electrode to work via th
plasma. Temperatures up to 17 000 °C can be obtained with this arc. T
shape the arc two auxiliary gas passages on each side of the main orific
may be included in the nozzle design. The flow of cooler gas throug
these, squeezes the circular pattern of the jet into oval form, giving
narrower heat-affected zone and increased welding speed. If a coppe
electrode is used instead of tungsten as in the welding of zirconium, it
made the anode. The low-current arc plasma with currents in the rang
0.1–15 A has a longer operating length than the TIG arc with muc
greater tolerance to change in arc length without significant variation i

the heat energy input into the weld. This is because it is straight, of narrow diameter, directional and cylindrical, giving a smaller weld pool, deeper penetration and less heat spread whereas the TIG arc is conical so that small changes in arc length have much more effect on the heat output. Fig. 6.24 compares the two arcs. Since the tungsten electrode is well inside the nozzle (about 3 mm) in plasma welding, tungsten con-tamination by touchdown or by filler rod is avoided, making welding easier.

Equipment. In the range 5–200 A a d.c. rectifier power unit with drooping characteristic and an OCV of 70 V can be used for argon and argon–hydrogen mixtures. If more than 5% hydrogen is used, 100 V or more is required for pilot arc ignition. This arc may be left in circuit with the main arc to give added stability at low current values. Existing TIG power sources such as that in Fig. 6.10b may be used satisfactorily, reducing capital cost, the extra equipment required being in the form of a console placed on or near the power unit. Input is 380–440 V, 50 Hz, a.c. single-phase with approximately 3.5 A full load current. It houses relays and solenoid valves controlling safety interlocks to prevent arc initiation unless gas and water pressures are correct; flowmeters for plasma and shielding gases, gas purge and post-weld gas delay and cooling water controls. For low-current welding in the range 0.1–15 A, often referred to in Europe as micro-plasma and in America as needle plasma welding, the power unit is about 3 KVA, 200–250 V, 50 Hz single-phase input, fan cooled with OCV of 100 V nominal and 150 V peak, d.c., for the main arc, and output current ranges 0.1–2.0 A and 1.0–15 A. The pilot arc takes 6 A at 25 V start and 2.5 A at 24 V running, the main arc being cylindrical and only 0.8 mm wide. Tungsten electrodes are 1.6, 2.4 and 3.2 mm diameter, depending upon appli-cation.

Gases. Pure argon is used for plasma and shield when welding reactive metals such as titanium and zirconium which have a strong affinity for hydrogen. For stainless steel and high-strength nickel alloys argon, or argon–hydrogen mixtures are used. Argon–5% hydrogen mix-tures are supplied for this purpose (cylinders coloured blue with a wide red band around the middle) and argon–8% hydrogen and even up to 15% hydrogen are also used. With these mixtures the arc voltage is increased giving higher welding speeds (up to 40%) and the thin oxide film present even on stainless and alloy steels is removed by hydrogen reduction giving a clean bright weld, and the wetting action is improved. De-ionized cooling water should be used, and gas flow rates at 2.0 bar pressure are about 0.25–3.3 litres per minute plasma and 3.8–7 litres per

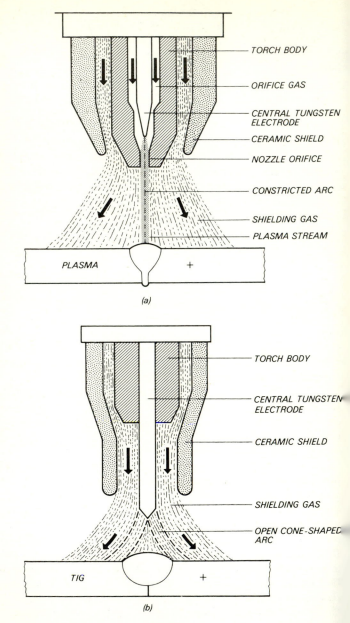

Fig. 6.24. (*a*) Arc plasma. (*b*) TIG arc.

minute, shielding in the 5–100 A range. Cleaning and preparation are similar to those for TIG, but when welding very thin sections, oil films or even fingerprints can vary the degree of melting, so that degreasing should be thorough and the parts not handled in the vicinity of the weld after cleaning.

Technique. Using currents of 25–100 A square butt joints in stainless steel can be made in thicknesses of 0.8–3.2 mm with or without filler rod, the angle of torch and rod being similar to that for TIG welding. The variables are current, gas flow and welding rate. Too high currents may break down the stabilizing effect of the gas, the arc wanders and rapid nozzle wear may occur. In 'double arcing' the arc extends from electrode to nozzle and then to the work. Increasing plasma gas flow improves weld appearance and increases penetration, and the reverse applies with decreased flow. A turbulent pool gives poor weld appearance.

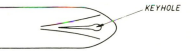

KEYHOLE

Fig. 6.25. Surface tension causes molten metal to flow and close keyhole.

In 'keyhole' welding of thicknesses of 2.5–6.5 mm a hole is formed in the square-edge butt joint at the front edge of the molten pool, with the arc passing through the section (Fig. 6.25). As the weld proceeds, surface tension causes the molten metal to flow up behind the hole to form the welding bead, indicating complete penetration, and gas backing for the underbead is required. When butt welding very thin sections, the edges of the joint must be in continuous contact so that each edge melts and fuses into the weld bead. Separation of the edges gives separate melting and no weld. Holding clamps spaced close together near the weld joint and a backing bar should be used to give good alignment, and gas backing is recommended to ensure a fluid pool and good wetting action. Flanging is recommended for all butt joints below 0.25 mm thickness and allows for greater tolerance in alignment. In the higher current ranges joints up to 6 mm thick can be welded in one run. Lap and fillet welds made with filler rod are similar in appearance and employ a similar technique to those made with the manual TIG process. Edge welds are the best type of joint for foil thickness, an example being a plug welded into a thin-walled tube.

In tube welding any inert gas can be used within the tube for underbead protection.

Faults in welding very thin sections are: excessive gaps which the metal cannot bridge, poor clamping allowing the joint to warp, oil films varying the degree of melting, oxidation or base metal oxides preventing good 'wetting', and nicking at the ends of a butt joint. This latter can be avoided by using run-off plates or by using filler rod for the start and finish of the weld.

PLASMA CUTTING

This process gives clean cuts at fast cutting speeds in aluminium, mild steel and stainless steel. All electrically conducting materials can be cut using this process, including the rare earth metals such as tungsten and titanium (Fig. 6.26*a*, *b*).

The arc is struck between the central tungsten electrode and a copper nozzle body of a water-cooled torch. A gas mixture passes under pressure through the restricted nozzle orifice and around the arc, emerging as a high-temperature (up to 16 000 °C) ionized plasma stream, and the arc is transferred from the nozzle and passes between the electrode and work.

Great improvements have been made in plasma cutting torches by greatly constricting the nozzle and thus narrowing the plasma stream, giving a narrower and cleaner cut with less consumption of power. In certain cases 'double arcing' may occur, in which the main arc jumps from tungsten electrode to nozzle and then to work, damaging both nozzle and electrode.

Power unit. A typical 20 kVA unit has input voltages of 220, 380 or 415 V, is fan cooled with a d.c. output of 200 V open circuit and a current range of 10–100 A. The auxiliary (or striking) arc operates with a lower current than the main arc, thus reducing nozzle erosion.

Gas mixtures of 10% Ar and 90% N_2 are used for stainless steel up to and including 6.5 mm thick. Above this, up to 25 mm thickness an 80% Ar 20% H_2 mixture is used for stainless steel and aluminium alloys. A typical cut in 13 mm thick stainless steel at 100 A uses 20 litres of Ar and 5 litres of H_2 per minute at a cutting speed of 0.4 m/min. Thicker sections require higher currents.

Torch. Torches can be either hand or machine operated and are supplied with spare electrodes, cutting tips and heat shields (Fig. 6.26*c*). In certain cases the presence of water can contribute to an electrolytic action in the torch which quickly causes sufficient corrosion in the waterways of the torch to give water leaks, ruining the torch. This

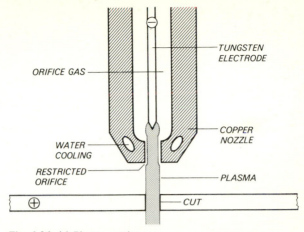

TUNGSTEN
ELECTRODE

ORIFICE GAS

COPPER
NOZZLE

WATER
COOLING

RESTRICTED
ORIFICE

PLASMA

⊕

CUT

Fig. 6.26. (*a*) Plasma cutting.

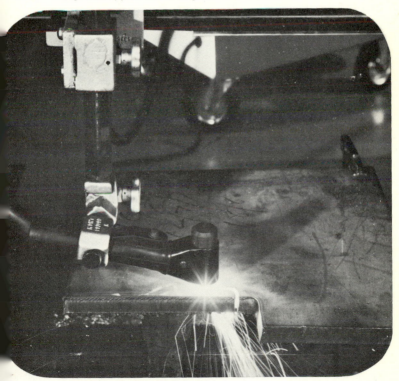

Fig. 6.26. (*b*) Mechanized plasma cutting.

electrolytic action occurs as a result of the presence of dissolved minerals in the water, and potential differences between the electrode and nozzle. To prevent this a supply of de-ionized water is used (about 2 litres per minute) and this prevents the electrolytic action.

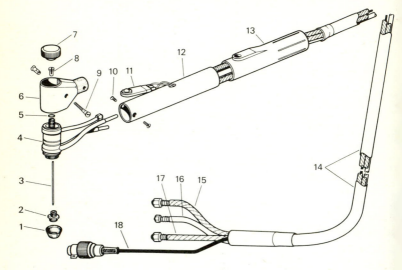

Key:

1. Heat shield.
2. Cutting tip.
3. Tungsten electrode 3.2 mm diameter.
4. Torch body.
5. O ring.
6. Torch head cover.
7. Torch cap.
8. Collet.
9. Cover retaining nuts and screw.
10. Screw.
11. Torch switch.
12. Handle.
13. Switch boot.
14. Sheath.
15. Main arc cable.
16. Gas hose.
17. Pilot arc cable.
18. Control cable with plug.

(*c*) Plasma cutting torch.

Fig. 6.26. (*c*) Plasma cutting torch.

De-ionized water

Water taken directly from a tap contains dissolved carbon dioxide (CO_2) with possibly hydrogen sulphide (H_2S) and sulphur dioxide (SO_2) together with mineral salts such as calcium bicarbonate and magnesium sulphate which have dissolved into the water in its passage through various strata in the earth. These minerals separate into ions in the water, for example magnesium sulphate $MnSO_4$ separates into Mg^{2+} ions (positively charged) and sulphate ions SO_4^{2-} which are negatively charged. Similarly with calcium bicarbonate ($Ca(HCO_3)_2$) which gives metallic ions, Ca^{2+} and bicarbonate ions HCO_3^-.

By boiling or distilling the water the minerals are not carried over in the steam (the distillate) so that the distilled water is largely free of mineral salts but may contain some dissolved CO_2. Distillation is however rather slow and costly in fuel and a more efficient method of obtaining de-ionized water is now available and uses ion exchange resins. The resins are organic compounds which are insoluble in water. Some resins behave like acids and others behave like alkalis and the two kinds can be mixed without any chemical change taking place.

When water containing dissolved salts is passed through a column containing the resins the metal ions change places with the hydrogen ions of the acidic resin so that the water contains hydrogen ions instead of metallic ions. Similarly the bicarbonate ions are replaced by hydroxyl (OH) ions from the alkaline resin. During this exchange insoluble metallic salts of the acid resin are formed and the insoluble alkali resin is slowly converted into insoluble salts of the acids corresponding to the acid radicals previously in solution.

The water emerging from the column is thus completely de-ionized and now has a greater resistivity (resistance) than ordinary tap water so that it conducts a current less easily. A meter can be incorporated in the supply of de-ionized water to measure its resistivity (or conductivity) and this will indicate the degree of ionization of the water in the cooling circuit. When the ionization rises above a certain value the resins must be regenerated. This is done by passing hydrochloric acid over the acidic resin, so that the free hydrogen ions in the solution replace the metallic ions (Ca and Mg). This is followed by passing a strong solution of sodium hydroxide (NaOH) through the column, when the hydroxyl ions displace the sulphate and bicarbonate ions from the alkaline resin.

This process produces de-ionized water, purer than distilled water, easily and quickly.

Water injection plasma cutting

In this process, water is injected through four small-diameter jets tangentially into an annular swirl chamber, concentric with the nozzle, to produce a vortex which rotates in the same direction as the cutting gas (Fig. 6.27). The water velocity is such as to produce a uniform and stable film around the high-temperature plasma stream, constricting it and reducing the possibility of double arcing. Most of the water emerges from the nozzle in a conical jet which helps to cool the work surface.

The cut produced is square within about 2 ° on the right-hand side (viewed in the direction of cutting), whilst the other side is slightly bevelled, caused in general by the clockwise rotation of the cutting gas, which is commercial purity nitrogen. The use of nitrogen reduces cutting costs as it is cheaper than gas mixtures.

The process gives accuracy of cut at high cutting speeds with very smooth cut surfaces. There is little or no adherent dross and the life of the cutting nozzles is greatly increased; mild and carbon steels, alloy and stainless steels, titanium and aluminium are among the metals which can be efficiently cut, in thicknesses from 3 to 75 mm for stainless steel.

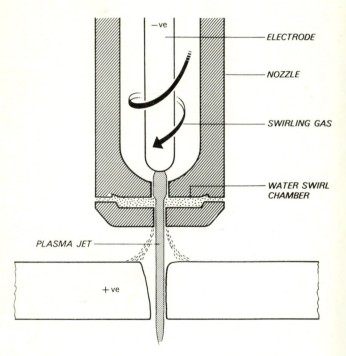

Fig. 6.27. The principle of water-injection plasma cutting.

The noise and fumes associated with plasma cutting can be reduced by the use of a water muffler fitted to the torch to reduce the noise and by the use of a water table which replaces the normal cutting table and which removes up to 99.5% of the particles and fumes by scrubbing, using the kinetic energy of the hot gases and molten metal stream from the kerf.

This cutting equipment is currently fitted to existing cutting installations up to the largest sizes used in shipyards and including those with numerical control.

PLASMA HOT WIRE SURFACING

This is an automated process developed to deposit overlays of nickel, inconel, other nickel alloys, austenitic stainless steel, etc. on to thick walled vessels for the chemical, petroleum and nuclear industries.

The process uses a transferred-arc plasma head mounted at right angles to the surface to be overlaid and fed from its own d.c. power source. Twin filler wires are fed by motor-driven, constant-speed rollers from wire spools making a small angle with the work surface and with each other so that the wires meet where the molten pool is produced by the plasma on the work surface. Because the stand-off distance from work to plasma head is about 20 mm with the tubular plasma there is room for the wires and for subsequent positioning of them. A separate a.c. power source is used to pass a current of the order of 200 A along the two wires which are in series connexion and there is considerable heating effect where each wire meets the molten pool, so that they melt and are fused into the surface by the plasma (which takes currents of 450–600 A for the transferred arc), with minimum dilution.

The gas for the plasma is a 75–25 helium–argon mixture with a shielding gas of argon fed to the outer nozzle. A further argon supply is

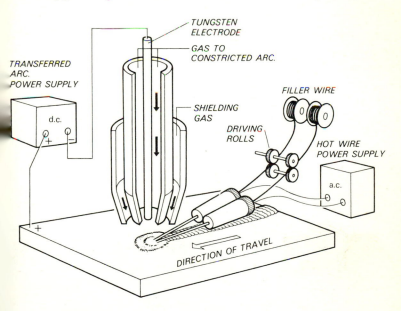

Fig. 6.28

fed to a water-cooled trailing shield about 150 × 100 mm in area to provide a protection over a large area of the hot deposit. The wires project about 35–40 mm from their respective guides, and a gas lens is fitted to ensure adequate protection for the molten pool (Fig. 6.28).

The head can be oscillated to a width of 40 mm at a frequency of 35–40 cycles per minute to give a deposit width of about 50 mm. Overlap between runs is 3–6 mm giving a regular surface and uniform deposit about 5 mm thick.

Deposition rates are from 15–27 kg per hour with dilution varying between 4 and 7% depending upon the type of overlay and the transferred-arc plasma current used.

7

Gas shielded metal arc welding

METALLIC INERT GAS (MIG), CO_2 AND MIXED GAS PROCESSES

It is convenient to consider, under this heading, those applications which involve shielding the arc with argon, carbon dioxide (CO_2) and mixtures of argon with oxygen and/or CO_2, since the power source and equipment is essentially similar except for gas supply.

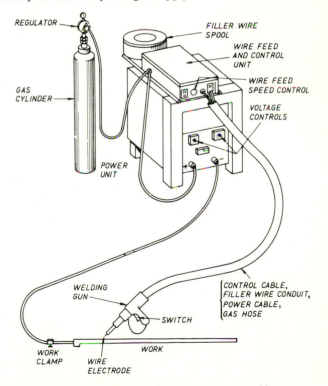

Fig. 7.1. Components of gas shielded metal arc welding process.

With the tungsten inert gas shielded arc welding process, inclusions of tungsten become troublesome with currents above 300 A. The MIG process does not suffer from these disadvantages and larger welding currents giving greater deposition rates can be achieved. The process is suitable for welding aluminium, magnesium alloys, plain and low-alloy steels, stainless and heat-resistant steels, copper and bronze, the variation being filler wire and type of gas shielding the arc.

The consumable electrode of bare wire is carried on a spool and is fed to a manually operated or fully automatic gun through an outer flexible cable by motor-driven rollers of an adjustable speed, and rate of burn-off of the electrode wire must be balanced by the rate of wire feed. Wire feed rate determines the current used.

In addition, a shielding gas or gas mixture is fed to the gun together with welding current supply, cooling water flow and return (if the gun is water cooled) and a control cable from gun switch to control contractors. A d.c. power supply is required with the wire electrode connected to the positive pole (Fig. 7.1).

Spray transfer

In manual metal arc welding, metal is transferred in globules or droplets from electrode to work. If the current is increased to the continuously fed, gas shielded wire, the rate at which the droplets are projected across the arc increases and they become smaller in volume, the transfer occurring in the form of a fine spray.

The type of gas being used as a shield greatly affects the values of current at which spray transfer occurs. Much greater current densities are required with CO_2 than with argon to obtain the same droplet rate. The arc is not extinguished during the operation period so that arc energy output is high, rate of deposition of metal is high, penetration is deep and there is considerable dilution. If currents become excessively high, oxide may be entrapped in the weld metal, producing oxide enfoldment or puckering (in Al). For spray transfer therefore there is a high voltage drop across the arc (30–45 V) and a high current density in the wire electrode, making the process suitable for thicker sections, mostly in the flat position.

The high currents used produce strong magnetic fields and a very directional arc. With argon shielding the forces on the droplets are well balanced during transfer so that they move smoothly from wire to work with little spatter. With CO_2 shielding the forces on the droplet are less balanced so that the arc is less smooth and spatter tendency is greater (Fig. 7.2). The power source required for this type of transfer is of the constant voltage type described later. Spray transfer is also termed free flight transfer.

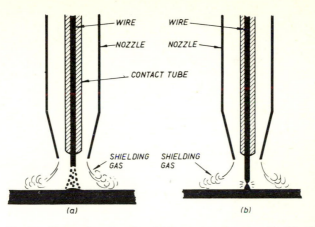

Fig. 7.2. Types of arc transfer. (*a*) Spray transfer: arc volts 27–45 V. Shielding gases: argon, argon–1 or 2% oxygen, argon–20% CO$_2$, argon–2% oxygen–5% CO$_2$. High current and deposition rate, used for flat welding of thicker sections. (*b*) Short-circuit or dip transfer: arc volts 15–22 V. Shielding gases as for spray transfer. Lower heat output and lower deposition rate than spray transfer. Minimizes distortion, low dilution. Used for thinner sections and positional welding of thicker sections.

Short circuit or dip transfer

With lower arc volts and currents transfer takes place in globular form but with intermittent short-circuiting of the arc. The wire feed rate must just exceed the burn-off rate so that the intermittent short-circuiting will occur. When the wire touches the pool and short-circuits the arc there is a momentary rise of current which must be sufficient to make the wire tip molten, a neck is then formed in it due to magnetic pinch effect and it melts off in the form of a droplet being sucked into the molten pool aided by surface tension. The arc is then re-established, gradually reducing in length as the wire feed rate gains on the burn-off until short-circuiting again occurs (Fig. 7.2). The power source must supply sufficient current on short-circuit to ensure melt-off or otherwise the wire will stick into the pool, and it must also be able to provide sufficient voltage immediately after short-circuit to establish the arc. The short-circuit frequency depends upon arc voltage and current, type of shielding gas, diameter of wire, and the power source characteristic. The heat output of this type of arc is much less than that of the spray transfer type and makes the process very suitable for the welding of thinner sections and for positional welding, in addition to multi-run thicker sections, and it gives much greater welding speed than manual arc on light gauge steel, for example. Dip transfer has the lowest weld metal dilution value of all the arc processes.

Semi-short-circuiting arc

In between the spray transfer and dip transfer ranges is an intermediate range in which the frequency of droplet transfer is approaching that of spray yet at the same time short-circuiting is taking place, but is of very short duration. This semi-short-circuiting arc has certain applications, as for example the automatic welding of medium-thickness steel plate with CO_2 as the shielding gas.

d.c. power supply and arc control

There are two methods of automatic arc control:

(1) Constant voltage or potential, known as the self-adjusting arc.
(2) Drooping characteristic or controlled arc (constant current).

The former is more usually employed both on MIG and CO_2 welding plant though the latter may be used with larger diameter wires and higher currents and with the flux cored welding process.

Constant voltage d.c. supply

Power can be supplied from a welding generator with level characteristic (q.v.) or from a natural or forced draught cooled three-phase (balanced load) or one-phase transformer and rectifier arranged to give output voltages of approximately 11–50 V and ranges of current according to the output of the unit (Fig. 7.3a, b, c and d).

The voltage–current characteristic curve, which should be flat or level in a true constant voltage supply, is usually designed to have a slight droop as shown in Fig. 7.4a. Evidently this unit maintains an almost constant arc voltage irrespective of the current flowing. The wire feed motor has an adjustable speed control with which the wire feed speed must be pre-set for a given welding operation. Once pre-set the motor feeds the wire to the arc at constant speed. Supplies for the auxiliaries are generally at 110 V a.c. for the wire feed motor and 25 V a.c. for the torch switch circuit, but some units have a 120 V d.c. supply. For the arc to function correctly the rate of wire feed must be exactly balanced by the burn-off rate to keep the arc length constant. Suppose the normal arc length is that with voltage drop V_M indicated in Fig. 7.4a at M and the current for this length is I_M amperes. If the arc shortens (manually or due to slight variation in motor speed) to S (the volts drop is now V_S) the current now increases to I_S, increasing the burn-off rate, and the arc is lengthened to M. Similarly if the arc lengthens to L, current decreases to I_L and burn-off rate decreases, and the arc shortens to M.

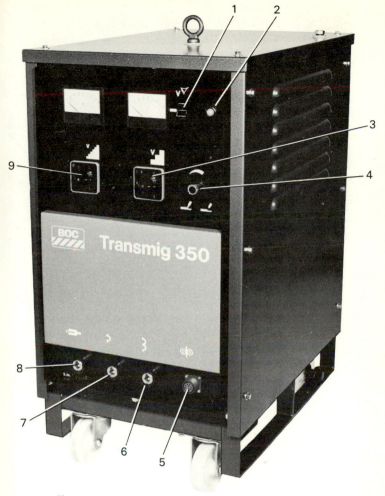

Key:
1. 'Press to set voltage' switch.
2. Indicator lamp.
3. Coarse voltage selector switch.
4. 'Burn-off' control.
5. Wire feed unit socket.
6. High inductance output socket.
7. Low inductance output socket.
8. 'Work' output socket.
9. Fine voltage selector switch.

Fig. 7.3

(a) Fan-cooled transformer–rectifier power unit for inert, mixed and CO₂ gas welding by dip and spray transfer. Input; 3-phase; 50 Hz, 220 or 380–440 V. Output 60 A at 14 V to 400 A at 35 V with 45% duty cycle, 15–53 OCV. The unit incorporates the auxiliary supplies for operating the wire feed unit and a CO₂ heater.

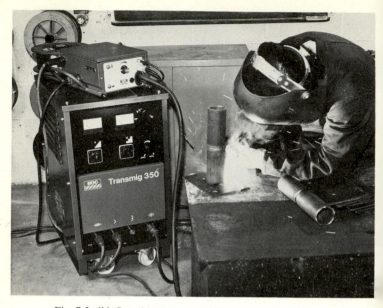

Fig. 7.3. (*b*) Gas-shielded arc welding process (semi-automatic).

Evidently the gradient or slope of the output curve affects the welding characteristics and slope-controlled units are now produced in which the gradient or steepness of the slope can be varied as required and the correct slope selected for given welding conditions.

Drooping characteristic d.c. supply

With this system the d.c. supply is obtained from a welding generator with a drooping characteristic (see p. 259) or more usually from a transformer–rectifier unit. If a.c. is required it is supplied at the correct voltages from a transformer.

The characteristic curve of this type of supply (Fig. 7.4*b*) shows that the voltage falls considerably as the current increases, hence the name. If normal arc length M has volts drop V_M and if the arc length increases to L, the volts drop increases substantially to V_L. If the arc is shortened the volts drop falls to V_s while the current does not vary greatly, hence the name constant current which is often given to this type of supply. The variations in voltage due to changing arc length are fed through control gear to the wire feed motor, the speed of which is thus varied so as to keep a constant arc length, the motor speeding up as the arc lengthens and slowing down as the arc shortens. With this system, therefore, the

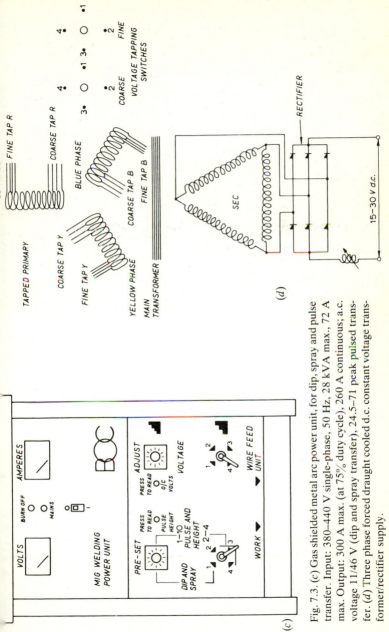

Fig. 7.3. (c) Gas shielded metal arc power unit, for dip, spray and pulse transfer. Input: 380–440 V single-phase, 50 Hz, 28 kVA max., 72 A max. Output: 300 A max. (at 75% duty cycle), 260 A continuous; a.c. voltage 11/46 V (dip and spray transfer), 24.5–71 peak pulsed transfer. (d) Three phase forced draught cooled d.c. constant voltage transformer/rectifier supply.

welding current must be selected for given welding conditions and the control circuits are more complicated than those for the constant voltage method.

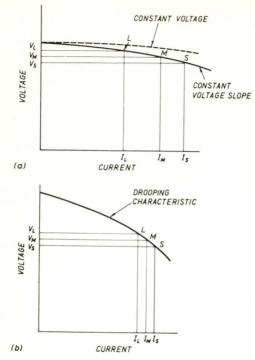

Fig. 7.4. Volt–ampere curves for constant voltage and drooping characteristic sources.

Power source–dip transfer

In order to keep stable welding conditions with a low voltage arc (17–20 V) which is being rapidly short-circuited, the power source must have the right characteristics. If the short-circuit current is low the electrode will freeze to the plate when welding with low currents and voltages. If the short-circuit current is too high a hole may be formed in the plate or excessive spatter may occur due to scattering of the arc pool when the arc is re-established. The power supply must fulfil the following conditions:

(1) During short-circuit the current must increase enough to melt the wire tip but not so much that it causes spatter when the arc is re-established.

(2) The inductance in the circuit must store enough energy during short-circuit to help to start the arc again and assist in maintaining it during the decay of voltage and current. If an inductive reactor or choke (see p. 274) is connected in the arc circuit when the arc is short-circuited the current does not rise to a maximum immediately, so the effect of the choke is to limit the rate of rise of current, and the amount by which it limits it depends upon the inductance of the choke. This limitation is used to prevent spatter in CO$_2$ welding. When the current reaches its maximum value there is maximum energy stored in the magnetic field of the choke. When the droplet necks off in dip transfer and the arc is re-struck, the current is reduced and hence the magnetic field of the choke is reduced in strength, the reduction in energy being fed into the circuit helping to re-establish the arc. If the circuit is to have variable inductance so that the choke can be adjusted to given conditions the coil is usually tapped to a selector switch and by varying the number of turns in circuit the inductive effect is varied. The inductance can also be varied by using a variable air gap in the magnetic circuit of the choke. On some units the fine control of the arc is made by this variable series inductor.

To summarize: the voltage of a constant voltage power source remains substantially constant as the current increases. In the case of a welding power unit the voltage drop may be one or two volts per hundred amperes of welding current, and in these circumstances the short-circuit current will be high. This presents no problem with spray transfer where the current adjusts to arc length and thus prevents short-circuiting, but in the case of short-circuiting (dip) transfer, excessive short-circuit currents would cause much spatter.

The steeper the slope of the power unit volt–ampere characteristic curve the less the short-circuit current and the less the 'pinch effect' (which is the resultant inward magnetic force acting on the molten metal in transfer) so that spatter is reduced. Too much reduction of the short-circuit current, however, may lead to difficulty in arc initiation and stubbing. Power units are available having slope control so that the slope can be varied to suit welding conditions, and the control can be by tapped reactor or by infinitely variable reactor, the power factor of the latter being better than that of the former.

Three variables can thus be provided on the power unit – slope control, voltage and inductance. Machines with all three controls give the most accurate control of welding conditions but are more expensive and require more setting than those with only two variables, slope–voltage or voltage–inductance. In general, units with voltage–inductance control offer better characteristics for short-circuit transfer than those with slope–voltage control. For spray transfer conditions all types perform well, with the proviso that for aluminium welding the unit should have sufficient slope.

Wire feed and control cabinet

The wire of diameter 0.8, 1.2, 1.6, 2.4 mm hard, 1.2, 1.6, 2.4 mm soft and 1.6, 2.0, 2.4 and 3.2 mm flux cored is supplied on reels. The wire passes between motor-driven rollers which may have serrations or grooves to provide grip and which drive the wire at speeds between 2.5 and 15 m per min. The pressure on this drive can be varied and care must be taken that there is enough pressure to prevent slipping, in which case the arc lengthens and may burn back to the contact tube, and on the other hand that the pressure is not so great as to cause distortion of the wire or the flaking off of small metal particles with consequent increased wear on the guide tubes and possibility of jamming. Some machines have a small

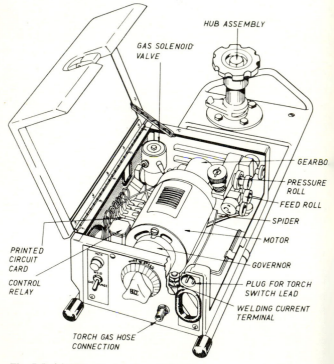

Fig. 7.5. (*a*) Wire feed and control unit.

removable magnet fixed after the wire drive to pick up such particles when ferrous wires are used. The flexible outer cables through which the wire is fed to the gun may have nylon liners for smoother feeding of fine wire sizes. The cabinet houses the wire drive motor and assembly, gas and cooling water solenoids and valves and main current contactor (controlled from the switch on the gun or automatic head) and the wire speed control which is also the current control is pre-set manually (Fig. 7.5a, b

Fig. 7.5. (b) Combined MIG power unit and wire feed unit. Unit comprises transformer, rectifier, inductor for 3-phase 50 Hz supply at 220, 380, 415 and 500 V with 28 OCV. 0.6, 0.8 and 1.0 mm diameter steel wire with current range 35–320 A for CO_2 and 80% Ar, 20% CO_2. One control sets the voltage and current and a potentiometer gives fine welding adjustment.

and c). There are also gas purging and inching switches, and in some units regenerative braking on the wire drive motor prevents over-run at the end of a weld.

Fig. 7.5. (c) Reverse side of power unit in Fig. 7.5b showing transformer (above) rectifier (below) and inductor (lower right).

Torch (Fig. 7.6a, b, c, d, e and f).

To the welding torch of either gooseneck or pistol type or an automatic head are connected the following supplies:

(1) Flexible cable through which the wire electrode is fed.
(2) Tubes carrying shielding gas and cooling water flow and return (if water cooled).
(3) Cable carrying the main current and control wire cable.

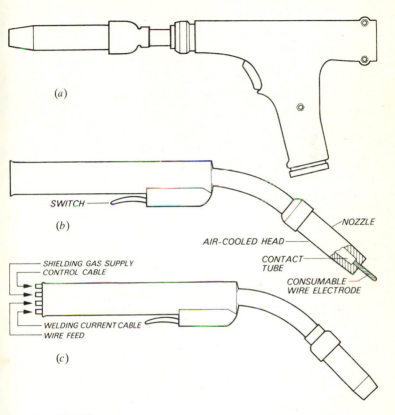

Fig. 7.6
(a) Straight-necked, air-cooled torch, variable in length by 38 mm. Current: at 75% duty cycle, 300 A with argon-rich gases, 500 A with CO$_2$; at 50% duty cycle, 350 and 550 A respectively. Wires: hard and soft, 1.2 and 1.6 mm, flux cored 1.6, 2.0 and 2.4 mm.
(b) Air-cooled torch, 45° neck angle. Current: at 75% duty cycle 300 A with argon-rich gases, 350 A with CO$_2$; at 50% duty cycle 350 A and 400 A respectively. Wires: hard 0.8, 1.0, 1.2 and 1.6 mm; soft 1.2 and 1.6 mm, flux cored, 1.6 mm.
(c) Air-cooled torch, 60° neck angle, Current: at 75% duty cycle 300 A with argon-rich gases, 350 A with CO$_2$; at 50% duty cycle 350 and 400 A respectively. Wires: hard 0.8, 1.0, 1.2 and 1.6 mm.

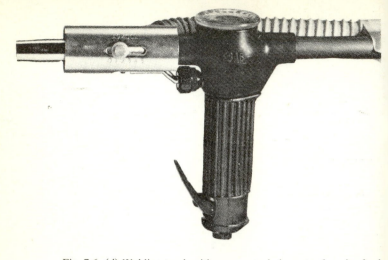

Fig. 7.6. (*d*) Welding torch with compressed air motor for wire feed drive built into the handle. Speed of motor is set by means of a knob which controls an air valve. Guns can be fitted with swan necks rotatable through 360°. Available in sizes up to 400 A, these latter have a heat shield fitted to protect the hand. A fume extraction unit is also fitted.

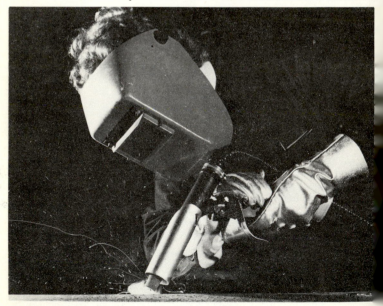

Fig. 7.6. (*e*) Torch with fume extraction unit.

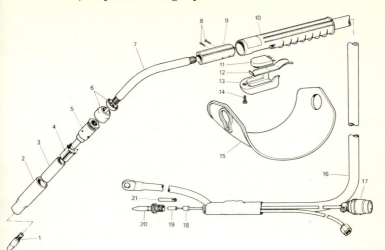

Fig. 7.6. (f) Welding torch or gun.

Key:
1. Contact tip for 1.2 and 1.6 mm hard and soft wire; 1.6, 2.0 and 2.4 mm tubular wire.
2. Nozzle.
3. Nozzle insulator.
4. Nozzle spring clip.
5. Torch head.
6. Head insulator and O clip.
7. Neck.
8. Self-tapping screw.
9. Spider.
10. Handle mounting.
11. Microswitch.
12. Switch lever.
13. Switch housing.
14. Screw.
15. Heat shield.
16. Integrated cable.
17. Plug; 7 pin.
18. Basic liner.
19. Liner; 1.2, 1.6 mm soft wire, 1.2, 1.6 mm hard wire, 1.6, 2.0 mm tubular wire.
20. Outlet guide; 1.2, 1.6 mm soft wire, 1.2, 1.6 mm hard wire, 1.6 2.0, 2.4 mm tubular wire.
21. Collet (for soft wire outlet guides).

A centrally placed and replaceable contact tube or tip screws into the torch head and is chosen to be a sliding fit on the diameter of wire being used (Fig. 7.6f). Contact from power unit to welding wire is made at the contact tube, which must be removed and cleaned at intervals and replaced as required. A metal shield or nozzle surrounds the wire emerging from the tube through which the shielding gas flows and surrounds arc and molten pool. Air cooled torches are used up to 400 A and water cooled up to 600 A. In the latter type the cooling water return flows around the copper cable carrying the welding current and thus this cable can be of smaller cross-sectional area and thus lighter and more flexible. A water cooled fuse in the circuit ensures that the water cooling flow must be in operation before welding commences and thus protects the circuit.

The wire feed can be contained in the head (Fig. 7.6d) only when the feed is by 'pull' or there can be a 'pull' torch unit, a 'push' unit, and a control box. These units handle wires of 0.8, 1.2 or 1.6 mm diameter in soft aluminium or 0.8 and 1.2 mm in hard steel wire, the unit being rated at 300 A. Pre-weld and post-weld gas flow are operated by the gas trigger and operate automatically. They operate from the standard MIG or CO_2 rectifier units and are suitable for mixed gas or CO_2 shielding, the very compact bulk greatly adding to their usefulness since they have a very large working radius.

Gases

Since CO_2 and oxygen are not inert gases, the title metallic inert gas is not true when either of these gases is mixed with argon or CO_2 is used on its own. The title *metallic active gas* (MAG) is sometimes used in these cases. The gases used in this process are shown in the accompanying table.

Argon, Ar. Commercial grade purity argon (99.996%) is obtained by fractional distillation of liquid air from the atmosphere, in which it is present to about 1% by volume. It is supplied in blue-painted cylinders containing 8500 litres at a pressure of 17.5 N/mm^2 (175 bar) or from bulk supply. It is used as a shielding gas because it is chemically inert and forms no compounds.

Carbon dioxide, CO_2. This is produced as a by-product of industrial processes such as the manufacture of ammonia, from the burning of fuels in an oxygen-rich atmosphere or from the fermentation processes in alcohol production, and is supplied in black-painted steel cylinders containing up to 35 kg of liquid CO_2.[1] To avoid increase of water vapour above the limit of 0.015% in the gas as the cylinder is emptied, a dip tube or syphon is fitted so that the liquid CO_2 is drawn from the cylinder, producing little fall in temperature. An electric vaporizer or heater of up to 250 W loading, fitted between the cylinder and

[1] The cylinder pressure depends upon the temperature being approximately 33 bar at 0°C and 50 bar at 15°C.

regulator, vaporizes the liquid. The heater contains a mains-operated heating element and a warning light on it is extinguished when the heater has warmed up sufficiently. The syphon cylinder may have a grey stripe running down its length to identify it. Manifold cylinders can be fed into a single vaporizer, and if the supply is in a bulk storage tank, this is fed into an evaporator and thence fed to the welding points at correct pressure as with bulk argon and oxygen supplies.

Gas or gas mixture	Application
Pure argon (99.996%)	Aluminium and alloys, magnesium and alloys, titanium and alloys nickel and alloys, copper and alloys. 9% Nickel steel.
Argon + 1% oxygen	Stainless steel (spray and pulse).
Argon + 2% oxygen	Stainless steel. Used for dip transfer if increased oxidation can be tolerated.
CO_2 (carbon dioxide)	Mild and low-alloy steel, dip and spray; bare and flux cored wire. Single-run stainless steel.
Argon + 5% CO_2	Mild and low-alloy steel. Stainless steel. Dip, spray and pulse (surge). Cannot be used on stainless steel of less than 0.06% C.
Argon + 20% CO_2	Mild and low-alloy steel. Similar to the above but not for pulsed arc.
Argon + 5% hydrogen	Nickel and its alloys. Dip, spray and pulse.
Argon + 15% nitrogen	Copper and its alloys. Nitrogen reduces pre-heat requirements.
Argon + 15/20% helium	Special applications such as for nickel 200.

Application of gases

1. Argon. Although argon is a very suitable shielding gas for the non-ferrous metals and alloys, if it is used for the welding of steel there exists an unstable negative pole in the work-piece (the wire being positive) which produces an uneven weld profile. Mixtures of argon and oxygen are selected to give optimum welding conditions for various metals.

2. Argon + 1% or 2% oxygen. The addition of oxygen as a small percentage to argon gives higher arc temperatures and the oxygen acts as a wetting agent to the molten pool, making it more fluid and stabilizing the arc. It reduces surface tension and produces good fusion and penetration. The argon + 1% oxygen mixture is used for stainless steels with spray and pulse methods. Dip transfer is better with argon + 2% oxygen as long as the increased oxidation can be tolerated. The addition of 5% hydrogen is the maximum for titanium-stabilized stainless steel and larger amounts than this increase porosity.

3. Helium. If helium is used as the shielding gas it requires about two or more times the gas flow of argon. It is usually used mixed with argon, e.g. argon–15–20% helium for welding certain high nickel content alloys – up to 50% helium for copper welding and argon–helium mixtures as the plasma gas in plasma welding.

4. Carbon dioxide. Pure CO_2 is the cheapest of the shielding gases and can be used as a shield for welding steel up to 0.4 % C and low-alloy steel. Because there is some dissociation of the CO_2 in the arc resulting in carbon monoxide and oxygen being formed, the filler wire is triple deoxidized to prevent porosity, and this adds somewhat to its cost and results in some small areas of slag being present in the finished weld. The droplet rate is less than that with pure argon, the arc voltage drop is higher, and the threshold value for spray transfer much higher than with argon. The forces on the droplets being transferred across the arc are less balanced than with argon–oxygen so that the arc is not as smooth and there is some spatter, the arc conditions being more critical than with argon–oxygen.

Using spray transfer there is a high rate of metal deposition with excellent low-hydrogen properties of the weld metal. The dip transfer process is especially suited for positional work. The process has not displaced the submerged arc and electro-slag methods for welding thick steel sections, but complements them and competes in some fields with manual metal arc process using iron powder electrodes. If offers the most competitive method for repetitive welding operations, and the use of flux cored wire greatly increases the scope of the CO_2 process. Thicknesses up to 75 mm can be welded in steel using fully automatic heads.

With stainless steel, because of the loss of stabilizers (titanium and niobium) in the CO_2 shielded arc there is some carbon pick-up resulting in some precipitation of chromium carbide along the grain boundaries and increased carbon content of the weld, reducing the corrosion resistance. Multi-pass runs result in further reduction in corrosion resistance, but with stabilized filler wire and dip transfer on thinner sections satisfactory single-pass welds can be made very economically. Non-toxic, non-flammable paste is available to reduce spatter problems.

5. Argon + 5% CO_2, Argon + 20% CO_2. The addition of CO_2 to argon for the welding of steel improves the 'wetting' action, reduces surface tension and makes the molten pool more fluid. Both mixtures give excellent results with dip and spray transfer, but the 20% mixture gives poor results with pulse while the 5% mixture gives much better results. The mixtures are more expensive than pure CO_2 but give a smoother, less critical arc with reduced spatter and a flatter weld profile, especially on fillets (Fig. 7.7a). The current required for spray transfer is less than for similar conditions with CO_2 and the 5% mixture is suitable for single-run welding of stainless steel with the exception of those with extra low carbon content (ELC), below 0.06%. Both mixtures contain a small amount of oxygen.

If the CO_2 content is increased above 20% the mixture behaves more and more like pure argon. If argon and CO_2 are on bulk supply, mixers enable the percentage to be varied as required.

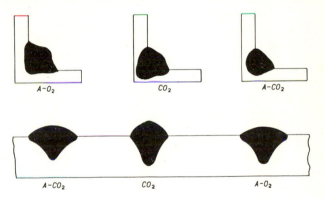

Fig. 7.7. (a) Variation of fillet weld shapes in dip transfer arc, and of bead shapes in spray transfer arc.

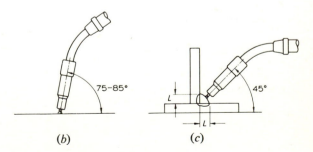

Fig. 7.7. (b) and (c). Angle of welding.

6. Argon + 15–20% nitrogen. This mixture can be used instead of pure argon for copper welding. Arc voltages are higher, giving greater heat output for a given current value, thus reducing the pre-heating requirements. If pure nitrogen is used, the droplets are of coarse size and there is much spatter and porosity with poor weld appearance.

Filler wires

Note. BS 2901 (1970) gives specifications for *Filler rods and wires for inert gas welding. Part 1, Ferritic steels; Part 2, Austenitic stainless steels; Part 3, Copper and copper alloys; Part 4, Aluminium and aluminium alloys and magnesium alloys; Part 5, Nickel and nickel alloys.*

Filler wires are supplied on convenient reels of 300 mm or more diameter and of varying capacities with wire diameters of 0.6, 0.8, 1.0, 1.2, 1.6, and 2.4 mm. The bare steel wire is usually copper coated to improve conductivity, reduce friction at high feed speeds and minimize corrosion while in stock. Manganese and silicon are used as deoxidizers in many cases but triple deoxidized wire using aluminium, titanium and zirconium gives high-quality welds and is especially suitable for use with CO_2. The following are examples of available steel wires and can be used with argon–5% CO_2, argon–20% CO_2 and CO_2, the designation being to BS 2901 Pt 1 1970.

A 18. General purpose mild steel used for mild and certain low alloy steels. Analysis: 0.12% C; 0.9–1.6% Mn; 0.7–1.2% Si. 0.04% max S and P.

A 15. Triple deoxidized steel wire recommended for pipe welds and root runs in heavy vessel construction. Analysis 0.12% max C; 0.9–1.6% Mn; 0.3–0.9% Si; 0.04–0.4% Al; 0.15% max Ti; 0.15% max Zr; 0.04% max S and P.

A 31. Molybdenum bearing mild steel wire. Used for most mild steel applications requiring extra strength; high tensile and quench and tempered steel; and suitable for offshore pipeline applications and root runs in thick joints. Analysis: 0.14% max C; 1.6–2.1% Mn; 0.5–0.9% Si; 0.4–0.6% Mo; 0.03% max S and P.

Flux cored and powder cored filler wires

Flux cored wires used with a CO_2 shield are now extensively used for fabrications giving rapid deposition of weld metal and high-quality welds in steel. The wire is manufactured from a continuous narrow flat steel strip formed into a U shape which is then filled with flux and formed into a tube. It is then pulled through reducing dies which reduce the diameter and compress the flux uniformly and tightly into a centre core (Fig. 7.8).

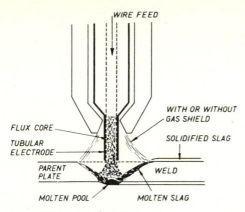

Fig. 7.8. Flux cored process.

The wires are supplied on reels as for MIG welding in diameters 1.6, 2.0, 2.4, 3.2 mm and fed to the feed rollers of a MIG type unit. The welding gun can be straight or swan-necked and should be water cooled for currents above 400 A, and the power unit is the same as for MIG welding.

There is deep penetration with smooth weld finish and mimimum spatter, and deposition rate is of the order of 10 kg/h using 2.4 mm diameter wire. The deep penetration characteristics enable a narrower V preparation to be used for butt joints resulting in a saving of filler metal, and fillet size can be reduced 15–20%.

Some of the filler wires available are:

(1) Rutile type, steel, giving a smooth arc and good weld appearance with easy slag removal. Its uses are for butt and fillet welds in mild and medium tensile steel in the flat and horizontal–vertical position.

(2) Basic hydrogen-controlled type, steel. This is an all-positional type and gives welds having good low-temperature impact values.

(3) Basic hydrogen-controlled type, steel, with $2\frac{1}{2}\%$ Ni for applications at -50 °C.

(4) Basic hydrogen-controlled type for welding 1% Cr, 0.5% Mo steels.

(5) Basic hydrogen-controlled type for welding 2.25% Cr, 1% Mo steels.

Self-shielded flux cored electrode wires are used without an external gas shield and are useful in outdoor or other draughty situations where an external gas shield would be blown away. Wires are available for mild steel and certain higher strength steels, giving welds of good appearance and quality with easily detachable slag.

Metal cored filler wire

This wire has a core containing metallic powders with minimal slag-forming constituents and there is good recovery rate (95%) with no interpass deslagging. It is used with CO_2 or argon–CO_2 gas shield to give welds with low hydrogen level (less than 5 ml/100 g weld metal). The equipment is similar to that used for MIG welding, and deposition rates are higher than with stick electrodes especially on root runs.

Safety precautions

The precautions to be taken are similar to those when metal arc welding, given on p. 323. The BS 679 recommended welding filters are up to 200 A, 10 or 11 EW; over 200 A, 12, 13, or 14 EW. When welding in dark surroundings choose the higher shade number, and in bright light the lower shade number. Because there is greater emission of infra-red energy in this process a heat absorbing filter should be used. The student should consult the following publications for further information: BS 679, *Filters for use during welding,* British Standards Institution; *Electric arc welding,* new series no. 38, *Safety, health and welfare.* Published for the Department of Employment by HMSO.

TECHNIQUES

There are three methods of initiating the arc. (1) The gun switch operates the gas and water solenoids and when released the wire drive is switched on together with the welding current. (2) The gun switch operates the gas and water solenoids and striking the wire end on the plate operates the wire drive and welding current (known as 'scratch start'). (3) The gun switch operates gas and water solenoids and wire feed with welding current, known as 'punch start'.

The table on p. 443 indicates the various gases and mixtures at present in use. As a general rule dip transfer is used for thinner sections up to 6.4 mm and for positional welding, whilst spray transfer is used for thicker sections. The gun is held at an angle of 80 ° or slightly less to the line of the weld to obtain a good view of the weld pool, and welding proceeds from right to left with the nozzle held 6–12 mm from the work (see Fig 7.8). The further the nozzle is held from the work the less the efficiency of the gas shield, leading to porosity. If the nozzle is held too close to the work spatter may build up, necessitating frequent cleaning of the nozzle while arcing between nozzle and work can be caused by a bent wire guide tube allowing the wire to touch the nozzle, or by spatter build-up short-circuiting wire and nozzle. If the wire burns back to the guide tube it may be caused by a late start of the wire feed, fouling of the wire in the feed conduit or the feed rolls being too tight. Intermittent wire feed is generally due to insufficient feed roll pressure or looseness due to wear in the rolls. Excessively sharp bends in the flexible guide tubes can also lead to this trouble.

Root runs are performed with no weave and filler runs with as little weave as possible consistent with good fusion since excessive weaving tends to promote porosity. The amount of wire projecting beyond the contact tube is important because the greater the projection, the greater the I^2R effect and the greater the voltage drop which may reduce the welding current and affect penetration. The least projection commensurate with accessibility to the joint being welded should be aimed at. Backing strips which are welded permanently on to the reverse side of the plate by the root run are often used to ensure sound root fusion. Backing bars of copper or ceramics with grooves of the required penetration bead profile can be used and are removed after welding. It is not necessary to back-chip the root run of the light alloys but with stainless steel this is often done and a sealing run put down. The importance of fit-up in securing continuity and evenness of the penetration bead cannot be over-emphasized.

Flat welds may be slightly tilted to allow the molten metal to flow against the deposited metal and thus give a better profile. If the first run has a very convex profile poor manipulation of the gun may cause cold laps in the subsequent run.

Positional welding

This is best performed by the dip transfer method since the lower arc energy enables the molten metal to solidify more quickly after deposition. Vertical welds in thin sections are usually made downwards with no weave. Thicker sections are welded upwards or with the root run downwards and subsequent runs upwards, weaving as required. Overhead welding, which is performed only when absolutely necessary, is performed with no weave.

Fillet welds are performed with the gun held backwards to the line of welding, as near as possible to the vertical consistent with a good view of the molten pool, bisecting the angle between the plates and with a contact tip-to-work distance of 16–20 mm (Fig. 7.7c). On unequal sections the arc is held more towards the thicker section. The root run is performed with no weave and subsequent runs with enough weave to ensure equal fusion on the legs. Tilted fillets give better weld profile and equal leg length more easily.

Aluminium alloys

When fabricating aluminium alloy sections and vessels, the plates and sections are cut and profiled using shears, cold saw, band saw or arc plasma and are bent and rolled as required and the edges cut to the necessary angle for welding where required. (*Note.* As the welding currents employed in the welding of aluminium are high, welding lenses of a deeper than normal shade should be used to ensure eye protection.)

The sections are degreased using a degreaser such as methyl chloride and are tack welded on the reverse side using either TIG or MIG so as to give as far as possible a penetration gap of close tolerance. Where this is excessive it can be filled using the TIG process, and pre-heating may be performed for drying. The areas to be welded are stainless steel wire brushed to remove all oxide and the root run is made. Each successive run is similarly wire brushed and any stop–start irregularities should be

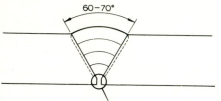

Fig. 7.9. (*a*) Flat butt weld preparation, aluminium plate MIG, semi-automatic process argon shield, back-chip and sealing run. Work negative.

Plate thickness (mm)	Number of runs	Approximate current (A)		Root face (mm)
		Root	Subsequent	
6 and 8	2	220	250–280	1.6
9.5 and 11	3	230	260–280	1.6
12.7 and 16	4	230–240	270–290	3.2
19	5	240–250	280–310	3.2
22 and 25	6	240–260	280–330	4.8

Refer also to BS 3571 *General recommendations for manual inert gas metal arc welding,* Part 1, *Aluminium and aluminium alloys.*

Fig. 7.9. (b) MIG weld in aluminium NS8 plate, 8 mm thick with NG61 wire. Stainless steel backing bar.
1st run; 250–270 A, 26–27 V, speed 9–10 mm/second.
2nd run; 260–280 A, 28–29 V, speed 8–10 mm/second.
Preparation: 90 ° V, 0–1 mm root gap, 3–4 mm root face.

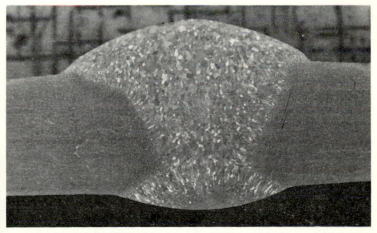

Fig. 7.9. (c) TIG weld on penetration bead of the above weld. No back chipping, no filler wire (can be added if required). 350 A, 32 V, speed 3–4 mm/second.
Section X 0, etched with cupric chloride $CuCl_2$ followed by a 50% nitric acid wash.

removed. It is important that the tack welds should be incorporated into the overbead completely so that there is no variation in it. Back-chipping of the root run, where required for a sealing run, should be performed with chisels, routers or saws and not ground, as this can introduce impurities into the weld. Fig. 7.9*a* indicates typical butt preparation, using 1.6 mm filler wire: flat.

Copper and copper alloys

Many of the problems associated with the welding of copper and copper alloys have been discussed in the chapter on the TIG process (Chapter 6) and apply equally to the MIG process. Because of the high thermal conductivity of copper and to reduce the amount of pre-heating, it is usual in all but the thinnest sections to use high currents with spray transfer conditions which are obtainable with argon and argon–helium mixtures. The addition of nitrogen to argon destroys the spray transfer condition but arc conditions are improved with up to 50% helium added to argon and results in an increased heat output. For thin sections, fine feed wires can be used giving spray transfer conditions with lower current densities, thus preventing burn-through. Fig. 7.10 shows edge preparation for TIG and MIG butt welds in copper.

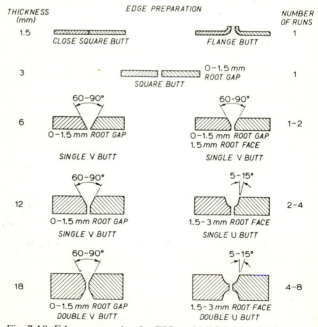

Fig. 7.10. Edge preparation for TIG and MIG butt welds in copper.

MIG process. Recommended filler wires for copper welding

Type (BS 2870–2875)	Grade	Filler wire	
		Argon or helium shield	Nitrogen shield
C106	Phosphorus deoxi-dized, non-arsenical	C7, C8, C21	Not recommended
C107	Phosphorus deoxi-dized, arsenical		
C101	Electrolytic tough pitch, high conductivity	C7, C8, C21	Not recommended
C102	Fire refined tough pitch, high conductivity		
C103	Oxygen-free, high conductivity	C7, C21	Not recommended

Stainless steels

Preparation angle is usually similar for semi-automatic and automatic welding for butt welds in stainless steel, being 70–80 ° with a 0–1.5 mm gap (Fig. 7.11). Torch angle should be 80–90 ° with an arc length long enough to prevent spatter but not long enough to introduce instability. All oxides should be removed by stainless steel wire-brushing. Back chipping is usually performed by grinding. With spray transfer using high currents there is considerable dilution effect and welds can only be made flat, but excessive weaving should be avoided as this increases dilution. Direct current with the torch (electrode) positive is used with a shield of argon + 1 % oxygen. If a strip of stainless sheet is tack welded to the back side of the weld it assists back-purging by using argon from the torch and reduces the oxide formation on the penetration bead, but in general because of the variation in penetration in long seams the root run is back-chipped and a sealing run laid. For stainless steels, root runs are of the order of 200–250 A and subsequent runs 230–300 A according to the thickness of the plate.

In the welding of dissimilar metals to stainless steel the short-circuiting arc (dip transfer) gives much less dilution with lower heat input and is generally preferred. The arc should be kept on the edge of the parent metal next to the molten pool and not in the pool itself, to reduce the tendency to the formation of cold laps.

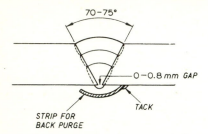

Fig. 7.11. Butt welding stainless steel MIG, 99% A–1% O_2, d.c. wire positive. Automatic or semi-automatic.

Argon–1% oxygen is used for spray transfer and argon–2% oxygen for dip transfer. Filler wires available include the following:

(1) Low carbon (0.03% max) 19.5–22% Cr, 9–11% Ni suitable for welding 18% Cr, 8% Ni, including low carbon types.
18% Cr, 8% Ni, Nb stabilized for use up to 400 °C operating temperatures and for chromium steels except where there is risk of sulphur corrosion.

(2) 0.12% C, 25% Cr, 12–14% Ni, for welding mild and low alloy to stainless steel: for ferritic to austenitic steels up to 300 °C and for similar steels (25/12).

(3) Low carbon type, for welding stainless steels where high impact strength at low temperatures is required with resistance to corrosion. Analysis: 0.03% max C, 18–20% Cr, 11–14% Ni 2–3% Mo (19/12/3), suitable for welding 18% Cr, 8% Ni, Mo: 18% Cr, 8% Ni, Mo, Nb: 18% Cr, 8% Ni not in strongly oxidizing conditions.

(4) 19% Cr, 12% Ni, 3% Mo, for steels of the same composition.

(5) 20% Cr, 10% Ni, niobium stabilized for welding similar steels.

Nickel alloys

The welding of nickel and nickel alloys can be done using spray transfer and short circuit (dip transfer) arc methods.

Spray transfer with its higher heat input gives high welding speeds and deposition rate and is used for thicker sections, usually downhand because of the large molten pool. Argon is used as the shielding gas but 10–20% helium can be added for welding Nickel 200 and Inconel 600, giving a wider and flatter bead with reduced penetration. The gun should be held as near to 90 ° to the work as possible, to preserve the efficiency of the gas shield, and the arc should be kept just long enough to prevent spatter yet not long enough to affect arc control.

Short circuiting arc conditions are used for thinner sections and positional welding, the lower heat input giving a more controllable molten pool and minimum dilution. The shielding gas is pure argon but the addition of helium gives a hotter arc and more wetting action so that the danger of cold laps is reduced. To further reduce this danger the gun should be held as near as possible to 90 ° to the work and moved so as to keep the arc on the plate and not on the pool. High-crowned profile welds increase the danger of cold laps.

The table on p. 404 indicates materials and filler metal suitable for welding by the MIG process and Fig. 6.20 the recommended methods of preparation.

SG cast irons

Pearlitic and ferritic cast irons are very satisfactorily welded using dip transfer conditions (e.g. 150 A, 22 V with 0.8 mm diameter wire) to give low heat input using filler wire of Nickel 6 (93% Ni) or Monel 60 (62/69 Ni 21–28% Cu). Carbide precipitation in the HAZ is confined to thin envelopes around some of the spheroids, unlike the continuous film associated with MMA welding. SG iron can be welded to other metals and Monel 60 can be used to give a corrosion-resistant surface on SG iron castings or as a buffer layer for other weld deposits. Cleaning and degreasing should be performed before welding, and pre-heat is only required for heavy pearlitic section or joints under heavy restraint when pre-heat of 200 °C is suitable. Minimum-heat input compatible with adequate fusion should always be used.

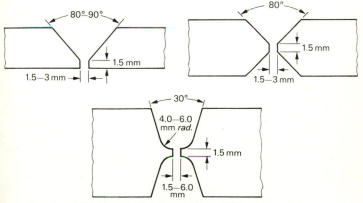

Fig. 7.12. Preparation for welding of SG cast irons.

CO₂ WELDING OF MILD STEEL

There are four controls to enable optimum welding conditions to be achieved: (1) wire feed speed which also controls the welding current, (2) voltage, (3) choke or series inductance and (4) gas flow (Fig. 7.13).

For a given wire diameter the wire feed rate must be above a certain minimum valve to obtain a droplet transfer rate of above about 20 per second, below which transfer is unsatisfactory. With increasing wire feed rate the droplet transfer rate and hence the burn-off increases and the upper limit is usually determined by the capacity of the wire feed unit. The voltage setting also affects the droplet frequency rate and determines the type of transfer, about 15–20 V for short-circuit or dip, and 27–45 V for spray, with an intermediate lesser used zone of about 22–27 V for the semi-short-circuiting arc.

Fig. 7.13. Gas-shielded welding unit for dip transfer, spray transfer and flux-cored CO₂ welding. Constant potential d.c. source (transformer–rectifier). Input 3-phase, 415 V, 50 Hz. 27 kVA at maximum continuous welding current. Output; 75 A at 17 V to 480 A at 40 V. Wire diameters; solid 0.8–1.6 mm flux cored 1.6–2.4 mm. A boom can be fitted for easy cable manipulation and the unit is readily converted from semi to fully automatic operation.

The choke, which limits the rate of current rise and decay, is also an important control because too low a setting can give a noisy arc with much spatter and poor weld profile, while too high a setting can give unstable arc conditions with more difficult start and even occasional arc extinguishing. Between these limits there is a setting which with correct arc voltage and wire feed rate gives a smooth arc with minimum spatter and good weld profile. Penetration is also affected by the choke setting.

As stated before, the short-circuiting arc is generally used for welding thinner sections, positional welding, tacking and on thicknesses up to 6.5

mm. In positional welding the root run may be made downwards with no weave and subsequent runs upwards. The lower heat output of this type of arc reduces distortion on fabrications in thinner sections and minimizes over-penetration. The spray-type arc is used for flat welding of thicker sections and gives high deposition rates.

Gas flow rate can greatly affect the quality of the weld. Too low a flow rate gives inadequate gas shielding and leads to the inclusion of oxides and nitrides, while too high a rate can introduce a turbulent flow of the CO$_2$ which occurs at a lower flow rate than with argon. This affects the efficiency of the shield and leads to porosity in the weld. The aim should be to achieve an even non-turbulent flow and for this reason spatter should not be allowed to accumulate on the nozzle which should be directed as nearly as possible at 90° to the weld, again to avoid turbulence.

The torch angle is, in practice, about 70–80° to the line of travel consistent with good visibility and the nozzle held about 10–18 mm from the work. If the torch is held too close, excess spatter build-up necessitates frequent cleaning, and in deep U or V preparation the angle can be increased to obtain better access. Weaving is generally kept as low as convenient to preserve the efficiency of the gas shield and reduce the tendency to porosity. Wide weld beads can be made up of narrower 'stringer' runs, and tilted fillets compared with HV fillets give equal leg length more easily, with better profile.

Economic considerations

Although filler wire for the CO$_2$ process, together with the cost of the shielding gas, is more expensive than conventional electrodes, other factors greatly affect the economic viability of the process. The deposition rate governs the welding speed which in turn governs the labour charge on a given fabrication.

The deposition rate of the filler metal is a direct function of the welding current. With metal arc welding the upper limit is governed by the overheating of the electrode. The current I amperes flows through the electrode, the wire of which has an electrical resistance R ohms, so that the heating effect is $\propto I^2R$. The resistance of any metallic conductor increases with the rise in temperature, so that, as the electrode becomes hotter, the resistance and hence the I^2R loss increases so that with excessive currents, when half of the electrode has been consumed, the remaining half has become red hot and the coating ruined. Iron powder electrodes have a greater current-carrying capacity due to the conductivity of the coating, the electrical resistance being reduced. With the CO$_2$ process the distance from contact tube to wire tip is of the order of 20 mm so that the electrical resistance is greatly reduced even though the wire diameter is smaller. The current can thus be increased greatly, resulting in higher deposition rates, greater welding speeds and reduced labour charges. In addition the duty cycle is increased since there is no constant electrode change and need for deslagging, but there may be greater spatter loss. Economically therefore the CO_2 process shows an advantage in very many applications, though the final choice of process is governed by the application and working conditions.

Details of plate preparation are given as a guide only, since there are too many variables to give generalized recommendations (see Figs. 7.14 and 7.15).

Mild steel sheet, butt welds, CO_2 shielding, flat, 0.8 mm diam. wire (approximate values)

Thickness (mm)	Gap (mm)	Wire feed (m/minute)	Arc (volts)	Current (A)
1	0	2.8–3.8	16–17	65–80
1.2	0	3.2–4.0	18–19	70–85
1.6	0.5	4.0–4.8	19–20	85–95
2.0	0.8	5.8–7.0	19–20	110–125
2.5	0.8	7.0–8.4	20–21	125–140
3.0	1.5	7.0–8.4	20–21	125–140

AUTOMATIC WELDING

Automatic welding, by MIG, pulse and CO_2 processes now plays an important part in welding fabrication practice. It enables welds of consistently high quality and accuracy to radiographic standard to be performed at high welding speeds because of the close degree of control over the rate of travel and nozzle-to-work distance. It is less tolerant than semi-automatic welding to variations of root gap and fit-up but reduces the number of start–stop breaks in long sequences. The choice between semi-automatic and automatic process becomes a question of economics, involving the length of runs, number involved, volume of deposited metal if the sections are thick, method of mechanization and set-up time. The torches are now usually air cooled even for current up to 450 A and are carried on welding heads fitted with controls similar to those used for semi-automatic welding, and may be remote controlled.

The head may be; (1) fixed, with the work arranged to move or be rotated beneath it, (2) mounted on a boom and column which can either be of the positioning type in which the work moves, or the boom can

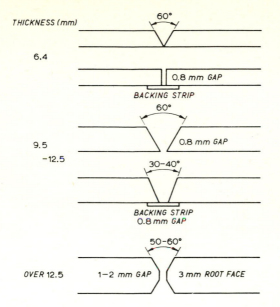

Fig. 7.14. Various preparations for thicker plate (varying with applications).

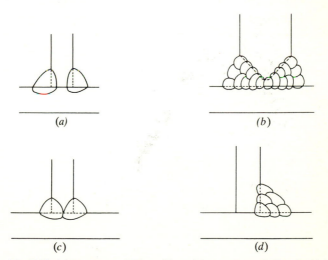

Fig. 7.15. (a) Unprepared fillets, (b) multi-run prepared fillets in thick plate, (c) deep preparation fillets, (d) multi-run unprepared fillet.

traverse over the work (Figs. 7.16 and 7.17), (3) gantry mounted so as to traverse over the stationary work, (4) tractor mounted, running on guide rails to move over the fixed work, (5) mounted on a special machine or fixture designed for a specific production. A head may carry two torches arranged to weld simultaneously, thus greatly reducing the welding time.

Fig. 7.16. A ram-type boom with welding head.

Fig. 7.17. High lift boom with CO_2 welding head and traversing roller bed.

For the CO_2 process in steel a typical example would be with wire of 1.2 mm diameter using 150–170 A on thinner sections and multiple passes up to 30 on thicknesses up to 75 mm with current in the 400–500 A range and 2.4 mm wire. With automatic surge (pulse) arc welding on stainless steel, accurate control of the underbead is achieved, obviating the necessity for back chipping and sealing run. In the case of aluminium welding on plate above 10 mm thickness the accuracy of the underbead produced with the MIG automatic process results in more economical welding than by the double-operator vertical TIG method. In general the full automation of these processes results in greater productivity with high-quality welds.

MIG spot welding

Spot welding with this process needs access to the joint on one side only and consists of a MIG weld, held in one spot only, for a controlled period of time. Modified nozzles are fitted to the gun, the contact tip being set 8–12 mm inside the nozzle, and the timing unit can either be built into the power unit or fitted externally, in which case an on–off switch does away with the necessity of disconnecting the unit when it is required for continuous welding. The timer controls the arcing time, and welds can be made in ferrous material with full or partial penetration as required. A typical application of a smaller unit is that of welding thin sheet as used on car bodies. The spot welding equipment timer is built into a MIG unit and is selected by a switch. Wire of 0.6 mm diameter is used with argon + CO_2 5% + O_2 2% as the shielding gas. Currents of 40–100 A are available at 14–17 V for continuous welding with a maximum duty cycle of 60%. The spot welding control gives a maximum current of 160 A at 27 V, enabling spot welds to be made in material down to 0.5 mm thickness, the timing control varying from 0.5–2.0 seconds.

On larger units arcing time can be controlled from 0.3 to 4.5 seconds with selected heavier currents enabling full penetration to be achieved on ferrous plate from 0.7 to 2.0 mm thick and up to 2.5 mm sheet can be welded on to plate of any thickness.

Cycle arc welding is similar to spot welding except that the cycle keeps on repeating itself automatically as long as the gun switch is pressed. The duration of the pause between welds is constant at about 0.35 seconds while the weld time can be varied from 0.1 to 1.5 seconds. This process is used for welding light-section components which are prone to burn through. In the pause between the welding period, the molten pool cools and just solidifies, thus giving more accurate control over the molten metal.

PULSED ARC WELDING

Pulsed arc welding is a modified form of spray transfer in which there is a controlled and periodic melting off of the droplets followed by projection across the arc. A pulse of current is applied for a brief duration at regular frequency and thus results in a lower heat output than with pure spray transfer, yet greater than that with dip transfer. Because of this, thinner sections can be welded than with spray transfer and there is no danger of poor fusion in a root run as sometimes occurs with dip transfer. There is regular and even penetration, no spatter and the welds are of high quality and appearance.

To obtain these conditions of transfer it is necessary to have two currents fed to the arc, (a) a background current which keeps the gap ionized and maintains the arc and (b) the pulsed current which is applied at 50 or 100 Hz and which melts off the wire tip into a droplet which is then projected across the arc gap. These two currents, which have critical values if satisfactory welding conditions are to be obtained, are supplied from two sources, a background source and a pulse source contained in one unit, and their voltages are selected separately. The background current, of much lower value than the pulse, is half a cycle out of phase with it (see Figs. 7.18 and 7.19). A switch enables the pulse source to be used for dip and spray transfer methods as required. The power supply is a silicon rectifier with constant voltage output and a maximum current value of about 350 A at open-circuit voltages of 11–45 V similar to those already described.

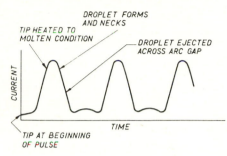

Fig. 7.18. Wave shape of pulse supply.

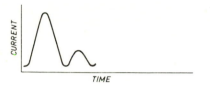

Fig. 7.19. Modified pulse supply with background pulse.

To operate the unit, the 'pulse height' (the value of pulse current) and the background current are selected on separate switches, the wire is adjusted to protrude about 12 mm beyond the nozzle and welding is commenced moving from right to left down the line of the weld with the gun making an angle of 60–70 ° with the line of weld. When welding fillets the gun is usually held at right angles to the line of weld with the wire pointing directly into the joint, and thus an excellent view is obtained of the degree of fusion into the root; welding is again performed from right to left. The process gives good root fusion with even penetration and good fill-in and is especially efficient when used fully automatically. As in all other welding processes, welding is best performed flat but positional welding with pulsed arc is very satisfactory and relatively easy to perform. Thicknesses between 2 mm and 6.5 mm which fall intermediate in the ranges for dip and spray transfer are easily welded. One of the drawbacks to pulse arc is the necessity to ensure accurate fit-up. If there are any sizeable gaps a keyhole effect is produced and it is impossible to obtain a regular underbead. The gap should be of the order of from 0 to about 1.0 mm max. The shielding gas is argon

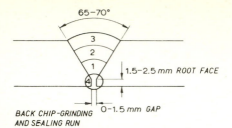

Fig. 7.20. Automatic pulsed arc welding of stainless steel, flat, 99% argon–1% oxygen.

with 1% or 2% oxygen or argon with 5% CO_2 and 2% oxygen for welding mild and low-alloy steels, stainless steel and heat resistant steel, and pure argon for aluminium and its alloys, 9% nickel steel and nickel alloys. Pulse arc is especially useful when used automatically for stainless steel welding, since the accurately formed under- or penetration bead obviates the expensive operation of back-chipping a sealing run and there is little carbon pick-up and thus little increase of carbon content in the weld. Aluminium requires no back purge and for the back purge on stainless steel a thin strip of plate can be tack welded on the underside of the joint (see Fig. 7.20) and removed after welding. The argon from the torch supplies the back purge and prevents oxidation of the underside of the weld.

There is good alloy recovery when welding alloy steel and because of the accurate heat control, welding in aluminium is consistently good without porosity and a regular underbead so that it can be used in place of double argon TIG with a saving of time and cost.

Fabrications and vessels can be fully tack weld fabricated with TIG on the underside of the seam and pulse arc welded, greatly reducing the tendency to distortion. The torch should be held at 75–80 ° to the line of the weld and good results are also obtained fully automatically by welding from left to right with the torch held vertically to the seam. This method allows greater tolerance in fit-up and preparation with better penetration control. Because this process is sensitive to the accuracy of fit-up and preparation, care should be taken to work to close tolerances and excessive gaps should be made up from the reverse side with, for example, the TIG process and then carefully cleaned.

The table gives a typical preparation for automatic flat butt welds in stainless steel using 99% A, 1% O_2 as shielding gas and 1.6 mm diam filler wire. For the alloys of nickel the recommended shielding gas is pure argon. A similar technique is used as for the MIG welding of stainless steel, with a slight pause each side of the weave to avoid undercut.

Plate thickness (mm)	Runs	Current (A) Roots and subsequent	Volts (approx.)
4.8, 6.4, 8	2	180–200	27–28
9.5	3	180–210	27–28

There appears to be no advantage in using the pulse method over the normal shielded metal arc with unpulsed wave when using CO_2 as the shielding gas.

Plasma MIG process. A process is being developed in which MIG filler wire is fed through a contact tube situated centrally in the torch head. A tungsten electrode is set at an angle to this contact tube and projects nearly to the mouth of the shielding nozzle through which the plasma gas issues. An outer nozzle provides the gas shield. The current through the wire assists the melting and gives good starting characteristics and a stable arc, so that thin plate is weldable at high welding speeds.

8

Resistance welding and flash butt welding

SPOT WELDING

In this method of welding use is made of the heating effect which occurs when a current flows through a resistance. Suppose two plates A and B Fig. 8.1 are to be welded together at X. Two copper electrodes are pressed against the plates squeezing them together. The electrical resistance is greatest at the interface where the plates are in contact, and if a large current at low voltage is passed between the electrodes through the plates, heat is evolved at the interface, the heat evolved being equal to I^2Rt Joules, where I is current in amperes, R the resistance in ohms and t the time in seconds. A transformer supplies a.c. at low voltage and high current to the electrodes (Figs. 8.1 and 8.2).

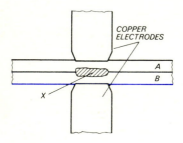

Fig. 8.1

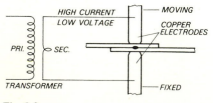

Fig. 8.2

For any given joint between two sheets a suitable time is selected and the current varied until a sound weld is obtained. If welds are made near each other some of the current is shunted through the adjacent weld (Fig. 8.3.) so that a single weld is not representative of what may occur when several welds are made. Once current and time are set, other welds will be of consistent quality. If the apparatus is controlled by a pedal as in the simplest form of welder, then pressure is applied mechanically and the time for which the current flows is switch controlled, as for example in certain types of welding guns used for car body repairs. This method has the great disadvantage that the time cannot be accurately controlled so that if it is too short there is insufficient heat and there is no fusion between the plates, whilst if the time is too long there is too much heat generated and the section of the plates between the electrodes melts, the molten metal is spattered out due to the pressure of the electrodes and the result is a hole in the plates. In modern spot-welding machines the pressure can be applied pneumatically or hydraulically or a combination of both and can be accurately controlled. The current is selected by a tapping switch on the primary winding of the transformer and the time is controlled electronically, the making and breaking of the circuit being performed by thyristors or ignitrons.

When the current flows across the interface between the plates, the heating effect causes melting and fusion occurs at A (Fig. 8.4). Around this there is a narrow heat-affected zone (HAZ) since there is a quench-ing effect due to the electrodes, which are often water cooled. There are equi-axed crystals in the centre of the nugget and small columnar cry-stals grow inwards towards the centre of greatest heat.

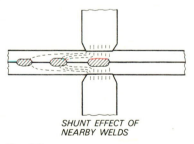

SHUNT EFFECT OF
NEARBY WELDS

Fig. 8.3

FUSION ZONE A

HAZ

Fig. 8.4

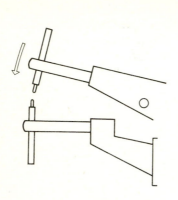
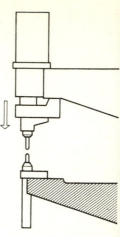

(a) PIVOTTING ARMS   (b) VERTICAL TRAVEL

Fig. 8.5. (a) Pivoting arms, (b) Vertical travel.

Types of spot welders

In the pedestal type there is a fixed vertical pedestal frame and integral transformer and control cabinet. The bottom arm is fixed to the frame and is stationary during welding and takes the weight of the workpiece. The top arm may be hinged so as to move down in the arc of a circle (Fig. 8.5a) or it may be moved down in a straight line (Fig. 8.5b). Pivoting arms are adjustable so as to have a large gap between the electrodes, the arms are easily adjusted in the hubs and various length arms are easily fitted giving easy access to difficult joints. The vertical travel machine has arms of great rigidity so that high pressures can be applied, and the electrode tips remain in line irrespective of the length of stroke, so that the machine is easily adapted for projection welding. Additionally the spot can be accurately positioned on the work, and since the moving parts have low inertia, high welding speeds can be achieved without hammering (see Figs. 8.6a and b).

Welding guns

Portable welding guns are extensively used in mass production. The equipment consists of the welding station often with hydro-pneumatic booster to apply the pressures, a water manifold, one or two welding guns with balancers, cable between transformer and guns and a control station. In modern machines the composite station with built-in cabinet comprises sequence controls with integrated circuits, thyristor con

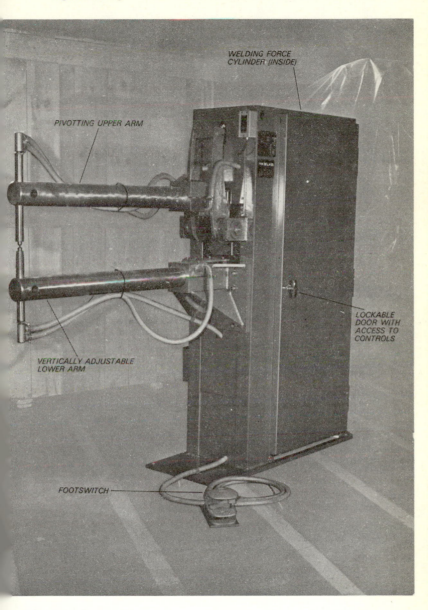

Fig. 8.6. (*a*) Air-operated spot welding machine with pivoting arms.

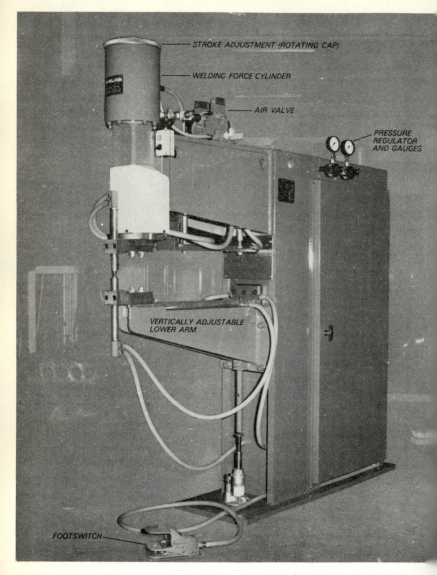

Fig. 8.6. (*b*) Air-operated spot welding machine with vertical action.

tactors and disconnect switch. Articulated guns (Figs. 8.7a and b) have both arms articulated as in a pair of scissors, giving a wide aperture between electrodes, and are used for welding joints difficult of access. Small articulated guns are used for example in car body manufacture for welding small flanges, and in corners and recesses. C guns (Fig. 8.7c) have the piston-type ram of the pressure cylinder connected to the moving electrode, which thus moves in a straight line. There is great rigidity and a high working speed because of the low inertia of the moving parts. The electrodes are always parallel and the precision motion is independent of the arm length with easily determined point of welding contact. They are used for quality welds where their height does not preclude them. Integral transformer guns are used to reduce hand-ling costs of bulky fabrications, and the manual air-cooled gun and twin spot gun are used in repair work on car bodies and agricultural equip-ment and general maintenance, the latter being used where there is access for only one arm as in closed box sections and car side panels.

The welding cycle

The cycle of operations of most modern machines is completely auto-matic. Once the hand or foot switch is pressed, the cycle proceeds to completion, the latest form having digital sequence controls with printed circuits, semi-conductors, digital counting and air valve operation by relays. The simplest cycle has one function, namely weld time, the electrode force or pressure being pre-set, and it can be best illustrated by two graphs, one the electrode pressure plotted against time and the other the current plotted against the same time axis (Fig. 8.8). The time axis is divided into 100 parts corresponding to the 50 Hz frequency of

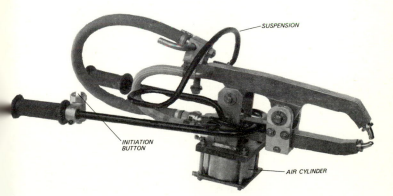

Fig. 8.7. (a) Articulated type welding head; air operated.

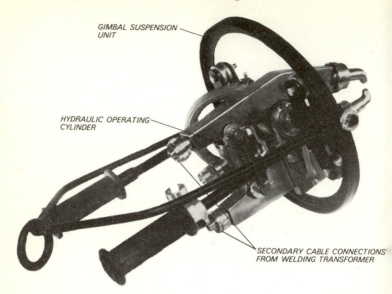

Fig. 8.7. (*b*) Articulated type welding head; hydraulically operated.

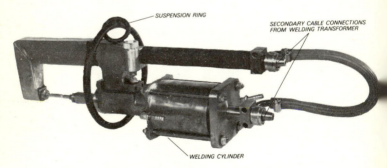

Fig. 8.7. (*c*) Air-operated C type welding head.

the power supply. Fig. 8.8 shows a simple 4-event control. First the squeeze is applied and held constant (*a–b* on the graph). The current is switched on at *x*, held for seven complete cycles (7/50 s) and switched off at *y*. The pressure is held until *c* on the graph, the weld cooling in this period. Finally the squeeze is released and the repeat cycle of operation begins again at a_1. The switches comprise repeat–non-repeat, ignitron on–off and heat control in–out.

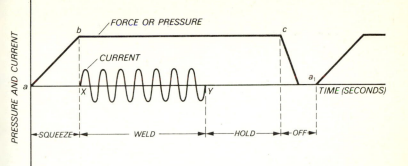

Fig. 8.8. Standard four-event sequence for spot welding.

For more complex welding cycles, necessary when welding certain metals to give them correct thermal treatment, the graphs are more complex with up to ten functions with variable pressure cycle, and these machines are often operated from three phases. Machines for the aeronautical and space industries are specially designed and incorporate variable pressure heads and up to ten functions. Most aluminium alloys lose the properties given to them by work hardening or heat treatment when they are heated. To enable the correct thermal treatment of heating and cooling to be given to them during the welding cycle three-phase machines are often employed.

The transformer

The transformer steps the voltage down from that of the mains to the few volts necessary to send the heavy current through the secondary welding circuit. When the current is flowing the voltage drop across the secondary may be as low as 3–4 volts and in larger machines the current can be up to 35 000 A. Because of these large currents there is considerable force acting between the conductors due to the magnetic field so transformers must be robustly constructed or movement may cause breakdown of the insulation and in addition they may be water cooled because of the heating effect of the current. The secondary winding consists of one or two turns of copper strip or plates over which the primary coils fit, the whole being mounted on the laminated iron core. Because any infiltration of moisture may lead to breakdown, modern transformers have primary winding and secondary plates assembled as a unit which is then vacuum encapsulated in a block of epoxy resin. The primary winding has tappings which are taken to a rotary switch which selects the current to be used.

The current flowing in the secondary circuit of the transformer provides the heating effect and depends upon (1) the open circuit voltage, (2) the impedance of the circuit, which depends upon the gap, throat depth and magnetic mass introduced during welding and the resistance of the metal to be welded. The duty cycle is important, as with all welding machines, as it affects the temperature rise of the transformer. For example, if a spot welder is making 48 spot welds per minute, each of 0.25 seconds duration, the duty cycle is (48 × 0.25 × 100) ÷ 60 = 20%. Evidently knowing the duty cycle and welding time, the number of welds that can be made per minute can be calculated.

ELECTRODES

The electrode arms and tips (Fig. 8.9a) which must carry the heavy currents involved and apply the necessary pressure must have the following properties; (1) high electrical conductivity so as to keep the I^2R loss (the heating due to the resistance) to a minimum, (2) high thermal conductivity to dissipate any heat generated, (3) high resistance to deformation under large squeeze pressures, (4) keep their physical properties at elevated temperatures, (5) must not pick up metal from the surface of the workpiece, (6) must be of reasonable cost. The following types of electrodes are chiefly used, and are given in the table. Electrolytic copper (99.95% Cu) has high electrical and thermal conductivity and when work hardened, resists deformation but at elevated temperatures that part of the electrode tip in contact with the work becomes annealed due to the heating and the tip softens and deforms. Because of this it is usual to water-cool the electrodes to prevent excessive temperature rise.

Water cooling
Adequate electrode cooling is the most essential factor to ensure optimum tip life; the object is to prevent the electrode material from reaching its softening temperature, at which point it will lose its hardness and rapidly deteriorate. The normal cooling method is by internal water circulation where the water is fed via a central tube arranged to direct the water against the end of the electrode cooling hole (Fig. 8.9b). In some cases a short telescopic extension tube is used inside the main tube and must be adjusted to suit the length of the electrode used.

Electrodes are available with a taper fit to suit the electrode arms and also as tips to fit on to a tapered shank body which makes them easily interchangeable. The face that makes contact with the metal to be welded can be a truncated cone with central or offset face, flat or domed (Fig. 8.9) and tip diameters can be calculated from formulae which depend upon the plate thickness, but in all cases these are approximate only and do not replace a test on the actual part. Electrodes are subject to great wear and tear in service due to the constant heating and cooling and varying pressure cycles. The chief causes of wear are: electrical; wrong electrode material, poor surface being welded, contacts not in line, mechanical; electrode hammering, high squeeze, weld and forge pressures, abrasion in loading and unloading and tearing due to the parting of the electrodes.

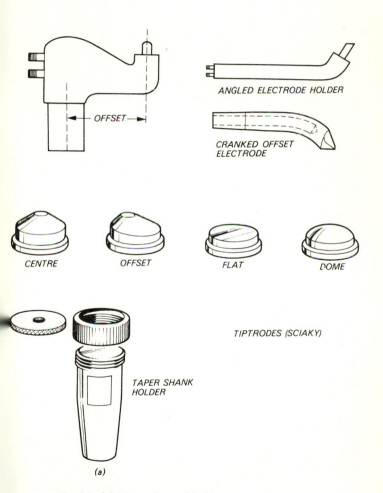

Fig. 8.9. (a) Electrodes and holders.

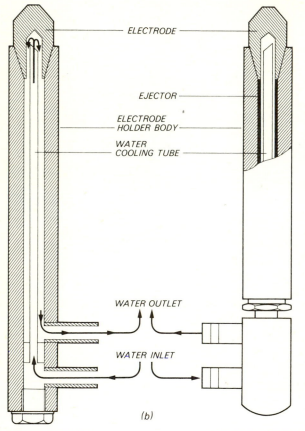

ELECTRODE

EJECTOR

ELECTRODE
HOLDER BODY

WATER
COOLING TUBE

WATER OUTLET

WATER INLET

(b)

PLAIN ELECTRODE HOLDER

'EJECTO' ELECTRODE HOLDER

Fig. 8.9(b)

Electrode	Properties	Uses
(1) Copper, 1% silver	High conductivity, medium hardness.	Light alloys, coated sheets, scaly steel.
(2) Copper, 0.6% Cr or 0.5% Cr and Be	Best electrical conductivity with greatest hardness.	Clean or lightly oxidized steel, brass and cupro-nickel. Used for the arms of the machine and electrical conductors.
(3) Copper, 2.5% Co and 0.5% Be	Poor conductivity but very hard.	For welding hard metals with high resistivity, e.g. stainless steel, heat-resistant and special steels.
(4) Copper, beryllium.	Poor conductivity, great hardness.	For clamping jaws of flash butt welders.
(5) Molybdenum		Drawn bar or forged buttons used as pressed fit inserts in supports of (2) above. For welding thin sheet or wires or electro-brazing silver-based metals.
(6) Sintered copper–tungsten	Fair conductivity, very hard.	Keeps its mechanical properties when hot.

Note. The addition of various elements increases the hardness but reduces the conductivity.

SEAM WELDING

In a seam welding machine the electrodes of a spot welder are replaced by copper alloy rollers or wheels which press on the work to be welded. (Fig. 8.10a and b). Either one or both are driven and thus the work passes between them. Current is taken to the wheels through the rotary bearings by silver contacts with radial pressure and the drive may be by knurled wheel or the more usual shaft drive which enables various types of wheel to be easily fitted. If the current is passed continuously a continuous seam weld results but, as there is a shunt effect causing the current to flow through that part of the weld already completed, over-heating may occur resulting in burning of the sheets. To avoid this the current can be pulsed, allowing sufficient displacement of the already welded portion to take place and thus obviating most of the shunt effect. For materials less than 0.8 mm thick or at high welding speeds (6 m/min), no pulsing is required, the 50 Hz frequency of the supply providing a natural pulse. Above 2 × 0.8 mm thickness pulsation is advisable, and essential above 2 × 1.5 mm thickness, while for pressure-tight seams the welds can be arranged to overlap and if the seams are given a small overlap with wide-faced wheels and high pressure a mash weld can be obtained.

By the use of more complex electro-mechanical bearing assemblies, longitudinal and circumferential welds can be made (Fig. 8.10c).

Seam welding guns are extremely useful for fabricating all types of tanks, exhaust systems, barrels, drip-mouldings on car body shells, etc. They have electrode drive which automatically propels the gun along the seam so that it only requires guidance, and they are operated in the same way as spot welding guns.

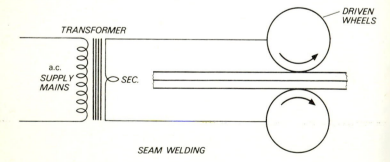

Fig. 8.10 (a)

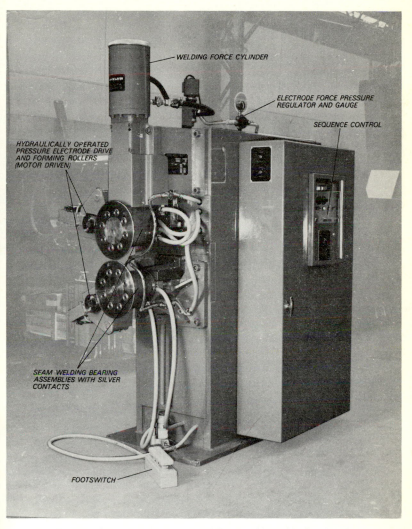

Fig. 8.10. (*b*) 'Thin wheel' seam welding machine with silver bearings.

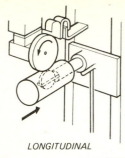
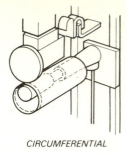

LONGITUDINAL   *CIRCUMFERENTIAL*

SEAM WELDING

Fig. 8.10 (*c*)

Direct current spot welding

We have seen that when iron and steel to be welded is placed in a spot or seam welder the impedance of the secondary circuit is increased and the secondary current varies according to the mass introduced. By using d.c. this loss is obviated, the power consumed is reduced and the electrode life increased because skin effect is eliminated. This enables coated steels, stainless and other special steels in addition to aluminium alloys to be welded to high standards. The machine is similar in appearance to the a.c. machine but has silicon diodes as the rectifying elements.

Three-phase machines

Three-phase machines have been developed to give impulses of current in the secondary circuit at low frequency, with modulated wave form to give correct thermal treatment to the material being welded. These machines have greatly increased the field of application of spot welding into light alloys, stainless, and heat-resistant steels, etc. A typical sequence of operations is: squeeze, which multiplies the number of high spots between the contact faces; welding pressure, which diminishes just before the welding current flows; immediately after the passage of the current, forging or recompression pressure is applied which is above the elastic limit of the material and completes the weld. Setting of the welding current can be done by welding test pieces with increasing current until the diameter of the nugget, found by 'peeling' the joint is $(2t + 3)$, where t is the thickness of the thinner sheet. This current is noted. It is then increased until spatter of the nugget occurs and the welding current is taken as midway between these two values with final adjustments made on test pieces.

Nickel and nickel alloys have higher electrical resistivity and lower thermal conductivity than steel and are usually welded in thin sections as lap joints, although the crevice between the sheets may act as a stress raiser and affect the corrosion and fatigue resistance. High pressures are required for the high nickel alloys so as to forge the solidifying nugget, and the machine should have low inertia of the moving parts so that the electrode has rapid follow-up during welding. Current may be set as in the previous section, and in some cases it is advantageous to use an initial squeeze followed by an increasing (up-sloping) current. When the nugget is just beginning to form with diffusion of the interface, the squeeze force is reduced to about one-third of its initial value and held until the current is switched off; this reduces danger of expulsion of molten metal and gives better penetration of the weld into the two sheets.

PROJECTION WELDING

Protrusions are pressed on one of the sheets or strips to be welded and determine the exact location of the weld (Fig. 8.11). Upon passage of the current the projection collapses under the electrode pressure and the sheets are welded together. The machines are basically presses, the tipped electrodes of the spot welder being replaced by flat platens with T slots for the attachment of special tools, and special platens are available which allow the machine to be used as a spot welder by fitting arms and electrodes (Fig. 8.12) and automatic indexing tables can be used to give increased output. Protection welding is carried out for a variety of components such as steel radiator coupling elements, brake shoes, tin-plate tank handles and spouts, etc. The press type of machine is also used for resistance brazing in which the joining of the parts is achieved with the use of an alloy with a lower melting point than the parent metal being welded so that there is no melting involved.

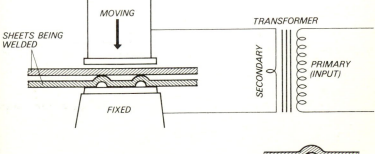

Fig. 8.11. Projection welding.

SHAPE OF PROJECTION

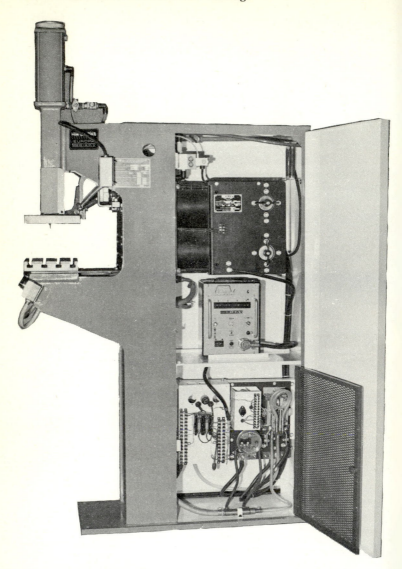

Fig. 8.12. Projection welder.

Cross wire welding is a form of projection welding, the point of contact of the two wires being the point of location of the current flow. Low carbon mild steel, brass, 18/8 stainless steel, copper-coated mild steel and galvanized steel wire can be welded but usually the bulk of the work is done with clean mild steel wire, bright galvanized, or copper coated as used for milk bottle containers, cages, cooker and refrigerator grids, etc., and generally several joints are welded simultaneously.

Modern machines are essentially spot welders in ratings of 25, 70, or 150 kVA fitted with platens upon which suitable fixtures are mounted and having a fully controlled pneumatically operated vertical head, and the electrical capacity to weld as many joints as possible simultaneously. Large programmed machines are manufactured for producing rein-forcement mesh automatically.

RESISTANCE BUTT WELDING

This method is similar to spot welding except that the parts to be butt welded now take the place of the electrodes. The two ends are prepared so that they butt together with good contact. They are then placed in the jaws of the machine, which presses them close together end to end (Fig. 8.13). When a given pressure has been reached, the heavy current is switched on, and the current flowing through the contact resistance between the ends brings them to welding heat. Extra pressure is now applied and the ends are pushed into each other, the white hot metal welding together and an enlargment of section taking place. The section may be machined to size after the operation if necessary.

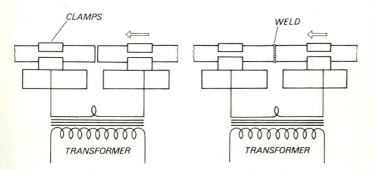

Fig. 8.13. Resistance butt welding.

FLASH BUTT WELDING

Although this is not a resistance welding process it is convenient to consider it here. Flash butt welding machines must be very robustly made and have great rigidity because considerable pressures are exerted and exact alignment of the components is of prime importance. The clamping dies of copper alloy, which carry the current to the components and hold them during butting up under high pressure, should grip over as large an area as possible to reduce distortion tendencies. The clamping pressure, which is about twice the butting pressure, is usually done pneumatically or hydraulically, and current is of the order of 7–10 A/mm² of joint area. The following are the stages in the welding cycle.

Pre-heat. The components are butted together and a current passing across the joint heats the ends to red heat.

Flash. The parts are separated and an arc is established between them until metal begins to melt, one of the components being moved to keep the arc length constant.

Upset. The parts are butted together under high pressures with the current still flowing. Impurites are forced out of the joint in the butting process and an impact ridge or flash is formed (Fig. 8.14). Post-heat treatment can be given by a variation of current and pressure after welding. For welding light alloys, pre-heating is generally dispensed with and the flashing is of short duration.

As an example of currents involved, a butt weld in 6 mm thick 18/8/3 stainless steel with a cross-section of 600 mm² involves currents of 20 000 A with a 9 second flashing time.

Flash butt welding is used very extensively by railway systems of the world to weld rails into continuous lengths. British Rail for example weld rails approximately 18.3 m long into continuous lengths from 91 to

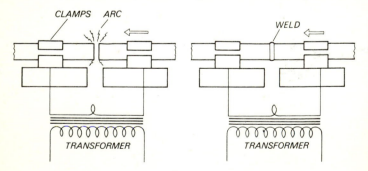

Fig. 8.14. Flash butt welding.

Fig. 8.15. Machine for welding rail sections. 1000 kVA, 3-phase, 440 V. Welding sequence: (1) Both rails securely clamped in welder. (2) Rails aligned vertically and horizontally under full clamping pressure. (3) Weld initiated by push button control. (4) Moving head moves forward on the burn-off or pre-flashing stroke used to square up the ends of the rails if necessary. On completion of this, pre-heating begins until the requisite number of pre-heats have lapsed. At this point the rails should be in a suitably plastic state to allow for straight flashing and finally forging.

366 m and conductor rails are welded in the same way. Fig. 8.15 shows a modern rail welding machine. Hydraulic rams, equally spaced on each side of the rail section apply the forging load of 200–400 kN to the rails of section approximately 7200 mm². The rail is clamped by two vertically acting cylinders and horizontally acting cylinders align each rail to a common datum and an anti-twist device removes axial twist. When welding long lengths of rail it is more convenient to move the machine to the exact position for welding rather than the rail and for this the machine can be rail mounted. Machines of this type can make 150–200 welds per 8 hour shift. In this case the sequence of operations is preflash, pre-heat, flashing and forging.

The upset is removed by a purpose-designed machine with hydraulic power-shearing action usually mounted in line with the welding machine.

The flash butt welded lengths are welded *in situ* on the track by the thermit process.

9

Additional processes of welding

There has been a great increase in the number of automatic processes designed to speed up welding production. Automatic welding gives high rates of metal deposition because high currents from 400 to 2000 A can be used, compared with the limit of about 600 A with manual arc welding. Automatic arc control gives uniformly good weld quality and finish and the high heat input reduces distortion and the number of runs for a given plate thickness is reduced. Twin welding heads still further reduce welding time, and when used, for example, one on each side of a plate being fillet welded, distortion is reduced. The welding head may be:

(1) Fixed, with the work arranged to move beneath it.
(2) Mounted on a boom and column which can either be of the positioning type in which the work moves or the boom can traverse at welding speed over the fixed work.
(3) Gantry mounted so that it can traverse over the stationary work.
(4) Self propelled on a motor-driven carriage.

The processes which have been described previously in this book, namely TIG, MIG and CO_2 (gas shielded metal arc) with their modifications are extensively used fully automatically. Heads are now available which, by changing simple components, enable one item of equipment to be used for MIG (inert gas), CO_2 and tubular wire, and submerged arc processes.

SUBMERGED ARC WELDING

In this automatic process the arc is struck between bare or flux cored wire and the parent plate, the arc, electrode end and the molten pool are submerged or enveloped in an agglomerated or fused powder which turns into a slag in its lower layers under the heat of the arc and protects the weld from contamination. The wire electrode is fed continuously to the arc by a feed unit of motor-driven rollers which is voltage-controlled in the same way as the wire feed in other automatic processes and ensures an arc of constant length. The flux is fed from a hopper fixed to the welding head, and a tube from the hopper spreads the flux in a continuous mound in front of the arc along the line of weld and of sufficient depth to completely submerge the arc, so that there is no spatter, the weld is shielded from the atmosphere and there are no radiation effects (UV and IR) in the vicinity. (Fig. 9.1*a* and *b*).

Welding heads

Fully automated welding heads for this process can also be used with modifications for gas shielded metal arc welding including CO_2, solid and flux cored, thus greatly increasing the usefulness of the equipment. The head can be stationary and the work moved below it as for example in the welding of circumferential and longitudinal seams, or the head may be used with positioners or booms or incorporated into custom-built mass production welding units for fabricating such components as brake shoes, axle housings, refrigerator compressor housings, brake vacuum cylinders, etc., and hard surfacing can also be carried out. The unit can also be tractor mounted (cf. Fig. 9.3) and is self-propelled, with a range of speeds of 100 mm to 2.25 m per minute and arranged to run on guide bars or rails. Oscillating heads can be used for root runs on butt joints to maintain a constant welding bead on the underside of the joint, and two and even three heads can be mounted together or the heads can be arranged side by side to give a wide deposit as in hard surfacing. Fillet welding can be performed by inclining two heads, one on each side of the

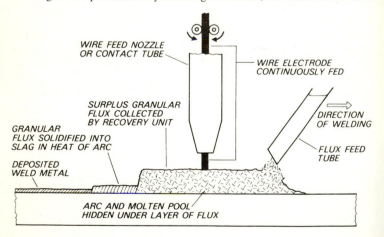

Fig. 9.1 (*a*)

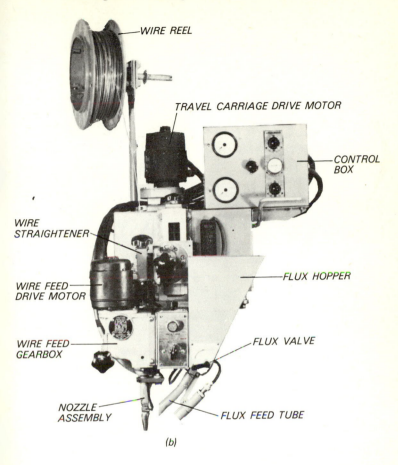

WIRE REEL

TRAVEL CARRIAGE DRIVE MOTOR

CONTROL BOX

WIRE STRAIGHTENER

WIRE FEED DRIVE MOTOR

FLUX HOPPER

WIRE FEED GEARBOX

FLUX VALVE

NOZZLE ASSEMBLY

FLUX FEED TUBE

(b)

joint with flux feeds and recovery, the heads being mounted on a carriage which travels along a gantry over the work (Fig. 9.2). Two heads mounted in tandem and travelling either along a guide rail or directly on the workpiece are used for butt joints on thick plate, and both can operate on d.c. or the leading head can operate on d.c. and the trailing head on a.c.

Three electrode heads can be gantry mounted on a carriage, the leading electrode being d.c. operated with the trailing electrodes a.c. This method gives high deposition rates with deep penetration. Special guide units ensure in all cases that the electrode is correctly positioned relative to the joint.

The main components in the control box are: welding voltage and arc

(a)

(b)

(c)

Fig. 9.2. Various moutings for automatic welding equipment.

(*a*) An automatic welding machine in which the head is mounted on a carriage which travels along a beam.

(*b*) An automatic welding head as in (*a*) designed for stationary mounting on a manipulator column or boom in order to be an integral part of a mechanized welding system.

(*c*) A tractor-mounted automatic welding head as in (*a*). The machine has a single welding head and is designed for welding butt joints and for making fillet welds in the flat or horizontal–vertical position.

current controls and wire feed controlled by a thyristor regulator which maintains set values of arc voltage. The head is accurately positioned by slide adjusters for horizontal and vertical movement and has angular adjustment also. The wire feed motor has an integral gearbox and wire-straightening rolls give smooth wire feed. The gear ratio for the metal arc process is much higher than for submerged arc and each wire diameter usually requires its own feed rolls, which are easily interchanged. For fine wires less than 3 mm diameter a fine-wire-straightening unit can be fitted.

Current is passed to the electrode wire through a contact tube and jaws which fit the wire diameter being used, and the contact tube is used for the shielding gas when gas shielded welding is being performed and is water cooled. The coil arm holder has a brake hub with adjustable braking effect and carries 300 mm i.d. coils, and a flux hopper is connected to a flux funnel attached to the contact tube by a flexible hose.

A guide lamp which is attached to the contact tube provides a spot of light which indicates the position of the wire, thus enabling accurate positioning of the head along the joint, and a flux recovery unit collects unfused flux and returns it to the hopper.

A typical sequence of operations for a boom-mounted carriage carrying multiple welding heads is: power on, carriage positioned, welding heads 1, 2 etc. down, electrode feed on, wire tips set, flux valve open, welding speed set, welding current set, welding voltage set, flux recovery on; press switch to commence welding.

Power unit

The power unit can be a motor- or engine-driven d.c. generator or transformer-rectifier with outputs in the 30–55 V range and with currents from 200 to 1600 A with the wire generally positive. In the case of multiple head units in which the leading electrode is d.c. and the trailing electrode is a.c. a transformer is also required. In general any power source designed for automatic welding is usually suitable when feeding a single head.

Wires are available in diameters of; 1.6, 2.0, 2.4, 3.2, 4.0, 5.0, 6.0 mm on plastic reels or steel formers. Wrappings should be kept on the wire until ready for use and the reel should not be exposed to damp or dirty conditions.

A variety of wires are available including the following (they are usually copper coated). For mild steel types with varying manganese content, e.g. 0.5%, 1.0%, 1.5%, 2.0% manganese, a typical analysis being 0.1% C, 1.0% Mn, 0.25% Si, with S and P below 0.03%. For low alloy steels, 1.25% Cr, 0.5% Mo; 2.25% Cr, 1.0% Mo; 0.2% Mn, 0.5% Mo; and 1.5% Mn 0.5% Mo are examples, whilst for the stainless steel range there are: 20% Cr, 10% Ni for unstabilized steels; 20% Cr, 10% Ni, 0.03% C, for low-carbon 18/8 steels; 19% Cr, 12% Ni, 3% Mo for similar steels; 20% Cr, 10% Ni, niobium stabilized; 24% Cr, 13% Ni for steels of similar composition and for welding mild and low alloy steels to stainless steel.

Many factors affect the quality of the deposited weld metal: electrode wire, slag basicity, welding variables (process), cleanliness, cooling rate, etc. For hardfacing, the alloy additions necessary to give the hard surface usually come from the welding wire and a neutral flux, and tubular wire with internal flux core is also used in conjunction with the external flux. Hardness values as welded, using three layers on a mild steel base, are between 230 and 650 HV depending upon the wire and flux chosen.

Flux

Fluxes are suitable for use with d.c. or a.c. They are graded according to their form, whether (1) fused or (2) agglomerated. Fused fluxes have solid glassy particles, low tendency to form dust, good recycling properties, good slag–flux compatibility, low combined water and little sensitivity to humid conditions. Agglomerated fluxes have irregular-sized grains with low bulk density, low weight consumption at high energy inputs with active deoxidizers and added alloying elements where required.

Fluxes are further classified as to whether they are acidic or basic, the basicity being the ratio of acidic oxides to basic oxides which they contain (see p. 59). In general the higher the basicity the greater the absorption of moisture and the more difficult it is to remove.

The general types of flux include manganese silicate, calcium manganese aluminium sulphate, rutile, zirconia and bauxite, and the choice of flux affects the mechanical properties of the weld metal. Manufacturers supply full details of the chemical composition and mechanical properties of the deposited metal when using wires of varying compositions with various selected fluxes (i.e. UTS, % elongation and Charpy impact values at various temperatures).

Fluxes that have absorbed moisture should be dried in accordance with the makers' instructions, as the presence of moisture will affect the mechanical properties of the deposited metal.

Joint preparation

Joint edges should be carefully prepared and free from scale, paint, rust and oil, etc. and butt seams should fit tightly together. If the fit-up has gaps greater than 0.8 mm these should be sealed with a fast manual weld

When welding curved circumferential seams there is a tendency for the molten metal and slag, which is very fluid, to run off the seam. This can be avoided or reduced by having the welding point 15–65 mm before top dead centre in the opposite direction to the rotation of the work and in some cases the speed of welding and current can be reduced. Preparation of joints is dependent upon the service to which the joint is to be put and the following preparations are given as examples only (Fig. 9.3).

JOINT BUTTED TIGHTLY. GAPS ABOVE 0.8 mm SEALED WITH MANUAL WELD

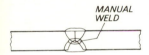

JOINT BUTTED TIGHTLY. GAPS SEALED WITH MANUAL WELD

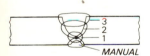

MANUAL WELD MUST HAVE 50% PENETRATION MINIMUM

PLATE 10–20 mm THICK

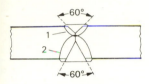

PLATE 20 mm THICK

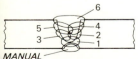

MANUAL
PLATE 25 mm THICK

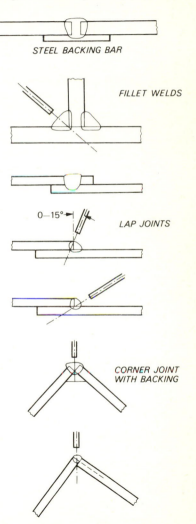

STEEL BACKING BAR

FILLET WELDS

LAP JOINTS

CORNER JOINT WITH BACKING

CORNER JOINT WITHOUT BACKING
50–70% PENETRATION

Fig. 9.3. Types of butt welds.

ELECTROSLAG WELDING (Fig. 9.4)

Developed in Russia, this process is used for butt welding steel sections usually above 60 mm in thickness although plates down to 10 mm thick can be welded. The sections to be welded are fixed in the vertical position and part of the joint line, where welding is to commence, is enclosed with water-cooled copper plates or dams which serve to position the molten weld metal and slag. The dams are pressed tightly against each side of the joint to prevent leakage. There may be from one to three electrode wires depending upon the thickness of the section and they are fed continuously from spools. The self-adjusting arc is struck on to a run-off plate beneath a coating of powder flux which is coverted into a liquid in about half a minute. The current is then transferred, not as an arc but through the liquid slag which gives the same order of voltage drop as would the arc. During welding some slag is lost in forming a skin

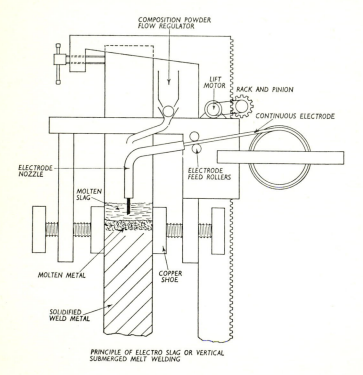

PRINCIPLE OF ELECTRO SLAG OR VERTICAL
SUBMERGED MELT WELDING

Fig. 9.4. Principle of electroslag or vertical submerged melt welding.

between the molten metal and the copper dams, and a flow of powder, carefully metered to avoid disturbing the welding conditions, is fed in to make good wastage. The vertical traverse may be obtained by mounting the welding head on a carriage which is motor-driven and travels up a rack on a vertical column in alignment with the joint to be welded. The rate of travel is controlled so that the electrode nozzle and copper dams are in the correct position with regard to the molten pool, and since the electrode is at right angles to the pool, variations to fit-up are not troublesome. For thick sections the electrodes are given a weaving motion across the metal.

The welds produced are free from slag inclusions, porosity and cracks, and the process is rapid, preparation is reduced, and there is no deslag-ging. Composite wires containing deoxidizers and alloying elements can be used when required. A variation of the process uses a CO_2 shield instead of the flux powder, the CO_2 being introduced through pipes in the copper dams just above the molten metal level.

Electroslag welding with consumable guides or nozzles

Consumable guide welding is a development of the electroslag process for welding straight joints in thick plate in the vertical or near vertical position, in a range of 15–40 mm thick plate and joints up to 2 m long. The set-up gap between plates is 25–30 mm, but when welding thick-nesses less than 20 mm the joints can be reduced to 18–24 mm, the gap ensuring that the guide tube does not touch the plate edges. Water-cooled copper shoes act as dams and position the molten metal and also give it the required weld profile. As with electroslag welding the current passing through the molten slag generates enough heat to melt electrode end, guide and edges of the joint, ensuring a good fusion weld. If a plain uncoated guide tube is used, flux is added during welding, but if a coated guide tube is used flux need only be added to cover electrode and guide end before welding commences. To start the process the arc is struck on the work. It continues burning under the slag with no visible arc or spatter. The slag should be viewed through dark glasses as in gas cutting because of its brightness when molten.

An a.c. or d.c. power source in the range 300–750 A is suitable, such as is used for automatic and MMA processes. Striking voltage is of the order of 70–80 V, with arc voltages of 30–50 V higher with a.c. than d.c.

The advantages claimed for the process are: relatively simpler, cheaper, and more adaptable than other similar types, faster welding than MMA of thick plate, cheaper joint preparation, even heat input into the joint thus reducing distortion problems, no spatter losses, freedom from weld metal defects and there is a low consumption of flux. Fig. 9.5 illustrates the layout of the machine.

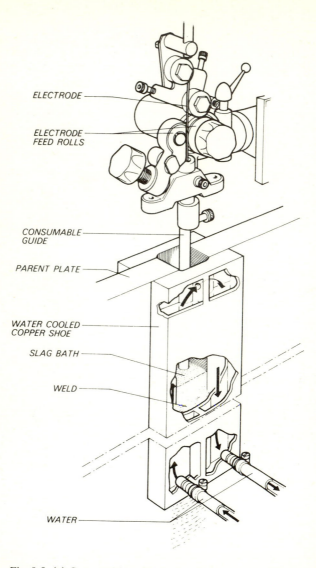

ELECTRODE

ELECTRODE
FEED ROLLS

CONSUMABLE
GUIDE

PARENT PLATE

WATER COOLED
COPPER SHOE

SLAG BATH

WELD

WATER

Fig. 9.5. (*a*) Consumable guide layout showing water-cooled dams.

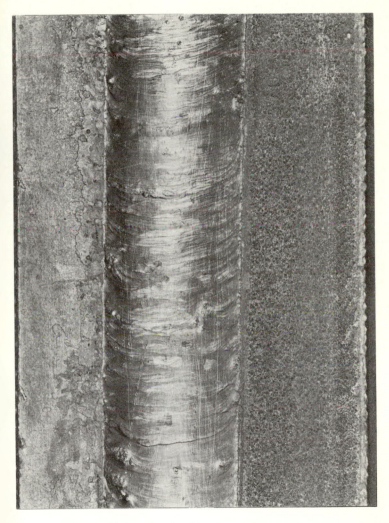

Fig. 9.5. (*b*) Completed weld.

Fig. 9.5. (*c*) The consumable guide welding process using, in this case, twin wire/tube system. The dams have been removed to show the position of the guide tubes.

MECHANIZED WELDING WITH ROBOTS

Fully automatic welding using the gas shielded metal arc or submerged arc processes in conjunction with columns and booms, positioners, rotators and other equipment is extensively used for the making of long welds either straight, circumferential or circular. This has resulted in increased speed of fabrication with a reduction in monotony and fatigue for the welder.

The fabrication of large numbers of similar components of relatively simple shape with regularly placed welds on straightforward joints with continuous or intermittent welds becomes exceedingly monotonous to a welder, and robots are now available which can perform this repetitive work with increased speed and uniformly high quality welds.

Welding equipment

The robots are used with standard gas shielded metal arc adapted for the purpose in conjunction with rotators and positioners (Fig. 9.6). One type of robot has a mini-computer programmed by push buttons on a control console. Another type is hydraulically operated and is programmed by the operator making the first welding sequence manually, the position of the torch head being recorded with high resolution on a cassette tape recorder.

Components in a given batch must be similar to each other or the welds must be in a similar position with respect to a given reference line or plane. Material should be clean, with fit-up similar to that usual for gas shielded metal arc semi-automatic welding.

The equipment comprises a power unit with stepless control of welding current which is programmed in advance by a programming unit which regulates arc current and voltage and adapts the robot's signals to the welding plant. The welding positioner on which the component is mounted is connected to the robot control equipment through a relay cabinet, and thus all positioning sequences are made prior to each welding run.

ATOMIC HYDROGEN ARC WELDING

In this method, an alternating arc is maintained between two tungsten electrodes. Hydrogen is fed into this arc, and the energy of the arc, together with the action of the tungsten electrodes, splits up or dissociates the molecules of hydrogen into atoms. The atoms recombine when they reach the slightly cooler regions outside the arc and the heat liberated by this recombination results in temperatures of up to 4000 °C being attained.

Hydrogen, as previously mentioned, is a strong reducing agent, especially in its atomic state, and as a result the weld metal is protected by a strongly reducing gas and no oxidation or other atmospheric absorption can take place, and in addition the electrodes are prevented from

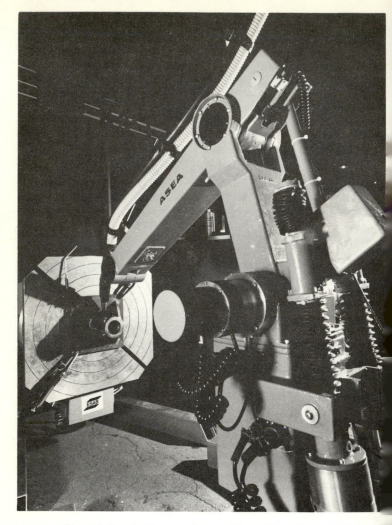

Fig. 9.6. Gas-shielded metal arc welding with a robot, with the work being held on a positioner.

burning away rapidly by oxidation, since no oxygen can possibly be present near the arc. Any oxygen present in the surrounding region combines with the hydrogen, forming water, which is immediately converted into steam, while the remaining hydrogen burns beyond the region of recombination into molecules, in the usual way.

Equipment

Atomic hydrogen sets are designed in various sizes, such as $7\frac{1}{2}$ to 35 A, 15 to 75 A, and a transformer supplies the current for the arc. The arc is struck at 300 V, falling to 70 to 90 V when operating (since a much higher voltage is required to maintain an arc in hydrogen than in air). The current is, as usual, adjusted according to the thickness of the work. The blowpipe or torch is fitted with a cable supplying current to the electrodes and with a metallic hose which supplies the hydrogen gas to the arc. The hydrogen supply may either be from cylinders containing the gas in compressed form, through a reducing valve, or it may be from equipment by which ammonia gas (NH_3) is cracked or broken down into hydrogen and nitrogen, the hydrogen being fed to the arc.

The set is fitted with a start and stop button, connected to the control cabinet by a flexible cable. With this the welder can control the gas and electrical supply from where he is welding.

Welding process

The start button is pressed and the tungsten electrodes are drawn apart from each other by a lever on the torch striking the arc. The gas supply is then adjusted, and the bottom of the elliptical or fan-shaped arc is brought on to the metal to be fused. Filler rod may be used as in oxy-acetylene welding, or the arc may be used simply to fuse two surfaces together. The arc is extinguished by drawing the electrodes well apart or by pressing the stop button, which cuts off the current and shuts off the gas.

Automatic atomic arc welding machines are available in which the electrodes are fed automatically through a water-cooled nozzle, through which the gas is also fed. These machines are mounted on frames, so that their rate of travel along the seam to be welded is automatic and at a constant rate, and they are especially useful for welding long seams, such as those in transformer cooling radiators, the resulting weld being absolutely watertight.

The atomic hydrogen welding process gives excellent welds entirely free from inclusions of any kind, and is applicable chiefly to steel plate, although non-ferrous metals may also be welded, taking the same precautions as for oxy-acetylene welding.

The two electrodes burn away evenly and the consumption is very slow (e.g. with 3 mm electrodes at 50 A the rate is 35 mm per hour); this is an important point, since they are expensive. Tungsten may be absorbed into the weld, strengthening it and making up for any deficiency that may occur. In addition, a certain amount of hydrogen may be absorbed.

Fig. 9.7 shows a typical blowpipe, while Fig. 9.8 shows the general

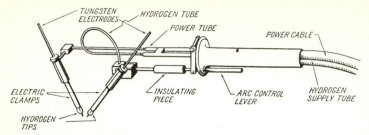

Fig. 9.7. Atomic hydrogen arc welding blowpipe.

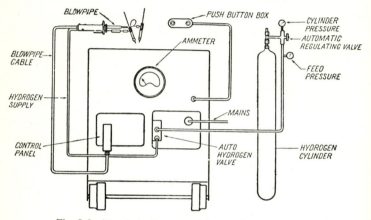

Fig. 9.8. Atomic hydrogen welding equipment.

layout of the equipment, the current being adjusted on the control panel, while the hydrogen supply pressure is adjusted at the reducing valve on the cylinder.

PRESSURE WELDING

This is the joining together of metals in the plastic condition (not fusion) by the application of heat and pressure as typified by the blacksmith's weld. In general the process is confined to butt welding. The parts (tubes are a typical example) are placed on a jig which can apply pressure to force the parts together.

The faces to be welded are heated by oxy-acetylene flames, and when the temperature is high enough for easy plastic flow to take place heating ceases and the tubes are pushed together causing an upset at the welded face. The welding temperature is about 1200 °C for steel. It is considered that atoms diffuse across the interfaces and recrystallization takes place, the grains growing from one side to the other of the welded faces since they are in close contact due to the applied pressure. Any oxide is completely broken up at a temperature well below that of fusion welding but due to the heating time concerned, grain growth is often considerable. Steel, some alloy steels, copper, brass, and silver can be welded by this process.

Cold pressure welding

This is a method of joining sections of metal together by application of pressure alone using no heat or flux.

Pressure is applied to the points to be welded at temperatures usually below the recrystallization temperatures of the metals involved. This applied pressure brings the atoms on the interfaces to be welded into such close contact that they diffuse and weld is thereby made.

The most important factor is cleanliness. Oxide and grease films are very troublesome, even when they are down to microscopic thinness, since they are difficult to break up by applied pressure and acting as a barrier they prevent contact between metal and metal. Although all types of cleaning are used, scratch brushing is still considered the most efficient. This type of welding can be applied satisfactorily to the joining of dissimilar metals such as aluminium and copper. In a variation of this method the faces at the point of welding are made to slide when the pressure is applied, the sliding action removing oxide and grease film and presenting clean, metal-to-metal faces.

Ultrasonic welding

Ultrasonic vibrations of several megacycles per second (the limit of audibility is 20 000–30 000 Hz) are applied to the region of the faces to be welded. These vibrations help to break up the grease and oxide film and heat the interface region. Deformation then occurs with the result that welding is possible with very greatly reduced pressure compared with an ordinary pressure weld. Very thin section and dissimilar metals can be welded and because of the reduced pressure there is reduced deformation (Fig. 9.9.)

FRICTION WELDING

The principle of operation of this process is the changing of mechanical energy into heat energy. One component is gripped and rotated about its axis while the other component to be welded to it is gripped and does not rotate but can be moved axially to make contact with the rotating

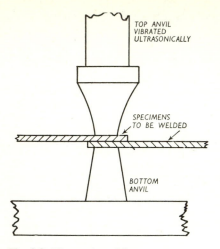

Fig. 9.9. Ultrasonic welding.

component. When contact is made between rotating and non-rotating parts heat is developed at the contact faces due to friction and the applied pressure ensures that the temperature rises to that required for welding. Rotation is then stopped and forging pressure applied, causing more extrusion at the joint area, forcing out surface oxides and impurities in the form of a flash (Fig. 9.10a). The heat is concentrated and localized at the interface, grain structure is refined by hot work and there is little diffusion across the interface so that welding of dissimilar metals is possible.

In general at least one component must be circular in shape, the ideal situation being equal diameter tubes, and equal heating must take place over the whole contact area. If there is an angular relationship between the final parts the process is not yet suitable.

The parameters involved are; (1) the power required, (2) the peripheral speed of the rotating component, (3) the pressure applied and (4) the time of duration of the operation. By adjusting (1), (2) and (3), the time can be reduced to the lowest possible value consistent with a good weld.

Power required. When the interfaces are first brought into contact, maximum power is required, breaking up the surface film. The power required then falls and remains nearly constant while the joint is raised to welding temperature. The power required for a given machine can be chosen so that the peak power falls within the overload capacity of the driving motor. It is the contact areas which determine the capacity

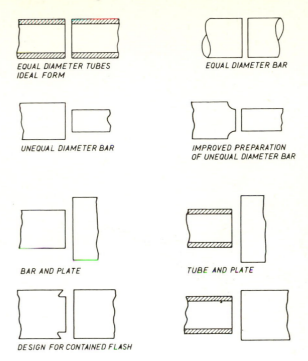

EQUAL DIAMETER TUBES
IDEAL FORM

EQUAL DIAMETER BAR

UNEQUAL DIAMETER BAR

IMPROVED PREPARATION
OF UNEQUAL DIAMETER BAR

BAR AND PLATE

TUBE AND PLATE

DESIGN FOR CONTAINED FLASH

Fig. 9.10. (*a*) Friction welding.

of a machine. The rotational speed can be as low as 1 metre per second peripheral and the pressure depends upon the materials being welded, for example for mild steel it can be of the order of 50 N/mm² for the first part of the cycle followed by 140 N/mm² for the forging operation. Non-ferrous metals require a somewhat greater difference between the two operations. The faster the rotation of the component and the greater the pressure, the shorter the weld cycle, but some materials suffer from hot cracking if the cycle is too short and the time is increased with lower pressure to increase the width of the HAZ. At the present time most steels can be welded including stainless, but excluding free cutting. Non-ferrous metals are also weldable and aluminium (99.7% Al) can be welded to steel.

There are various control systems: (1) Time control, in which after a given set time period after contact of the faces, rotation is stopped and forge pressure is applied. There is no control of length with this method. (2) Burn-off to length: parts contact and heating and forging take place within a given pre-determined length through which the axially moving

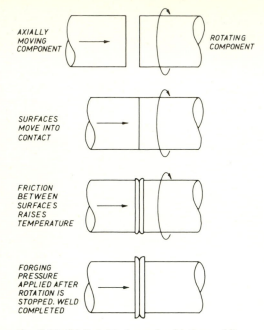

AXIALLY
MOVING
COMPONENT

ROTATING
COMPONENT

SURFACES
MOVE INTO
CONTACT

FRICTION
BETWEEN
SURFACES
RAISES
TEMPERATURE

FORGING
PRESSURE
APPLIED AFTER
ROTATION IS
STOPPED. WELD
COMPLETED

Fig. 9.10. (*b*) Suitable forms for friction welding.

component moves. (3) Burn-off control: a pre-determined shortening of the component is measured off by the control system when minimum pressures are reached. Weld quality and amount of extrusion are thus controlled, but not the length. (4) Forging to length: the axially moving work holder moves up to a stop during the forging operation irrespective of the state of the weld and generally in this case extrusion tends to be excessive.

The extrusion or flash can be removed by a subsequent operation or, for example for tubes of equal diameter, a shearing unit can be built into the machine operating immediately after forging and while the component is hot, thus requiring much less power. Fig. 9.10*b* illustrates joints designed to contain the flash.

In the process known as inertia welding the rotating component is held in a fixture attached to a flywheel which is accelerated to a given speed and then uncoupled from its drive. The parts are brought together under high thrust and the advantage claimed is that there is no possibility of the driving unit stalling before the flywheel energy is dissipated.

Friction welding machines resemble machine tools in appearance, as illustrated in Fig. 9.11.

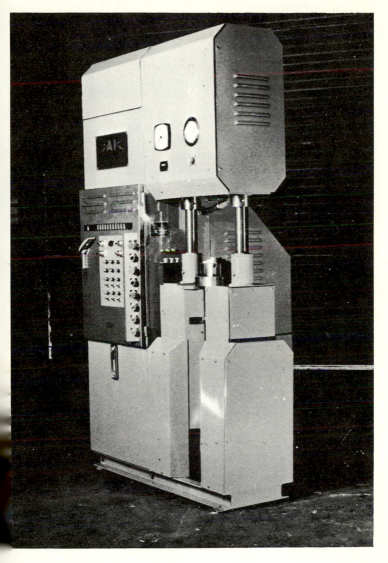

Fig. 9.11. Friction welder.

ELECTRON BEAM WELDING

If a filament of tungsten or tantalum is heated to high temperature in a vacuum either directly by means of an electric current or indirectly by means of an adjacent heater, a great number of electrons are given off from the filament which slowly evaporates. This emission has been mentioned previously in the study of the tungsten arc welding process. The greater the filament current the higher the temperature and the greater the electron emission, and if a metal disc with a central hole is placed near the filament and charged to a high positive potential relative to the filament, so that the filament is the cathode and the disc the anode, the emitted electrons are attracted to the disc and because of their kinetic energy pass through the hole as a divergent beam. This can then be focused electrostatically, or magnetically, by means of coils situated adjacent to the beam and through which a current is passed. The beam is now convergent and can be spot focused. This basic arrangement, an electron 'gun', is similar to that used for television tubes and electron microscopes (Fig. 9.12).

If the beam is focused on to a metal surface the beam can have sufficient energy to raise the temperature to melting point, the heating effect depending upon the kinetic energy of the electrons. The kinetic energy of an electron is $\frac{1}{2}mV^2$ where m is the mass of an electron (9.1×10^{-28} g) and V its velocity. The electron mass is small, but increasing the emission from the filament by raising the filament current increases the number of electrons and hence the mass effect. Because the kinetic

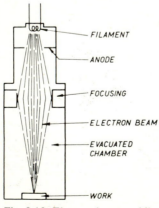

FILAMENT

ANODE

FOCUSING

ELECTRON BEAM

EVACUATED CHAMBER

WORK

Fig. 9.12. Electron beam welding.

energy varies directly as the square of the velocity, accelerating the electrons up to velocities comparable with the velocity of light by using anode voltages (up to 200 kV) greatly increases the beam energy. The smaller the spot into which the beam is focused the greater the energy density but final spot size is often decided by working conditions, by aberration in the focusing system, etc., so that spot size may be of the order of 0.25–2.5 mm.

When the beam strikes a metal surface X-rays are generated, so that adequate precautions must be taken for screening personnel from the rays by using lead or other metal screens or making the metal walls of the gun chamber sufficiently thick. If the beam emerges into the atmosphere the energy is reduced by collision of the electrons with atmospheric molecules and focus is impaired. Because of this it has been the practice to perform many welding operations in a vacuum, either in the gun chamber in which case each time the component is loaded the chamber must be evacuated to high vacuum conditions, thus increasing the time and cost of the operation, or in a separate steel component chamber fixed to the gun chamber. This can be made of a size to suit the component being welded and is evacuated to a relatively low vacuum after each loading. In either case welds suffer no contamination because of vacuum conditions. Viewing of the spot for set-up, focusing and welding is done by various optical arrangements.

Welding in non-vacuum conditions requires much greater power than for the preceding method because of the effects of the atmosphere on the beam and the greater distance from gun to work, and a shielding gas may be required around the weld area. Research work is proceeding in this field involving guns of higher power consumption. Difficulties may also be encountered in focusing the beam if there is a variation in the gun-to-work distance as on a weld on a component of irregular shape.

Welds made with this process on thicker sections are narrow with deep penetration with minimum thermal disturbance and at present welds are performed in titanium, niobium, tungsten, tantalum, beryllium, nickel alloys (e.g. nimonic), inconel, aluminium alloys and magnesium mostly in the aero and space research industries. The advantages of the process are that being performed in a vacuum there is no atmospheric contamination and the electrons do not affect the weld properties, accurate control over welding conditions is possible by control of electron emission and beam focus, and there is low thermal disturbance in areas adjacent to the weld (Fig. 9.13). Because of the vacuum conditions it is possible to weld the more reactive metals successfully. On the other hand the equipment is very costly, production of vacuum conditions is necessary in many cases and there must be protection against radiation hazards.

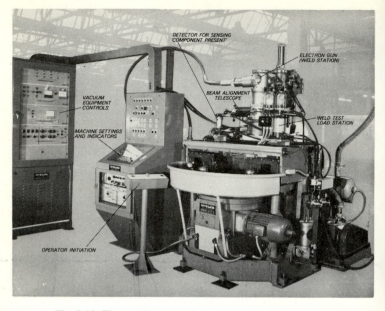

Fig. 9.13. Electron beam welding machine with indexing table, tooled for welding distributor shafts to plates at a production rate of 450 per hour. The gun is fitted with optical viewing system. Power 7.5 kW at 60 kV, 125 mA. Vacuum sealing is achieved by seals fitted in the tooling support plate and at the bottom of the work chamber. As the six individual tooling stations reach the welding station they are elevated to the weld position and then rotated by an electronically controlled d.c. motor.

LASER BEAM WELDING

Radio waves, visible light, ultra-violet and infra-red radiations are electro-magnetic radiations which have two component fields, one electric and the other magnetic. If either one of these components is suppressed the resulting radiations are said to be polarized and the direction in which the resultant electric or magnetic forces act is the plane of polarization. Light from a source such as a tungsten filament electric light consists of several frequencies involving various shades of colour. These waves are not in phase and are of various amplitudes and planes of polarization, and the light is said to be non-coherent. Light of a single wave-length or frequency is termed monochromatic. The wave-length of light is measured in metres or micro-metres, termed microns (μm) visible light being in the range 0.4–0.7 μm. The frequency is related to the wave-length by the expression $V = n\lambda$ where n is the frequency in Hz, λ the wave-length in metres and V the velocity of electro-magnetic radiation, 3×10^8 m/s, so that the frequency range of visible light is in the range 430–750×10^{12} Hz.

Atoms of matter can absorb and give out energy and the energy of any atomic system is thereby raised or lowered about a mean or 'ground' level. Energy can only be absorbed by atoms in definite small amounts (quanta) termed photons, and the relationship between the energy level and the frequency of the photon is $E = hv$, where E is the energy level, h is Planck's constant and v the photon frequency so that the energy level depends upon the frequency of the photon. An atom can return to a lower energy level by emitting a photon and this takes place in an exceedingly short space of time from when the photon was absorbed, so that if a photon of the correct frequency strikes an atom at a higher energy level, the photon which is released is the same in phase and direction as the incident photon.

The principle of the laser (Light Amplification by Stimulated Emission of Radiation) is the use of this stimulated energy to produce a beam of coherent light, that is one which is monochromatic and the radiation has the same plane of polarization and is in phase. At the present time lasers operate with wave-lengths in the visible and infra-red region of the spectrum. When the beam is focused into a small spot and there is sufficient energy, welding, cutting and piercing operations can be performed on metals.

The ruby laser has a cylindrical rod of ruby crystal (Al_2O_3) in which there is a trace of chromium as an impurity. An electronic flash gun, usually containing neon, is used to provide the radiation for stimulation of the atoms. This type of gun can emit intense flashes of light of one or two milliseconds' duration and the gun is placed so that the radiations impinge on the crystal. The chromium atoms are stimulated to higher energy levels, returning to lower levels with the emission of photons. The stimulation continues until an 'inversion' point is reached when there are more chromium atoms at the higher levels than at the lower levels, and photons impinging on atoms at the higher energy level cause them to emit photons. The effect builds up until large numbers of photons are travelling along the axis of the crystal, being reflected by the ends of the crystal back along the axis, until they reach an intensity when a coherent pulse of light, the laser beam (of wave-length about 0.63μm), emerges from the semi-transparent rod end. The emergent pulses may have high energy for a short time period, in which case vaporization may occur when the beam falls on a metal surface, or the beam may have lower energy for a longer time period in which case melting may occur, while a beam of intermediate power and duration may produce intermediate conditions of melting and vaporization, so that control of the time and energy of the beam and focusing of the spot exercise control over the working conditions.

Developments of the ruby laser include the use of calcium tungstate and glass as the 'host' material with chromium, neodymium, etc., as impurities, a particular example being yttrium–aluminium–garnet with neodymium (YAG), used for operations on small components.

Gas lasers operate on the same basic principles as the solid state type. The gas is contained in a long tube of quartz or pyrex with end windows, and specially designed mirrors are arranged to reflect the beam back along the tube axis. Neon with a trace of helium was first used and an electro-magnetic radiation at chosen frequency is applied from a radio-frequency generator to electrodes around the tube. The helium atoms are stimulated and their energy level raised. Collision with neon atoms causes energy to be transferred to the neon atoms until inversion occurs, the radiation stimulating release of photons from atoms faster than by normal emission and a coherent beam is emitted. The CO_2 laser uses CO_2 with some nitrogen and/or helium added, in a tube some metres long, the wavelength of the beam ($10.6\,\mu$m) being longer than that of the solid state lasers, and either continuous wave or pulsed, the power increasing as the length of the tube increases.

A pulsed or continuous wave laser beam can produce enough energy to heat, melt and vaporize a metal surface and refractory metals such as niobium, tatalum and tungsten can be welded in thin sections, and welds made between dissimilar metals, the operations being performed within an inert gas shield when required. Because of the vaporization which can occur, holes can be pierced even in diamonds.

2 kW CO_2 lasers can be used to weld up to 3 mm thick material and are an alternative to the electron beam for thin gauge material. The width of the weld may be increased at speeds below 12 mm/s due to interaction between the beam and an ionized plasma which occurs near the work. At speeds of 20 mm/s and over, laser and electron beam welds are practically indistinguishable from one another.

Lasers of 10–20 KW are now being developed and, applied to welding and cutting (with oxygen) of thicker sections, will greatly widen the scope of the process.

STUD WELDING

This is a rapid, reliable and economical method of fixing studs and fasteners of a variety of shapes and diameters to parent plate. The studs

may be of circular or rectangular cross-section, plain or threated internally or externally and vary from heavy support pins to clips used in
component assembly.

There are two main methods of stud welding: (1) arc (drawn arc), (2)
capacitor discharge, and the process selected for a given operation
depends upon the size, shape and material of the stud and the composition and thickness of the parent plate, the arc method generally
being used for heavier studs and plate, and capacitor discharge for
lighter gauge work.

Arc (drawn arc) process

This is used in both engineering and construction work and the equipment consists of a d.c. power source, controller and a hand-operated or
bench-mounted tool or gun.

A typical unit consists of a forced-draught-cooled power source with a
380/440 V, 3-phase, 50 Hz transformer, the secondary of which is
connected to a full-wave, bridge-connected silicon rectifier (p. 255),
with tappings giving ranges of 120–180 A, 170–320 A, and 265–1500 A
approximately (equipment is also available up to 1800 A), at about 65
V. The two lower ranges can also be used for MMA welding and two or
more units can be connected in parallel if required, the loading being
shared automatically.

The power controller contains a main current contactor and timer
relays and a pilot arc device, energized from the d.c. source. The solid-
state timer has a multi-position switch or switches for the selection of
operating time and current for the diameter of stud in use. Work is
usually connected to the positive pole with the stud negative except for
aluminium, when the polarity is reversed.

The hand gun (Fig. 9.14) has the operating solenoid and return spring
within the gun body, which also carries the operating switch and adjustable legs to accommodate varying lengths of studs, an interchangeable
chuck for varying stud diameters and a foot adaptor to maintain con-
centricity between the stud and ferrule or arc shield, this latter being
held by ferrule grips. Studs are fluxed on the contact end which is slightly
pointed (Fig. 9.15) and are supplied with ferrules.

To operate the equipment, the welding current and time for the
diameter of stud in use are selected, the stud is loaded into the appropriate chuck, the legs adjusted for length and the stud positioned on the
plate. A centre punch can be used for locating the stud point and the
plate should be free of contamination.

Sequence of operations. The gun switch is pressed, a low current flows between pointed stud end and workpiece and immediately the

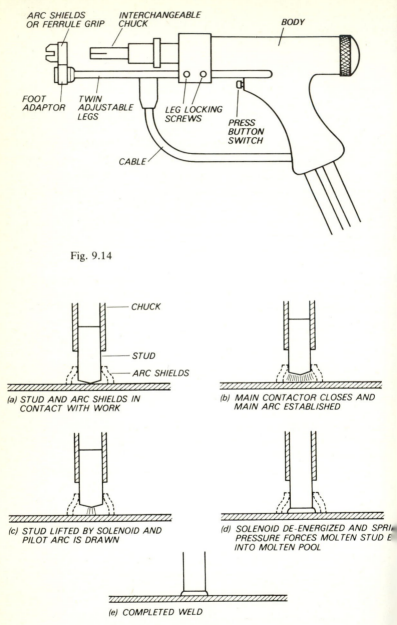

Fig. 9.14

(a) STUD AND ARC SHIELDS IN CONTACT WITH WORK

(b) MAIN CONTACTOR CLOSES AND MAIN ARC ESTABLISHED

(c) STUD LIFTED BY SOLENOID AND PILOT ARC IS DRAWN

(d) SOLENOID DE-ENERGIZED AND SPRI PRESSURE FORCES MOLTEN STUD E INTO MOLTEN POOL

(e) COMPLETED WELD

Fig. 9.15. Arc stud welding, cycle of operations.

stud is raised, drawing an arc and ionizing the gap. The main current contactor now closes and full welding current flows in an arc, creating a molten state in plate and stud end. The solenoid is de-energized and the stud is pushed under controlled spring pressure into the molten pool in the plate. Finally the main current contactor opens, the current is switched off, and the operation cycle is complete, having taken only a few hundredths of a second (Fig. 9.15).

This method, which is usually employed, whereby the welding current is kept flowing until the return cycle is completed, is termed 'hot plunge'. If the current is cut off just before the stud enters the molten pool it is termed 'cold plunge'. The metal displaced during the movement of the stud into the plate is moulded into fillet form by the ceramic ferrule or arc shield held against the workpiece by the foot. This also protects the operator, retains heat and helps to reduce oxidation.

Studs from 3.3 mm to 20 mm and above in diameter can be used on parent plate thicker than 1.6 mm, and the types include split, U shaped, J bent anchors, etc. in circular and rectangular cross-section for the engineering and construction industries, and can be in mild steel (low carbon), austenitic stainless steel, aluminium and aluminium alloys (3–4% Mg). The rate of welding varies with the type of work, jigging, location, etc., but can be of the order of 8 per minute for the larger diameters and 20 per minute for smaller diameters.

CAPACITOR DISCHARGE STUD WELDING

In this process a small projection on the end of the stud makes contact with the workpiece and the energy from a bank of charged capacitors (pp. 231–4) is discharged across the contact, melting the stud projection, ionizing the zone and producing a molten end on the stud and a shallow molten pool in the parent plate. At this time the stud is pushed into the workpiece under controlled spring pressure, completing the weld (Fig. 9.16).

If C is the capacitance of the capacitor, Q the charge and V the potential difference across the capacitor, then $C = Q/V$ or capacitance = *charge/potential*. The definition of the unit of capacitance, the farad, is given on p. 233. The energy of a charged capacitor is $\frac{1}{2} QV$, and since $Q = CV$, the energy is $\frac{1}{2} CV^2$, so that the energy available when a capacitor is discharged is dependent upon: (1) the capacitance – the greater the capacitance the greater the energy; (2) the square of the potential difference (voltage) across the capacitance. The greater this voltage, the greater the energy. Thus the energy required for a given welding operation is obtained by selection of the voltage across the

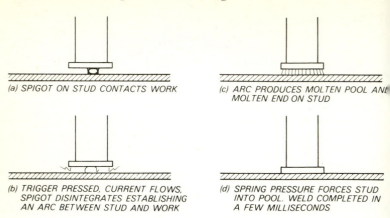

(a) *SPIGOT ON STUD CONTACTS WORK*

(c) *ARC PRODUCES MOLTEN POOL AND MOLTEN END ON STUD*

(b) *TRIGGER PRESSED, CURRENT FLOWS, SPIGOT DISINTEGRATES ESTABLISHING AN ARC BETWEEN STUD AND WORK*

(d) *SPRING PRESSURE FORCES STUD INTO POOL. WELD COMPLETED IN A FEW MILLISECONDS*

Fig. 9.16. Capacitor discharge stud welding, sequence of operations.

capacitor, i.e. that to which it is charged, and the total capacitance in the circuit.

In the contact method of capacitor discharge welding the small projection on the stud end is placed in contact with the workpiece as explained above. In the hold-off method, useful for thin gauge plate to avoid reverse marking, a holding coil in the hand gun is energized when the welding power is selected in the 'hold-off' position on the capacitor switch. When the stud is pushed into the gun-chuck both stud and chuck move into the hold-off position giving a pre-set clearance between the projection on the stud end and the workpiece. When the gun switch is pressed the hold-off coil is de-energized and spring pressure pushes the stud into contact with the workpiece, the discharge takes place and the weld is made.

This process minimizes the depth of penetration into the parent metal surface and is used for welding smaller diameter ferrous and non-ferrous studs and fasteners to light gauge material down to 0.45 mm thickness in low carbon and austenitic stainless steel, and 0.7 mm in aluminium and its alloy and brass, the studs being from 2.8 to 6.5 mm diameter. Studs are usually supplied with a standard flange on the end to be welded (Fig. 9.16) but this can be reduced to stud diameter if required and centre punch marks should not be used for location. The studs can be welded on to the reverse side of finished or coated sheets with little or no marking on the finished side. The studs are not fluxed and no arc shield or ferrule is required.

Typical equipment with a weld time cycle of 3–7 milliseconds consisted of a control unit and a hand- or bench-mounted tool or gun.

The control unit houses the banks of capacitors of 100 000–200 000 μF capacitance depending upon the size of the unit, the capacitance required for a given operation being selected by a switch on the front panel. The control circuits comprise a charging stage embodying a mains transformer and a bridge-connected full-wave silicon rectifier and the solid-state circuits for charging the capacitors to the voltage pre-determined by the voltage sensing module. Interlocking prevents the energy being discharged by operating the gun switch until the capacitors have reached power pre-set by voltage and capacitance controls.

The printed circuit voltage sensing module controls the voltage to which the capacitors are charged and switches them out of circuit when they are charged to the selected voltage, and is controlled by a voltage dial on the panel. A panel switch is also provided to discharge the capacitors if required.

Solid-state switching controls the discharge of the capacitors between stud and work, so there are no moving contactors. The gun, similar in appearance to the arc stud welding gun, contains the adjustable spring pressure unit which enables variation to be made in the speed of return of the stud into the work-piece, and a chuck for holding the stud. Legs are provided for positioning or there can be a nosecap gas shroud for use when welding aluminium or its alloys using an argon gas shield. The weld cannot be performed until legs or shroud are firmly in contact with the work-piece.

Welding rates attainable are 12 per minute at 6.35 mm diameter to 28 per minute at 3.2 mm diameter in mild steel. Similar rates apply to stainless steel and brass but are lower for aluminium because of the necessity for argon purging. Partial scorch marks may indicate cold laps (insufficient energy) with the possibility of the stud seating too high on the work. The weld should show even scorch marks all round the stud, indicating a sound weld. Excessive spatter indicates the use of too much energy.

Automatic single- and multiple-head machines, pneumatically operated and with gravity feed for the fasteners, are currently available.

EXPLOSIVE WELDING

This process is very successfully applied to the welding of tubes to tube plates in heat exchangers, feedwater heaters, boiler tubes to clad tube plates, etc.; and also for welding plugs into leaking tubes to seal the leaks.

The welds made are sound and allow higher operating pressures and temperatures than with fusion welding, and the tubes may be of steel, stainless steel or copper; aluminium brass and bronze tubes in naval brass tube plates are also successfully welded.

The tube and tube plate may be parallel to each other (parallel geometry) with a small distance between them and the tube plate can be counterbored as in Fig. 9.17a. The explosive (e.g. trimonite) must have a low detonation velocity, below the velocity of sound in the material, and there is no limit to the joint area so that the method can be used for cladding surfaces. For tube plate welds several charges can be fired simultaneously, the explosive being in cartridge form.

In the oblique geometry method now considered (YIMpact patent) the two surfaces are inclined at an angle to each other, Fig. 9.17b, the tube plate being machined or swaged as shown. As distance between tube and plate continuously increases because of the obliquity, there is a limit to the surface area which can be welded because the distance between tube and plate becomes too great.

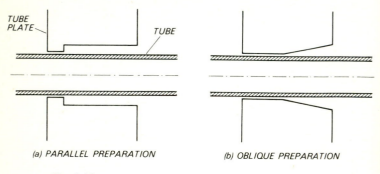

(a) PARALLEL PREPARATION   (b) OBLIQUE PREPARATION

Fig. 9.17.

The detonation value of the explosive can be above the velocity of sound in the material and the explosive (e.g. PETN) can be of pre-fabricated shape and is relatively cheap. The charge is fired electrically from a fuse head on the inner end of the charge and initiates the explosion, the detonation front then passing progressively through the charge.

The size of the charge depends upon the following variables; surface finish, angle of inclination of tube and plate, yield strength and melting point of the materials, the tube thickness and diameter; and the upper limit of the explosive is dependent upon the size of ligament of the tube plate between tubes, this usually being kept to a minimum in the interest of efficient heat exchange.

The tube plate is tapered towards the outer surface otherwise there would be a bulge in the tube on the inner side after welding and the tube would be difficult to remove. The charge must be fired from the inner end so that the weld will progress from contact point of tube and plate and thus the detonator wires must pass backwards along the charge to the outer end of the tube, the charge being within a polythene insert (Fig. 9.18*a*).

Upon initiation of the explosion, tube and tube plate collide at the inner end of the taper and, due to the release of energy, proceed along it, and ideally the jet of molten metal formed at collision point is ejected at the tube mouth. In effect the welded surfaces assume a sinusoidal

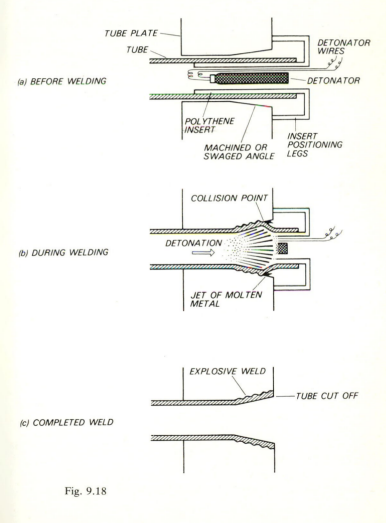

Fig. 9.18

wave form, some of the molten metal, which is rather porous and brittle, being entrapped in the troughs and crests of the wave. This entrapment can be reduced to a minimum by having a good surface finish (e.g. of the order of 0.003 mm) and the angle between tube and tube plate from 10° to 15°. At the lower angle the waves are pronounced and of shorter wavelength while at the higher angle the wavelength is longer and the waves more undulating so that 15° is the usual angle (Fig. 9.18*b* and *c*).

Surface oxide and impurities between the surfaces increase the charge required compared with a smooth surface and the positioning of the charge is important. If it is too far in, the energy at the mouth of the tube is not sufficient to produce a weld in this region, while if the charge is not far enough in the tube, welding is not commenced until some distance along the taper. To position the charge correctly and quickly a polythene insert has been developed to contain the charge and is positioned by a brass plug.

At present tubes of any thickness and of diameters 16–57 mm are weldable, with plate thickness greater than 32 mm and 9.5 mm plate ligaments.

The plug for explosive plugging of leaking tubes is of tubular form, the end of which is swaged to give the necessary taper, the polythene insert protruding beyond the open end of the plug to allow for extraction in case of misfire, thus increasing the safety factor (Fig. 9.19).

Since all the configuration is confined to the plug the tube plate hole is cleaned by grinding, the plug with explosive charge is inserted and the detonator wires connected. The plugs are available in diameters with 1.5 mm increments.

This method has proved very satisfactory in reducing the time and cost in repair work of this nature.

Personnel can be trained by the manufacturers or trained personnel are available under contract.

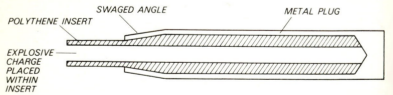

Fig. 9.19. A swaged explosive plug.

GRAVITY WELDING

This is a method of economically welding long fillets in the flat position using gravity to feed the electrode in and to traverse it along the plate. The equipment is usually used in pairs, welding two fillets at a time, one on each side of a plate, giving symmetrical welds and reducing stress and distortion. The electrode holder is mounted on a ball-bearing carriage and slides smoothly down a guide bar, the angle of which to the weld can be adjusted to give faster or slower traverse and thus vary the length of deposit of the electrode and the leg length of the weld.

The special copper alloy electrode socket is changed for varying electrode diameters and screws into the electrode holder, the electrode being pushed into a slightly larger hole in the socket and held there by the weight of the carriage. Turning the electrode holder varies the angle of electrode inclination. The base upon which the guide and support bar is mounted has two small ball bearings fitted so that it is easy to move along the base plate when resetting. If the horizontal plate is wider than about 280 mm a counterweight can be used with the base, otherwise the base can be attached to the vertical plate by means of two magnets or a jig can be used in place of the base to which the segment is fixed. A flexible cable connects the electrode holder to a disconnector switch carried on the support arm. This enables the current to be switched on and off so that electrodes can be changed without danger of shock. A simple mechanism at the bottom of the guide bar switches the arc off when the carriage reaches the bottom of its travel (Fig. 9.20).

At the present time electrodes up to 700 mm long are available in diameters of 3.5, 4.0, 4.5, 5.0 and 5.5 mm using currents of 220–315 A with rutile, rutile-basic and acid coatings suitable for various grades of steel.

Gravity welding is usually used for fillets with leg lengths of 5–8 mm,

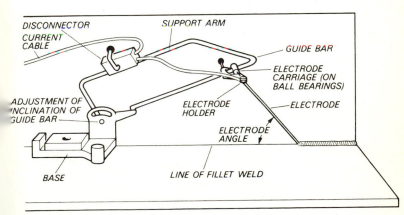

Fig. 9.20. Gravity welding.

the lengths being varied by altering the length of deposit per electrode. An a.c. power source is used for each unit with an OCV of 60 V and arc volts about 40 V with currents up to 300 A. Sources are available for supplying up to 6 units (three pairs) manageable by one welding operator and so arranged that when the current setting for one unit is chosen, the remaining units are supplied at this value. In general gravity welding is particularly suitable for welding, for example, long parallel stiffeners on large unit panels, enabling one operator to carry out three or four times the deposit length as when welding manually.

THERMIT WELDING

Thermit (or alumino-thermic) is the name given to a mixture of finely divided iron oxide and powdered aluminium. If this mixture is placed in a fireclay crucible and ignited by means of a special powder, the action, once started, continues throughout the mass of the mixture, giving out great heat. The aluminium is a strong reducing agent, and combines with the oxygen from the iron oxide, the iron oxide being reduced to iron (see p. 43).

The intense heat that results, because of the chemical action, not only melts the iron, but raises it to a temperature of about 3000 °C. The aluminium oxide floats to the top of the molten metal as a slag. The crucible is then tapped and the superheated metal runs around the parts to be welded, which are contained in a mould. The high temperature of the iron results in excellent fusion taking place with the parts to be welded. Additions may be added to the mixture in the form of good steel scrap, or a small percentage of manganese or other alloying elements, thereby producing a good quality thermit steel. The thermit mixture may consist of about 5 parts of aluminium to 8 parts of iron oxide, and the weight of thermit used will depend on the size of the parts to be welded. The ignition powder usually consists of powdered magnesium or a mixture of aluminium and barium peroxide.

Preparation
The ends which are to be welded are thoroughly cleaned of scale and rust and prepared so that there is a gap between them for the molten metal to penetrate well into the joint. Wax is then moulded into this gap, and also moulded into a collar round the fracture. This is important, as it gives the necessary reinforcement to the weld section. The moulding box is now placed around the joint and a mould of fireclay and sand made, a riser, pouring gate and preheating gate being included. The ends to be welded are now heated through the pre-heating gate by means of a flame and the

wax is first melted from between the ends of the joint. The heating is continued until the ends to be welded are at a red heat. This prevents the thermit steel being chilled, as it would be if it came into contact with cold metal. The pre-heating gate is now sealed off with sand and the thermit process started by igniting the powder. The thermit reaction takes up to about 1 minute, depending upon the size of the charge and the additions that have been made in the form of steel scrap, etc. When the action is completed the steel is poured from the crucible through the pouring gate, and it flows around the red hot ends to be welded, excellent fusion resulting. The riser allows extra metal to be drawn by the welded section when contraction occurs on cooling, that is, it acts as a reservoir. The weld should be left in the mould as long as possible (up to 12 hours), since this anneals the steel and improves the weld.

Thermit steel is pure and contains few inclusions. It has a tensile strength of about 460 N/mm^2. The process is especially useful in welding together parts of large section, such as locomotive frames, ships' stern posts and rudders, etc. It is also being used in place of flash butt welding for the welding together of rail sections into long lengths.

At the present time British Rail practice is to weld 18 m long rails into lengths of 91 to 366 m at various depots by flash butt welding. These lengths are then welded into continuous very long lengths *in situ* by means of the thermit process and conductor rails are welded in the same way. Normal running rails have a cross-sectional area of 7184 mm^2 and two techniques are employed, one requiring a 7 minute pre-heat before pouring the molten thermit steel into the moulds and the other requiring only $1\frac{1}{2}$ minutes pre-heat, the latter being the technique usually employed. Excess metal can be removed by pneumatic hammer and hot chisel or by portable trimming machine.

10

Cutting processes

GAS CUTTING OF IRON AND STEEL

Iron and steel can be cut by the oxy-hydrogen, oxy-propane, oxy-natural gas and oxy-acetylene cutting blowpipes with ease, speed and a cleanness of cut (see also 'Cutting with carbon arc', p. 343, and arc plasma cutting, Chapter 6).

Principle of cutting operation
There are two operations in gas cutting. A heating flame is directed on the metal to be cut and raises it to bright red heat or ignition point. Then a stream of high-pressure oxygen is directed on to the hot metal. The iron is immediately oxidized to magnetic oxide of iron (Fe_3O_4) and, since the melting point of this oxide is well below that of the iron, it is melted immediately and blown away by the oxygen stream.

It will be noted that the metal is cut entirely by the exothermic chemical action and the iron or steel itself is not melted. Because of the rapid rate at which the oxide is produced, melted and blown away, the conduction of the metal is not sufficiently high to conduct the heat away too rapidly and prevent the edge of the cut from being kept at ignition point.

The heat to keep the cut going once it has started is provided partly by the heating jet, and partly by the heat of the chemical action.

The cutting torch or blowpipe (Fig. 10.1)
Cutting blowpipes may be either high or low pressure. The high pressure pipe, using cylinder acetylene or propane as the fuel gas,[1] can have the mixer in the head (Fig. 10.2), or in the shank, while the low pressure pipe with injector mixing can be used with natural gas at low pressure.

In the high pressure pipe, fuel gas and heating oxygen are mixed in the

---

[1] Proprietary gases are available and contain mixtures of some of the following: methyl acetylene, propadiene, propylene, butane, butadiene, ethane, methane, diethyl ether, dimethyl ether, etc.

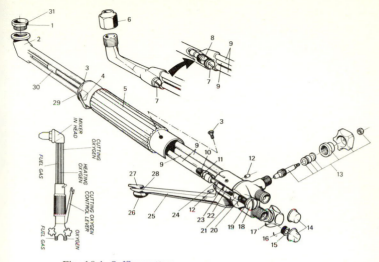

Fig. 10.1. Saffire cutter.

Key:

1. Nozzle nut.
2. Head 90°.
   Head 75°.
   Head 180°.
3. Screw.
4. Bracket latch.
5. Handle.
6. Nozzle nut.
7. Injector cap.
8. Injector assembly.
9. Tube.
9a. Tube.
10. Push rod.
11. Lock nut.
12. Pivot pin.
13. Control valve fuel gas.
14. Red cap.
15. Spring clip.
16. Filter.
17. Control valve oxygen.
18. Rear cap.
19. Cap washer.
20. Valve spring disc.
21. Rear valve.
22. Plunger.
23. O ring.
24. Lever pin.
25. Lever.
26. Button.
    Grub screw.
27. Lever latch.
28. Spring washer.
29. Forward tube.
30. Tube support.
31. Cutogen nut.

head (Fig. 10.2), and emerge from annular slots for propane or holes for acetylene (Fig. 10.4).

The cutting oxygen is controlled by a spring loaded lever, pressure on which releases the stream of cutting oxygen which emerges from the central orifice the diameter of which increases as thickness of plate to be cut increases. An oxygen gauge giving a higher outlet pressure (up to 6.5 bar) than for welding is required, but the gauge for cylinder acetylene can be the same with pressures up to 0.28 bar. Propane pressure is usually up to 0.6 bar and if natural gas is used no regulator is required but a non-return valve should be fitted in the supply line to prevent flash back.

The size of blowpipe used depends upon whether it is for light duty or

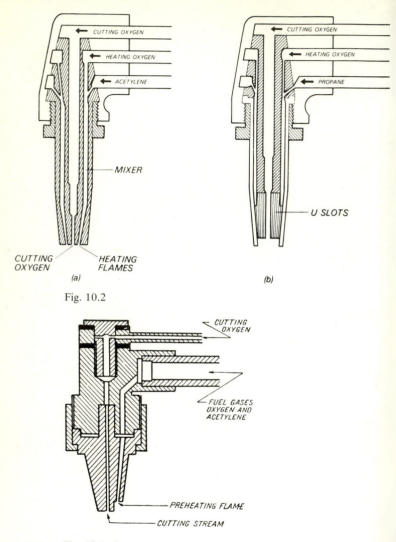

Fig. 10.2

Fig. 10.3. Cutting head for thin steel sheet.

heavy continuous cutting and the volume of oxygen used is much greater than of fuel gas (measured in litres per hour, 1/h).

Fig. 10.3 shows a stepped nozzle used for cutting steel sheet up to 4 mm thick and this type is also available with head mixing.

The size of torch varies with the thickness of work it is required to cut

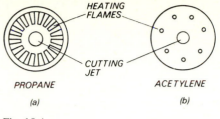

HEATING
FLAMES

CUTTING
JET

PROPANE   ACETYLENE

(a)   (b)

Fig. 10.4

and whether it is for light duty, or heavy, continuous cutting. Fig. 10.3 shows a specially designed head for cutting steel sheet up to 4 mm thick.

Adjustment of flame

Oxy-hydrogen and Oxy-propane. The correctly adjusted pre-heating flame is a small non-luminous central cone with a pale blue envelope.

Oxy-natural gas. This is adjusted until the luminous inner cone assumes a clear, definite shape, that may be up to 8–10 mm in length for heavy cuts.

Oxy-acetylene. This flame is adjusted until there is a circular short blue luminous cone, if the nozzle is of the concentric ring type, or until there is a series of short, blue, luminous cones (similar to the neutral welding flame), if it is of the multi-hole type (Fig. 10.5a and b). The effect of too much oxygen is indicated in Fig. 10.5c.

It may be observed that when the cutting valve is released the flame may show a white feather, denoting excess acetylene. This is due to the

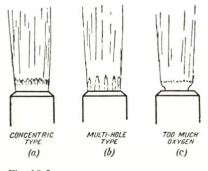

CONCENTRIC
TYPE
(a)

MULTI-HOLE
TYPE
(b)

TOO MUCH
OXYGEN
(c)

Fig. 10.5

slightly decreased pressure of the oxygen to the heating jet when the cutting oxygen is released. The flame should be adjusted in this case so that it is neutral when the cutting oxygen is released.

Care should be taken to see that the cutter nozzle is the correct size for the thickness to be cut and that the oxygen pressure is correct (the nozzle sizes and oxygen pressures vary according to the type of blowpipe used).

The nozzle should be cleaned regularly, since it becomes clogged with metallic particles during use. In the case of the concentric type of burner, the outer ring should be of even width all round, otherwise it will produce an irregular-shaped inner cone, detrimental to good cutting (Fig. 10.6).

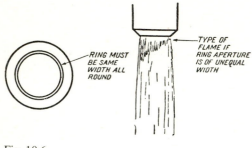

Fig. 10.6

Technique of cutting

The surface of the metal to be cut should be free of grease and oil, and the heating flame held above the edge of the metal to be cut, farthest from the operator, with the nozzle at right angles to the plate. The distance of the nozzle from the plate depends on the thickness of the metal to be cut, varying from 3 to 5 mm for metal up to 50 mm thick, up to 6 mm for metal 50 to 150 mm thick. Since the oxide must be removed quickly to ensure a good, clean cut, it is always preferable to begin on the edge of the metal.

The metal is brought to white heat and then the cutting valve is released, and the cut is then proceeded with at a steady rate. If the cutter is moved along too quickly, the edge loses its heat and the cut is halted. In this case, the cutter should be returned to the edge of the cut, the heating flame applied and the cut restarted in the usual manner. Round bars are best nicked with a chisel, as this makes the starting of the cut much easier. Rivet heads can be cut off flush by the use of a special type nozzle while if galvanized plates are to be cut for any length of time, a respirator is advisable, owing to the poisonous nature of the fumes.

To cut a girder, for example, the cut may be commenced at *A* and taken to B (Fig. 10.7). Then commenced at C and taken to B, that is, the flange is cut first. Then the bottom flange is cut in a similar manner. The cut is then commenced at B and taken to E along the web, this completing the operation. By cutting the flanges first the strength of the girder is altered but little until the web itself is finally cut.

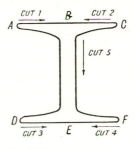

Fig. 10.7

Rollers and point guides can be affixed to the cutter in order to ensure a steady rate of travel and to enable the operator to execute straight lines or circles, etc., with greater ease (Fig. 10.8a).

The position of the flame and the shape of the cut are illustrated in Fig. 10.8b.

To close down, first shut off the cutting stream, then the propane or acetylene and then the oxygen valve. Close the cylinder valves and release the pressure in the tubes by momentarily opening the cutter valves.

To cut holes in plates a slightly higher oxygen pressure may be used.

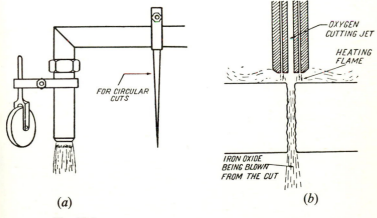

(a)   *(b)*

Fig. 10.8

The spot where the hole is required is heated as usual and the cutting valve released gently, at the same time withdrawing the cutter from the plate. The extra oxygen pressure assists in blowing away the oxide, and withdrawing the nozzle from the plate helps to prevent oxide from being blown on to the nozzle and clogging it. The cutting valve is then closed and the lower surface now exposed is heated again, and this is then blown away, these operations continuing until the hole is blown through. The edges of the hole are easily trimmed up afterwards with the cutter.

When propane or natural gas is used instead of acetylene, the flame temperature is lower; consequently it takes longer to raise the metal to ignition point and to start the cut, and it is not suitable for hand cutting over 100 mm thick nor for cast iron cutting. Because of this lower temperature, the speed of cutting is also slower. The advantages, however, are in its ease of adjustment and control; it is cheap to instal and operate and, most important of all, since the metal is not raised to such a high temperature as with the oxy-acetylene flame, the rate of cooling of the metal is slower and hence the edges of the cut are not so hard. This is especially so in the case of low-carbon and alloy steels. For this reason, oxy-propane or natural gas is often used in works where cutting machines are operated (see later).

The oxy-hydrogen flame can also be used (the hydrogen being supplied in cylinders, as the oxygen). It is similar in operation to oxy-propane and has about the same flame temperature. It is advantageous where cutting has to be done in confined spaces when the ventiliation is bad, since the products of combustion are not so harmful as in the case of oxy-acetylene and oxy-propane, but is not so convenient and quick to operate.

The effect of gas cutting on steel

It would be expected that the cut edge would present great hardness, owing to its being raised to a high temperature and then subjected to rapid cooling, due to the rapid rate at which heat is dissipated from the cut edges. Many factors, however, influence the hardness of the edge.

Steels of below 0.3% carbon can be easily cut, but the cut edges will definitely harden, although the hardness rarely extends more than 3 mm inwards and the increase is only 30 to 50 points Brinell.

Steels of 0.3% carbon and above and also alloy steels are best preheated before cutting, as this reduces the liability to crack.

Nickel, molybdenum, manganese and chrome steels (below 5%) can be cut in this way. Steels having a high tungsten, cobalt or chrome content, however, cannot be cut satisfactorily. Manganese steel, which is machined with difficulty owing to the work hardening, can be cut without any bad effects at all.

The oxy-acetylene flame produces greater hardening effect than the oxy-propane flame, as before mentioned, owing to its higher temperature. Excessive cutting speeds also cause increased hardness, since the heat is thereby confined to a narrower zone near the cut and cooling is thus more rapid. Similarly, a thick plate will harden more than a thin one, owing to its more rapid rate of cooling from the increased mass of metal being capable of absorbing the heat more quickly. The hardening effect for low carbon steels, however, can be removed either by preheating or heat treatment after the cut. The hardening effect in mild steel is very small. On thicknesses of plate over 12 mm it is advisable to grind off the top edge of the cut, as this tends to be very hard and becomes liable to crack on bending.

The structure of the edges of the cut and the nearby areas will naturally depend on the rate of cooling. Should the cutting speed be high and the cooling be very rapid in carbon steels, a hard martensitic zone may occur, while with a slower rate of cutting and reduced rate of cooling the structure will be softer. A band of pearlite is usually found, however, very near to the cut edge and because of this, the hardness zone, containing increased carbon, is naturally very narrow. When the cut edge is welded on directly, without preparation, all this concentration of carbon is removed.

Thus, we may say that, for steels of less than 0.3% carbon, if the edges of the cut are smooth and free from slag and loose oxide, the weld can be made directly on to the gas-cut edge without preparation.

OXYGEN OR THERMIC LANCE

This is a method of boring holes in concrete, brick, granite, etc., by means of the heat generated by chemical reactions.

The lance consists of a tube about 3 m long and 6.5–9.5 mm diameter which is packed with steel wires. One end of the tube is threaded and is connected by means of a flexible hose to an oxygen supply. To operate the lance the free end is heated and oxygen passed down the tube. Rapid oxidation of the wires begins at the heated end with great evolution of heat. Magnesium and aluminium are often added to the packing to increase the heat output. The operator can be protected by a shield and protective clothing should be worn. The exothermic reaction melts concrete and other hard materials to a fluid slag and cast iron is satisfactorily bored. Standard gas pipe can be used for thick steel sections.

As an example of boring speed and oxygen consumption a hole 30–40 mm diameter can be bored through 300 mm thick concrete with a 6.4

mm lance in 63 seconds using 310 litres of oxygen at 1.0 N/mm² pressure and consuming 1.2 m of lance.

Cutting machines

Profiles cut by hand methods are apt to be very irregular and, where accuracy of the cut edge is required, cutting machines are used. The heating flame is similar to that used in the hand cutter and is usually oxy-propane or oxy-acetylene (either dissolved or generated), while the thickness of cut depends on the nozzle and gas pressures.

The mechanical devices of the machines vary greatly, depending upon the type of cut for which they are required. In many types a tracing head on the upper table moves over the drawing or template of the shape to be cut. Underneath the table or on the opposite side of the machine (depending on the type) the cutting head describes the same motion, being worked through an intermediate mechanism. The steel being cut is placed on supports below the table (Fig. 10.9).

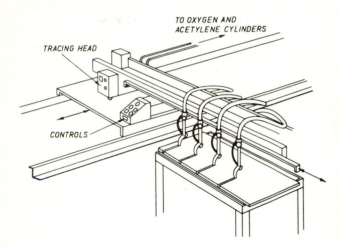

Fig. 10.9. Cutting machine fitted with four cutting heads.

Simpler machines for easier types of cuts, such as straight line and circles, bevels, etc., are also made.

A typical machine is capable of cutting from 1.5 m to 350 mm thick, 3 m in a straight line and up to 1.5 m diameter circle. The machine incorporates a magnetic tracing roller, which follows round a steel or iron template the exact shape of the cut required, while the cutting head cuts the replica of this shape below the table.

Stack cutting

Thin plates which are required in quantities can be cut by clamping them tightly in the form of a stack and, due to the accuracy of the modern machines, this gives excellent results and the edges are left smooth and even. Best results are obtained with a stack 75–100 mm thick, while G clamps can be used for the simpler types of stack cutting.

Cast iron cutting

Cast iron cutting is made difficult by the fact that the graphite and silicon present are not easily oxidized. Reasonably clean cuts can now be made, however, using a blowpipe capable of working at high pressure of oxygen and acetylene. Cast iron cannot be cut with hydrogen.

Since great heat is evolved in the cutting process, it is advisable to wear protective clothing, face mask and gloves.

The oxygen pressure varies from 7 N/mm² for 35 mm thick cast iron to 11 N/mm² for 350 to 400 mm thick, while the acetylene pressure is increased accordingly.

The flame is adjusted to have a large excess of acetylene, the length of the white plume being from 50–100 mm long (e.g. 75 mm long for 35–50 mm thick plate). The speed of cutting is slow, being about 2.5 m per hour for 75–125 mm thick metal.

Technique. The nozzle is held at 45° to the plate with inner cone 5–6 mm from the plate, and the edge where the cut is to be started is heated to red heat over an area about 12–18 mm diameter. The oxygen is then released and this area burnt out. The blowpipe is given a zig-zag movement, and the cut must not be too narrow or the slag and metal removed will clog the cut. About 12 to 18 mm is the normal width. After the cut is commenced the blowpipe may be raised to an angle of 70–80°, which will produce a lag in the cut, as shown in Fig. 10.10.

Owing to the fact that high pressures are used in order to supply

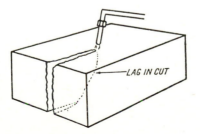

Fig. 10.10

sufficient heat for oxidation, large volumes of gas are required, and this is often obtained by connecting several bottles together.

FLAME GOUGING BY THE OXY-ACETYLENE PROCESS

Flame gouging is an important extension of the principle of oxy-acetylene cutting by which grooves with very smooth contours can be cut easily in steel plates without the plate being penetrated.

Principle of operation
This is the same as that in the oxy-acetylene process, except that a special type of nozzle is used in the standard cutting blowpipe. A pre-heating flame heats the metal to red heat (ignition temperature), the cutting oxygen is switched on, oxidation occurs and the cut continues as previously explained.

Equipment
The cutting torch may have a straight 75° or 180° angle head together with a range of special gouging nozzles, as shown in Fig. 10.11. The nozzle sizes are designated by numbers and they are bent at an angle which is best for the gouging process. Regulators and other equipment are as for cutting.

Fig. 10.11. Flame gouging nozzle.

Operation
There are two main techniques:

    (1) progressive,
    (2) spot.

In the former the groove is cut continuously along the plate – it may be started at the plate edge or anywhere in the plate area. It can be used for removing the underbeads of welds prior to depositing a sealing run, or it may be used for preparing the edges of plates. Spot gouging, however, is used for removing small localized areas such as defective spots in welds.

To start the groove at the edge of a plate for continuous or progressive gouging the nozzle is held at an angle of about 20° so that the pre-heating flames are just on the plate and when this area gets red hot the cutting oxygen stream is released and at the same time the nozzle is brought at a shallower angle to the plate as shown, depending upon depth of gouge required. The nozzle is held so that the nozzle end clears the bottom of the cut and the pre-heating flames are about 1.5 mm above the plate.

The same method is adopted for a groove that does not start at the plate edge. The starting point is pre-heated with the nozzle making a fairly steep angle with the plate at 20–40°. When the pre-heated spot is red hot the cutting oxygen stream is released and the angle of the nozzle reduced to 5–10° depending upon the depth of gouge required (Fig. 10.12).

To gouge a single spot, it is pre-heated as usual where required, but when red hot and the cutting stream of oxygen is turned on, the angle of the nozzle is *increased* (instead of, as previously, decreased) so as to make the gouge deep.

The depth of groove cut depends upon nozzle size, speed of progress and angle between nozzle and plate (i.e. angle at which the cutting stream of oxygen hits the plate). The sharper this angle, the deeper the

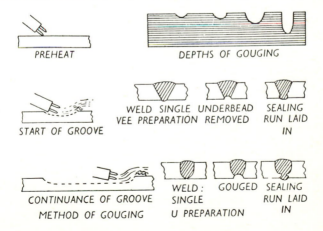

Fig. 10.12. Method and uses of flame gouging.

groove. If the cutting oxygen pressure is too low ripples are left on the base of the groove. If the pressure is too high the cut at the surface proceeds too far in advance of the molten pool, and eventually the cut is lost and must be restarted.

Uses of flame gouging

Certain specifications, such as those for fabrication of butt welded tanks, etc., stipulate that the underbead (or back bead) should be removed and a sealing run laid in place. This can easily and efficiently be done by flame gouging, as also can the removal of weld defects, tack welds, lugs, cleats, and also the removal of local areas of cracking in armour plate, and flashing left after upset welding.

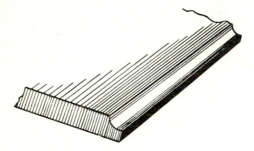

Fig. 10.13. Flame-gouged edge of plate, prepared for welding.

POWDER INJECTION CUTTING

Prior to the advent of powder cutting it was necessary to cut stainless steel always by mechanical means since it could not be cut by the gas cutter. Powder cutting enables stainless steel to be cut, bevelled and profiled with much the same ease and facility as with the oxygen cutting of low-carbon steels.

In cutting low-carbon steels a pre-heating flame raises the temperature of a small area to ignition point. This is the temperature at which oxidation of the iron occurs, and iron oxide is formed when a jet of oxygen is blown on to the area. The heat of chemical combination, together with that of the pre-heating flame enables the oxidation, and hence the cut to continue, the oxde being removed by being blown away by the oxygen jet, resulting in a narrow cut.

For this sequence to occur, the melting point of the oxide formed must be lower than that of the metal being cut. This is the case in low-carbon steel. In the case of stainless steels and non-ferrous metals the oxides formed have a melting point higher than that of the parent metal.

When attempting to cut stainless steel with the ordinary oxy-acetylene equipment the chromium combines with oxygen at high temperature and forms a thin coating of oxide which has a melting point higher than that of the parent steel, and since it is difficult to remove, further oxidation does not occur and the cut cannot continue.

In the powder-cutting process a finely divided iron powder is sprayed by compressed air or nitrogen into the cutting oxygen stream on the line of the cut. The combustion of this iron powder so greatly increases the ambient temperature that the refractory oxides are melted, fluxed, and to a certain extent eroded by the action of the particles of the powder, so that a clean surface is exposed on to which the cutting oxygen impinges and thus the cut continues. The quality of the cut is very little inferior to that of a cut in a low-carbon steel.

Equipment

For hand or machine cutting the powder is delivered to the reaction zone of the cut by means of an attachment fitted to the cutting blowpipe (Fig. 10.15). The attachment consists of powder valve, powder nozzle and tubing. The nozzle is fitted over the standard cutting nozzle and the powder valve is clamped near the gas valves. The iron powder is carried down the outside of the cutting nozzle and after passing through inclined ports, is injected through the heating flame into the cutting stream of oxygen which it meets at approximately 25 mm below the end of the cutting nozzle.

The nozzle is normally one size larger than for cutting the same thickness of low-carbon steel and is held as for normal cutting except that a clearance of 25 to 35 mm is given between nozzle and plate to be cut to allow the powder to burn in the oxygen stream. The great heat produced makes pre-heating unnecessary on stainless steel and what may be termed a 'flying start' can be made.

The powder dispenser unit (Fig. 10.14) is of the injector type and is a pressure vessel which incorporates:

(1) A hopper for filling.
(2) An air filter.
(3) An air pressure regulator.
(4) A dryer.
(5) An injector unit.

The removable cover enables the hopper to be filled and is fitted with a pressure relief valve which lifts at 0.15 N/mm^2. A screen for removing

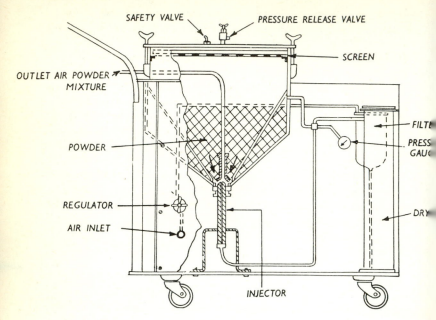

Fig. 10.14. Powder dispenser unit.

over-large particles from the powder and a tray for the drying agent are fitted.

Compressed air flows into the dispenser, picks up powder and carries it along a rubber hose to the cutting nozzle. Nitrogen may be used instead of air but never use oxygen, as it is dangerous to do so. The powder flow from the dispenser is regulated by adjusting the nozzle or by varying the air pressure.

The dispenser should be fed from an air supply of 1.5 m³ per hour at a pressure of 0.3 N/mm². The usual pressure for operating is from 0.02 to 0.03 N/mm².

Since the whole process depends upon a uniform and smooth flow of powder every care must be taken to ensure this. Any moisture in either the powder or in the compressed air can cause erratic operation and affect the quality of the cut.

Silica gel (a drying agent) is incorporated in the dispenser to dry the powder but the amount is not sufficient to dry out the compressed air and a separate drying and filtering unit should be installed in the air line to ensure dry, clean air. The oxygen and acetylene is supplied through not less than 9 mm bore hose with 6 mm bore for the powder supply.

The single-tube attachment discharges a single stream of powder into the cutting oxygen and is used for straight line machine cutting of stainless steel. The multi-jet type (Fig. 10.15) has a nozzle adaptor which fits over a standard cutting nozzle and has a ring of ports encircling it. The powder is fed through these ports and passes through the heating flame into the cutting oxygen. This type is recommended for hand cutting and for profile, straight and bevel machine cutting. Special cutting nozzles are available for this process. These give a high velocity parallel cutting oxygen stream their bore being convergent–divergent. The pre-heater holes are smaller in number but more numerous and are set closer to the cutting oxygen orifice than in the standard nozzle. This gives a soft narrow pre-heating flame, giving a narrower cut and better finish with a faster cutting speed. Since the iron powder is very abrasive, wear occurs in ports and passages through which the powder passes. By using stainless tube where possible, avoiding sharp internal bends and reinforcing certain parts, wear is reduced to a minimum.

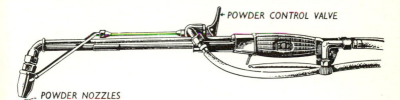

Fig. 10.15. Cutter equipped for powder cutting.

Technique
In cutting any metal by this process, correct dispenser setting, dry air and powder, clean powder passages and leak-tight joints all help towards ensuring a good quality cut. The rate of powder flow is first adjusted to the correct amount for the particular work in hand by trial cuts.

Stainless steel. Size of nozzle is one size larger than for the same thickness low-carbon steel. For thicknesses up to 75 or 100 mm the nozzle is held about 25 mm from the plate, for thicknesses up to 150 mm the distance is increased to about 35 mm, while for heavier sections it can be about 50 mm distant. No pre-heating is necessary – an immediate or flying start can be made. Scrap plate may be loosely put together and a single cut can be made, the great heat enabling this to be done in spite of gap between the plates.

Cast iron and high-alloy steels. Technique for cast iron is similar to that for stainless steel but cutting speed is up to 50% slower.

With high-alloy steels for example, a 25/20 nickel–chrome steel takes 30–40% longer than for the 18/8 nickel–chrome steel. Pre-heating of high-alloy tool steels is advisable to prevent risk of cracking due to localized heat.

Round carbon steels bars are easily cut by laying them side by side. The cut is done from bar to bar, powder being switched on to start the cut on each bar giving no interruption in cut from bar to bar.

Nickel and nickel alloys. Pure nickel is extremely difficult to cut because it does not readily oxidize and it can only be cut at slow speed in thicknesses up to 25 mm section. Alloys like Inconel, Nimonic and Nichrome can be cut up to 125 mm thick, speeds being up to 60% slower than for the same thickness of stainless steel.

Copper and copper alloys. In this operation copper is melted (and not cut) the particles of iron powder removing the molten metal and eroding the cut.

Much pre-heating is required due to the high conductivity of copper and copper alloys – for example – a powder cutting blowpipe capable of cutting 180 mm thick stainless steel can only deal with copper up to 25 mm thick and brass and bronze up to 100 mm thick. A lower cutting oxygen pressure than normal is used and the nozzle moved in a forwards, upwards, backwards, downwards motion in the line of the cut, thus helping to remove the molten metal and avoiding cold spots.

Aluminium and aluminium alloys. The quality of the cut is poor and of ragged profile. Alloys containing magnesium such as MG7 develop a hard surface to the cut due to formation of oxide and this may extend to a depth of 6 mm. As a result powder cutting is limited to scrap recovery. Cutting (or melting) speeds on thinner sections are much the same as for stainless steel, but with thicker sections molten metal chokes the cut, so that larger nozzles with reduced cutting oxygen pressures should be used.

Powder cutting can be incorporated on cutting machines and gouging can be carried out as with the normal blowpipe with the same ease and speed, the nozzle being again held further away from the work, to allow space for the combustion of the powder.

OXY-ARC CUTTING PROCESS

In this process the electric arc takes the place of the heating flame of the oxy-acetylene cutter. The covered electrodes of mild steel are in 4 sizes and are of tubular construction, the one selected depending upon whether it is required for cutting, piercing, gouging, etc. They are about 5 to 6 mm outside diameter with a fine hole about 1.5 mm diameter or more through which the cutting oxygen stream passes, down the centre. The gun type holder secures the electrode by means of a split collet and has a trigger controlling the oxygen supply. A d.c. or a.c. supply of 100–300 A is suitable.

The arc is struck with the oxygen off, the oxygen valve is released immediately and cutting begins, the electrode being held at an angle of 60° to the line of cut, except at the finish when is is raised to 90°, and is consumed in the process. The oxygen pressure varies with the thickness of steel and with the size of electrode being used, about 4 N/mm² for mild and low alloy steel plate of 8 to 10 mm thick and about 5–6 N/mm² for 25 to 25 mm thick. In addition to mild and low-alloy steel, and stainless steel, copper, bronze, brass, monel and cupro-nickels can be roughly cut by this process.

ARC-AIR CUTTING AND GOUGING PROCESS

In this process a carbon arc is used with a d.c. supply from a welding generator, or rectifer, together with a compressed air supply at 5–8 N/mm²

The equipment comprises a holder for the carbon electrode to which is supplied the direct current with *electrode +ve*, and the compressed air which is controlled by a lever-operated valve on the electrode holder.

The jaws of the electrode holder are rotatable enabling the carbon electrode to be held in any position and twin jets of compressed air emerge from the head on each side of the carbon and parallel to it, the two jet streams converging at the point where the arc is burning. Air is also circulated internally in the holder to keep it cool.

Operation

The carbon electrode is placed in the holder with 75–150 mm pro-jecting, with the twin jet holes pointing towards the arc end of the carbon.

The carbon is held at approximately 45° to the job to be cut while for shallow gouging the angle may be reduced to 20°. This angle together with the speed of travel affects the depth of cut. The carbons have been specially designed for the process and are a mixture of carbon and graphite covered over all with a thin sheath of copper. The copper coating prevents tapering and ensures a cut of regular width in addition to enabling higher current to be used and consequently greater speeds of cutting to be obtained.

The carbons range in diameter from 4 mm (75–150 A) to 16 mm (550–700 A) and the higher the current density the more efficient the operation. Too high a current for a given size of electrode destroys the copper coating and burns the carbon at an excessive rate. Normally the copper coating burns away about 20mm from the arc.

The process is applicable to work in all positions and is used for removing defects in castings, in addition to cutting and gouging. A reasonably clean surface of cut is produced with no adverse effect so that welding can be carried out on the cut surface without further grinding.

11

Inspection and testing of welds

During the process of welding, faults of various types may creep in. Some, such as those dealing with the quality and hardness of the weld metal, are subjects for the chemist and research worker, while others may be due to lack of skill and knowledge of the welder. These, of course, can be overcome by correct training (both theoretical and practical) of the operator.

In order that factors such as fatigue may not affect the work of a skilled welder, it is evidently necessary to have means of inspection and testing of welds, so as to indicate the quality, strength and properties of the joint being made.

Visual inspection, both while the weld is in progress and afterwards, will give an excellent idea of the probable strength of the weld, after some experience has been obtained.

Inspection during welding

Metal arc welding. The chief items to be observed are: (1) rate of burning of rod and progress of weld, (2) amount of penetration and fusion, (3) the way the weld metal is flowing (no slag inclusions), (4) sound of the arc, indicating correct current and voltage for the particular work.

Oxy-acetylene welding. The chief items are: (1) correct flame for the work on hand, (2) correct angle of blowpipe and rod, depending on method used, (3) depth of fusion and amount of penetration, (4) rate of progress along the joint.

The above observations are a good indication to anyone with experience what quality of weld is being made, and this method furnishes one of the best ways of observing the progress of welders when undergoing training.

Inspection after welding

Examination of a weld on completion will indicate many of the following points:

(1) Has correct fusion been obtained between weld metal and parent metal?

(2) Is there an indentation, denoting undercutting along the line where the weld joins the parent metal (line of fusion)?

(3) Has penetration been obtained right through the joint, indicated by the weld metal appearing through the bottom of the V on a single V or U joint?

(4) Has the joint been built up on its upper side (reinforced), or has the weld a concave side on its face, denoting lack of metal and thus weakness?

(5) Does the metal look of close texture or full of pinholes and burnt, denoting incorrect flame?

(6) In arc welding has spatter occurred, indicating too high a current or too high a voltage across arc or too long an arc?

(7) Are the dimensions of the weld correct, tested, for example, by gauges such as shown in Fig. 11.1?

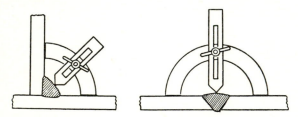

Fig. 11.1. Weld test gauges.

A study of the above will indicate to an experienced welder what faults, if any, exist in the work and then provide a rapid and useful method of ensuring that the right technique of welding is being followed.

A very useful multi-purpose pocket-size welding gauge has been designed by The Welding Institute. It is of stainless steel and enables the following measurements to be taken in either metric or Imperial units: material thickness up to 20 mm, preparation angle 0–60°, excess weld metal capping size, depth of undercut and of pitting, electrode diameter, fillet weld throat size and leg length and high-low misalignment.

Visual inspection, however, has several drawbacks. Take, for example, the double V joint shown in Fig. 11.2. It will obviously be impossible to observe by visual means whether penetration has occurred at the bottom of the V except at the two ends.

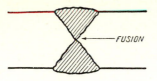

Fig. 11.2

A great variety of methods of testing welds are now available and, for convenience, we can divide them into two classes: (1) non-destructive, (2) destructive.

Destructive tests are usually carried out either on test specimens made specially for the purpose, or may even be made on one specimen taken as representative of several similar ones.

Destructive tests are of greatest value in determining the ultimate strength of a weld and afford a check on the quality of weld metal and skill of the operator. (Visual inspection obviously falls under the heading of non destructive tests.)

NON-DESTRUCTIVE TESTS

(1) Visual inspection (including use of a penetrant fluid).
(2) Magnetic: (*a*) iron filings in paraffin (magnetic fluid), (*b*) Search coil.
(3) Sound (acoustic) methods: (*a*) hammer, (*b*) hammer and stethoscope.
(4) X-ray.
(5) Gamma ray.
(6) Ultrasonic.
(7) Application of load.

Visual inspection (penetrant fluid)

Surface defects can be identified by painting or aerosol spraying on a penetrant fluid, wiping clean, and then a developer which shows up, visually or by ultra-violet light, defects found by the fluid.

Magnetic tests

Method (a). Iron filings in an extremely finely divided state (colloidal) are suspended (or mixed) in paraffin (often termed magnetic fluid).

The specimen under test is highly magnetized, usually by magnetizing coils, or by being placed in a strong magnetic field, and the fluid is then painted on the weld metal, which must have a machined or polished surface. If there is any crack in the metal, an alteration in the magnetic field (or flux) occurs at the crack, which is in reality a minute air gap. As a result, the finely divided particles of iron cling to the edges of the crack and show it up as a dark hair.

For rapid inspection, the fluid can be contained in a circular celluloid container, called a detector (Fig. 11.3), with a thin base. The specimen is magnetized and the container well shaken so as to mix up the iron. Upon placing the detector on the surface of the specimen, any crack will again be shown up as a dark hair. Its advantage is that it can be used continually with no deterioration.

The fluid method is extensively used at the present time in detecting cracks in every type of metal product from the steel bar to the finished part.

Its drawbacks are that:

(1) It can only be applied to iron and steel, as these are the only magnetic substances.
(2) It only shows up surface cracks.
(3) The specimen must be machined or polished and magnetized.

Fig. 11.3. Portable magnetic crack detector.

Method (b). The specimen is magnetized as before or by having a heavy current passed through it. Search coils connected to a galvanometer (an instrument which will measure small currents) are then moved over the specimen. If a crack exists in the specimen, the change of magnetic field or flux across it will cause a change in the current in the search coil, and this is indicated by fluctuations of the galvanometer needle. This method has the advantage over method (*a*) that its surface need not be machined and that defects below the surface are also indicated.

Hammer or acoustic method

If metal is struck by a hammer, the note given out will depend on the metal and its size. If, however, there is a crack in the metal, the note is quite different. This is a well-known fact and can easily be verified experimentally. After considerable experience, it is possible to become reasonably proficient in detecting in this way metal which is cracked, by the note it gives out on being struck.

By the use of a stethoscope, different regions of the weld can be explored just as a doctor uses the same instrument to discover weak areas in a patient's chest. In this way, the note given out by any particular part of the weld can be separated from the note given out by the surrounding plate, and this enables the crack to be more easily located. Great experience is necessary, however, to identify the different notes emitted and to understand fully exactly to what each refers.

X-ray method

X-rays are produced in the following way: a glass tube with two arms and a central bulb is exhausted of air. In one arm is a small electric light filament F which can be heated to white heat. In the other arm is a thick copper stem S ending in a target T made of platinum and inclined at an angle of 45 ° to the axis of the tube (Fig. 11.4). A high voltage of between 60 000 and 180 000 volts is placed across the ends of the tube, the end A being positive (called the *anode*). The fact that the filament is white hot, and emits negatively charged particles, enables the high voltage to send a current through the tube, and the current causes a stream of negatively charged particles, called *cathode rays*, to move from the filament F to the

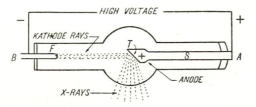

Fig. 11.4

positively charged target *T*, as shown. On hitting the target, they are reflected as shown and are then termed X-rays.

They are electro-magnetic radiations of short wave-length and they can penetrate solid substances, but, in doing so, a certain proportion of the rays is absorbed. The amount of absorption depends on the thickness of the substance and on its density. The denser and thicker the substance, the smaller the proportion of rays which get through. Certain substances, such as calcium tungstate and barium platino-cyanide, become fluorescent or luminous when X-rays strike them. If a screen is coated with one of these substances and the rays fall on it, the sensitized screen becomes brightly illuminated. If now an object, such as a piece of steel, is placed in the path of the rays in front of the screen, a shadow of the object will be thrown on the screen, and if the object is of the same thickness and density throughout, the shadow will also be of the same degree of darkness over all its surface. If, however, any holes or cavities exist in the steel, the rays will be less absorbed at these points and the shadow will be more brightly illuminated at these spots.

In the same way, films covered with silver halides are affected by X-rays in the same way that ordinary photographic films are affected by light. Thus if an X-ray film is put in place of the sensitive screen in the above, a 'shadowgraph' of the object will appear when the film is developed. Films are used more than the screen method, because they provide a permanent record of the shadow which can be carefully studied (Fig. 11.5).

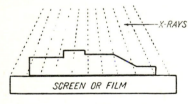

Fig. 11.5

Now, just as in photography an incorrectly exposed film will possess no detail, so with an X-ray photograph no detail of defects in the object can be observed unless correct exposure is given. This is entirely a matter of practice.

In order to make sure that we are getting a correctly exposed negative, so that even the smallest defects will be shown up, it is usual to place a penetrometer, which is a small strip of cutting of the same material as the object (steel for example) and about $\frac{1}{100}$ to $\frac{2}{100}$ of its thickness, on the upper surface next to the X-ray tube, as shown in Fig. 11.6*a*.

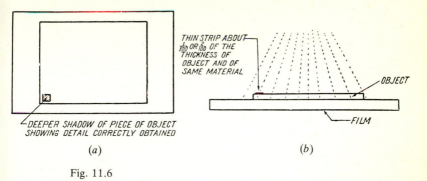

DEEPER SHADOW OF PIECE OF OBJECT
SHOWING DETAIL CORRECTLY OBTAINED

(a)

THIN STRIP ABOUT
$\frac{1}{100}$ OR $\frac{2}{100}$ OF THE
THICKNESS OF
OBJECT AND OF
SAME MATERIAL

OBJECT

FILM

(b)

Fig. 11.6

If this now appears as shown in Fig. 11.6b, as a shadow on the negative, we are sure that any defects or holes of the size $\frac{1}{100}$ to $\frac{2}{100}$ of the thickness of the object will be indicated.

Great practice is necessary to interpret the X-ray films of welds correctly and to distinguish between various defects shown up as shadows. Gas holes causing porosity are usually regular in shape, while any included slag is usually very irregular. In this way, we can determine whether penetration to the full depth required has been obtained, whether correct fusion between parent metal and weld metal or between layer and layer in a multi-layer weld has been obtained, and whether there are regions of entrapped slag, blowholes or other porous defects. In addition, any defects, such as contraction cracks, will also be shown up clearly.

The X-raying of butt welds is comparatively straightforward, but with fillet welds special methods have to be adopted, and often more than one photograph taken in different directions. Figure 11.7 indicates some of the defects mentioned above and how they appear in the X-ray photograph.

Finally, the method of X-ray inspection of pressure vessels, such as boilers, may be mentioned. During the welding operation on the boiler, one or more test pieces are made by the welder, using the same technique exactly as on the boiler, the test pieces being of the same metal and thickness as the boiler itself. The weld on the boiler is X-rayed along its whole length (Fig. 11.8) and the test pieces are also X-rayed. A comparison of the negatives will indicate how closely the structure of the test pieces compares with that of the boiler itself and if (as is usually the case) this is very close, it can reasonably be supposed that the boiler weld would behave as the test pieces under test. The test pieces are then tested by the usual methods and finally tested to destruction. This affords an excellent indication as to the probable behaviour of the boiler

(*a*) Tungsten inert gas (argon) weld on an aluminium plate illustrating tungsten pick-up due to the current being too high for the diameter of tungsten electrode being used.

(*b*) Metallic inert gas (argon) weld with 1.5 mm diameter wire on 12 mm aluminium plate showing puckering due to the current (380 amperes) being too high.

(*c*) Metallic inert gas (argon) weld on aluminium plate showing linear porosity due to moisture. This can be prevented by pre-heating.

(*d*) Metallic inert gas weld with copper wire on a copper plate to show various faults–undercut, porosity, unequal reinforcement, etc.

Fig. 11.7. X-rays of welds.

Fig. 11.8. X-ray equipment photographing welded seams on pressure vessels.

weld under test. The American Society of Mechanical Engineers (ASME) specify the above test in their Boiler Code, as does Lloyds.

Ultrasonic testing

Ultrasonic testing employs waves above the frequency limit of human audibility and usually in the frequency range 0.6 to 5 MHz. A pulse consisting of a number of these waves is projected into the specimen under test. If a flaw exists in the specimen an echo is reflected from it and from the type of echo the kind of flaw that exists can be deduced.

The equipment comprises an electrical unit which generates the electrical oscillations, a cathode ray tube on which pulse and echo can be seen, and probes which introduce the waves into the specimen and receive the echo. The electrical oscillations are converted into ultrasonic waves in a transducer which consists of a piezo-electric element mounted in a perspex block to form the probe, which, in use, has its one face pressed against the surface of the material under test. When a pulse is injected into the specimen a signal is made on the cathode ray tube. The echo from a flaw is received by another probe, converted to an electrical e.m.f. (which may vary from microvolts to several volts) by the

Inspection and testing of welds

transducer and is applied to the cathode ray tube on which it can be seen as a signal displaced along the time axis of the tube from the original pulse (Fig. 11.9a).

The first applications of ultrasonics to flaw detection employed longitudinal waves projected into the specimen at right angles to the surface (Fig. 11.9b). This presented problems because it meant that the weld surface had to be dressed smooth before examination, and more often than not the way in which the flaw oriented, as for example, lack of penetration, made detection difficult with this type of flaw. The type of wave used to overcome these disadvantages is one which is introduced into the specimen at some distance from the welded joint and at an angle to the surface (e.g. 20°) and is known as a shear wave. The frequency of the waves (usually 2.5 and 1.5 MHz for butt welds), the angle of incidence of the beam, the type of surface and the grain size, all affect the intensity of the echo which is adjustable by means of a sensitivity control. The reference standard on which the sensitivity of the instrument can be checked consists of a steel block 300 × 150 × 12.7 mm thickness with a 1.6 mm hole drilled centrally and perpendicularly to the largest face 50.8mm from one end. Echoes are obtained from the hole after 1, 2 or 3 traverses of the plate (Fig. 11.9c) and from the amplitude of the echo the sensitivity from a hole of known size can be checked.

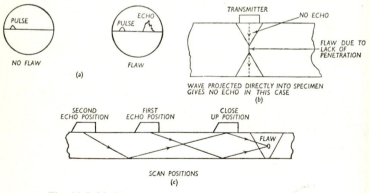

Fig. 11.9 (a) (b) (c)

Three types of probe are available:

> (1) A single probe which acts as both transmitter and receiver, the same piezo-electric elements transmitting the pulse and receiving the echo. The design of the probe is complicated in order to prevent reflections within the perspex block confusing the echo.

(2) The twin transmitter–receiver probe in which transmitter and receiver are mounted together either side by side or one in front of the other but are quite separate electrically and ultrasonically so that there is no trouble with interference with the echo. This type is the most popular.

(3) The separate transmitter and receiver each used independently (two-handed operation) (Fig. 11.9d).

To make a 'length scan' of the weld the transmitter–receiver unit is moved continuously along a line parallel to the welded seam so that all points of the whole area of the welded joint are covered by the scanning beam and care must be exercised that by the use of too high a spread of the beam, double echoes are not obtained from a single flaw. It is evident that varying the distance from the weld to the probe varies the depth at which the main axis of the beam crosses the welded joint and moving the probe at right angles to the line is thus known as depth scan. A spherical flaw will have no directional characteristics and a wave falling upon its centre will be reflected along the incident path, the amplitude of the echo depending upon the size of the flaw. Cylindrical flaws behave in the same way but in the case of a narrow planar flaw it is evident that optimum echo will be received when the crack is at right angles to the wave and there will be no echo when the crack lies along the wave, but if the probe is moved to the first echo position the crack is no longer lying along the beam.

The probes must make good contact with the specimen and on slightly curved surfaces a thin film of oil is used to improve the contact. On surfaces with greater curvature, as for example when investigating circumferential welds on drums, curved probes are used.

We have only considered the essential points of ultrasonic testing and it must be emphasized that there is a considerable amount of theory involved in the connexion between distance of transmitter from the weld in terms of the beam angle, etc., and that a great amount of practice is required to interpret correctly the echoes received and from them

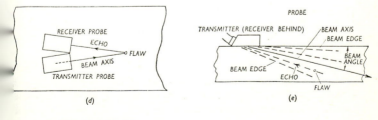

Fig. 11.9 (d) (e)

decide the nature and position of the flaw. See also BS 3923, *Methods for ultrasonic examination of welds*, Parts 1 and 2.

Gamma ray method

If certain elements are irradiated or bombarded with neutrons, protons, and other radiations, artificial radioactive elements are formed. These emit electrons, positrons and X-rays as they decompose back to their original state and the period for which they can emit radiation after being irradiated is given in terms of the 'half-life' which is the time taken for disintegration of one-half of the atoms of the substance. For a given element the radioactive sample may have the same atomic number but a different mass number (the mass number is the sum of the atomic number and the neutron number) from the element and these are termed isotopes. Gamma rays are electromagnetic radiation of very short wavelength and high frequency and they can penetrate solid matter to a high degree. Like X-rays they can show a shadowgraph on a sensitized film and are interpreted in the same way.

In the gamma-ray testing of welds the isotopes generally in use are cobalt 60, iridium 192, caesium 137, and thulium 170. Using a radioactive isotope, radiographic pictures or gammagraphs similar to X-ray pictures can be taken without an electricity supply, with a great range of penetration and at low cost. The small size of the isotope enables work that is inaccessible to an X-ray unit to be examined. The isotopes mentioned above are transported from the source such as Harwell in lead transport containers and are kept in the laboratories in lead storage containers in which different isotopes can be accommodated. They can be inserted in the container which is used for the picture taking by means of a handling rod the screw end of which fits the end of the isotope rod. The film is placed in a cassette as for X-ray examination and the source is suitably placed so as to throw a shadow on the film.

The exposure timer is set and the apparatus can then be left while the exposure is taking place. Cobalt 60, specific gravity 8.9, melting point 1480 °C, half-life 5.2–5.3 years is the most widely used isotope. It penetrates all sections of metal up to 280 mm thickness of steel. Above 50 mm of steel the gammagraph is comparable to those obtained on a 2 000 000-V X-ray unit. Below 20 mm thickness there is a loss of contrast and iridium 192 gives better results. The cobalt is available in solid cylinders 6 × 6 mm, 4 × 4 mm, and 2 × 2 mm, the largest ones being the strongest. Iridium 192, specific gravity 22.4, melting point 235 °C, gives good contrast on welded joints in 6–40 mm thickness of steel. The radiation is softer than from cobalt. Its half-life is 74 days and at this interval exposure times are doubled. Two sources a year would be enough for the average radiographic requirements and the sources can be returned for re-irradiation. A source strength decay clock indicates the source strength at any time. Typical sources used at present are cobalt 60 up to 100 curie and iridium 192 up to 120 curie.

The caesium 137 radioactive pellet is sealed in an iridio-platinum liner and encased in a stainless steel shell and sealed by silver soldering. It is used for a range of 20–100 mm steel and because its half-life is 33 years it has stable exposure times over many months but at present it is expensive.

The sources are designed and priced according to their strength and curies. A curie is the unit of radioactivity and is the quantity of any radioactive nuclide (that is a nucleus that is capable of existing for a measurable time) in which the number of disintegrations per second is 3700×10^{10}. Tests can be performed on site or in a specially constructed test room. Various fixtures are available to hold and position the source for examination of welds on storage tanks, pressure vessels, pipes etc. (Fig. 11.10a). Stringent precautions are taken to ensure that capsules are stored and used under strict safety conditions to ensure the safety of personnel against radiation.

In a typical test room the source may be stored below ground-level, with wall or stand-mounted adjustable supports to carry the isotope heads. Pre-exposure flashing lights, and 'exposure in progress' red light and emergency switches are wall-mounted and a radiation monitor, wall-fitted in an ante room, controls a pneumatically operated sliding door and lock which operates whenever radiation is present, giving greatest safety. An electro-pneumatic control and a control panel with an exposure timer (e.g., $4\frac{1}{2}$ seconds to 45 minutes) is mounted on an outside wall (Fig. 11.10b).

To operate the equipment the parts, source in its container and film are positioned within the room which is then evacuated by personnel. The start button is pressed, a relay closes and after a 5 second delay a warning system is energized after which the gamma ray capsule is projected by air pressure from its container into the exposure head. The timer controls the exposure time, after which the source is projected back into its container to complete the cycle.

Safety precautions

Factory regulations demand that personnel operating X-ray or gamma-ray equipment must either wear film badges or carry dosemeters. Film badges are supplied by a special service which processes them after they have been worn for a specified time and reports on the amount of radiation that the wearer has received. One type of dosemeter indicates the radiation dose received by checking it on a master control. Another type indicates the dose visibly on a built-in optical system. A geiger

(*a*) Isotope positioner.

Fig. 11.10

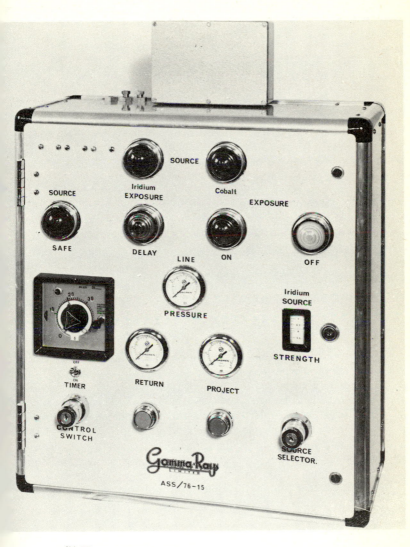

(*b*) Timer control panel.

counter indicating the intensity of radiation is used to indicate the effectiveness of shielding and the limits of the area in which it is safe for personnel to work.

Application of load

An illustration of this method is furnished by the hydraulic test on boilers. Water is pumped into the welded boiler under test (the safety valve if fitted having been clamped shut) to a pressure usually $1\frac{1}{2}$ to 2 times the working pressure. Should a fault develop in a joint, the hydraulic pressure rapidly falls without danger to persons near, such as there would have been if compressed air or steam had been used.

In the same way, partial compressive or tensile loads may be applied to any welded structure to observe its behaviour. The method adopted will, of course, depend on the nature of the work under test.

DESTRUCTIVE   TESTS

These may be divided as follows:

(1) Test capable of being performed in the workshop.
(2) Laboratory tests, which may be divided as follows: (1) micro-scopic and macroscopic, (2) chemical, analytical and corrosive, (3) mechanical.

Workshop tests

(The student is advised to study BS 1295, *Tests for use in the training of welders*, which gives standard workshop tests for butt and fillet welds in plates, bars and pipes.)

These are usually used to break open the weld in the vice for visual inspection. When operators are first learning to weld, this method is very useful, because as a rule the weld contains many defects and, when broken open, these can quickly be pointed out. Little time is thus lost in finding out the faults and rectifying them. As the welding technique of the beginner improves, however, this test becomes of much less value. Obviously much will depend on the actual position of the specimen in the vice, whether held on the joint or just below it. Also on the ham-mering, whether heavy erratic blows are used or a medium-weight, even hammering is given. In addition, if the weld metal is stronger than the parent metal, fracture may occur in the parent metal and thus the weld itself has hardly been tested. We can make sure that the specimen will break in the weld and afford us opportunity for examination by making a nick with a hacksaw as shown on each end of the weld, having previously filed or ground the ends square (see Fig. 11.11).

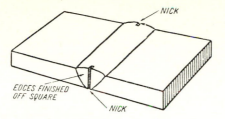

Fig. 11.11. Specimen for 'nick bend' or 'nick break' test.

Another useful method for determining the ductility of the weld is to bend the welded specimen in the vice through 180° with an even bending force. Any cracks appearing on the weld face will indicate lack of ductility. A better method of conducting this test will be described later (see Fig. 11.12).

Fig. 11.12

A useful workshop test, for use in the case in which the welded parts have to be heated up or even forged after welding, consists of actually forging a test specimen after welding. It is always advisable to apply the tests given later also, such as tensile, in order to obtain the ultimate strength of the weld.

Workshop tests are very limited, and their chief advantage is the little time taken to perform them. They are useful during training of welders, but little knowledge of the weld can be gained from them. The visual method, as previously explained, is a valuable addition to the workshop methods given above.

Microscopic and macroscopic tests

Microscopic tests. The use of the microscope is very important in determining the actual structure of the weld and parent metal. When a polished section of the weld is observed with the eye, it will look completely homogeneous if no blowholes or entrapped slag are present. On the other hand, if a section is broken open, as in the nick bend test, it may be found that there is a definite crystal-like structure. Since, however, this type of section may have broken at the weakest line, we must take a section across any desired part of the weld in order to have a typical example to examine. Specimens to be microscopically examined are best cut by means of a hacksaw. Any application of heat, as for example with gas cutting, may destroy part of the structure which it is desired to examine. If this specimen was polished by means of abrasives in the usual commercial way, when observed under the microscope it would be found to be covered with a multitude of scratches.

The best method of preparation is to grind carefully the face of the specimen after cutting on a water-cooled slow-running fine grinding wheel of large diameter, care being taken to obtain a flat face. Polishing can then be continued by hand, using finer abrasives, finally polishing by the polishing wheel, using rouge or aluminium oxide as the abrasive.

In order to bring out the structure of the section of metal clearly, the surface must now be 'etched'. This consists of coating it with a chemical, which will eat away and dissolve the metal. Since the section is a definite structure consisting of composite parts, some are more easily dissolved than others, and thus the etching liquid will bring up the pattern of the structure very clearly when observed under the microscope.

The etching liquids employed depend on the metal of the specimen. For iron and steel, a 1 or 2% solution of nitric acid in alcohol, or picric acid in alcohol, is used. For copper, either ammonium persulphate or ferric chloride acidified with hydrochloric acid. For aluminium and aluminium alloys, either caustic soda or dilute hydrofluoric acid and nitric acid.

Most of the microphotographs in this book were prepared by etching with the 2% nitric acid solution in alcohol.

The length of time for which the etching liquid remains on the metal depends on the detail and the magnification required. After etching is complete, the liquid is washed off the surface of the specimen to prevent further action. For example, if steel etched with picric acid is to be examined at 100 diameters, etching could be carried out from 25 to 35 seconds, giving a clear well-cut image. If this, however, was observed under the high-power glass of 1000 diameters, it would be found that picric acid had eaten deeply into the surface, and the definition and result would be extremely poor. Thus, for high magnification, the etching would only need carrying out for 5 to 10 seconds. Naturally, however, the time will vary entirely with the etching liquid used, the power of magnification and the detail required.

When the section is prepared in this way and the whole crystal structure is visible, the exact metallic condition of the weld can be examined, together with that of the surrounding parent metal. For example, examination of microphotographs of steel at 150 to 200 diameters will indicate the size of the grain, the arrangement of pearlite and ferrite. Increasing magnification to 1000 diameters will indicate the presence of oxides or nitrides, oxides being shown up as fine cracks between the crystals (producing weakness) (Fig. 2.19), and iron nitrides as needle-like crystals (producing brittleness) (Fig. 2.21). From this, the metallurgist can tell the suitability of the weld metal, how well the structure compares with that of the parent metal, and its probable strength. This study or test plays an important part in the manufacture of new types of welding rods. Microphotographs of varying magnifications are used in various parts of this book to illustrate the structures referred to.

Macroscopic tests. This method consists as before of preparing a cross-section of the weld by polishing and etching. It is then examined either by a low-power microscope magnifying 3 to 20 diameters or even with a magnifying glass. This will show up any cracks, entrapped slag, pin-size blow or gas holes, and will also indicate any coarse structure present (Fig. 11.13).

The etching fluids most suitable for macroscopic examination are:

Steel and iron. 10% iodine, 20% potassium iodine and 70% distilled water; 10 to 20% nitric acid in water; 8% cuprous ammonium chloride in water.

Copper. 25% solution of nitric acid in water; ammonium hydrate; nitric acid in alcohol.

Brass and bronze. 25% solution of nitric acid.

Aluminium and aluminium alloys. 10% solution of hydrofluoric acid in water.

The macrographic examination of welds can easily be undertaken in the workshop, using a hand magnifying glass, and the degree of polish required is not so high as for microscopic examination. The microscope, however, will obviously bring out defects and crystal structures which will not be apparent in the macrograph.

Sulphur prints. This is an easy method by which the presence of sulphur, sulphides and other impurities can be detected in steel. It is not suitable for non-ferrous metals or high alloy steels.

The principle of sulphur printing is that a dilute acid such as sulphuric will attack sulphur and sulphides, liberating a gas, hydrogen sulphide (H_2S), which will stain or darken bromide or gaslight photographic paper.

(*a*) Fillet weld, single-run each side, with good penetration and no undercut.

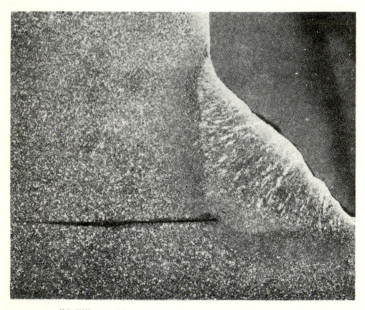

(*b*) Fillet weld, two runs on one side. (× 3.) Note good fusion.

Fig. 11.13

To make a sulphur print, the specimen is first prepared by filing or machining and then by rubbing by hand or machine to obtain a scratch-free surface (O grade emery). A piece of photographic paper is soaked in dilute sulphuric acid for about 3 to 4 minutes and then after excess acid has been sponged off, the paper is laid carefully on the prepared surface of the steel specimen and pressed down perfectly flat on its surface. It is left on the specimen for about 4 to 5 minutes, the edge of the paper being lifted at intervals to ascertain how it is staining. After removing it, the paper is treated as in photographic printing, namely rinsed, then immersed in a 20% hypo solution for few minutes and then again thoroughly washed. The darker the stains on the paper the higher the sulphur content.

Chemical tests

Analytical tests are used to determine the chemical composition of the weld metal. From its composition, the physical properties of the metal can be foretold. The addition of manganese increases the toughness of steel, uranium increases its tensile strength, and these are indicated fully in the chapter on metallurgy.

Corrosive tests are used to foretell the behaviour of the weld metal under conditions that would be met with in years of service.

The action of acids and alkalis, present in the atmosphere of large industrial areas and which may have a marked effect on the life of the welded joints, can be observed, the effect in the laboratory being concentrated so as to be equal in a few days to years of normal exposure. From these tests, the most suitable type of weld metal is indicated. The following examples will serve as illustrations.

Along the sea coast, greatest corrosion takes place to those metal parts which are subject to the action both of the salt water and the atmosphere, that is, the areas between high and low tide; for example, pier and landing-stage supports and caissons, and railings and structures exposed to the sea spray. By dipping welded specimens alternately in and out of a concentrated brine solution corrosion effects equal to years of exposure are produced.

Suppose it is required to compare the resistance to acid or alkaline corrosion of plates welded together with different types of electrodes. The specimens are polished and marked and then photographed. They are then rotated in a weak acid or alkaline solution. The specimens are photographed at given intervals and the degree of corrosion measured in each case. From the results it is evident which electrode will give the best resistance to this particular type of corrosion.

In the chemical industry, tanks are required for the storing of corrosive chemicals. It is essential that the welded joints should be just as proof against corrosion as the metal of the tank itself. Corrosive tests undertaken as above in the laboratory will indicate this, and will enable a correct weld metal to be produced, giving proof against the corrosion.

Evidently, then, these tests are specialized, in that they reproduce as nearly as possible, in the laboratory, conditions to which the weld is subjected.

Mechanical tests

These may be classified as follows:

    (1) Tensile.
    (2) Bending.
    (3) Impact: Charpy and Izod.
    (4) Hardness: Brinell, Rockwell, Vickers Diamond Pyramid (Hardness Vickers HV) and Scleroscope.
    (5) Fatigue: Haigh and Wöhler.
    (6) Cracking: Reeve.

Tensile test

As stated previously, a given specimen will resist being pulled out in the direction of its length and up to a point (the yield point) will remain elastic, that is, if the load is removed it will recover its original dimensions. If loaded beyond the yield point or elastic limit the deformation becomes permanent.

Preparations of specimens. In order to tensile test a welded joint, specimens are cut from a welded seam and one specimen from the plate itself. This latter will give the strength of the parent metal plate. The specimens are machined or filed so as to have all the edges square, and the face can be left with the weld built up or machined flat, depending on the test required. It is usual, in addition, to cut specimens for bend testing from the same plate, and these are usually cut alternately with the tensile specimens, as shown in Fig. 11.14.

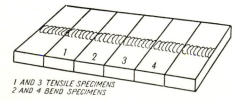

I AND 3 TENSILE SPECIMENS
2 AND 4 BEND SPECIMENS

Fig. 11.14

If the elongation is required, it is usual to machine the specimen flat on all faces and to make two punch marks 50 mm apart on each side of the weld, as shown in Fig. 11.15. Fig. 16*a* and *b* shows two specimens prepared for tensile test.

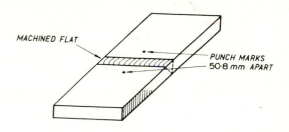

MACHINED FLAT

PUNCH MARKS
50·8 mm APART

Fig. 11.15

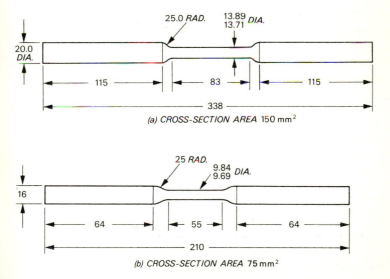

25.0 RAD.

13.89
13.71 DIA.

20.0
DIA.

115

83

115

338

(a) CROSS-SECTION AREA 150 mm²

25 RAD.

9.84
9.69 DIA.

16

64

55

64

210

(b) CROSS-SECTION AREA 75 mm²

ALL DIMENSIONS IN MILLIMETRES

Fig. 11.16. The preparation of two specimens for tensile test with cross-sectional areas of 75 mm² and 150 mm².

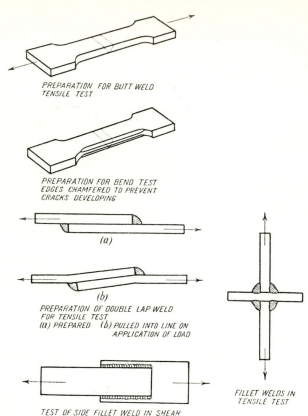

PREPARATION FOR BUTT WELD
TENSILE TEST

PREPARATION FOR BEND TEST
EDGES CHAMFERED TO PREVENT
CRACKS DEVELOPING

(a)

(b)

PREPARATION OF DOUBLE LAP WELD
FOR TENSILE TEST
(a) PREPARED   *(b)* PULLED INTO LINE ON
APPLICATION OF LOAD

FILLET WELDS IN
TENSILE TEST

TEST OF SIDE FILLET WELD IN SHEAR

Fig. 11.17. Preparations of specimens for test.

Preparation of all-weld metal specimens

Two steel plates approximately $200 \times 100 \times 20$ mm thick are prepared with one face at an angle of $80°$ as in Fig. 11.18. The plates are set about 16 mm apart on a steel backing strip about 10 mm thick and are welded in position. The groove is built up with the weld metal under test and with a top reinforcement of about 3 mm. The welded portion is then cut out (thermally) along a line about 20 mm each side of the weld line. A tensile specimen is prepared from the all-weld metal as in Fig. 11.19 and a Charpy specimen as in Fig. 11.28.

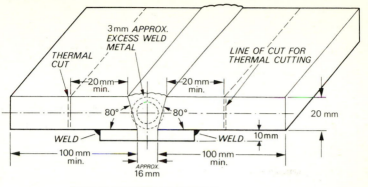

VIEW IN DIRECTION OF ARROW X

Fig. 11.18. Preparation of an all-weld metal test piece.

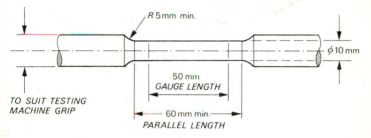

Fig. 11.19. An all-weld tensile test specimen.

The specimen is prepared from deposit well away from the parent plate as there will be effects of dilution on the two or three initial layers.

Testing machines

Present-day testing machines are available in a variety of designs suitable for specialized testing. A typical universal machine for tests in tension, cold bend, compression, double shear, transverse, cupping and punching has four ranges, 0–50, 100, 250 and 500 kN (50 tonf).

It comprises an hydraulic pumping unit (Fig. 11.20) with a multi-piston pump of variable displacement enabling the movement of the straining rams to be closely controlled. Hydraulic pressure is applied to the pistons of the rams R which move the cross beam H to which the straining wedgebox is attached. The specimen under test is gripped between this and an upper wedge box connected to a series of lever balances A, B and C, the movement of which when load is applied is indicated on the figure. These balances or beams are mounted on

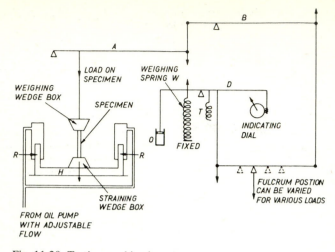

Fig. 11.20. Testing machine layout.

hardened steel knife edges and are of deep section to prevent deflexion. The balance arm D is attached at one end to the weighing spring W and the other end actuates the dial pointer which moves over a scale graduated in kN. An additional spring T helps to keep the knife edges in contact and an oil dash pot O acts as a damper. The arm C has four fulcra, any of which can be selected by movement of a hand-operated cam dependent upon the range of the test required, and at the same time the range selected is indicated on the dial.

Tensile test of a welded joint. It will be evident that a tensile test on a welded joint is not quite similar to a test on a homogeneous bar, and the following considerations will make this clear. The steel weld metal may be strong, yet brittle and hard. When tested in the machine, the specimen would most probably break outside the weld, in the parent metal, whereas in service due to its brittleness, failure might easily occur in the weld itself. The result of this test gives the tensile strength of the bar itself and indicates that the weld is sound. It does not indicate any other condition.

If the weld metal is softer than the parent metal, when tested the weld metal itself will yield, and fracture will probably occur in the weld. Because of this, the elongation of the specimen will be small, since the parent bar will have only stretched a small amount, and this would lead to the belief that the metal had little elasticity. Quite on the contrary, however, the weld metal may have elongated by a considerable amount, yet because of its small size in comparison to the length of the specimen the actual elongation observed in small. Great care must therefore be taken to study carefully the results and to interpret them correctly, bearing in mind the properties which it is required to test.

A tensile test on an all-weld metal specimen prepared as previously explained indicates the strength and ductility of the metal in its deposited condition and is a valuable test.

A very useful form of test which is used nowadays is that known at the longitudinal test. In this test the weld runs along the length of the test piece (Fig. 11.21). As the load is applied, if the weld metal is ductile, it will elongate with the parent metal and is placed in the machine so that the load is applied longitudinally to help to share the load. If, on the other hand, it is brittle, it will not elongate with the parent metal but will crack. Should the parent metal be of good quality and structure, the cracks will be confined to the weld metal mostly and will merely increase in width. If the parent metal is not of such good quality, the cracks will extend into the parent metal and breakage will occur with little elongation of the specimen. This test therefore indicates the quality of the parent metal as well as that of the weld metal.

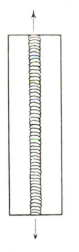

Fig. 11.21. Longitudinal tensile test of welded specimen.

Torsion test. This test is useful to test the uniformity of work turned out by welders. A weld is made between steel plates V'd in the usual manner, and a cylindrical bar is turned out of the deposited metal. This specimen is then gripped firmly at one end, while the other end is rotated in a chuck or other similar device, until breakage occurs. The degree of twist which occurs before breakage will depend upon the type of metal under test.

Bend test (for ductility of a specimen)

In this test the bar is prepared by chamfering the edges to prevent cracking (if it is of rectangular section), and is then supported on two edges and loaded at the centre (Fig. 11.22a).

As the load is applied the bar first bends elastically, and in this state if the load is removed it would regain its original shape. On increasing the load a point will be reached when the fibres of the beam at the centre are no longer elastically deformed, i.e. they have reached their yield point, and the bar deforms plastically at the centre (Fig. 11.22c).

Further increase of load causes yielding to occur farther and farther from the centre, while at the same time the stress at the centre increases ultimately, when maximum stress is reached, fracture of the bar will occur. If this maximum stress is not reached, fracture of the bar will not occur for any angle of bend. The method of determining the ductility of the bar from the above test is as follows. Lines are scribed on the machined or polished face of the specimen parallel and equidistant to each other across a width of about 150 to 250 mm. As the load is applied, the increase in distance between the scribed lines is measured, and this increase is plotted vertically against the actual position of the lines horizontally. When the bar deforms elastically the result is a triangle (Fig. 11.22b), termed the stress diagram, since it represents graphically

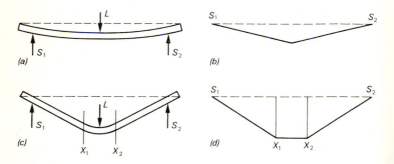

Fig. 11.22. Uniform bar supported at S_1 and S_2 and loaded at the centre. (a) Beam deflected elastically. (b) Bottom layer stress diagram for elastic deflexion, i.e. extension of bottom surface layers. (c) Beam deformed or yielded plastically between X_1 and X_2 but with elastic deflexion between X_2S_2 and X_1S_1. (d) Bottom layer stress diagram when yield commences.

the stress at these points. The stress diagram is shown for plastic deformation of the bar in Fig. 11.22*d*.

Now consider the test applied to a welded joint and let the weld be placed in position under the applied load. There are now two different metals to be considered, since the weld metal might have quite different properties from those of the parent metal (Fig. 11.23).

If the load is applied and the yield point of the weld metal is greater than that of the parent bar, plastic yield or bend will occur in the bar, and as the load is increased the bar bends plastically, as in Fig. 11.23*b*. During this bending, if the yield point of the weld metal is reached, the weld metal will flow or yield somewhat, but in any case most of the bend is taken by the bar. If the yield point of the weld metal is not reached, then all the bend will be taken by the bar.

If however, the yield point of the weld metal is lower than that of the bar, the weld metal will first bend plastically and will continue to do so, plastic deformation occurring long before the yield point of the bar is

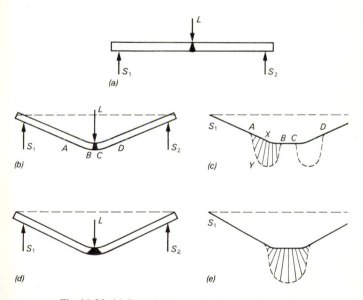

Fig. 11.23. (*a*) Bar with inserted edge of weld metal having mechanical properties differing from that of the bar itself, supported at S_1 and S_2 and loaded at the centre. (*b*) Load applied. Yield point of weld not reached. Bend taking place in the bar. (*c*) Bottom layer stress diagram. Length of XY represents extension at point X. (*d*) Load applied. Yield point of *bar* not reached. Bend taking place in the weld. (*e*) Bottom layer stress.

reached (Fig. 11.23d). On such a small area as the weld metal has, therefore, fracture will occur in the weld metal at a small angle of bend. The stress diagrams given in Fig. 11.23c and e indicate where the greatest elongations of the fibres occur in each case.

Since the weld metal is almost always harder or softer than the parent metal the bending will not occur, therefore, equally in the weld and parent metal, and as a result the chief value of this test is to determine whether any flaws exist in the weld. Otherwise its value as a test of ductility of a welded specimen is very limited.

If the weld is placed so that its face is under the central applied load, fracture will occur at the root of the weld if the penetration is imperfect.

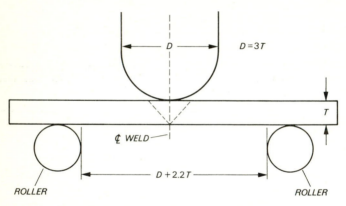

TRANSVERSE BEND TEST ON BUTT WELD SPECIMEN

Fig. 11.24. A typical transverse bend test. Upper and lower surfaces are ground or machined flat. The specimen is about 30 mm wide. The bending should be through an angle of 180° over a former with a diameter three times that of the plate thickness. Test should be made with: (1) the weld face in tension, and (2) the root of the weld in tension.

Impact tests

We have seen when discussing notch brittleness in steel that localized plastic flow at a notch may cause cracking and that the transition from ductile to brittle state is affected by temperature, strain rate and the occurrence of notches. It should be noted that as there is no ductile–brittle transition with aluminium, impact tests are performed to a lesser degree with aluminium than with steel. Serrations, tool marks, changes of section and other discontinuities on the surface of metals that are met with in service reduce their endurance so that the term 'notch sensitivity' is applied to the degree to which these discontinuities reduce the mechanical properties. This is an important consideration in welding because for example any reduction in section due to undercut along the toes of butt welds and in the vertical plate in HV fillets reduces the mechanical properties of the structure.

To determine the notch brittleness (or notch toughness), impact tests are performed on specimens prepared with a notch of precise width, depth and shape, and the resistance which the specimen offers to breaking at the notch when hit by a striker moving at a given velocity and having a given energy is a measure of the notch brittleness.

The two main tests, Charpy and Izod, employ a swinging pendulum to which a slave pointer is attached. This moves over a scale calibrated in joules as the pendulum swings and stays at the impact value of the test, being afterwards reset by hand. The pendulum tup (or bob) incorporating the striker hits the notched specimen at a given velocity and with a given energy (measured in joules). If no specimen were present the pendulum would swing unhindered to the zero position on the scale, but since energy is lost in breaking the specimen the pointer will take up a position say x joules on the scale. This is the impact value for the specimen at the particular temperature and represents the energy lost by the pendulum in breaking the specimen.

Impact tests are being increasingly used at sub-zero temperatures in order to give indications and possibilities of brittle fracture. Diethyl ether and liquid nitrogen are used to obtain temperatures down to $-196\,°C$ using a copper–constantan thermo-couple for temperature measurement.

Charpy and Izod machines

Machines can be Charpy and Izod combined or Charpy only or Izod only and can be manually or pneumatically operated. In manual machines the pendulum is lifted physically to the start position where it is held in position by a release box which has a self-setting catch. There is a pendulum release lever and a safety lock lever which prevents accidental release of the pendulum. Pneumatically operated machines operate in a similar manner to the manual machines except that the pendulum is lifted under power to the Charpy or Izod start position and can be set for automatic release.

Machine capacities are 0–150 J (striker velocity 3–4 m/s) for the Izod machine and 0–300 J (striker velocity 5–5.5 m/s) with an optional 0–150 J for the Charpy machine. On the combined machines, tups and strikers are changed for the different tests and gauges are provided for the Charpy machine to check that the striker hits the specimen centrally between the anvils.

Charpy test

This test may be either with a V or a U section notch, the specimen and notch sizes being shown in Fig. 11.26. The V notch test is becoming increasingly used in Britain and is the test required for impact values in BS 639 – *Covered electrodes for the MMA welding of carbon and carbon–manganese steels.*

Fig. 11.25*a* and *b* indicates the method of operation of the machine.

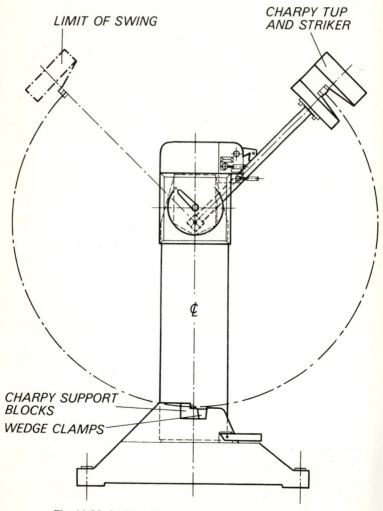

Fig. 11.25. (*a*) The Charpy machine; 150 and 300 joules.

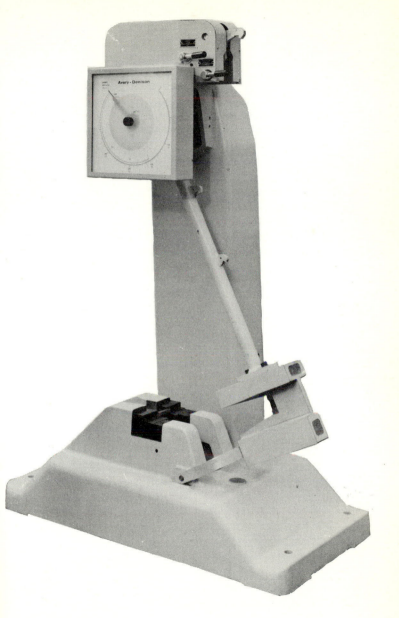

Fig. 11.25 (b)

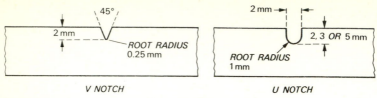

V NOTCH   U NOTCH

SPECIMENS 55 mm *LONG AND* 10 mm *SQUARE SECTION*

NOTCHES CENTRALLY PLACED AND AT RIGHT ANGLES TO THE LONGITUDINAL AXIS

Fig. 11.26. Preparation of V and U notches.

The specimen is supported squarely at its two ends by machine supports, the notch being centrally placed by means of small tongs (Fig. 11.27). The pendulum is raised to the test height and the pointer indicates 300 J on the scale. A hand lever is operated, the pendulum swings and the striker hits the specimen exactly on the side behind the notch. Energy is absorbed in fracturing the specimen and the pointer swings to say x joules, on the scale this being the Charpy value at the particular temperature on the 300 J scale for either V or U notch, whichever was chosen. Fig. 11.28a shows the preparation of an all-weld metal test piece and Fig. 11.28b shows the preparation for impact testing a weld. (See also BS 131 Pt 2 – *Charpy* V *test* and Pt 3, *Charpy* U *notch test*.)

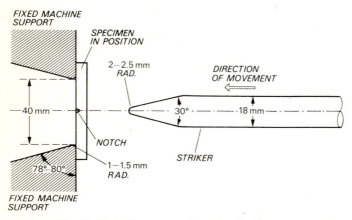

Fig. 11.27. Striker and specimen for the Charpy test.

Izod test

This test is performed on a specimen with a V notch and of dimensions as in Fig. 11.29a. The specimen is mounted vertically in a groove in the vice wedge block assembly which is tightened by handwheel.

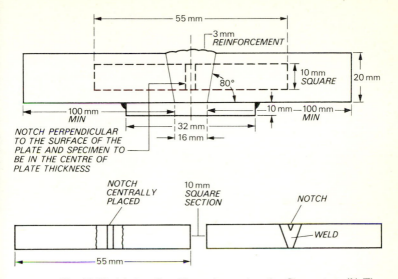

Fig. 11.28. (*a*) An all-weld metal test piece for Charpy test. (*b*) The preparation for impact testing.

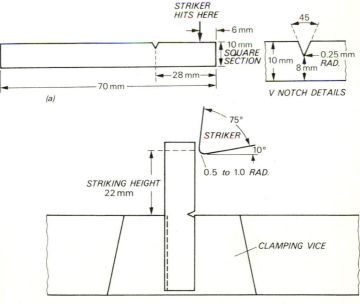

Fig. 11.29. The Izod test (single notch).

The V notch is located facing the striker and with the base of the V exactly in line with the top edge of the vice, and is lined up with a small hand jig. The striker hits the specimen at the striking height 6 mm above the V notch (Fig. 11.20*b*).

To operate the machine the pendulum is raised to the 150 J position and upon release swings as in the Charpy test and the pointer indicates the impact value in joules. Fig. 11.30 shows a combination machine with Izod release box in position and set for the Izod test and also with the Charpy release box in position. (See also BS 131 Pt 1.)

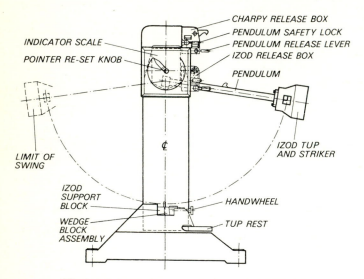

Fig. 11.30. Machine for the Izod test; 0–150 joules.

Hardness tests

These are useful to indicate the resistance of the metal to wear and abrasion, and give a rough indication of the weldability of alloy steels. If parts, such as tramway crossings, dredger bucket lips, plough shares or steel gear wheels, have been reinforced or built up, it is essential to know the degree of hardness obtained in the deposit. This can be determined by portable hardness testers of the following types.

The chief methods of testing are: (*a*) Brinell, (*b*) Rockwell, (*c*) Vickers Diamond Pyramid, (*d*) Scleroscope.

The **Brinell test** consists in forcing a hardened steel ball, 10 mm diameter, hydraulically into the surface under test. The area in square millimetres, of the indentation (calculated from the diameter measured by a microscope) made by the ball, is divided into the pressure in kilograms, and the result is the Brinell hardness number or figure. Figure 11.31 shows an indentation of 4.2 mm diameter when measured on the microscope scale.

Fig. 11.31. Indention made by Brinell ball in hard surface and measured by microscope scale.

For example: If the load was 3000 kg, and the area of indentation 10 mm², Brinell number equals 3000 divided by 10, which is 300.

The Brinell figure can be calculated from the following:

$$\text{Brinell figure} = \frac{P}{\frac{\pi D}{2}(D - \sqrt{(D^2 - d^2)})}$$

where P = load in kg, D = diameter of ball in mm, d = diameter of indentation in mm.

The tensile strength of mild steel in N/mm² is approximately 3.4 times the Brinell hardness value.

Evidently the ball must be harder than the metal under test or the ball itself will deform and as a result it is used only up to a figure of about 500. See also BS 240 Pt 1 and 2.

Rockwell hardness test

There are three standard indenters, a diamond cone with an included angle of 120° and with the tip rounded to a radius of 0.2 mm; a 1.6 mm diameter hardened steel ball and a 3.2 mm diameter hardened steel ball. The diamond cone or steel ball is first pressed into the clean surface under test with a load of F_1 kgf. to a point B distant AB above a reference line at A, (Fig. 11.32a). A further load F_2 is now applied making a total load of $F_1 + F_2$ kgf and the indenter is pushed further into the surface to a point C, distant AC above A. The major load is then released leaving only the initial load F_1, and there is some recovery to the point D distant AD above A. Then BD represents the permanent depth of indentation due to the additional load. This indentation is automatically measured on the dial of the machine and indicates the Rockwell hardness HR, the number having an added letter indicating the scale of hardness used.

In each case in the following typical scales the initial load is 10 kgf (F_1).

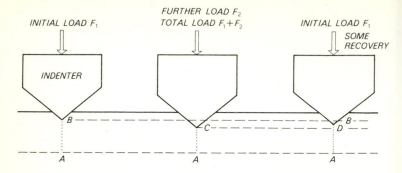

Fig. 11.32. (*a*) *BD* represents the permanent indentation due to the additional load. The reference plane at *A* represents the zero of the particular hardness scale, *AB* having a constant value of 100 units for the diamond indenter and 130 for the steel ball indenter.

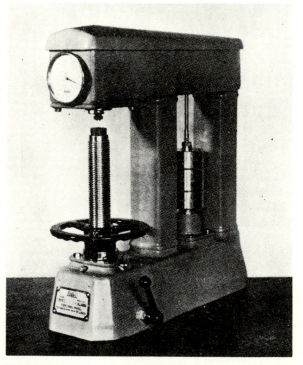

Fig. 11.32. (*b*) Direct reading hardness testing machine.

Scale A, additional load 50 kgf, total load 60 kgf, diamond indenter.

Scale B, additional load 90 kgf, total load 100 kgf, 1.6 mm diameter steel ball indenter.

Scale C, additional load 140 kgf, total load 150 kgf, diamond indenter.

BD represents the permanent indentation due to the additional load. The reference plane at A represents the zero of the particular hardness scale, AB having a constant value of 100 units for the diamond indenter and 130 for the steel ball indenter. To use a testing machine the particular indenter in use is fitted to the machine and the scale selected. Loads are applied automatically, the hardness number appearing on the dial (Fig. 11.32b) so that routine hardness tests are easily and quickly performed. See also BS 891 Pt 1 and 2.

Vickers hardness test (BS 427)

The Vickers hardness test is similar to the Brinell test but uses a square base right pyramidal diamond with an angle of 136° between opposite faces as the penetrator. The two diagonals d_1 and d_2 mm of the indentation are measured and the average calculated. The Hardness Vickers is obtained thus:

$$HV = 1.854 \frac{F}{d^2},$$

where F = load in kgf and d is the diameter of the diagonal (or the average of the diameters). To avoid having to perform this calculation for each reading, the diameter is obtained from an ocular reading fitted to the measuring microscope and the HV is obtained from the ocular reading with the use of a table.

The pressure which is applied for a short time can be varied from 1 to 120 kgf according to the hardness of the specimen under test. The Brinell number and the HV number are practically the same up to 500, the Brinell number being slightly lower (see table).

The **Scleroscope test** consists of allowing a hard steel cylinder, called the hammer, having a pointed end, to fall from a certain height on to the surface under test. The height to which it will rebound will depend upon the hardness of the surface, and the rebound figure is taken as the hardness figure. The fall is about 250 mm, giving with a hard steel surface a rebound height of about 150 mm, this being 90 to 100 on the Scleroscope scale.

Fatigue test

If a specimen is subjected to a continuously alternating set of push and pull forces operating for long periods, the specimen may fail due to

Brinell number	Hardness Vickers	Rockwell	
		Scale A	Scale C
59	100	43	
235	240		20
380	400		40
430	460		45
460	500		48
535	600		54
595	700		58
	800		62
	1000		68

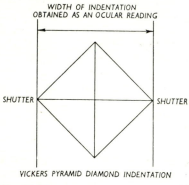

WIDTH OF INDENTATION
OBTAINED AS AN OCULAR READING

SHUTTER

SHUTTER

VICKERS PYRAMID DIAMOND INDENTATION

Fig. 11.33

fatigue of the molecules, and the magnitude of the force under which it may fail will be much less than its maximum tensile or compressive strength. The forces applied rise to a maximum tension, decrease to zero, rise to a maximum compression and decrease again to zero. This is termed a cycle of operations and may be written 0–maximum tension–0–maximum compression–0 and so on. Fatigue tests are based on this phenomenon exhibited by metals.

In the **Haigh tests,** a soft-iron core or armature vibrates between the poles of an electro-magnet carrying alternating current, and is connected to the specimen under test. As alternating current is passed through the coil, the armature vibrates at the frequency of the supply (usually 50 Hz) between the poles, and the welded specimen is thus subjected to alternating push-pull forces at this frequency. The alternating current, and therefore the force on the armature, rises as above: 0–maximum in one direction–0–maximum in opposite direction–0; this being, as before, one cycle of operations.

The drawback to this test is that at 3000 reversals per minute an endurance test of 10 000 000 reversals would take about 56 hours and a complete endurance test will take many days.

The latest type of electromagnetic fatigue tester gives approximately 17 000 reversals per second, and thus the required tests can be performed in a fraction of the time and the machine automatically shuts off when failure occurs.

The pick-up which causes the vibration is controlled from an oscillator and non-magnetic metals can also be tested.

In the **Wöhler test,** the specimen is gripped at one end in a device like a chuck and the load is applied at the other end by fixing it to a bearing, as shown in Fig. 11.34. When the chuck rotates at speed, the specimen is continuously under an alternating tension and compression, tension when the face of the weld is uppermost, as shown, and compression when it is below. If the load applied is great, difficulty is experienced by it pulling the specimen out of balance. These out-of-balance forces then increase the forces on the specimen, and we are unable to tell the load under which the weld failed. This can be overcome, however, by a slight modification of the machine having a bearing at each side of the joint under test and the load applied between the bearings, but the test remains the same. In conducting a fatigue test, a certain load is placed on the specimen, and this produces a certain stress. Suppose the stress produced is 140 N/mm²; this stress varies from zero to 140 N/mm² tensile stress, then back to zero and to 140 N/mm² compressive stress and back to zero. This is a complete cycle.

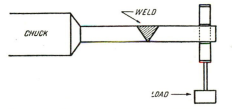

Fig. 11.34. Illustrating principle of Wöhler test.

Fatigue tests are extremely useful for observing the resistance to fatigue of welded shafts, cranks and other rotating parts, which are subjected to varying alternating loads. They also provide a method of comparing the resistance to fatigue of solid drop forged and welded fabricated components.

Cracking (Reeve) test

This is used in the study of the hardening and cracking of welds and is of especial value in ascertaining the weldability of low-alloy structural steels and high tensile steels, which as before mentioned are prone to harden and develop cracks on cooling. A 150 mm square plate of the metal to be welded is placed on another larger plate of the same metal and the two are firmly secured to a heavy bed plate, 50 mm or more in thickness, by means of bolts as shown in Fig. 11.35.

Edges *a*, *b* and *c* are then welded with any selected electrode, thus firmly welding the two plates together, and they are then allowed to cool off. Edge *d* is the one on which the test run is to be deposited using the electrode under test, and evidently since the two plates are completely restrained in movement, any tendency to crack on cooling will show in the weld on the edge *d*.

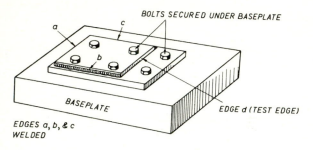

Fig. 11.35. Reeve test.

After cooling, the bolts are removed and the weld examined by previously described methods for cracks. Sections can then be sawn off from the plate, the hardness of the weld tested at various points, and sections etched and examined microscopically.

It can be seen from this outline of available tests for welds that the particular test chosen will depend entirely upon the type of welded joint and the conditions under which it is to operate. These conditions will govern the tests which must be applied to indicate the way in which the weld will behave under actual service conditions.

Erichsen test (cupping)

This is used for determining the suitability of a metal for deep drawing and pressing. A punch with a rounded head is pushed into the surface of the metal, the depth and appearance of the indentation before cracking occurs being an indication of the suitability for drawing and pressing.

12

Engineering drawing and welding symbols

ENGINEERING DRAWING

During periods of training, and certainly in the course of his work, the welder will find that an elementary knowledge of technical drawing is of great help, both from the point of view of his being able to read and understand the conventional blue print or working drawing, and being able to make sketches of parts to scale. To enable the welder to have some knowledge of the subject it will be well to consider the way in which drawings are made, the standard methods used, and then to proceed to the consideration of a few special examples of drawings of welded construction with brief descriptions.

The principal method usually adopted in the making of machine drawings is known as orthographic projection.

Suppose the part under construction is shown in Fig. 12.1a. This 'picture' is known as an isometric view. It is of small use to the engineer, since it is difficult to include on it all the details and dimensions required, especially those on the back of the picture, which is hidden.

Imagine that around the object a box is constructed (o being the corner farthest from the observer) having the sides ox, oy, oz all at right angles to each other. The plane or surface of the box bounded by ox and oy is the *vertical* plane, indicated by vp; that bounded by oy, oz is the side vertical plane, svp, and that bounded by ox and oz, the horizontal plane, hp, these three planes being the three sides of the box farthest from the observer. Lines are projected, as shown, on to these planes from the object under consideration, and the view projected on the vertical plane is the side elevation, and is the view obtained when looking at the object in the direction of the arrow A. The end elevation is the view obtained by projection on to the side vertical plane, while the plan is the view obtained by projection on to the horizontal plane. The arrow B shows the direction in which the object is viewed for the side elevation and C the direction for the plan.

Now imagine the sides vp, svp and hp opened out on their axes ox, oy and oz. The three projections will then be disposed in position, as shown in Fig. 12.1b, i.e. the end elevation is to the *right* of the elevation and the plan is *below* the elevation.

On these three projections, which are those used by the engineer, almost all the details required during manufacture can be included, and hence they are of the greatest importance.

This method of projection is known as First Angle Projection, and is

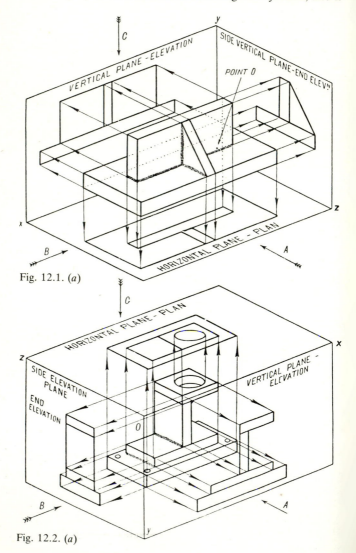

Fig. 12.1. (a)

Fig. 12.2. (a)

that usually adopted in British Engineering circles, the projection lines being clearly indicated in Fig. 12a.

A second method, called Third Angle Projection, is extensively used in the USA, and can be understood by reference to Fig. 12.2a.

From this it will be seen that the corner of the box *o* is chosen to be that nearest the observer, and the three planes are those sides of the box also nearest to the observer, the part under consideration being seen through these planes of projection. The elevation is again that view formed by

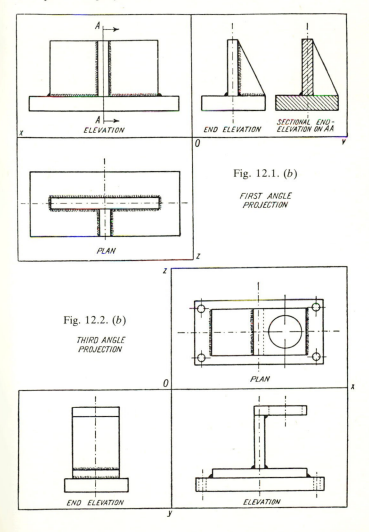

Fig. 12.1. (*b*)

FIRST ANGLE
PROJECTION

Fig. 12.2. (*b*)

THIRD ANGLE
PROJECTION

projection on to the vertical plane *ox, oy*, the end elevation that formed by projection on the side vertical plane *oy, oz* and the plan formed by projection on the horizontal plane *ox, oz*. Owing, however, to the change in the axes when they are unfolded, the projections are disposed differently, the plan now being *above* the elevation and the side elevation being to the *left* of the elevation (Fig. 12.2*b*). (N.B. The object is viewed in the same direction as previously, as indicated by the arrows.)

By noting the above difference between the two methods the welder can immediately tell which method has been used. Sometimes a combination of these two methods may be encountered but need not be considered here.

Scales

Engineering or working drawings are usually drawn to a definite scale. Small parts may be drawn full size (scale 1:1) and its choice is limited usually by the size of drawing paper used. Larger parts may be drawn 1 : 2; 1 : 5; 1 : 10, etc., but the measurements or dimensions given on the drawing will represent the *true* size of the parts.

Tracings of these drawings are then made, and these are used as negatives in the production of blue, brown or white prints. In the process of making the print (which is chemical) a change of size may occur and therefore lines on the blue print may not be to the scale of their correct length. For this reason measurements should not be made by means of a ruler or other measuring instrument from the print – the dimension printed on the print gives the correct lengths and distances and should be followed. A good working drawing is always fully dimensioned, and if a dimension does not appear on one view it will always be found on one or other of the remaining views. In a case where the actual measurement on a drawing is evidently not equal to the dimensioned size after allowing for the scale of the drawing, the dimension will be underlined thus: 50 mm, and this means that the true size which the part must be is 50 mm irrespective of its apparent size on the drawing. The engineer or welder should accustom himself to the addition and subtraction of dimensions to give him those not indicated, e.g. a tube of inner radius 75 mm, outer radius 100 mm, evidently has a wall thickness of 25 mm. Hidden parts are usually denoted by dotted lines, as in Fig. 12.2*b*.

Sections

A section may be considered as the view on an imaginary plane which cuts through the object under consideration at any given point, and these are used to give further detailed information about the part. When the section plane, as it is termed, coincides with part of the true surface of the object it is no longer a true section. It is usual to indicate the different materials used by means of a schedule in the bottom right-hand corner of the drawing.

Another case in which cross-hatching is omitted is that of a rib. This is illustrated in the sectional end elevation (Fig. 12.1*b*), the rib being left plain. Figure 12.3 indicates a simple bearing with an oil hole drilled in it. Sections on *AA* and *BB* are, as indicated by the arrows, to the left of the part, while those in *CC* and *DD* are to the right. *A, B* and *D* are true sections and are therefore cross-hatched, but at *CC* it will be noticed that the section plane coincides with surface of the flange and this portion is therefore not cross-hatched. Evidently any number of section planes may be chosen, but usually each is chosen so as to give some additional information essential to the manufacture of the part.

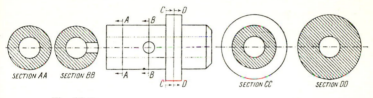

Fig. 12.3

In an elevation and section welds are usually blacked out thus:⎯▼⎯, while in a plan they are indicated by a series of parallel lines thus: ////////////, or by × × × ×.

Elevations and plans of butt and fillet welds are shown in Fig. 12.4.

WELDING SYMBOLS

A weld is indicated on a drawing by (1) a symbol (Fig. 12.4*a*) and (2) an arrow connected at an angle to a reference line (Fig. 12.4*b*). The side of the joint on which the arrow is placed is known as the 'arrow side'.

The method of use of the symbol and the meaning of reference line and position of the symbol regarding it are as follows.

If the weld symbol is placed *below* the reference line, the weld is to be made on the arrow side (Fig. 12.4*c*).

If the weld symbol is placed *above* the reference line, the weld is to be made on the other side from the arrow.

If the symbols are both above and below the reference line the welds are to be made on both sides (Fig. 12.4*d*).

If one plate only is to be prepared, the arrow points to that plate and the positioning of the symbol regarding the reference line indicates the side to be welded and thus the widest part of the preparation (Fig. 12.4*e*).

A circle at the junction of arrow and reference line indicates that the weld should be made 'all round' while a smaller filled-in circle in the same place indicates that the weld should be made 'on site' as opposed to 'in the shop' (Fig. 12.4*f*).

Intermittent runs of welding are indicated by figures denoting the welding portion and figures in brackets denoting the unwelded portions (Fig. 12.4*g*). A figure before the symbol for a fillet weld indicates the

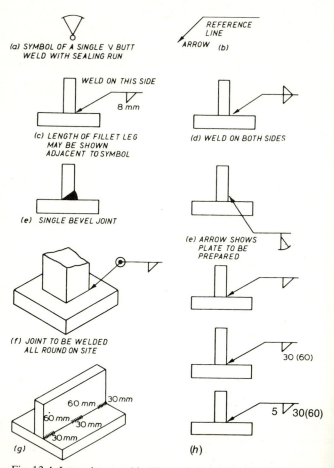

Fig. 12.4. Intermittent welds. Figures in brackets indicate the lengths not welded.

leg length (Fig. 12.4.*h*). If the intermittent welds are to be made regularly along the joint only the two figures, welded and not welded, need be given. Figure 12.5*a* and *b* gives some typical examples and the student should study BS 499, Part 2, *Symbols for welding*, for a complete account of this subject.

The following points will be found to be very helpful when making sketches of parts for fabrication or welding:

FORM OF WELD	SECTIONAL REPRESENTATION	APPROPRIATE SYMBOL	FORM OF WELD	SECTIONAL REPRESENTATION	APPROPRIATE SYMBOL
FILLET			DOUBLE J BUTT		
SQUARE BUTT			EDGE		
SINGLE V BUTT			SEALING RUN		
DOUBLE V BUTT			BACKING STRIP		
SINGLE U BUTT			SPOT		
DOUBLE U BUTT			SEAM		
SINGLE BEVEL BUTT			PROJECTION	BEFORE AFTER	
DOUBLE BEVEL BUTT			FLASH	ROD OR BAR TUBE	
SINGLE J BUTT			BUTT (RESISTANCE OR PRESSURE) WELDING	ROD OR BAR TUBE	

Fig. 12.5. (*a*) Welding symbols (BS 499).

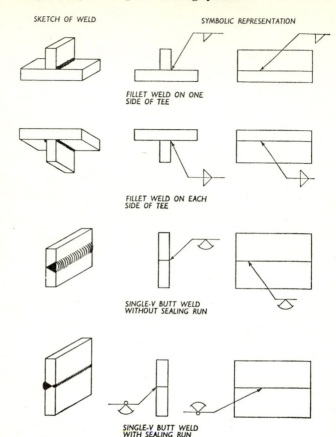

SKETCH OF WELD

SYMBOLIC REPRESENTATION

FILLET WELD ON ONE
SIDE OF TEE

FILLET WELD ON EACH
SIDE OF TEE

SINGLE-V BUTT WELD
WITHOUT SEALING RUN

SINGLE-V BUTT WELD
WITH SEALING RUN

Fig. 12.5. (*b*) Welding symbols (BS 499).

(1) Draw a centre line for each part and build up each part of the drawing from these centre lines.

(2) Always make them in orthographic projection (elevation, side elevation and plan), because they are simpler, clearer and easier to dimension.

(3) Make the sketch as large as possible and include all detail in its correct proportion, i.e. diameter of holes, radii of fillets etc.

(4) Dimension the sketch fully, as a missing dimension may hold up production, and run dimensions from line to line, all holes having centre lines.

(5) Use a hard pencil and good paper to ensure clarity.

13

Metallic alloys and equilibrium diagrams[1]

METALLIC ALLOYS

The metals with which a welder may have to deal are often not pure metals but alloys consisting of a parent metal with one or more alloying elements added in various proportions, e.g. mild steel is an alloy mainly of iron with a small amount of carbon, while brasses are alloys of copper containing up to 45% of zinc.

The properties of such alloys vary according to the form in which the alloying element is present. There are several possible forms:

(1) The alloying element may be present in an unchanged form in a state of fine mechanical mixture with the parent metal so that it can be seen as a separate particle or crystal (constituent) under the microscope.

Examples of this are:

- (a) Carbon present as flakes of graphite in grey cast irons.
- (b) Silicon present as fine silicon crystals in the aluminium silicon alloys.
- (c) Lead present as round particles in free cutting brasses. Alloying elements present in this condition do not generally produce a great increase in strength, but they increase or decrease the hardness, reduce the ductility and improve the machinability of the parent metal.

(2) The alloying element may be present in solid solution in the parent metal, i.e. actually dissolved as salt or sugar dissolves in water, so that under the microscope only one constituent, the solid solution, can be seen, similar in appearance to the parent metal except that the colour may be changed.

Examples of this are:

(a) Up to 37% of zinc can be present in solid solution in copper,

---

[1] This chapter will give the welding engineer an introduction to equilibrium diagrams and their uses.

causing its colour to change gradually from red to yellow as the percentage of zinc increases. These alloys are known as alpha brasses.

(*b*) Up to about 8% of tin or aluminium can be present in solid solution in copper giving alloys known as alpha tin bronzes or alpha aluminium bronzes respectively.

(*c*) Copper and nickel in any proportion form a solid solution of the same type. Examples are cupro-nickels containing 20–30% nickel and monel metal containing about 70% nickel. All have a similar appearance under the microscope showing a simple structure similar to that of a pure metal.

Elements in solid solution improve strength and hardness without making the metal brittle.

(3) The alloying element may form a compound with the parent metal which will then have properties different from either. Such compounds are generally hard and brittle and can only be present in small amounts and finely distributed without seriously impairing the properties of the alloy.

These appear under the microscope as distinct new constituents often of clearly crystalline shape. A second type of inter-metallic compound (a new constituent) having ductility and properties more like a solid solution forms in some alloys at certain ranges of composition. An example is the β constituent in brasses containing 40–50% zinc.

(4) While some pairs of metals are soluble in each other in all proportions, e.g. copper and nickel, other pairs can only retain a few per cent of each other in solid solution. If one of the alloying elements is present in a greater amount than the maximum which can be retained in solid solution in the parent metal, then part of it will be in solid solution and part will be present as a second constituent, i.e. either:

(*a*) as crystals of the alloying element which may themselves have a small amount of the parent metal in solid solution, or

(*b*) as an inter-metallic compound.

The state in which an alloying element is present can vary with the temperature so that the constituents present in an alloy at ordinary temperatures may be different from those present at the temperature of the welding operation. For all alloys used in practice, however, the constituents present at any temperature, from that of the welding range down to that of the room, have been determined by quenching the alloys from the various temperatures and examining them under the microscope. Each constituent which can be recognized as separable distinct particles or crystals is called a phase, and an alloy is said to be single-phase, two-phase or three-phase, etc., according to the number of phases present.

For any pair of metals, the different combinations of phases which may be produced by varying either the composition of the alloy, or the temperature, can be shown in an *equilibrium* or constitutional diagram for that pair of metals. In these diagrams composition is plotted horizontally and temperature vertically. The diagram is subdivided into a number of areas called phase-fields each of which is labelled with the phase or phases which occur within its limits. On the diagram are also plotted curves showing how the melting point and freezing point changes with the composition.

Thus a glance at the diagram will show for any composition, the phases present at any chosen temperature. Furthermore the diagram shows what changes will occur in any composition if it is cooled slowly from, say the welding temperature to normal air temperature.

EQUILIBRIUM DIAGRAMS AND THEIR USES

Two metals soluble in each other in all proportions

In Fig. 13.1 A is the melting point of pure copper and C that of pure nickel. The percentage composition (Cu and Ni) is plotted horizontally and the temperature vertically.

Above the curve ABC which is called the liquidus any mixture of copper and nickel will be a single liquid solution. This region is called a single-phase field. Below the curve ADC called the solidus any mixture of copper and nickel will consist of a single solid solution and under the microscope will show no difference between one composition and another except for a gradual change in colour from the red colour of copper to the white of nickel as the percentage of nickel is increased. The region below the curve ADC is therefore a single-phase field.

In the region between the curves ABC, ADC any alloy contains a mixture of:

(1) solid solution crystals, and
(2) liquid solution.

This region is therefore a two-phase field. There is a simple rule for determining the compositions and proportions of the liquid and solid solutions which are present together at any selected temperature in a two-phase field. Let X be the composition of any selected alloy (Fig. 13.2) and let $t°$ be selected temperature.

Draw a horizontal line at temperature $t°$ to cut the liquidus at O and the solidus at Q. Then O is the composition of the liquid solution (say approx. 28% Ni, 72% Cu) and Q the composition of the solid solution

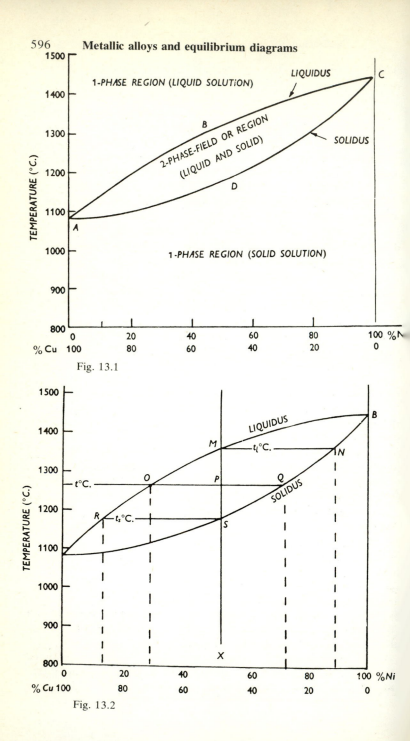

Fig. 13.1

Fig. 13.2

(say 73% Ni, 27% Cu) present in the alloy X at temperature $t°$. The relative amounts of liquid O and solid P are given by the lengths of PQ and OP respectively.

When the alloy of composition X cools from the liquid condition to normal temperatures, the changes involved may be summarized as follows. No change occurs until the alloy cools to the liquidus temperature $t_1°$. When $t_1°$ is reached a small amount of solid solution of composition N (say 90% Ni, 10% Cu) is formed. On further cooling these crystals absorb more copper from the liquid and more crystals separate out but their composition changes progressively along the solidus line towards S getting richer in copper, while the composition of the remaining liquid changes progressively along the liquidus line towards composition R. The amount of solid present therefore increases and the amount of liquid decreases until when temperature $t_s°$ is reached (this is the solidus temperature of our alloy of composition X) the last drop of liquid solidifies and the whole alloy is then a uniform solid solution of composition X. No further changes occur on cooling to normal temperatures, and under the microscope the cooled alloy will appear to consist of polygonal grains of one kind only, differing from a pure metal only in the colour.

This simplified description applies only when the alloy solidifies very slowly, for time is required for the first crystals deposited (composition N) to absorb copper by diffusion so that they change progressively along NS as solidification occurs. When cooling is rapid there is not sufficient time for these changes to occur in the solidified crystals and we get the first part of each crystal having composition N so that the centre of each crystal or grain is richer in nickel than the average while as we progress from the centre to the outside the composition gradually changes, becoming richer in copper until at the boundaries of each grain they are richer in copper than the average composition X. This effect is known as *coring* and is shown as a gradual change in colour of the crystals when the alloy is etched with a suitable chemical solution and examined under the microscope. All solid solution alloys show coring to a greater or lesser degree when they are solidified at normal rates, for example in a casting or welding operation.

Coring can be removed and the composition of each grain made uniform throughout by re-heating or annealing at a temperature just below the solidus of the alloy.

Two metals partially soluble in each other in the solid state, which do not form compounds

This type of alloy system is very common, e.g. copper–silver, lead–tin, aluminium–silicon. It is called a *eutectic system* and at the eutectic composition, that is, at the minimum point on the liquidus curve, the alloy solidifies at a constant temperature instead of over a range of temperatures.

In the solid state there are three phase-fields (Fig. 13.3):

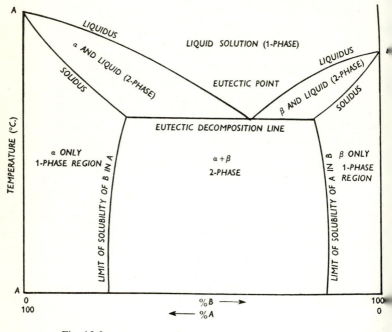

Fig. 13.3

(1) A single-phase region in which any composition consists of a solid solution of metal B in metal A (a only).
(2) A single-phase region in which any composition consists of a solid solution of metal A in metal B (β only).
(3) A two-phase region in which any alloy consists of a mechanical mixture of the two solid solutions mentioned above ($a + \beta$).

Solidification of alloys forming a eutectic system. (1) Alloys containing less than $X\%$ or more than $Z\%$ of metal B will solidify as solid solutions similar to those of the copper–nickel system. In Fig. 13.4 consider an alloy of composition at (1). As it cools, crystals of composition S_1 will begin to separate out at L_1. On further cooling the liquid will change in composition along the liquidus to L_2 and the solid deposited will change along the solidus to S_2. At S_2 the last drop of liquid will solidify and if cooling has been slow, the grains will be a uniform solid solution of composition (1); with rapid cooling the crystal grains will be cored, the centres richer in metal A and the boundaries richer in metal B.

(2) The eutectic alloy will cool as a liquid solution to Y. At Y, crystals of two solid solutions will separate simultaneously, their compositions being given by the points Z and X. Solidification will be completed at constant temperature and the solidified alloy will consist of a fine mechanical mixture of the two solid solutions, called a eutectic structure.

(3) An alloy of composition between X and Y, or between Y and Z will begin to solidify as a solid solution but when the composition of the liquid has reached the point Y, the remainder will solidify as a fine eutectic mixture of X and Z.

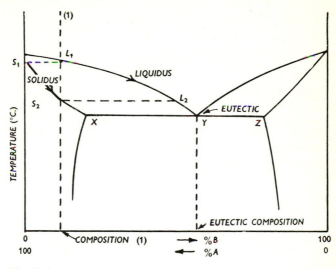

Fig. 13.4

Two metals which form intermetallic compounds

Compound forming a eutectic system with the parent metal. The equilibrium diagram for many alloying elements which form compounds with the parent metal, e.g. copper alloyed to aluminium, is of a simple eutectic form similar to that just considered for the important part of the diagram, and the changes occurring on cooling an alloy of such a system are similar to those given for Class 2. (In Fig. 13.5, which is for Al–Cu alloy, the intermetallic compound θ formed is $CuAl_2$.)

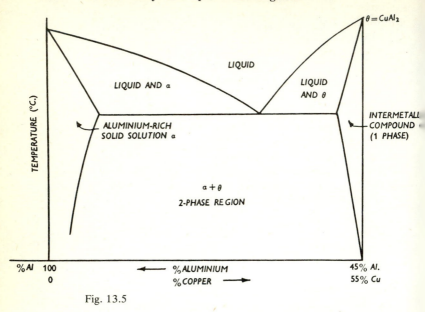

Fig. 13.5

System in which an intermetallic compound is formed by a reaction. In some alloy systems, notably those in which copper is the parent metal, an intermediate phase or inter-metallic compound is formed as a product of a reaction which occurs in certain compositions during solidification. An example is that of the copper–zinc system in which a β phase is formed by a peritectic reaction.

Solidification of alloys in a peritectic system. Referring to Fig. 13.6, alloys containing less than 32.5% of zinc at A, solidify as α solid solutions by the process described for Class 1. Alloys containing between 32.5% and 39% zinc (A and C in the figure) commence to solidify by forming crystals of the α solid solution, but when the temperature has fallen to 905 °C, the liquid is of composition at C (39% Zn) and the α crystals are of composition at A (32.5% Zn). At this temperature the α phase and the liquid react to form a new solution β having a composition at B (37% Zn). This reaction is termed a *peritectic reaction*.

If the mean composition of the alloy is B% zinc (37%) then all the liquid and the α crystals are converted to β solution and at the end of solidification the alloy is a uniform solid solution β.

If the alloy contains between A% and B% of zinc, the alloy after solidification consists of a mixture of α and β. If the alloy contains between B% and C% of zinc, all the α crystals are converted to β at the

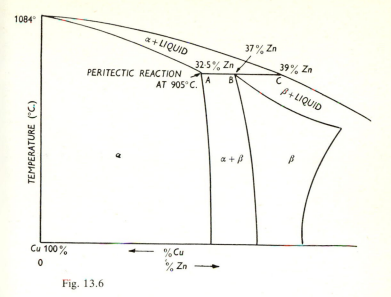

Fig. 13.6

peritectic reaction, but there is some liquid left which then solidifies as β by the usual mechanism of solidification for a solid solution.

Phase changes in alloys in the solid state
There are two important types of phase change which may occur in alloys while cooling in the *solid* state.

Simple solubility change. In many alloy systems the amount of the alloying element which can be kept in solid solution in the parent metal decreases as the temperature falls.

This effect is shown in the equilibrium diagram Fig. 13.7 by the slope of the solubility line, that is the line which marks the limit of the single-phase region of the α solid solution.

Thus in the copper–aluminium system, aluminium will dissolve 5.7% of copper at the eutectic temperature but the solubility falls to less than 0.5% copper at ordinary temperatures.

The alloys containing up to 5.7% copper are α solid solutions at temperatures above the line AB, but below the line AB they consist of α solid solution with varying amounts of an inter-metallic compound θ.

If the alloys of composition at X (say 4% copper) is annealed above the line AB say at 530 °C and quenched in water, a super-saturated solid solution is obtained. If, on the other hand, it is cooled slowly, on reaching the line AB, crystals of the inter-metallic compound θ begin to

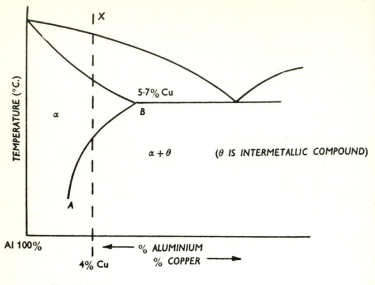

Fig. 13.7

separate out (or are precipitated) and increase in size and number on further cooling.

If a quenched specimen is retained at room temperature, the excess copper in solution tends to diffuse out to form separate θ crystals and this process produces severe hardening of the alloy known as *age hardening*. No visible change in the microstructure occurs.

If a quenched specimen is heated to 100–200 °C the diffusion and hardening occur more quickly. This is known as *temper hardening*, but if the heating is too prolonged, visible crystals of θ separate out and the alloy re-softens. This is known as *over-ageing*.

Duralumin, hiduminium and Y alloy are typical alloys which age harden.

Eutectoid change. In steels the remarkable changes in properties which can be obtained by different types of heat treatment are the result of a different type of transformation in the solid state. This is known as a *eutectoid transformation*.

Some solid solutions can only exist at high temperatures and on cooling they decompose to form a fine mixture of two other phases. This change, known as a *eutectoid* decomposition, is similar to a eutectic decomposition (see p. 598) except that the solution which decomposes is a solid one instead of a liquid one. The structure produced is also similar in appearance under a microscope to a eutectic structure but is in general finer.

In Fig. 13.8 the β phase undergoes eutectoid decomposition at temperature $t_1°$ to form a eutectoid mixture of α phase + Y phase.

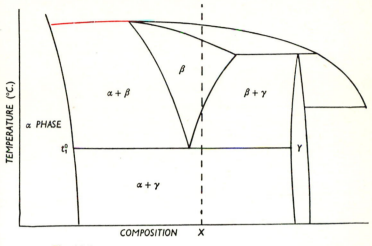

Fig. 13.8

If an alloy, which contains some β phase at high temperatures, say composition X in the figure, is quenched from a temperature higher than t_1, the decomposition of β is prevented and severe hardening is produced. If the alloy X is now heated to some temperature below t_1 the change of β to $\alpha + Y$ can occur and the hardening is removed. This is then a tempering process.

In some systems in which a change of this type occurs it is possible to obtain a wide range of hardness values in a single alloy according to the rate of cooling from above the decomposition temperature. The *slower* the cooling rate the coarser is the ($\alpha + Y$) eutectoid mixture and the lower the hardness, and vice versa, the quench giving the highest hardness obtainable.

Appendix

TABLES

Elements: their symbols, atomic weights and melting points

Element	Symbol	Atomic weight	Melting point (°C)
Actinium	Ac	227	—
Aluminium	Al	26.97	658.7
Americium	Am	241	—
Antimony	Sb	121.77	630
Argon	Ar	39.94	−188
Arsenic	As	74.96	850
Astatine	At	211	—
Barium	Ba	137.36	850
Berkelium	Bk	245	—
Beryllium	Be	9.02	1280
Bismuth	Bi	209.00	271
Boron	B	10.82	2200–2500
Bromine	Br	79.91	−7.3
Cadmium	Cd	112.41	320.9
Caesium	Cs	132.81	26
Calcium	Ca	40.07	810.0
Californium	Cf	246	—
Carbon	C	12.00	3600
Cerium	Ce	140.13	635
Chlorine	Cl	35.45	−101.5
Chromium	Cr	52.01	1615
Cobalt	Co	58.94	1480
Copper	Cu	63.57	1083
Curium	Cm	242	—
Dysprosium	Dy	162.5	—
Erbium	Er	167.64	—
Europium	Eu	152	—

Elements: their symbols, atomic weights and melting points (contd.)

Symbol	Element	Atomic weight	Melting point (°C)
Fluorine	F	19.0	−223
Francium	Fa	223	—
Gadolinium	Gd	157.26	—
Gallium	Ga	69.72	30.1
Germanium	Ge	72.60	958
Gold	Au	197.2	1063
Hafnium	Hf	179	2200
Helium	He	4.00	−272
Holmium	Ho	165	—
Hydrogen	H	1.0078	−259
Indium	In	114.8	155
Iodine	I	126.932	113.5
Iridium	Ir	193.1	2350
Iron	Fe	55.84	1530
Krypton	Kr	83.7	−169
Lanthanum	La	138.90	810
Lead	Pb	207.22	327.4
Lithium	Li	6.94	186
Lutecium	Lu	175	—
Magnesium	Mg	24.32	651
Manganese	Mn	54.93	1230
Mercury	Hg	200.61	−38.87
Molybdenum	Mo	96	2620
Neodymium	Nd	144.27	840
Neon	Ne	20.18	−253
Neptunium	Np	237	—
Nickel	Ni	58.69	1452
Niobium (Columbium)	Nb(Cb)	92.9	1950
Nitrogen	N	14.008	−210
Osmium	Os	190.8	2700
Oxygen	O	16.000	−218
Palladium	Pd	106.7	1549
Phosphorus	P	30.98	44
Platinum	Pt	195.23	1755
Plutonium	Pn	239	—
Polonium	Po	210	—
Potassium	K	39.1	62.3
Praseodymium	Pr	140.92	940
Promethium	Pm	147	—

Elements: their symbols, atomic weights and melting points (contd.)

Element	Symbol	Atomic weight	Melting point (°C)
Protactinium	Pa	231	—
Radon	Rn	222	−71
Radium	Ra	226.1	700
Rhenium	Re	186	3167
Rhodium	Rh	102.91	1950
Rubidium	Rb	85.44	38
Ruthenium	Ru	101.7	2450
Samarium	Sm	150.43	1300–1400
Scandium	Sc	45.10	1200
Selenium	Se	78.96	217–220
Silicon	Si	28.06	1420
Silver	Ag	107.88	960.5
Sodium	Na	22.997	97.5
Strontium	Sr	87.63	800
Sulphur	S	32.06	112.8
Tantalum	Ta	181.5	2900
Technetium	Tc	99	—
Tellurium	Te	127.5	452
Terbium	Tb	159.2	—
Thallium	Tl	204.39	302
Thorium	Th	232.12	1700
Tin	Sn	118.70	231.9
Thulium	Tm	169.4	—
Titanium	Ti	47.9	1800
Tungsten	W	184.0	3400
Uranium	U	238.14	1850
Vanadium	V	50.96	1720
Xenon	Xe	131.3	−140
Ytterbium	Yb	173.6	1800
Yttrium	Y	88.92	1490
Zinc	Zn	65.38	419.4
Zirconium	Zr	91.22	1700

Gauge table. Imperial standard

No.	Size (in)	Size (mm)
0	0.324	8.229
1	0.300	7.620
2	0.276	7.010
3	0.252	6.401
4	0.232	5.893
5	0.212	5.385
6	0.192	4.877
7	0.176	4.470
8	0.160	4.064
9	0.144	3.658
10	0.128 approx. $\frac{1}{8}$ in	3.251
11	0.116	2.946
12	0.104	2.642
13	0.092	2.337
14	0.080	2.032
15	0.072	1.829
16	0.064 approx. $\frac{1}{16}$ in	1.626
17	0.056	1.422
18	0.048	1.219
19	0.040	1.016
20	0.036	0.914
21	0.032 approx. $\frac{1}{32}$ in	0.813
22	0.028	0.711
23	0.024	0.610
24	0.022	0.559
25	0.020	0.508
26	0.018	0.457
27	0.0164	0.4166
28	0.0148	0.3759
29	0.0136	0.3454
30	0.0124	0.315

Millimetres to inches

mm	0	1	2	3	in 4	5	6	7	8	9
0	——	0.03937	0.07874	0.11811	0.15748	0.19685	0.23622	0.27559	0.31496	0.35433
10	0.39370	0.43307	0.47244	0.51181	0.55118	0.59055	0.62992	0.66929	0.70866	0.74803
20	0.78740	0.82677	0.86614	0.90551	0.94488	0.98425	1.02362	1.06299	1.10236	1.14173
30	1.18110	1.22047	1.25984	1.29921	1.33858	1.37795	1.41732	1.45669	1.49606	1.53543
40	1.57480	1.61417	1.65354	1.69291	1.73228	1.77165	1.81102	1.85039	1.88976	1.92913
50	1.96850	2.00787	2.04724	2.08661	2.12598	2.16535	2.20472	2.24409	2.28346	2.32283
60	2.36220	2.40157	2.44094	2.48031	2.51969	2.55906	2.59843	2.63780	2.67717	2.71654
70	2.75591	2.79528	2.83465	2.87402	2.91339	2.95276	2.99213	3.03150	3.07087	3.11024
80	3.14961	3.18898	3.22835	3.26772	3.30709	3.34646	3.38573	3.42520	3.46457	3.50394
90	3.54331	3.58268	3.62205	3.66142	3.70079	3.74016	3.77953	3.81890	3.85827	3.89764
100	3.93701	3.97638	4.01575	4.05512	4.09449	4.13386	4.17323	4.21260	4.25197	4.29134
110	4.33071	4.37008	4.40945	4.44882	4.48819	4.52756	4.56693	4.60630	4.64567	4.68504
120	4.72441	4.76378	4.80315	4.84252	4.88189	4.92146	4.96063	5.00000	5.03937	5.07874
130	5.11811	5.15748	5.19685	5.23622	5.27559	5.31496	5.35433	5.39370	5.43307	5.47244
140	5.51181	5.55118	5.59055	5.62992	5.66929	5.70866	5.74803	5.78740	5.82677	5.86614
150	5.90551	5.94488	5.98425	6.02362	6.06229	6.10236	6.14173	6.18110	6.22047	6.25984
160	6.29921	6.33858	6.37795	6.41732	6.45669	6.49606	6.53543	6.57480	6.61417	6.65354
170	6.69291	6.73228	6.77165	6.81102	6.85039	6.88976	6.92913	6.96850	7.00787	7.04724
180	7.08661	7.12598	7.16535	7.20472	7.24409	7.28346	7.32283	7.36220	7.40157	7.44094
190	7.48031	7.51969	7.55906	7.59843	7.63780	7.67717	7.71654	7.75591	7.79528	7.83465
200	7.87402	7.91339	7.95276	7.99213	8.03150	8.07087	8.11024	8.14961	8.18898	8.22835
210	8.26772	8.30709	8.34646	8.38583	8.42520	8.46457	8.50394	8.54331	8.58268	8.62205
220	8.66142	8.70079	8.74016	8.77953	8.81890	8.85827	8.89764	8.93701	8.97638	9.01575
230	9.05512	9.09449	9.13386	9.17323	9.21260	9.25197	9.29134	9.33071	9.37008	9.40945
240	9.44882	9.48819	9.52756	9.56693	9.60630	9.64567	9.68504	9.72441	9.76378	9.80315

Thousandths of an inch to millimetres

Mils*	0	1	2	3	mm 4	5	6	7	8	9
0	—	0.0254	0.0508	0.0762	0.1016	0.1270	0.1524	0.1778	0.2032	0.2286
10	0.2540	0.2794	0.3048	0.3302	0.3556	0.3810	0.4064	0.4318	0.4572	0.4826
20	0.5080	0.5334	0.5588	0.5842	0.6096	0.6350	0.6604	0.6858	0.7112	0.7366
30	0.7620	0.7874	0.8128	0.8382	0.8636	0.8890	0.9144	0.9398	0.9652	0.9906
40	1.0160	1.0414	1.0668	1.0922	1.1176	1.1430	1.1684	1.1938	1.2192	1.2446
50	1.2700	1.2954	1.3208	1.3462	1.3716	1.3970	1.4224	1.4478	1.4732	1.4986
60	1.5240	1.5494	1.5748	1.6002	1.6256	1.6510	1.6764	1.7018	1.7272	1.7526
70	1.7780	1.8034	1.8288	1.8542	1.8796	1.9050	1.9304	1.9558	1.9812	2.0066
80	2.0320	2.0574	2.0828	2.1082	2.1336	2.1590	2.1844	2.2098	2.2352	2.2606
90	2.2860	2.3114	2.3368	2.3622	2.3876	2.4130	2.4384	2.4638	2.4892	2.5146
100	2.5400	2.5654	2.5908	2.6162	2.6416	2.6670	2.6924	2.7178	2.7432	2.7686
110	2.7940	2.8194	2.8448	2.8702	2.8956	2.9210	2.9464	2.9718	2.9972	3.0226
120	3.0480	3.0734	3.0988	3.1242	3.1496	3.1750	3.2004	3.2258	3.2512	3.2766
130	3.3020	3.3274	3.3528	3.3782	3.4036	3.4290	3.4544	3.4798	3.5052	3.5306
140	3.5560	3.5814	3.6068	3.6322	3.6576	3.6830	3.7084	3.7338	3.7592	3.7846
150	3.8100	3.8354	3.8608	3.8862	3.9116	3.9370	3.9624	3.9878	4.0132	4.0386
160	4.0640	4.0894	4.1148	4.1402	4.1656	4.1910	4.2164	4.2418	4.2672	4.2926
170	4.3180	4.3434	4.3688	4.3942	4.4196	4.4450	4.4704	4.4958	4.5121	4.5466
180	4.5720	4.5974	4.6228	4.6482	4.6736	4.6990	4.7244	4.7498	4.7752	4.8006
190	4.8260	4.8514	4.8768	4.9022	4.9276	4.9530	4.9784	5.0038	5.0292	5.0546
200	5.0800	5.1054	5.1308	5.1562	5.1816	5.2070	5.2324	5.2578	5.2832	5.3086
210	5.3340	5.3594	5.3848	5.4102	5.4356	5.4610	5.4864	5.5118	5.5372	5.5626
220	5.5880	5.6134	5.6388	5.6642	5.6896	5.7150	5.7404	5.7658	5.7912	5.8166
230	5.8420	5.8674	5.8928	5.9182	5.9436	5.9690	5.9944	6.0198	6.0452	6.0706
240	6.0960	6.1214	6.1468	6.1722	6.1976	6.2230	6.2484	6.2738	6.2992	6.3246
250	6.3500	6.3754	6.4008	6.4262	6.4516	6.4770	6.5024	6.5278	6.5532	6.5786
260	6.6040	6.6294	6.6548	6.6802	6.7056	6.7310	6.7564	6.7818	6.8072	6.8326
270	6.8580	6.8834	6.9088	6.9342	6.9596	6.9850	7.0104	7.0358	7.0612	7.0866
280	7.1120	7.1374	7.1628	7.1882	7.2136	7.2390	7.2644	7.2898	7.3152	7.3406
290	7.3660	7.3914	7.4168	7.4422	7.4676	7.4930	7.5184	7.5483	7.5692	7.5946
300	7.6200	7.6454	7.6708	7.6962	7.7216	7.7470	7.7724	7.7978	7.8232	7.8486
310	7.8740	7.8994	7.9248	7.9502	7.9756	8.0010	8.0264	8.0518	8.0772	8.1026
320	8.1280	8.1534	8.1788	8.2042	8.2296	8.2550	8.2804	8.3058	8.3312	8.3566
330	8.3820	8.4074	8.4328	8.4582	8.4836	8.5090	8.5344	8.5598	8.5852	8.6106
340	8.6360	8.6614	8.6868	8.7122	8.7376	8.7630	8.7884	8.8138	8.8392	8.8646
350	8.8900	8.9154	8.9408	8.9662	8.9916	9.0170	9.0424	9.0678	9.0932	9.1186
360	9.1440	9.1694	9.1948	9.2202	9.2456	9.2710	9.2964	9.3218	9,3472	9.3726
370	9.3980	9.4234	9.4488	9.4742	9.4996	9.5250	9.5504	9.5758	9.6012	9.6266
380	9.6520	9.6774	9.7028	9.7282	9.7536	9.7790	9.8044	9.8298	9.8552	9.8806
390	99060	9.9314	9.9568	9.9822	10.0076	10.0330	10.0584	10.0838	10.1092	10.1346

* 1 Mil = 0.001 inch.

hbar to tonf/in², MN/m², lbf in² and kgf/mm²

hbar	tonf/m²	MN/m² N/mm²	lbf/in²	kgf/mm²	hbar	tonf/in²	MN/m² N/mm²	lbf/in²	kgf/mm²
0.5	0.3	5	700	0.5	30.5	19.7	305	44200	31.1
1	0.6	10	1500	1.0	31	20.1	310	45000	31.6
1.5	1.0	15	2200	1.5	31.5	20.4	315	45700	32.1
2	1.3	20	2900	2.0	32	20.7	320	46400	32.6
2.5	1.6	25	3600	2.5	32.5	21.0	325	47100	33.1
3	1.9	30	4400	3.1	33	21.4	330	47900	33.7
3.5	2.3	35	5100	3.6	33.5	21.7	335	48600	34.2
4	2.6	40	5800	4.1	34	22.0	340	49300	34.7
4.5	2.9	45	6500	4.6	34.5	22.3	345	50000	35.2
5	3.2	50	7300	5.1	35	22.7	350	50800	35.7
5.5	3.6	55	8000	5.6	35.5	23.0	355	51500	36.2
6	3.9	60	8700	6.1	36	23.3	360	52200	36.7
6.5	4.2	65	9400	6.6	36.5	23.6	365	52900	37.2
7	4.5	70	10200	7.1	37	24.0	370	53700	37.7
7.5	4.9	75	10900	7.6	37.5	24.3	375	54400	38.2
8	5.2	80	11600	8.2	38	24.6	380	55100	38.7
8.5	5.5	85	12300	8.7	38.5	24.9	385	55800	39.3
9	5.8	90	13100	9.2	39	25.3	390	56600	39.8
9.5	6.2	95	13800	9.7	39.5	25.6	395	57300	40.3
10	6.5	100	14500	10.2	40	25.9	400	58000	40.8
10.5	6.8	105	15200	10.7	40.5	26.2	405	58700	41.3
11	7.1	110	16000	11.2	41	26.5	410	59500	41.8
11.5	7.4	115	16700	11.7	41.5	26.9	415	60200	42.3
12	7.8	120	17400	12.2	42	27.2	420	60900	42.8
12.5	8.1	125	18100	12.7	42.5	27.5	425	61600	43.3
13	8.4	130	18900	13.3	43	27.8	430	62400	43.8
13.5	8.7	135	19600	13.8	43.5	28.2	435	63100	44.4
14	9.1	140	20300	14.3	44	28.5	440	63800	44.9
14.5	9.4	145	21000	14.8	44.5	28.8	445	64500	45.4
15	9.7	150	21800	15.3	45	29.1	450	65300	45.9
15.5	10.0	155	22500	15.8	45.5	29.5	455	66000	46.4
16	10.4	160	23200	16.3	46	29.8	460	66700	46.9
16.5	10.7	165	23900	16.8	46.5	30.1	465	67400	47.4
17	11.0	170	24700	17.3	47	30.4	470	68200	47.9
17.5	11.3	175	25400	17.8	47.5	30.8	475	68900	48.4
18	11.7	180	26100	18.4	48	31.1	480	69600	48.9
18.5	12.0	185	26800	18.9	48.5	31.4	485	70300	49.5
19	12.3	190	27600	19.4	49	31.7	490	71100	50.0
19.5	12.6	195	28300	19.9	49.5	32.1	495	71800	50.5
20	12.9	200	29000	20.4	50	32.4	500	72500	51.0
20.5	13.3	205	29700	20.9	50.5	32.7	505	73200	51.5
21	13.6	210	30500	21.4	51	33.0	510	74000	52.0
21.5	13.9	215	31200	21.9	51.5	33.3	515	74700	52.5
22	14.2	220	31900	22.4	52	33.7	520	75400	53.0
22.5	14.6	225	32600	22.9	52.5	34.0	525	76100	53.5
23	14.9	230	33400	23.5	53	34.3	530	76900	54.0
23.5	15.2	235	34100	24.0	53.5	34.6	535	77600	54.6
24	15.5	240	34800	24.5	54	35.0	540	78300	55.1
24.5	15.9	245	35500	25.0	54.5	35.3	545	79000	55.6
25	16.2	250	36300	25.5	55	35.6	550	79800	56.7
25.5	16.5	255	37000	26.0	55.5	35.9	555	80500	56.6
26	16.8	260	37700	26.5	56	36.3	560	81200	57.1
26.5	17.2	265	38400	27.0	56.5	36.6	565	81900	57.6
27	17.5	270	39200	27.5	57	36.9	570	82700	58.1
27.5	17.8	275	39900	28.0	57.5	37.2	575	83400	58.6
28	18.1	280	40600	28.6	58	37.6	580	84100	59.1
28.5	18.5	285	41300	29.1	58.5	37.9	585	84800	59.7
29	18.8	290	42100	29.6	59	38.2	590	85600	60.2
29.5	19.1	295	42800	30.1	59.5	38.5	595	86300	60.7
30	19.4	300	43500	30.6	60	38.8	600	87000	61.2

Factors

To convert	Multiply by	To convert	Multiply by
in to mm	25.4	mm to in	0.03937
in² to mm²	645.16	mm² to in ²	0.00155
in³ to cm³	16.387	cm³ to in³	0.061024
in⁴ to cm⁴	41.623	cm⁴ to in⁴	0.024025
ft to m	0.3048	m to ft	3.2808
ft² to m²	0.092903	m² to ft²	10.764
ft³ to m³	0.028317	m³ to ft³	35.315
yd to m	0.9144	m to yd	1.0936
yd² to m²	0.83613	m² to yd²	1.1960
yd³ to m³	0.76456	m³ to yd³	1.3080
lb to kg	0.45359	kg to lb	2.2046
cwt to kg	50.802	kg to cwt	0.019684
tons to kg	1016.1	kg to tons	0.00098421
tons to lb	2240	lb to tons	0.0004464
tons to short tons	1.12	short tons to tons	0.8929
tons to tonnes (metric)	1.0160	tonnes (metric) to tons	0.98421
lb/in² to kg/mm²	0.0007031	kg/mm² to lb/in²	1422.33
lb/in² to kg/cm²	0.07031	kg/cm² to lb/in²	14.2233
lb/in² to tons/in²	0.0004464	tons/in² to lb/in²	2240
lb/in³ to g/cm³	27.680	g/cm³ to lb/in³	0.036127
lb/ft to kg/m	1.4882	kg/m to lb/ft	0.67197
lb/ft² to kg/m²	4.8824	kg/m² to lb/ft²	0.2048
tonf/in² to MN/m² or MPa or N/mm²	15.444	MPa or MN/m² to tonf/in²	0.064749
tonf/in² to hbar	1.5444	hbar to tonf/in²	0.64749
tonf/in² to kgf/mm²	1.5749	kgf/mm² to tonf/in²	0.63497
lbf/ft² to N/m² or Pa	47.880	Pa or N/m² to lbf/ft²	0.02089
lbf/in² to N/m² or Pa	6894.8	Pa or N/m² to lbf/in²	0.00014504
lbf/in² to hbar or Pa	0.00068948	hbar to lbf/in²	1450.4
kgf/m² to N/m² or Pa	9.8067	Pa or N/m² to kgf/m²	0.10197
kgf/mm² to hbar	0.98067	hbar to kgf/mm²	1.0197
cal cm/cm² s°C to W/m °C	418.68	W/m °C to cal cm/cm² s °C	0.0023885
$\mu\Omega$ m to Ω m	10^{-6}	Ω m to $\mu\Omega$ cm	10^8

Other conversions

Multiply by

Density

Pounds/cubic inch to kilograms/cubic metre	27680
Pounds/cubic foot to kilograms/cubic metre	16.018
Tons/cubic yard to kilograms/cubic metre	1328.9

Force

Poundals to newtons	0.13825
Pounds force to newtons	4.448
Tons force to newtons	9964

Torque

Pounds force-inch to newton-metres	0.11298
Pounds force-feet to newton-metres	1.3558
Tons force-feet to newton-metres	3037
Inches of mercury to millibars	33.864
Inches of mercury to newtons per square metre or pascal	3386.4

Note. 1 bar = 10^5 newtons per square metre or 10 Pa

Work, Energy

Therms to mega-joules	105.5
Kilowatt-hours to mega-joules	3.6
British thermal unit to joules (metre-newtons)	1055.1
Centigrade heat unit to joules (metre-newtons)	1899.2
Foot pound f. to joules (metre-newtons)	1.3558

Power

Horse-power to watts (joules per sec)	745.7
Foot pound f. per sec to watts (joules per sec)	1.3558

Illumination

Lumens per square foot ⎫ to lumens per square	
Foot candles ⎭ metre (lux)	10.764

Angular measurement

Radians to degree	57.29
Degrees to radians	0.01745

Table giving some of the chief types of oxy-acetylene welding rods available for welding carbon and alloy steels

(See pp. 199–218 for technique)

Rod	Description and use
Low-carbon steel	A general purpose rod for mild steel. Easily filed and machined. Deposit can be case-hardened
High-tensile steel	Gives a machinable deposit of greater tensile strength than the previous rod. Can be used instead of the above wherever greater strength is required
High-carbon steel	Gives a deposit which is machinable as deposited, but which can be heat treated to give a hard abrasion-resisting surface. When used for welding broken parts, these should be of high carbon steel, and heat treatment given after welding

Table giving some of the chief types of oxy-acetylene welding rods available for welding carbon and alloy steels (contd.)

Rod	Description and use
High-nickel steel ($3\frac{1}{2}$–4%)	Produces a machinable deposit with good wear-resisting properties. Suitable for building up teeth in gear and chain wheels, splines and keyways in shafts, etc.
Wear-resisting steel (12–14% manganese)	Gives a dense tough unmachinable deposit which must be ground or forged into shape, and can be heat treated. Useful for building up worn sliding surfaces, cam profiles, teeth on excavators, tracks, etc.
Stellite	For hard surfacing and wear- and abrasion-resisting surfaces
Chrome–molybdenum steel (creep resisting)	High-tensile alloy steel deposit for pressure vessels and high-pressure steam pipes, etc. Rod should match the analysis of the parent metal
Chrome–vanadium steel	A high-tensile alloy rod for very highly stressed parts
Tool steel	Suitable for making cutting tools by tipping the ends of mild or low-carbon steel shanks
Stainless steel	Decay-proof. Rod should match the analysis of the parent metal

Table giving some of the chief types of metal arc welding electrodes available for welding alloy steels

Electrode	Description and uses
Mild steel heavy duty	Rutile or basic coated for medium and heavy duty fabrications. The latter suitable for carbon and alloy steels and mild steel under restraint and for thick sections and root runs in thick plate.
High tensile alloy steels	Austenitic rutile or basic coated for high tensile steels including armour plate and joints between low-alloy and stainless steel and in conditions where pre-heat is not possible to avoid cracking.
Structural steels	Basic coated for high strength structural steels 300–425 N/mm² tensile strength and for copper bearing weathering quality steels, e.g. Corten.
Notch ductile steels for low temperature service	Basic coated for steels containing 2–5% Ni, 3% Ni and carbon manganese steels. Also austenitic high nickel electrodes for 9% Ni steel for service to − 196 °C and for dissimilar metal welding and for high Ni–Cr alloys for use at elevated temperatures.
Creep-resisting steels	Ferritic, basic coating for (1) 1.25% Cr, 0.5% Mo, (2) 2.5% Cr, 1.0% Mo, (3) 4–6% Cr, 0.5% Mo, steels with pre- and post-heat. Austenitic ferrite controlled for creep-resistant steels and for thick stainless steel sections requiring prolonged heat treatment after welding.
Heat- and corrosion-resisting steels	(1) Rutile or basic coating 19% Cr 9% Ni for extra low carbon stainless steels. (2) Basic coating Nb stabilized for plain or Ti or Nb stabilized 18/8 stainless steels and for a wide range of corrosive- and heat-resisting applications. Variations of these electrodes are for positional welding, smooth finish, and for high deposition rates. (3) Rutile or basic coating Mo bearing for 18/10 Mo steels. (4) Rutile or basic coating, low-carbon austenitic electrodes for low-carbon Mo bearing stainless steels and for welding mild to stainless steel. (5) Basic coating austenitic for heat-resisting 25% Cr, 12% Ni steels, for welding mild and low-alloy steels to stainless steel and for joints in stainless-clad mild steel. (6) Rutile coating austenitic for 23% Cr, 11% Ni heat-resistant steels containing tungsten. (7) Basic coating 25% Cr, 20% Ni (non-magnetic) for welding austenitic 25/20 steels and for mild and low-alloy steels to stainless steel under mild restraint. (8) Basic coating 60% Ni, 15% Cr for high-nickel alloys of similar composition (incoloy DS, cronite, etc.) and for welding these alloys and stainless steel to mild steel in low restraint conditions. (9) A range of electrodes for high-nickel, monel, inconel and incoloy welding. These electrodes are also suitable for a wide range of dissimilar metal welding in these alloys.
Hard-surfacing and abrasion-resisting alloys	(1) High impact moderate abrasion-resistance rutile coating for rebuilding carbon steel rails, shafts, axles and machine parts subject to abrasive wear. 250 HV. (2) Medium impact medium abrasion-resistance rutile or basic coating for rebuilding tractor links and rollers, roller shafts, blades, punching die sets, reasonably machinable. The basic coated electrode gives maximum resistance to underbead cracking and is suitable for resurfacing low-alloy and hardenable steels. 360 HV. (3) High abrasion medium impact. (a) Rutile or basic coating for hard-surfacing bull-dozer blades, excavator teeth, bucket lips, etc. The basic coated electrode has greater resistance to underbead cracking and eliminates the need for a buffer layer on high-carbon steels. 650 HV. Unmachinable. (b) Tubular electrodes depositing chromium carbide. For worn carbon steel such as dredger buckets, excavator shovels etc. Matrix 560 HV. Carbide 1400 HV. (4) Severe abrasion moderate impact. (a) Tubular electrode depositing fused tungsten carbide giving highest resistance to abrasion with moderate impact resistance. Matrix 600 HV. Carbide 1800 HV. (b) High-alloy weld deposit for severe abrasion, suitable for use on sand and gravel excavators. Resistant to oxidation. Matrix 700 HV. Carbides 1400 HV.
12–14% Manganese steels	(1) Basic coating, 14% manganese steel (work-hardening) for 12–14% Mn steel parts – steel excavators and mining equipment. 240 HV. (2) Austenitic stainless steel weld deposit suitable for joining 12–14% Mn steels and for reinforcing and for buffer layer for 12–14% Mn electrodes. 250 HV, work-hardening to 500 HV. (3) Tubular-type electrodes for hardfacing 12–14 Mn steel parts with high resistance to abrasion ahd heavy impact. Matrix 640 HV. Carbides 1400 HV.

SELECTION OF BRITISH STANDARDS RELATING TO WELDING

Note. When consulting British Standards the engineer should make sure that the Standard concerned incorporates the latest amendments.

B.S. number

Welding terms and symbols:
    Part 1 Welding, brazing and thermal cutting glossary
    Part 2 Symbols for welding
    Part 3 Terminology of and abbreviations for fusion weld imperfect-
        ions as revealed by radiography   *499*

Arc welding plant, equipment and accessories   *638*

Approval testing of welders when welding procedure is not required   *4872*

Tests for use in the training of welders; manual metal arc and oxy-acetylene welding of mild steel   *1295*

Covered electrodes for the manual metal arc welding of carbon and carbon–manganese steels   *639*

Low-alloy steel electrodes for manual metal arc welding   *2493*
Chromium nickel austenitic and chromium steel electrodes for manual arc welding   *2926*

Class I welding of ferritic steel pipework for carrying fluids   *2633*
Class I arc welding of stainless steel pipework for carrying fluids   *4677*
Class II metal arc welding of steel pipelines and pipe and pipe assemblies for carrying fluids   *2971*

Field welding of carbon steel pipelines   *4515*

Specification for metal arc welding of carbon and carbon–manganese steels   *5135*

Weldable structural steels   *4360*

Fusion welding of steel castings:
    Part 1 Production, rectification and repair
    Part 2 Fabrication welding   *4570*

Fusion welding joints in copper   *1077*

General recommendations for manual inert gas tungsten arc welding:
    Part 1 Wrought aluminium, aluminium alloys and magnesium alloys
    Part 2 Austenitic stainless steels and heat resisting steels   *3019*

General recommendations for manual inert gas welding:
    Part 1 Aluminium and aluminium alloys   *3571*

Filler rods for gas shielded arc welding:
    Part 1 Ferritic steels
    Part 2 Austenitic stainless steels
    Part 3 Copper and copper alloys

Safety

Testing

Spot and seam welding, etc.

618   **Appendix**

City and Guilds of London Institute

Examination questions

RELATED STUDIES AND TECHNOLOGY

Science and calculations

Note. All dimensions are given in millimetres, unless otherwise stated.

1 State one important safety precaution that must be observed for each of the following:
   (a) before commencing welding repair work on a two-compartment tractor fuel tank, having one compartment used for petrol and the other for fuel oil,
   (b) when flame gouging in the vertical position a defective vertical joint out of a heavy fabrication in preparation for rewelding.
2 What is the purpose of (a) portable cylinder couplers in oxy-acetylene welding, (b) a brushless d.c. generator for metal arc welding?
3 State two advantages which may be obtained by the use of a U preparation instead of a V preparation for welded butt joints in thicker materials.
4 (a) Explain briefly how (1) carburizing and (2) oxidizing oxy-acetylene flame settings are obtained.
   (b) Give one important application of each flame setting.
5 (a) What is meant by 'all-position rightward welding'?
   (b) Give two important advantages which may be obtained by the use of this technique for the welding of pipe joints.
6 State the plate thickness limitations of
   (a) the upward–vertical and the downward–vertical metal arc welding techniques.
7 (a) Describe with the aid of sketches the technique for making a lap fillet weld in the horizontal–vertical position by the oxy-acetylene process.
   (b) Compare the respective advantages and limitations of the leftward and the lindewelding techniques when butt welding mild steel pipes.
8 (a) Sketch the joint preparation and set-up, and state the size of the electrodes, number of runs and current values to be used for making:
   (1) butt welds, without backing bar in the flat position,
   (2) tee fillet welds in the horizontal–vertical position, in each of the following thicknesses of mild steel: 6.4 mm and 12.7 mm.

(b) Describe the technique required when cutting mild steel by the oxygen arc process and explain the cutting mechanism that produces the severance of the metal.

9 Discuss briefly safety recommendations with regard to each of the following:
   (a) the type of current used in dangerous situations,
   (b) metal arc welding in confined spaces,
   (c) effects due to (1) arc radiations and (2) heat exhaustion,
   (d) earthing and conductivity of the welding return circuit.

10 (a) Describe with the aid of sketches, what effect each of the following would have on the depth of root penetration and the quality of the deposited metal:
     (1) the use of too high a current value,
     (2) incorrect angle of electrode slope,
     (3) too fast a speed of travel,
     (4) incorrect arc length.
   (b) Explain how measuring equipment could be used to check current and voltage values available in a metal arc welding circuit during welding.

11 State one safety precaution that must be carried out:
   (a) before commencing welding repair work on a tank which has contained acids,
   (b) when metal arc welding from a multi-operator set is carried out during the construction of multiple-storey steel framed structures.

12 What is the purpose of (a) a single-stage gas pipeline regulator, (b) a rectifier for metal arc welding?

13 State what is meant by:
   (a) deep-penetration coated electrode,
   (b) non-consumable electrode.

14 (a) Make a labelled section-sketch to show the arrangement of the nozzle assembly used for progressive flame gouging.
   (b) State two methods used for gas-cutting bevels on plate edges in preparation for welding.

15 Name one possible cause of each of the following, when oxy-acetylene welding butt joints in mild steel: (a) excessive penetration, (b) incomplete penetration, (c) adhesion.

16 (a) Name two functions of the filter glasses used when metal-arc welding.
   (b) Why should the cable used for the welding lead connexion of an arc welding circuit be flexible?

17 (a) Name two types of arc welding processes in which non-consumable electrodes are used.
   (b) Give one example of a semi-automatic arc welding process.

18 (a) What is the difference between a backfire and a flashback?
   (b) Name five possible causes of backfiring when using a gas welding blowpipe. In each case explain how backfiring could have been avoided.
   (c) State two difficulties which may be experienced in the efficient operation of gas pressure regulator.

19 (a) State the purpose of each part of the high pressure oxy-acetylene welding system.

(b) Explain what is meant by gas velocity and outline how this is controlled in welding practice.

(c) Describe, with the aid of a sectional sketch indicating the gas paths, the operation of an oxy-fuel gas cutting blowpipe.

20 (a) Sketch the joint preparation and set-up, and state the nozzle size (in cubic feet or litres per hour) and filler rod size to be used for making: (1) butt welds in the vertical position in each of the following thicknesses of mild steel: 3.2 mm (1/8 in) and 5 mm (3/16 in), (2) a tee fillet-weld in the horizontal–vertical position in 5 mm (3/16 in) thick mild steel plate.

(b) Describe in detail, with the aid of sketches, the fusion welding of a butt joint in cast iron 10 mm (3/8 in) thick.

21 (a) Name four different types of metal arc welding plant which fulfil the electric power supply requirements for welding. Describe with aid of a sketch one of the plants named.

(b) What would be the practical effect of each of the following during metal arc welding: (1) loose circuit connexions, (2) variations in the mains supply voltage?

22 (a) Describe in detail, with the aid of sectional sketches, three defects which may be produced during the metal arc welding in the horizontal–vertical position of close-square-tee joints in mild steel plate. In each case state the cause and explain how the defects should be avoided.

(b) Explain briefly the effect of any two of the following upon the production of an efficient joint in a partially chamfered butt weld: (1) depth of root face, (2) the angle of the vee, (3) the gap setting.

23 (a) What is meant by distortion of welded work? State four factors which may cause distortion during the welding of mild steel assemblies. Describe briefly one method used to control distortion when building up a short section of a 75 mm diameter steel shaft worn below the correct diameter.

(b) Describe, with the aid of a sketch, one manual metal arc welding technique used for making butt welds in mild steel pipelines.

24 Describe briefly the ultrasonic method of testing for weld defects.

25 Explain the effect of each of the following on grain structure of low-carbon steel weld metal: (a) fast cooling, (b) slow cooling.

26 Describe briefly how heat for welding is generated by each of the following: (a) combustion, (b) the electric arc.

27 Explain briefly how the following properties of a welded joint in low-carbon steel may be affected by welding: (a) ductility, (b) hardness.

28 State three functions of the flux coatings on manual metal arc welding electrodes.

29 What is the purpose of:
(a) a choke reactance used in manual metal arc welding,
(b) a gas economizer used in oxy-acetylene welding?

30 Outline one method which may be used during the oxy-acetylene welding of vertical butt joints, in 4 mm thick low-carbon steel, to ensure freedom from weld defects when
(a) starting the weld at the beginning of the joint,

   (*b*) restarting the weld at a stop-point along the joint.
31 (*a*) What is meant by arc blow in manual metal arc welding?
   (*b*) Arc blow may arise when manual metal arc welding with either direct current or alternating current supply. Give *two* possible causes.
32 (*a*) State what is meant by;
       (1) rutile-coated electrode,
       (2) hydrogen controlled coated electrode.
   (*b*) Give *one* defect which may arise when manual metal arc welding low-carbon steel, using an eccentrically coated electrode.
33 Give *one* reason for the use of *each* of the following in welded work:
   (*a*) a chill,
   (*b*) a heat retaining material
   (*c*) a fixture.
34 (*a*) Make labelled sketches to show the essential difference between the nozzle assemblies used with;
       (1) acetylene,
       (2) propane, for the oxy-fuel gas cutting of low-carbon steel.
   (*b*) After manual oxy-fuel gas cutting 50 mm thick low-carbon steel it is required to cut 6 mm thick low-carbon steel plate. State the adjustment which will need to be made for this operation to be efficiently carried out.
35 Make sectional sketches of the following weld joints and give their weld symbol in accordance with the appropriate BS 499: (*a*) close-square-tee fillet (weld both sides), (*b*) single bevel butt.
36 (*a*) What are meant by the following terms: (1) conduction, (2) convection, (3) radiation? Give one example of each in welding practice.
37 State *two* safety precautions which should be taken for *each* of the following:
   (*a*) storage or use of dissolved acetylene cylinders,
   (*b*) using gas pressure regulators,
   (*c*) oxy-acetylene cutting operations in a confined space,
   (*d*) using non-injector type gas welding blowpipes.
38 (*a*) Sketch the joint preparation and set-up and state the nozzle and filler rod sizes to be used for making;
       (1) a butt weld in the vertical position in 3 mm thick low-carbon steel,
       (2) a close-square-tee fillet weld in the horizontal–vertical position in 5 mm thick low-carbon steel.
       (3) a fusion welded butt joint in the flat position in 10 mm thick cast iron.
   (*b*) Describe in detail, with the aid of sketches, the procedure and the technique required for making any *one* of the joints in part (*a*) above.
39 (*a*) Explain the action which produces the severance of the metal when oxy-fuel gas cutting low-carbon steel and state the difference between the basic principles of gas cutting and flame gouging.
   (*b*) Describe, with the aid of a sectional sketch indicating the gas paths, the operation of an oxy-fuel gas cutting blowpipe.
40 (*a*) State the principles of a gas-shielded arc welding process and give *two* industrial applications of the process.
   (*b*) Describe with the aid of sketches the *technique* required when cutting

low-carbon steel plate in the flat position by the use of *each* of the following arc cutting processes:

(1) air-arc

(2) oxygen-arc

41 (*a*) Sketch and label *two* different types of edge preparation for butt joints other than close-square butt, suitable for manual metal arc welding.

(*b*) What is the included angle of preparation necessary for single-vee butt welded joints to be made by the oxy-acetylene process using (1) the leftward technique, and (2) the rightward technique?

42 Give *one* possible cause of *each* of the following defects when manual metal arc welding close-square-tee fillet joints in low-carbon steel in the vertical position by the upwards technique:

(*a*) incomplete root penetration,

(*b*) undercut,

(*c*) unequal leg length.

43 Explain, by means of sketches, what is meant by *each* of the following:

(*a*) the *slope and tilt* of the filler rod and blowpipe *when making* a butt weld in low-carbon steel, 2 mm thick, in the vertical position by the oxy-acetylene process,

(*b*) the *slope and tilt* of the cutting electrode *when making* a straight cut in 8 mm low-carbon steel plate in the flat position by the oxygen-arc process.

44 (*a*) State *three* factors which may influence slag control during manual metal arc welding.

(*b*) Give *one* example of when *each* of the following are used in manual metal arc welding:

(1) tong test ammeter,

(2) voltmeter.

45 (*a*) Unsuitable variations during oxy-fuel gas cutting of steel may lead to faults along the cut face. Show, by means of labelled sketches of the cut face, the faults which would be caused by

either (1) the speed of travel being too fast,

or (2) the nozzle being too high above the work surface.

(*b*) State *two* methods which may be used for back gouging the root, ready for a sealing run, on the reverse side of welded butt joints in low-carbon steel.

46 (*a*) What is meant by (1) open circuit voltage, and (2) arc voltage in manual metal arc welding?

(*b*) State the likely effect on root penetration and weld deposit when manual metal arc welding low-carbon steel 6 mm thick with too long an arc.

47 (*a*) Name *two* undesirable effects on welds which may arise when manual metal arc welding low-carbon steel on site, due to the effect of weather conditions.

(*b*) Name *two* arc welding processes which involve the use of inert gas shielding.

(*c*) Name *two* arc welding processes which involve the use of reducing gas shielding.

48 (*a*) Describe, with the aid of sketches, the all-position rightward technique to be used when making a butt joint in low-carbon steel pipe 100 mm diameter by 5 mm wall thickness, the pipe axis to be in the fixed vertical position throughout.

(b) Compare the respective advantages and limitations of the leftward and the rightward techniques when used for the butt welding of low-carbon steel plate, 5 mm thick, in the flat position.

49 (a) Describe briefly, with the aid of a sectional sketch and indicating the gas paths, the mode of operation of;

either (1) a non-injector type welding blowpipe,

or   (2) a single-stage gas pressure regulator.

(b) What precautions are necessary during the assembly of a high pressure oxy-acetylene cutting plant in order to ensure safe and effective operation?

(c) Explain why the specified discharge rate of a dissolved acetylene cylinder must not be exceeded when in use.

50 (a) Describe in detail, with the aid of sectional sketches, *four* defects which may occur during the oxy-acetylene welding of low-carbon steel, stating in *each* case its cause and explaining how it may be avoided.

(b) Explain the procedure which must be carried out in order to make an effective macroscopic examination of a transverse section through a welded joint in low-carbon steel, naming *four* defects which may be revealed by this method of examination.

51 State *two* safety precautions which should be observed with *each* of the following for manual metal arc welding:

(a) treating components prior to welding by the use of trichloroethylene degreasing plant,

(b) welding in confined spaces,

(c) welding in close proximity to glossy finished surfaces,

(d) preparing vessels which have contained liquids with flammable vapours for repair by welding.

52 (a) Sketch the joint set-up and state the diameter of electrodes and current values to be used when making butt welds in the flat position in *each* of the following thicknesses of low-carbon steel

(1) 3 mm,

(2) 6 mm,

(3) 10 mm,

(4) 14 mm.

(b) Discuss the effect of *each* of the following factors on depth of root penetration when making butt welded joints in the flat position:

(1) current,

(2) arc length,

(3) speed of travel,

(4) angle of electrode (slope and tilt).

53 A dye penetrant method may be used for detecting defects in a welded joint.

(a) Outline the principles of this method.

(b) What type of defects may be revealed?

54 State why pre-heating is to be recommended when welded joints are to be made in *each* of the following:

(a) low-alloy, high-tensile steel in cold weather,

(b) 50 mm thick low-carbon steel,

(c) 6 mm thick copper plate.

55 (a) Explain the principle of oxygen cutting.

(b) State the probable cause of oxidation in a low-carbon steel weld made by the oxy-acetylene process.

56 Explain briefly the meaning of *each* of the following electrical terms:

(a) voltage,

(b) current,

(c) resistance.

57 State *two* advantages in *each* case of (a) hot working, and (b) cold working a low-carbon steel.

58 A butt weld is to be made by the manu l metal arc process between two 2 m by 1 m by 10 mm thick low-carbon steel plates along the long side;

(a) explain briefly why the cooling rate should be controlled,

(b) state *three* factors which may influence the cooling rate of the weld.

59 (a) Describe briefly the effect of cold rolling on the grain structure of a metal.

(b) E plain what takes place when a metal that has been cold rolled is heated to its recyrstallisation temperature.

60 (a) Give *two* reasons why grain growth may take place in a metal.

(b) What effect will enlarged grain structures have on the mechanical properties of a metal?

61 Explain briefly why oxygen and nitrogen should be excluded throughout the welding operation.

62 Describe briefly the difference between the current pickup systems for the generation of alternating current and direct current.

63 In relation to manual metal arc welding state *two* functions in *each* case of:

(a) fluxes,

(b) slags.

64 The figure shows the dimensions of a gusset plate. Calculate the area of plate required to make 50, assuming no wastage.

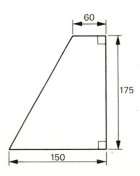

ALL DIMENSIONS IN MILLIMETRES

Question 64

65 Ten low-carbon steel plates each 3 m by 2 m by 20 mm thick are required to make an oil tank. Calculate the total mass of plate used if the metal density is 7830 kg/m³.

66 A number of butt welded joints have to be made between the ends of 100 mm diameter low-carbon steel pipe. Sketch a simple jig for holding them in position for tack welding.

67 Make a sectional sketch of *each* of the following types of welded joint, giving the appropriate weld symbol to show these joints in accordance with BS 499.
   (*a*) Single 'U' butt.
    b) Double-bevel butt.

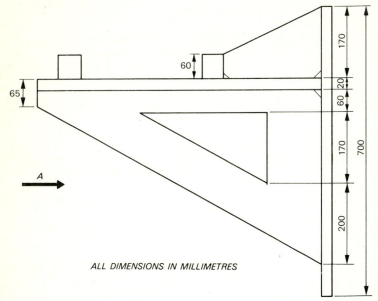

ALL DIMENSIONS IN MILLIMETRES

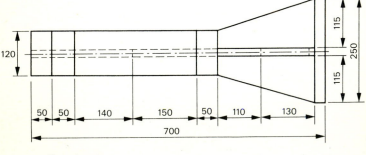

Question 68

68 Draw and fully dimension, in the direction of arrow A, an end elevation of the welded bracket shown in the figure. Insert the appropriate weld symbols according to BS 499 on the welded joints used.

69 Describe, with the aid of sketches where appropriate, how defects may be detected by using *each* of the following methods of testing:

(*a*) X-ray,

(*b*) ultrasonic,

(*c*) dye penetrant.

70 The figure shows a stool fabricated by manual metal arc welding. Each welded joint requires three runs of 4 mm diameter electrodes.

(*a*) Calculate the total length of weld deposited.

(*b*) Determine the number of electrodes used, to the nearest whole electrode, assuming 350 mm of weld is deposited by each electrode.

(*c*) Calculate the total cost of welding at £3.25 per metre of completed weld. Take π as 3.14 or 22/7.

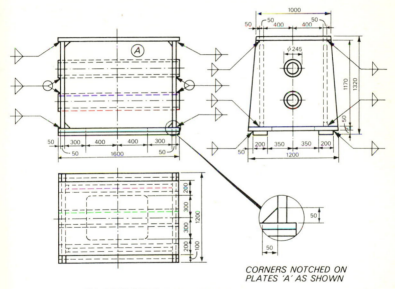

CORNERS NOTCHED ON
PLATES 'A' AS SHOWN

ALL DIMENSIONS IN MILLIMETRES

Question 70

71 A low-carbon steel open-top tank 7 m in mean diameter by 5 m deep by 20 mm thick is to be fabricated by welding. Calculate the total mass of plate if the metal density is 7830 kg/m³, taking π as 22/7.

72 A 90° segmental bend is to be made for a 150 mm diameter low-carbon steel pipeline. Sketch the completed bend.

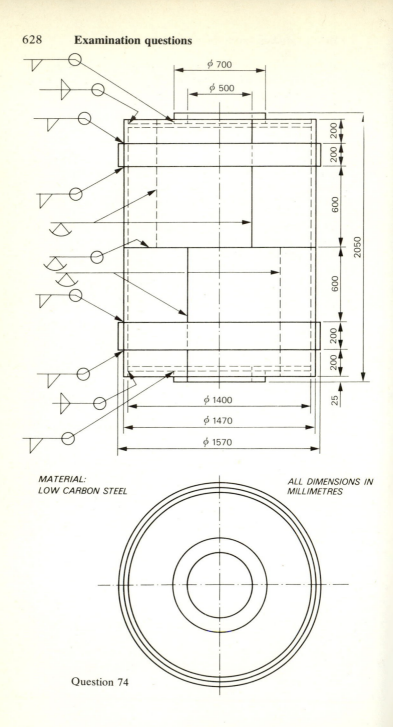

MATERIAL:
LOW CARBON STEEL

ALL DIMENSIONS IN
MILLIMETRES

ϕ 700

ϕ 500

200
200
600
600
200
200
2050
25

ϕ 1400
ϕ 1470
ϕ 1570

Question 74

73 Sketch cross sections of *each* of the following types of welded butt joint, giving
the appropriate weld symbol according to BS 499 for *each* on the drawings;
 (*a*) open single-bevel,
 (*b*) double U.
74 The figure shows a tank fabricated by manual metal arc welding.
 Each fillet-welded joint requires three runs to complete.
 (*a*) Calculate the total length of fillet welding.
 (*b*) To the nearest whole electrode, how many electrodes will be required to
 complete the fillet welding assuming 300 mm of weld is deposited per
 electrode?
 (*c*) Calculate the cost of butt welding at £4 per metre of completed weld.
 Take π as 22/7.
75 The figure shows a pictorial view of a bracket to be produced by welding.
 Sketch, approximately half full size, an elevation in direction to arrow '*A*' and
 a plan view in direction of arrow '*B*'.
 Insert on the sketch the appropriate weld symbols according to BS 499 in
 order to indicate the welded joints necessary to fabricate the bracket.

Question 75

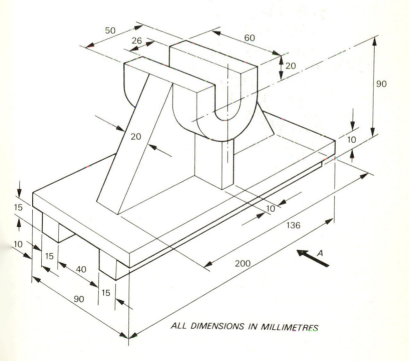

ALL DIMENSIONS IN MILLIMETRES

76 Draw an end elevation of the fabricated bracket shown in the figure. Dimension fully and insert the appropriate symbols according to BS 499 to indicate the welded joints used.

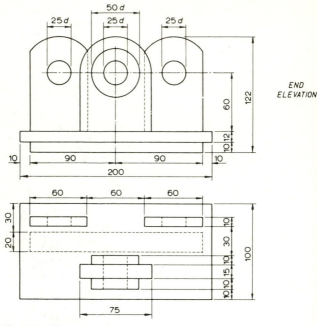

Question 76

Oxy-acetylene welding technology

1 Explain by means of a sketch what is meant by temperature gradient.

2 State *two* important functions of fluxes used during oxy-acetylene welding operations.

3 Name two main types of weld testing and give one example of each type.

4 Describe briefly the fundamental differences between the following: (*a*) the technique for the bronze welding of cast iron, and (*b*) the technique for the fusion welding of cast iron.

5 Stainless steel filler rod used for oxy-acetylene welding usually contains an element known as a stabilizer. Name one such element and state its main purpose.

6 What is the purpose of the arrow and the reference line for symbols used to indicate welded joints on drawings according to BS 499? A sketch *may* be used to answer the question.

7 Describe the technique necessary to form a tee-fillet joint in 1.6 mm commercially pure aluminium sheet by the flame brazing process.

State the type of filler wire and flame setting needed. You may use a sketch to illustrate your answer.

8 (a) State the volume of oxygen required for complete combustion of one volume of acetylene.

  (b) Give the proportions and sources of oxygen supply in a neutral oxy-acetylene flame.

9 Give *two* weld defects found by visual inspection which may influence the notch strength of welded joints.

10 Give *two* difficulties which may give rise to defects when welding grey cast iron by the oxy-acetylene process.

11 State *three* operations which are normally considered to be part of the total labour costs in oxy-acetylene welded repair work.

12 (a) Explain how the correct size of cutting nozzle to be used in the oxy-fuel gas cutting process should be determined.

  (b) Discuss the influence on the oxy-fuel gas cutting operation or any *two* alloying elements which may be present in steel.

  (c) Describe how cast iron can be cut by the oxy-fuel gas cutting process.

13 With reference to one particular application of each, describe briefly with the aid of sketches how any two of the following methods of weld inspection can be carried out: (1) radiographic, (2) magnetic crack detention, (3) ultrasonic.

14 (a) What is meant by 'post-heating'?

  (b) State two reasons why welded assemblies may be subjected to post-heating.

15 (a) Name three impurities found in acetylene gas immediately after generation.

  (b) Explain how acetylene gas may be tested for purity.

16 (a) What is meant by 'oxygen lance cutting'?

  (b) Give two examples of its use.

17 (a) Name three fuels used for pre-heating.

  (b) Why is coke an unsatisfactory pre-heating fuel?

18 Describe with the aid of a simple sketch what is meant by 'a pearlitic structure'.

19 Increasing the carbon content of a plain steel influences certain physical properties. Using one word only in each case, state the effect of increase in carbon content on the following properties: (a) tensile strength, (b) elongation, (c) hardness, (d) melting point.

20 (a) What is meant by 'hot-shortness'?

  (b) Give two causes of this weakness.

21 (a) What is meant by the term 'gas fluxing'?

  (b) Give one advantage of using this method.

22 Describe how a cemented tungsten carbide tip may be attached to a medium carbon steel shank using an oxy-acetylene flame.

23 (a) Describe briefly the operational principles of electric resistance spot welding.

  (b) Give one limitation of the process.

24 Give two examples of difficulties which may be encountered in the welding of high thermal conductivity materials.

25 (a) Summarize the problems encountered in the oxy-acetylene repair welding of each of the following cast materials: (1) zinc base die cast alloy, (2) magnesium alloy.

(b) Describe the preparation, welding technique and post-weld treatment necessary for each.

26 (a) State three reasons for using welding fixtures in welding fabrication.

(b) State three basic factors that must be considered for the effective operation of a welding fixture.

(c) Describe briefly, with the aid of a sketch, the principle of operation of any welding fixture with which you are acquainted.

27 (a) What is meant by 'a welding sequence'?

(b) What is the purpose of using a 'welding sequence'?

28 Explain what is meant by capillary attraction and describe how this affects the making of certain brazed joints.

29 (a) When would a carburizing flame be used for joining metals?

(b) What purpose is served by the use of such a flame adjustment?

30 Describe, with the aid of a simple outline sketch, one example of a repair welding operation where *studding* could be employed with advantage.

31 (a) Name two advantages of using a high silicon content in cast iron, or in cast iron filler rods for welding purposes.

(b) Give the approximate percentage of silicon contained in a 'super-silicon' cast iron filler rod.

32 (a) What is meant by the term 'hard-facing'?

(b) Name two types of filler rod that may be used for hard-facing operations.

33 Give one example of where each of the following flame gouging techniques would be used: (a) spot gouging, (b) progressive gouging.

34 (a) Name the two metals which form the basis of the alloy 'brass'.

(b) What is the recommended flame adjustment for the fusion welding of brass?

35 (a) State the approximate pressure of acetylene in a fully charged cylinder.

(b) What is the maximum permissible pressure at which acetylene may be used for welding or cutting operations?

36 (a) Name the fuel gas generally used for underwater flame cutting operations.

(b) State one good reason why this gas is used.

37 State two defects which may arise in a fusion weld made in low-carbon steel (a) from the use of a nozzle orifice which is too small, (b) from the use of a nozzle orifice which is too large.

38 (a) Describe, in detail, the preparation and procedure for making a controlled root bend test from a 5 mm thick low-carbon steel test piece taken from a butt welded joint.

(b) State two desirable features such a bend test should reveal.

39 (a) Explain, in detail, four safety measures that should be taken before carrying out welding repairs on a 150 litre (30 gallon) low-carbon steel petrol tank.

(b) Describe the hazards encountered when welding galvanized metals. State the precautions to be taken in the interests of health and safety.

40 The figure shows a cast iron wheel with two broken spokes. With the aid of a sketch, indicate how partial pre-heating may be used to minimize the risk of cracking.

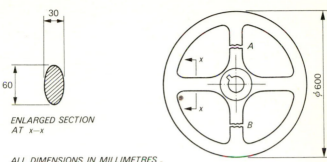

ENLARGED SECTION AT x—x

ALL DIMENSIONS IN MILLIMETRES

Question 40

41 State *four* advantages to be obtained by using the two-operator upward-vertical welding technique.

42 (*a*) What is meant by the term 'flame cleaning'?
(*b*) Give *two* advantages which may be obtained by the use of this process.

43 Give *two* reasons why oxygen should be excluded from molten low-carbon steel weld metal, stating *two* suitable precautions.

44 State briefly *two* possible causes of cracking of welded joints in carbon steel, and explain how this effect may be avoided.

45 Explain the action which causes the severance of the metal when using the cutting process for the cutting of austenitic stainless steel.

46 A fusion weld is to be carried out on a grey iron casting. State *two* precautions to ensure that the weld zone will be of grey cast iron after welding.

47 State *four* reasons why undercut should be avoided when fillet welds are made in low-carbon steel assemblies.

48 Briefly explain why a flux is unnecessary during the welding of a low-carbon steel.

49 Give *two* causes of porosity in the deposit, when hard-facing is carried out on low-carbon steel.

50 (*a*) What is meant by *stack cutting*?
(*b*) List *two* advantages of the use of stack cutting.

51 Why is grain growth more likely to occur in an oxy-acetylene welded joint than in a manual metal arc welded joint in low-carbon steel?

52 Give *two* advantages of the leftward technique over the rightward technique of oxy-acetylene welding.

53 (*a*) Give *two* difficulties encountered during the welding of zinc base alloys.
(*b*) How are these difficulties minimized in welding practice?

54 Two bars of low-carbon steel are to be butt welded by the oxy-acetylene welding process. Give *two* reasons why quenching must not be carried out immediately after welding.

55 Outline *two* workshop methods of distinguishing a grey iron casting from a malleable iron casting.

56 List any *four* safety precautions to be taken when flame gouging.

57 (*a*) State *four* problems associated with the welding of copper.
   (*b*) State what type of flame setting should be used and how it is attained.
   (*c*) Describe in detail the welding procedure.

58 What information is indicated by the weld symbols for the joint shown in the figure according to BS 499?

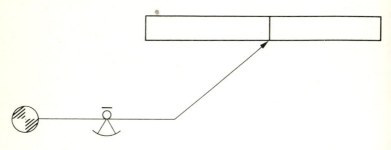

Question 58

59 Give *two* advantages in *each* case which may be obtained when oxy-fuel gas cutting;
   (*a*) manually,
   (*b*) by machine.

60 State briefly how any *four* of the following may arise in welding practice and explain how *each* may be counteracted:
   (*a*) grain growth in a brass,
   (*b*) over-ageing of precipitation hardenable aluminium alloys,
   (*c*) residual stresses in low-carbon steel,
   (*d*) intergranular corrosion of austenitic stainless steel,
   (*e*) cold cracking of low-alloy, high-tensile steel.

61 A cobalt based, hard-facing alloy is required to be deposited on to a low-carbon steel component.
   (*a*) State how the component could be prepared.
   (*b*) Name the type of flame setting required for a single layer deposit and give *two* reasons for your choice.
   (*c*) State *two* precautions to be taken to avoid defects.

62 A 6 mm thick magnesium alloy casting is cracked for a length of 150 mm and is to be repaired by oxy-acetylene welding.
   (*a*) Outline the preparation which may be required before welding.
   (*b*) Give *two* methods of indicating the correct pre-heat temperature.
   (*c*) Describe a suitable method of stress relieving after welding.
   (*d*) Explain how the flux residue should be removed.

Metal arc welding technology

(All dimensions in mm)

1 Why is alternating current potentially more dangerous than direct current at the same nominal voltage?

2 (*a*) In what type of weld would you expect to find a columnar structure?
 (*b*) With the aid of a sketch show in which part of the weld you would find this structure.

3 Explain how rapid cooling affects the microstructure of an 18% nickel–steel weld metal.

4 State *three* factors that may influence metal transfer phenomena when using a metal arc welding process.

5 Give *three* reasons why stray-arcing is undesirable.

6 Give three reasons why damp flux-coated electrodes should not be used for welding mild steel.

7 With the aid of a sketch, show the principle of the buttering technique.

8 What main advantage is claimed for the arc-on-gas technique?

9 (*a*) Name an arc welding process suitable for joining copper plates of thickness 6.4 mm.
 (*b*) Why is this process suitable?

10 State three ways in which weather conditions may adversely affect welding operations.

11 What is friction welding?

12 (*a*) Describe in detail the metal arc welding process commonly used for manually welding mild steel.
 (*b*) Give two examples showing how the process in (*a*), using standard electrodes, may be partially automated so as to reduce costs.

13 (*a*) Construct a table showing the ranges of welding currents typically used with 2.5 mm, 3.2 mm, 5.0 mm and 6.0 mm general-purpose, mild steel electrodes.
 (*b*) Plot a graph from the figures in the table in part (*a*).
 (*c*) From the graph in part (*b*). determine the probable current range to be used with a 4.0 mm electrode.

14 Manufacturers of metal arc welding electrodes take care to specify a minimum and maximum current for each size and type of mild steel electrode. Discuss in detail the important consequences of using (*a*) insufficient current, (*b*) excessive current.

15 Give four important functions of a slag.

16 (*a*) Briefly describe one test, other than visual inspection, used to reveal surface defects in welded joints.
 (*b*) Give one limitation of such a test.

17 (*a*) Name four constituents used in electrode coatings.
 (*b*) State the most important function of any one of these constitutents.

18 Give four differences in the behaviour of the metal that a welder will find between the metal arc welding of (*a*) aluminium and (*b*) low-carbon steel.

19 Explain the essential difference between macroscopic and microscopic examination.

20 (*a*) Explain what is meant by 'percentage metal recovery' in metal arc welding.

(*b*) If over 100% is claimed, what additions could have been made to the electrode coating?

21 (*a*) In what form would you expect the carbon to be present in (1) white cast iron, (2) grey cast iron?

(*b*) Which of these types of iron would most likely be formed in the heat affected zone if the cooling rate after welding is too fast?

22 Explain briefly why nickel is liable to crack when metal arc welded. State briefly how this defect may be avoided.

23 With the aid of sketches give two examples of distortion of welded work that may result from the application of metal arc welding.

24 Sketch the form of an as-welded joint produced by:
(*a*) electric resistance flash butt welding,
(*b*) electric resistance (upset) butt welding.

25 State three safety precautions to be observed when using engine-driven metal arc welding equipment.

26 Give any two advantages of using multi-operator a.c. metal arc welding equipment.

27 For each of three of the following give a description of (1) the principles of operation, (2) a typical application:
(*a*) submerged arc welding,
(*b*) electric resistance spot welding,
(*c*) electroslag welding,
(*d*) arc stud welding.

28 A vessel 4 m long and 1.2 m diameter has to be made from two dished end plates and six available plates rolled to form six semi-circular sections. The material is 9.5 mm thick low-alloy structural steel, and the longitudinal and circumferential joints are to be butt welds made by manual metal arc welding.
(*a*) Sketch a suitable form of plate edge preparation and describe the welding procedure to be used.
(*b*) Calculate how much it would cost to complete the welding of the vessel if the cost is £2.15 per metre of welded joint. Take π as 22/7.

29 The figure shows two views of a cast iron support bracket which has fractured in service.
(*a*) If the casting is to be replaced by a low-carbon steel welded fabrication:
    (1) make a sketch showing the complete bracket assembled ready for welding. By the use of weld symbols (BS 499) indicate the type and location of the weld joints;
    (2) detail freehand the steel parts that you would require.
(*b*) If the casting is to be repaired by manual metal arc welding state:
    (1) all preparations necessary to be made before welding,
    (2) the weld procedure, including details of electrode and current values,
    (3) any precautions to be carried out after welding.

30 (*a*) Give any three advantages obtained when using rectifier welding equipment.
(*b*) Explain what is meant by the terms arc voltage and open circuit voltage.

31 (*a*) Give three probable causes of poor-quality resistance spot welds.

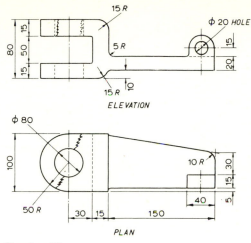

ELEVATION

PLAN

Question 29

   (*b*) With the aid of sketches show how resistance projection welding differs from resistance spot welding.

32 (*a*) Give two causes of hot cracking in carbon–steel fusion welds.

   (*b*) Name one impurity that may cause hot cracking in welded steel joints.

   (*c*) State one way in which hot cracking can be minimized by welding procedure.

33 (*a*) Explain briefly what is meant by solution treatment.

   (*b*) Give one example of an alloy that may be solution treated.

   (*c*) What effect will fusion welding have on the mechanical properties of a solution-treated alloy?

34 The table overleaf shows the electrode coverings, the performance of covered mild steel electrodes and the quality of the weld deposit.

   Indicate the class of electrode that will give the best result by placing one tick only in the appropriate space. The first line has been completed. You are required to complete the remaining six lines.

35 Describe briefly with the aid of a sketch a suitable arc cutting process for the cutting of ferrous and non-ferrous metals.

36 (*a*) Name *three* obnoxious fumes or poisonous gases which may be formed during metal arc welding operations.

   (*b*) Give *two* safety precautions to be taken in order to avoid personal injury from these fumes or gases.

37 Describe with the aid of sketches each of the following:

   (*a*) back-step welding,

   (*b*) backing strips,

   (*c*) backing bars.

38 Write short explanatory notes on any *four* of the following and in each case

Electrode coverings	Cellulosic	Rutile	Iron oxide, silicates	Hydrogen con- trolled
Current carrying capacity				√
Ductility				
Penetration				
Resistance to cracking				
Absence of spatter				
Deposition rate				
Ease of arc striking				

Question 34

describe how these problems are overcome in order to produce satisfactory joints in a welded fabrication:

(a) residual stress,
(b) hard-zone cracking,
(c) intergranular weakness,
(d) overheating of plain carbon steels,
(e) carbon pickup,
(f) dilution effects.

39 The figure shows details of a pipe assembly. Answer the following:

(a) Recommend a suitable method in each case for producing to correct shape and size (1) the flanges, (2) the main pipe and the branch pipe.

(b) Describe with the aid of sketches a suitable edge preparation that may be used to ensure satisfactory penetration at the following joints: (1) a flange to the main pipe, (2) branch pipe to the main pipe.

(c) Show with the aid of sketches the procedure for welding both of these joints, stating the sequence of runs, the current value and diameter of electrode used in each case.

(d) Describe briefly how the finished welds may be tested for surface defects.

40 The figure shows details of a pipe assembly. Answer the following:

(a) Show by means of a sketch how the main pipe and the flanges may have distorted as a result of welding.

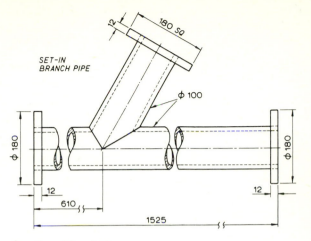

SET-IN
BRANCH PIPE

φ 100

φ 180

φ 180

12

12

610

1525

Questions 39 and 40

(b) Sketch a simple fixture that would be suitable for locating the branch to the main pipe and assist in controlling distortion.

(c) Describe briefly a suitable method for pressure testing the completed assembly.

(d) Calculate the total weight of the three flanges; 1 m³ of low-carbon steel weighs 7750 kg.

41 (a) Explain what is meant by the term 'arc eye'.

(b) What precautions should be taken to avoid 'arc eye'?

(c) What action should be taken in the case of severe 'arc eye'?

42 (a) What is meant by a metallic solid solution?

(b) Give one example of a solid solution using an alloy which is normally welded.

43 When each of the following materials is welded by the manual metal arc process state, in each case, two difficulties which may arise: (a) copper, (b) grey cast iron.

44 (a) State two difficulties which may be encountered when manual metal arc welding dissimilar metals.

(b) Explain how these difficulties can be overcome.

45 State the purpose of each of the following components of an arc welding plant: (a) rectifier, (b) transformer, (c) choke reactance.

46 Give three precautions which should be taken to produce acceptable joints when manual metal arc welding austenitic stainless steel.

47 Butt welds are to be made in the construction of a pipe line.

(a) Give two reasons why excessive penetration should be avoided.

(b) Describe briefly two methods of controlling root penetration.

48 Describe, with the aid of sketches, any four of the following. In each case state one typical application, one advantage and one limitation.

(a) Stove pipe technique, (b) tongue bend test, (c) manipulators, (d) arc on gas techniques, (e) 'studding' when welding cast iron.

49 Give *three* reasons why a craftsman welder, accustomed to welding low-carbon steel, may find aluminium difficult to weld.

50 The equivalent carbon content of an alloy steel can be found from the formula:

$$\text{Carbon equivalent} = \%C + \frac{\%Mn}{6} + \frac{\%Si}{24} + \frac{\%Ni}{40} + \frac{\%Cr}{5} + \frac{\%Mo}{4}$$

If a certain alloy steel alloy contains 0.25% C, 0.6% Mn, 0.2% Si, 2.4% Ni, 0.7% Cr, 0.6% Mo, find its equivalent carbon content.

51 Briefly explain the function of *each* of the following in manual metal arc welding:
(a) a low voltage safety device,
(b) a rectifier.

52 List *six* methods of testing welded joints, indicating clearly whether the methods are destructive or non-destructive.

53 (a) Give *three* reasons why pre-heating is sometimes necessary when manual metal arc welding.
(b) Briefly describe how pre-heating may be carried out by an electro-thermal method.

54 (a) Explain what is meant by dilution in weld deposits.
(b) List *three* factors which may influence the amount of dilution produced in a weld deposit.

55 (a) Briefly explain why notch effects must be avoided in stressed welded structures.
(b) Sketch *two* defects and *two* undesirable weld contours, *each* of which could create notch effects.

56 (a) What are the *three* basic types of wear to which hard-faced components are subjected?
(b) Select *one* of these types and suggest a suitable type of electrode to be used for surfacing to meet requirements.
(c) State *two* precautions which should be taken to minimize cracking.

57 An aluminium alloy casting is to be repaired by the manual metal arc welding process. Give *each* of the following
(a) a workshop method of indicating the pre-heating temperature,
(b) the type of current to be used,
(c) *Three* factors which make this material more difficult to weld than low-carbon steel.

58 Select a suitable electrode for welding *each* of the following low-carbon steel joints:
(a) a severely stressed single-vee butt joint welded in the flat position,
(b) a light stressed fillet welded joint in the vertical position,
(c) fillet welds in the flat position where a high metal recovery rate is required.

59 State *two* physical properties and *one* metallurgical problem that must be encountered when joining dissimilar materials by a fusion welding process.

60 Describe briefly the principles of operation, and give *one* suitable industrial application, for *three* of the following welding processes:

  (*a*) electroslag,
  (*b*) electric resistance flash butt,
  (*c*) electron beam,
  (*d*) arc stud,
  (*e*) friction.

61 In *each* case, list *six* factors that would require attention while carrying out visual inspection of metal arc welded work;

  (*a*) prior to welding,
  (*b*) during welding, and
  (*c*) after welding.

62 Describe the effects on the weldability of steel of *each* of the following elements and state in *each* case *one* method of overcoming difficulties that may arise.

  (*a*) Carbon.
  (*b*) Hydrogen.
  (*c*) Silicon.
  (*d*) Chromium.

63 When a fabrication requiring 10 metres length of welding was produced using manual metal arc welding, the cost of welding was £2.80 per metre. When the same fabrication was welded using CO_2 process, the cost per metre of welding was reduced to £2.20 per metre. If the CO_2 equipment costs £480, by means of graphs find:

  (*a*) the number of fabrications that would be required to be welded to cover the cost of the equipment,
  (*b*) the total number of fabrications that would be required to show a saving of £100.

64 The figure shows two views of a bracket which has to be fabricated by manual metal arc welding.

  (*a*) Make a pictorial sketch of the bracket and indicate, by the use of weld symbols, according to BS 499, the location and the type of weld joints required.
  (*b*) List *ten* items of information that could be included on a Welding Procedure Sheet for the bracket shown.

65 The figure shows two views of a cast iron bracket which has fractured in service.

  (*a*) If the casting is to be repaired by manual metal arc welding,
    (1) state *all* preparations necessary to be made before welding
    (2) describe a suitable procedure, including details of the type of electrodes and the current values
    (3) state *two* precautions to be carried out after welding.
  (*b*) Make a sketch to show the complete bracket assembled ready for welding, if the casting is to be replaced by a low carbon steel welded fabrication. By the use of weld symbols (BS 499), indicate the type and location of the weld joints to be used.

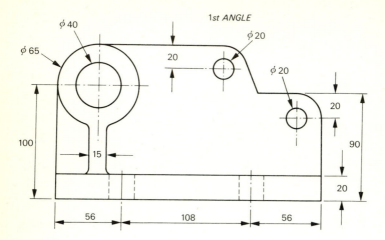

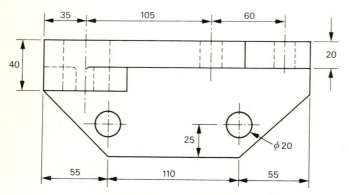

ALL DIMENSIONS IN MILLIMETRES

Question 64

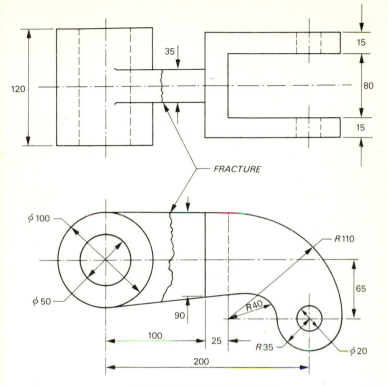

ALL DIMENSIONS IN MILLIMETRES

Question 65

Gas-shielded arc welding technology
(All dimensions in mm unless otherwise stated)

1 (a) If your welding generator caught fire and you could not switch off the supply current, what type of fire extinguisher would you use?
 (b) Is there any type of extinguisher that you should not use?
 (c) Why should you not use the type of extinguisher mentioned in (b)?
2 Explain why a welding generator neither blows its fuse nor burns out when a short-circuit occurs at the welding electrode.
3 (a) Explain why a destructive macrostructural examination of the cross section of a weld is often required.
 (b) Why is the welding craftsman not normally expected to make a macro-structural examination of a weld?
4 (a) Name any *two* limitations of the gas-shielded tungsten arc welding process.

    (*b*) Name any *two* limitations of the gas-shielded metal arc welding process.

    (*c*) Why is a gas shield needed in the two types of arc processes in which it is used?

5 For each of the following cases state which kind of cracking is most likely to occur in a fusion welded joint:

    (*a*) a weld highly stressed during the early stages of solidification,

    (*b*) a weld in a hardenable steel made without pre-heat,

    (*c*) a weld in an unstablized austenitic stainless steel.

6 In the welding of a solution-treatable type of aluminium alloy, describe any *two* weldability difficulties that you would expect to encounter.

7 (*a*) List the *three* main characteristic modes of metal transfer in metal arc welding.

    (*b*) State which mode is the most desirable of the three.

    (*c*) Name any common welding system in which the most desirable mode is unattainable.

8 (*a*) In tungsten arc welding how should the welding arc be initiated without contact between electrode and parent metal?

    (*b*) With the aid of a simple line diagram show all the essential equipment for the conditions that you outline in (*a*) and clearly label each part.

9 (*a*) With the aid of sketches show how a single Vee butt joint preparation for tungsten arc welding should differ from that of gas shielded metal arc welding in the same thicknesses of similar materials.

    (*b*) Give the main reason why they should differ.

10 (*a*) Name two practical difficulties likely to be encountered in inspecting a weld joint by radiographic means.

    (*b*) Why are magnetic crack detection methods not used for examining welds in copper alloys?

    (*c*) What crack detection method could be used for copper alloys?

11 (*a*) State the purpose of a surge injector unit as used in arc welding.

    (*b*) State the purpose of a suppressor unit as used in arc welding.

12 (*a*) For a plain carbon steel containing 0.4% C list three typical metallurgical states which might exist in the material in the vicinity of a fusion weld.

    (*b*) For each of the conditions under (*a*), outline the sequence of heating and cooling that would put the material in that particular condition.

13 State the main difficulty that you would expect to have to overcome in joining each of the following types of material by gas shielded tungsten arc welding: (*a*) malleable cast iron, (*b*) solution treatable aluminium alloy.

14 (*a*) Explain why, in spite of the overall efficiency of argon as a shielding gas, so much effort is spent in developing the use of gases such as carbon dioxide.

    (*b*) Name two difficulties likely to be encountered in using carbon dioxide as a shielding gas in arc welding.

    (*c*) For each of the difficulties given in your answer to (*b*) name one method used for overcoming the problem.

15 (*a*) What is arc plasma?

(*b*) By means of a simple labelled outline diagram, show one method of using arc plasma for cutting purposes.

16 Two small 60/40 brass castings are to be accurately butt welded to the ends of a 70/30 brass tube by means of gas shielded tungsten arc welding. Each casting incorporates an extension ring about 25 mm long machined to the cross-sectional dimension of the tube which is 200 mm long and 50 mm outside diameter, with a wall thickness of 6 mm. Two hundred of the assemblies are to be made.

(*a*) Giving reasons, explain (1) which type of joint preparation you would use, (2) which dimensions and angles you would recommend, and (3) the welding procedure that you would use.

(*b*) Outline the main features of any special equipment that you consider essential for making the welds.

(*c*) Explain any particular metallurgical problems that you would expect to have to overcome.

17 A cylindrical vessel 620 mm diameter by 2.5 metres long with low-carbon steel flanged ends 25 mm thick is to have two austenitic stainless steel pipes coming out at right angles to the vessel axis and to each other midway along the vessel axis. The pipes are 100 mm inside diameter with a wall thickness of 9.5 mm and the vessel is made of 12.5 mm thick low-carbon steel, clad inside with austenitic steel to a further 1.6 mm thickness. The pipes and flanges are to be joined to the vessel by gas shielded tungsten arc welding.

(*a*) Give particulars of (1) the type of joint, (2) the joint preparation and (3) the welding procedure that you recommend for welding the flanges to the vessel.

(*b*) Give particulars of (1) the type of joint, (2) the joint preparation and (3) the welding procedure that you recommend for welding the pipes into the vessel.

18 A special I-section girder 6 metres long has the following sectional dimensions: flanges 450 mm wide by 50 mm thick, web 19 mm thick, and overall depth 1 metre. Stiffening ribs 150 mm wide by 12.5 mm thick, running from flange to flange, are located opposite each other against the web at right angles to the girder axis at 1.2 m intervals. The girder is to be fabricated from high-tensile constructional steel of welding quality by gas shielded metal arc welding.

(*a*) Give particulars of (1) the type of joint, (2) the joint preparation, and (3) the welding procedure that you would use for joining the web to the flanges.

(*b*) Give particulars of (1) the type of joint, (2) the joint preparation and (3) the welding procedure that you would use for attaching the stiffening ribs.

19 There are particular economic and metallurgical problems associated with the effective use of gas-shielded metal arc welding in production applications. Outline these problems and explain how they may be overcome.

20 Austenitic heat-resisting steels containing higher nickel and chromium contents are used for fabricating structures for high temperature service.

(a) State *five* main welding problems that are likely to arise in the metal arc gas-shielded welding of this type of material.

(b) Explain how *each* of these problems may be effectively overcome to ensure efficient welded joints for service at high temperatures.

21 State the purpose of a drooping characteristic for arc welding.

22 Sketch the form of a drooping characteristic and indicate on it (1) the open circuit voltage, (2) the average arc voltage, (3) the average welding current.

23 (a) When is it necessary to use a flux when using the metal arc gas shielded process?

(b) Why is it necessary in this case?

(c) In what manner is the flux usually applied?

24 (a) Give one reason why the back-step welding procedure is not used with automatic metal arc welding.

(b) Does the gas shielded metal arc welding of a single-run V butt weld leave behind (1) any transverse residual stress, (2) any longitudinal residual stress?

25 If you are *tungsten arc* gas-shielded welding a butt joint in 0.4% C steel plate 12 mm thick with 0.1% C steel filler wire, estimate the approximate average carbon content of the deposit.

26 If you are *metal arc* gas shielded welding a butt joint in 0.4% C steel plate 12 mm thick with 0.1% C steel filler wire, estimate the approximate average carbon content of the deposit.

27 (a) When a d.c. arc is operating, what proportion of the heat may be generated at each side of the arc?

(b) Would the heating situation be the same when tungsten arc gas-shielded welding aluminium with the electrode positive?

28 (a) With what type of material would you expect to find equi-axial solidification occurring in a fusion welded deposit?

(b) Why is a colomnar growth almost invariably found in the structure of a progressive fusion weld in the as-welded state?

(c) State the type of grain structure which may be found in heat affected zone of a single-run weld made in normalized low-carbon steel.

29 On a labelled outline sketch (a) name the different types of structure that you would expect to find in the vicinity of a single-run single-vee butt weld made in an initially *annealed* solution-heat-treatable alloy, and (b) indicate the approximate areas in which you would expect to find each structure.

  Note. You are expected only to name the type of structure, not to show any details.

30 A gas pressure regulator and a flowmeter are each essential for the successful operation of a gas shielded arc welding process.

(a) With the aid of an outline diagram show the usual location of each in a conventional gas shielded tungsten arc welding arrangement.

(b) Why is it necessary to have a gas pressure regulator?

31 In gas shielded tungsten arc welding each of the following plays an important part: (1) length of electrode projecting beyond the nozzle, (2) length of arc, (3) angle of electrode relative to the workpiece.

(*a*) State the one factor that is influenced by all three.

(*b*) Why is it important that this factor should be controlled?

32 Give two important reasons why small diameter filler wire is used for gas-shielded metal arc welding.

33 In the cross-sectional shape of a fusion welded joint, sharp corners should be avoided.

(*a*) Give the most important reason for this precaution.

(*b*) State why this precaution is particularly important when welding structural steel for service in a cold atmosphere.

34 The support column shown in the figure is to be made up from 25 mm thick plates of high strength low-alloy structural steel mounted on a square mild steel base plate. The sizes are given on the diagram.

(*a*) Outline the main material problems to be overcome in welding this construction.

(*b*) Indicate where you would locate the four transverse joints required to make up the vertical member from the two plates.

35 For the support column shown in the figure and the materials quoted in question 34, give

(*a*) the details of the appropriate longitudinal joint preparation used to complete the hollow section,

(*b*) the number and sequence of deposition of the runs required to complete the section,

(*c*) any precautions necessary to ensure sound welds.

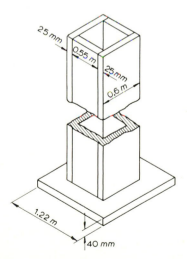

Questions 34–37. Overall height 5.6 m. Material available for main stem: plate 3.5 m × 1.2 m × 25 mm, plate 2.0 m × 1.2 m × 25 mm.
Note. Gas shielded metal arc welding to be used for all except tack welds.

36 For the support column shown in the figure and the materials quoted in question 34, give
  (a) the details of the appropriate joint preparations used to attach the vertical member to the base,
  (b) the number and sequence of deposition of runs used to complete this joint,
  (c) any precautions necessary to ensure sound welds.

37 Discuss the main factors affecting the desirability of making up a special welding fixture for use when making the longitudinal welds in the vertical member of the support column shown in the figure.

38 (a) Give *two* reasons why it is necessary to be particularly careful of plant insulation when using HF current.
  (b) Why may it be dangerous to weld with an exposed arc near to a tank filled with non-flammable degreasing liquid?

39 (a) Give *one* reason why, in particular situations, it might be difficult to control the arc when arc welding mild steel.
  (b) When would it not be safe to connect two welding generators in parallel to give increased power to a single arc?

40 (a) Give the principal reason why carbon dioxide gas is used for the gas shield in gas-shielded metal arc welding of mild steel.
  (b) Give *one* specific difficulty to be overcome when using carbon dioxide as a shielding gas for arc welding mild steel.

41 (a) Why is it that residual stress tends to become less of a problem the faster you are able to complete an arc-welded joint?
  (b) Give *one* reason why a particular material might be very liable to hot intergranular cracking during fusion welding.

42 (a) What is dilution in fusion welding?
  (b) What is pickup in fusion welding?
  (c) Can the atmosphere surrounding an arc affect any pickup that normally tends to occur?

43 Draw a simple outline sketch of the cross section of a two-run double V butt weld in a hardenable steel made without pre-heating and show (a) *three* different types of structure that might be found in the heat-affected zones, and (b) the most likely location(s) of each of the types you give.
   You are not expected to give details of the structure; a simple general word description will be sufficient if you do not know the technical name of a particular structure.

44 With the aid of an outline sectional sketch show the relative positions and proportions of (a) the electrode, (b) the collet, (c) the gas entry, and (d) the nozzle, in a typical gas-shielded tungsten arc welding torch head.

45 On a simple outline sketch locate and name the function of any *three* parts essential to the operation of the welding head of a fully automatic gas-shielded metal arc welding plant.

46 Give any *three* possible differences between the respective preparation and deposition techniques for making a flat single V butt weld in 12 mm thick material by (a) gas-shielded tungsten arc welding and (b) gas-shielded metal arc welding.

47 State briefly any *two* problems likely to be met in trying to weld an alloy containing one relatively low-melting-temperature constituent and with a wide solidification range of temperature.

48 What is meant by the term low-alloy steel?

49 Give what you consider to be *three* important factors likely to affect the cost of the welded fabrication of a very large one-off component.

50 Give details of;
- (*a*) the type of preparation,
- (*b*) the welding procedure,
- (*c*) particular problems encountered with the material, when making the joint between the flange and the body of the vessel shown in the figure.

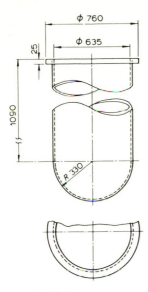

Questions 50–53. Processing vessel. Material: oxygen-free copper. Weld from forged flange ring, roll formed plate and pressed dome end by gas shielded tungsten arc process.

51 Large numbers of the vessel shown in the figure are to be made. Outline (*a*) the type, (*b*) the particular functions, of any welding fixture(s) you consider essential to facilitate welding.

52 If the hemispherical end of the vessel shown in the figure is to be made by fabrication from twelve equal segments of 12 mm plate, indicate clearly;
- (*a*) the approximate shapes of the segments that would be used,
- (*b*) the means to be used to prepare the joint faces.

53 Outline the particular welding problems you would expect to have to overcome in making the hemispherical fabrication of question 52.

54 Give details of (*a*) suitable joint preparation(s) and (*b*) suitable welding procedures for joining the flanges to the central web of the lightweight beam shown in the figure.

55 Give details of (a) the type or types of joint preparation and (b) the welding procedure for joining the thickening plates to the main flanges of the beam shown in the figure.

56 Give details of (a) suitable joint preparation(s) and (b) suitable welding procedure, for attaching the six stiffening ribs to the beam shown in the figure.

57 On a sketch, using the BS 499 (Part 2) system, give all the information needed to show the type and location of each welded joint.

58 (a) State one purpose of adding a proportion of oxygen to the argon for certain metal arc gas-shielded welding operations.

    (b) Approximately what proportion of oxygen is added?

59 (a) With the aid of an outline sketch show the effects of excessive current on the shape of cross section of a butt weld. Assume a tungsten arc gas-shielded, single-run, single-vee joint.

    (b) State two effects of using too long an arc on this type of joint.

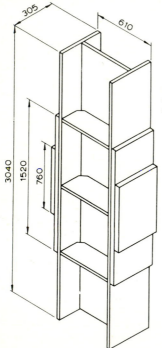

WEB AND RIB 16 mm THICK.
FLANGES AND THICKENING
PLATE 25 mm THICK

Questions 54–57. Lightweight beam. Material: commercial purity aluminium plate. All joints to be welded by gas-shielded metal arc process.

60 If the parent metal composition of a welded joint in plain carbon steel is 0.22% carbon, the all-weld metal deposit composition is 0.11% carbon and the cross-sectional area of the weld metal zone is 10 times the size of cross-sectional area of the fusion zone, estimate the approximate average carbon content of the weld deposit resulting from dilution.

61 (a) Give one reason why the hard brittle form of structure (martensite) is most likely to form close beside the fusion boundary in the heat affected zone of a welded joint in a hardenable steel.

    (b) Name the kind of structure to be found *just outside* a martensitic zone in the heat affected zone of a welded joint in a hardenable steel. (*Note.* If you do not know the technical name of the structure a simple word description will do.)

62 (a) State one difficulty with the dip-transfer mode of metal deposition as it is used in CO_2 shielded metal arc welding.

    (b) State briefly how this difficulty is overcome.

63 With the aid of a simple outline sketch show the crystal structure in a cross section of a single-pass single-vee butt weld in an unalloyed *non-ferrous* metal.

64 On a simple labelled block diagram show the name and purpose of each essential part of the plant and equipment needed for making tungsten arc gas-shielded welds in a variety of metals and a range of thicknesses.

65 Give two checks you should make on the electrode wire before fitting a new spool into a gas-shielded metal arc welding machine.

66 Why is a double-bevel or a double J butt welded tee joint preferable to a double fillet welded close-square-tee joint for welding two highly stressed members?

67 Suggest how two completely incompatible parent plant materials might be jointed to each other by fusion welding.

    What is meant by the term 'mild steel'?

68 If 500 of a small welded component are required, state any two measures that should be taken to keep the welding costs low.

69 State the sequence which could be used to complete the joints numbered 1, 2, 3, and 4 in the figure of a welded fabrication made from five parts as shown. Give reasons for your answer.

70 For joint 1 in the figure give details of (a) the type of preparation and (b) the welding procedure to be used.

71 For joint 2 in the figure give details of (a) the type of preparation and (b) the welding procedure to be used.

72 Describe briefly four problems which may be encountered in the welding of the fabrication shown in the figure.

73 With the aid of simple sketches show how you would prepare the joints for welding the top central part of the beam shown in the figure, where the central boss and the top flange join with each other and where the vertical web joins with the top plate.

74 State the welding sequences which should be used for welding the beam shown in the figure. Give reasons for each step.

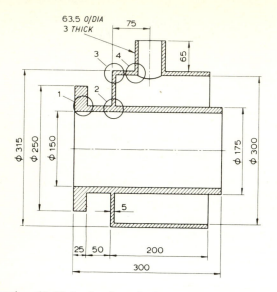

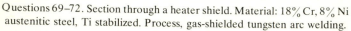

Questions 69–72. Section through a heater shield. Material: 18% Cr, 8% Ni austenitic steel, Ti stabilized. Process, gas-shielded tungsten arc welding.

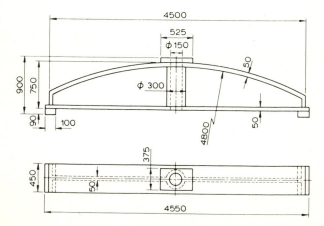

Questions 73–76. Support beam. Material: high yield stress structural steel (BS 4360 Grade 50) welded by CO_2 shielded metal arc process.

75 Describe four of the problems which are likely to arise in welding the ends of the beam shown in the figure.

76 On a sketch of the figure give all the information needed to show the type and location of each welded joint, using the system given in BS 499 (Part 2).

77 Give *three* reasons why a suitable shade of filter glass should be used for viewing the arc during gas-shielded arc welding.

78 (a) What is the purpose of a rectifier when used for welding from an a.c. power supply?
    (b) On a simple labelled graph show clearly the typical form of current flow likely to be obtained from a welding rectifier.

79 A metallic alloy may have a 'narrow' or a 'wide' solidification range.
    (a) State which type of solidification mode will give most difficulty in fusion welding.
    (d) Give *one* reason to justify your answer.

80 The following gas-shielding mixtures are used in gas-shielded arc welding.
    (a) Argon + carbon dioxide + oxygen.
    (b) Argon + oxygen.
    (c) Argon + hydrogen.
State, in *each* case,
        (1) the material for which the mixture is best suited.
        (2) the approximate percentages of gases in the mixture.

81 (a) What is the most commonly used ferrous alloy?
    (b) Name *two* different non-ferrous alloys used in welded fabrications.

82 (a) By means of a labelled sketch show what is meant by buttering.
    (b) Name *one* material on which buttering could helpfully be used to make an effective arc fusion welded joint.

83 State why a self-adjusting arc is 'not likely to operate effectively in CO_2 shielded metal arc welding'.

84 State *two* conditions essential for the formation of an equi-axial crystal structure, in an arc welded deposit in the as-welded condition.

85 If treated alloy plates, in the fully solution-treated and aged condition, are joined by fusion welding state whether:
    (a) the as-welded deposit will be harder or softer than the parent plate,
    (b) the heat affected zone will be harder or softer than the parent plate.

86 By means of a labelled block diagram show clearly the equipment required to give effective control of gas flow to the torch or welding head during gas-shielded arc welding.
Assume that a suitable gas is being used.

87 By means of a labelled sketch indicate any *three* types of weld defect likely to occur in gas-shielded arc welded joints.

88 What is the purpose of a suppressor unit in a gas-shielded arc welding circuit?

89 State *three* different types of material suitable for tungsten arc gas-shielded welding. In *each* case state
    (a) *Two* difficulties that may arise during the welding of the material
    (b) the methods used to overcome the difficulties encountered.

90 With the aid of a simple labelled block diagram show the name, location and

purpose of each part of the equipment essential for the effective tungsten arc gas welding of aluminium.

91 (a) Outline the typical main functions of a jig or fixture suitable for use in tungsten arc gas-shielded welding.

(b) Give *two* situations in which the use of a jig or fixture would be considered essential, in the manufacture of a component by tungsten arc gas-shielded welding.

92 Explain what is meant by gas-backing in tungsten arc welding and show *two* ways in which it may be applied.

93 List *two* advantages and *two* limitations of gas-shielded metal arc welding as a process for general-purpose welding repair work.

94 With the aid of a simple labelled block diagram show the name, location and purpose of each part of the equipment essential for the effective CO_2 shielded metal arc welding of low-carbon steels.

95 State *four* features which should be present in a manipulator intended for extensive gas-shielded metal arc welding of joints in a large component.

96 State the means by which it would be possible to control the irregular penetration profile made by a single run weld deposit in the gas-shielded metal arc welding of joints in aluminium.

97 (a) State the type of filter which should be used to provide eye and skin protection against radiation effects when gas-shielded arc welding with;
    (1) tungsten arc process,
    (2) the metal arc process.

(b) Give *one* reason, in *each* case, for the type of filter used.

98 (a) What is the purpose of a choke reactance when used for gas-shielded metal arc welding?

(b) State *two* effects which may be produced by using excessive electrode wire extension during welding.

99 (a) Give *two* advantages of tungsten alloyed electrodes over plain tungsten electrodes when used for tungsten arc welding.

(b) Show by means of simple labelled sketches what is meant by the electrode tip (vertex) angle when tungsten arc welding with;
    (1) alternating current.
    (2) direct current.

100 Show by means of a single labelled sketch the essential differences between the welding power source characteristics suitable for;
    (a) manual tungsten arc welding,
    (b) welding with a self-adjusting arc.

101 (a) State *two* factors which influence the amount of dilution of the weld deposit in gas-shielded arc welding.

(b) With the aid of an outline sketch, show how the level of dilution of the welded joint may be determined.

102 (a) State *two* modes of metal transfer, other than spray, used for gas-shielded metal arc welding.

(b) Give *one* application of *each* mode of transfer named.

103 (a) Give *two* important physical properties of ceramic gas nozzles.

(b) State *two* of the factors which govern the gas-shielding necessary to obtain efficient welded joints by tungsten arc welding.

104 (a) In *each* of the following cases state a possible cause of *one* type of cracking which may be produced in fusion welded joints made in;
    (1) low-alloy hardenable steels,
    (2) austenitic heat resisting steels.

(b) Outline why it is important to avoid sharp corners in fusion welded joints made in structural steel for service at low temperatures.

105 (a) Show by means of a labelled sketch *one* method of providing gas backing for the tungsten arc welding of butt joints in plate in the flat position.

(b) Why should syphon-type cylinders be used to supply carbon dioxide for shielding gas in welding?

106 (a) What is meant by the critical cooling rate of a plain carbon steel?

(b) State *two* detrimental effects which may be produced during the making of a welded joint in 0.4% carbon steel if the critical cooling rate is exceeded.

107 (a) Describe briefly the recommended procedure to follow for clearing a 'burn back' in gas-shielded metal arc welding.

(b) State *four* causes, other than joint preparation, of lack of root penetration in butt welded joints made by gas-shielded metal arc welding.

108 (a) List the following welding processes in their order of usefulness for welding steel components 50 mm thick, in the vertical position:
    (1) arc welding with coated electrodes,
    (2) electron beam welding,
    (3) electroslag welding,
    (4) laser beam welding.

(b) Give a brief outline of friction welding.

109 (a) State the type of current and tungsten alloyed electrode recommended to be used for tungsten arc welding *each* of the following materials:
    (1) low-carbon steel up to 2 mm thick
    (2) austenitic stainless steel up to 5 mm thick
    (3) aluminium over 3 mm thick
    (4) magnesium alloy up to 5 mm thick.

(b) List *three* gases or gas mixtures in tungsten arc welding and give one typical use of each.

110 (a) State *four* factors that have to be taken into account when costing for tungsten arc gas-shielded welding.

(b) Explain what is meant by:
    (1) arcing time,
    (2) floor to floor time.

111 (a) List the main welding problems when tungsten arc gas-shielded welding copper plates in thicknesses rangeing from 3–10 mm.

(b) Outline with the aid of sketches how each of these problems may be counteracted.

112 (a) Describe the effects on the weldability of low-alloy steel of any *two* of the following elements.
    (1) Nickel.

     (2) Hydrogen.

     (3) Chromium.

   (*b*) Explain how the difficulties which may arise from the presence of hydrogen may be overcome during the welding of low-alloy steels.

113 State *five* desirable features in jigs intended for gas-shielded metal arc welding of joints in components to be mass produced.

114 (*a*) With the aid of a labelled block diagram to show the arrangement, describe the purpose of *each* part of the equipment necessary for the effective argon-shielded metal arc welding of aluminium alloys.

   (*b*) State *one* precaution which must be taken in *each* case, with;

     (1) the shielding gas supply lines,

     (2) temporary backing bars,

     in order to assist in the production of efficient joints.

115 (*a*) Give *one* safety precaution to be taken before tungsten arc gas-shielded welding when the gas cylinder is located near the welding area.

   (*b*) Give *two* hazards present when metal arc gas-shielded welding equipment is inadequately earthed.

116 (*a*) State the purpose of a combined transformer-rectifier when used for tungsten arc welding.

   (*b*) Give *two* important reasons for the use of a welding contactor in tungsten-arc welding.

117 (*a*) Briefly describe the mode of solidification leading to columnar grain structure in an autogenous welded joint made by tungsten-arc welding.

   (*b*) Give *one* example of a type of material in which a band of refined grain structure may be expected nearest to the weld boundary in the heat affected zone of a fusion welded joint.

118 (*a*) The current in metal arc gas-shielded welding is indicated by an ammeter positioned in the power source. What is the difference between the current values shown by the ammeter during welding with *each* of the following modes of metal transfer?

     (1) Spray.

     (2) Dip.

   (*b*) What voltage is indicated by the voltmeter incorporated in the power source? State whether the voltage at the arc will remain constant at this same value during welding.

119 What is an alloy? Give *two* reasons why an alloying addition might be necessary in the composition of a metal intended for use as weld metal for joints to be made by gas-shielded arc welding.

120 (*a*) Sketch a tungsten arc welding torch fitted with a gas lens, indicating the pattern of gas flow which would be obtained from the torch.

   (*b*) State *two* advantages which may be obtained by the use of a gas lens in tungsten arc welding.

121 (*a*) Give *two* examples of when pre-heating is essential in the fusion welding of carbon steels.

   (*b*) Explain why the presence of moisture in any form should be avoided when gas-shielded arc welding low-alloy steels.

122 (*a*) Give *one* example of *either* dilution *or* pickup effects arising in gas-shielded arc welding practice.

(*b*) State the shielding gas or gas mixture which is best suited to obtain the required modes of metal transfer for the effective metal arc gas-shielded welding of *each* of the following
(1) low-carbon steel by dip transfer
(2) austenitic stainless steel by controlled spray (pulse) transfer.

123 (*a*) Why may magnetic particle inspection only be used for crack detection in carbon steel and some alloy steels, whereas dye penetrant methods can be applied to all metals?

(*b*) Name *four* of the most possible causes of cracking in welds made by metal arc gas-shielded welding.

124 (*a*) Explain the technique to be followed to avoid contamination of the electrode when tungsten arc welding butt joints in thicker section material in the flat position.

(*b*) Give *two* advantages which may be obtained by the use of the two-operator upward-vertical technique for tungsten arc welding.

125 (*a*) Give *two* reasons why flux-cored electrode wire is used with carbon dioxide shielding for the metal arc gas-shielded welding of steel.

(*b*) When using a gas-shielded arc welding process, on an exposed site, which *two* problems are most likely to be encountered as a result of weather conditions?

126 (*a*) What advantageous feature is provided by a constant potential power source when the arc is initiated to start the weld in metal arc gas-shielded welding?

(*b*) Outline the procedure for rectification when a fault condition has been indicated in a metal arc gas-shielded welding circuit, by the falling voltage reading at the power source, when the end of the electrode wire is touched on to the work with the torch control trigger actuated.

127 Name *four* factors which may affect the cost of metal arc gas-shielded welding a fabrication in a non-ferrous alloy.
With the aid of a sketch, explain briefly the principles of the tungsten arc spot welding process.

128 (*a*) Describe, with the aid of a sketch, the distortion effects which are likely to be produced when an unrestrained 300 mm long single-vee butt joint between 5 mm thick austenitic stainless steel plates, 150 mm wide, is made in two runs from one side by the tungsten arc process.

(*b*) Outline *two* methods used to control distortion in tungsten arc welding operations.

129 List *four* of the main welding problems encountered when tungsten arc gas-shielded welding magnesium alloys in thicknesses ranging from 2 mm to 10 mm, indicating how *each* of these problems is overcome, and mentioning the effects of surface preparation.

130 Twenty 2 metre pipe lengths are to be fabricated from existing stock of 1 metre lengths of 150 mm internal diameter low-carbon steel pipe of 10 mm wall thickness by metal arc gas-shielded welding. Describe, with the aid of sketches, *each* of the following for the welding of one joint:

(a) a suitable joint preparation and set-up,

(b) the assembly and tack welding procedure,

(c) the mode of metal transfer, shielding gas and electrode wire size to be used,

(d) an effective welding procedure.

131 (a) Compare the use of carbon dioxide with the use of argon for metal arc gas-shielded welding by listing *two* relative advantages and *two* relative disadvantages which may be obtained by the use of *each* gas.

(b) State *two* advantages which may be obtained by the use of a manipulator as an aid to fabrication by metal arc gas-shielded welding.

WELDING ENGINEERING CRAFT STUDIES

All dimensions are in millimetres

1 Spot welds are to be made by the tungsten arc gas-shielded welding process.

(a) Make a sectional sketch through one of these spot welds.

(b) State *one* advantage that this process has over the resistance spot welding process for making this type of weld.

(c) Name the additional equipment that would be necessary for making the spot welds using standard tungsten arc gas-shielded welding equipment.

2 (a) How is the size of a fillet weld with normal penetration determined in accordance to BS 499 in

(1) a convex fillet weld,

(2) a concave fillet weld?

(b) Make a sectional sketch of a mitre fillet weld.

3 State *five* factors that would influence the pre-heating temperature to be used for a welded steel fabrication.

4 (a) Give *two* reasons why hydrogen controlled electrodes are preferred for manual metal arc welding restrained joints in low-alloy steel.

(b) Explain the influence of the cellulose in the coating of an electrode on

(1) voltage

(2) penetration.

5 A circular low-carbon steel plate of 750 mm in diameter and 10 mm thick has to be fitted and welded into a deck plate.

(a) State why hot cracking is likely to occur in this particular type of assembly.

(b) Outline *either* a suitable weld sequence *or* a change in the form of the plate insert that could be used to avoid the occurrence of hot cracking.

6 A worn press die has to be built up using the oxy-acetylene flame powder spraying process.

(a) From the table opposite select the powder to be used for this repair.

(b) List *three* advantages that powder spraying has over arc welding for this type of repair.

(c) Name the type of flame setting to be used.

Table (Question 6)

Powder no.	Resistance to: Abrasion	Impact	Machinability
1	Fair	Excellent	Very Good
2	Very Good	Excellent	Very Good
3	Excellent	Poor	Grind Only

7 From the information given below, list in the correct order the welding sequence that should be carried out when friction welding two 20 mm diameter low-carbon steel bars.
(a) Place parts lightly in contact.
(b) Load machine.
(c) Apply axial force.
(d) Apply upset force.
(e) Rotate chuck and close gap.
(f) Release upset force.
(g) Arrest chuck movement.
(h) Remove specimen.

8 Steel cylindrical tanks 5 m long and rolled to 2 m internal diameter from 50 mm thick low-carbon steel plate are to be welded, using the electro-slag welding process for the longitudinal seams.
(a) With the aid of a sketch outline how the weld area is protected from atmospheric contamination.
(b) (1) State *three* advantages that the electroslag welding process has over arc welding processes for welding process for the longitudinal seams.
    (2) Give *two* limitations of the electroslag welding process.
(c) (1) What is the purpose of run-on and run-off plates when electroslag welding?
    (2) After electroslag welding has commenced small additions of flux must continue to be added to the weld pool. Why is this?
(d) If it was considered that the cylinder would be working under conditions that may cause stress corrosion cracking:
    (1) state what is meant by the term stress corrosion cracking;
    (2) state *two* precautions that could be taken to reduce the occurrence of failure from this form of attack.

9 The figure shows a low-carbon steel nut to be resistance projection welded to 2 mm thick low-carbon steel sheet.
(a) Explain the principles involved in making this welded connection using the resistance welding process.
(b) Make a sectional sketch through *A–A* of the welded assembly shown in the figure.
(c) State *two* advantages of resistance welding over manual metal arc welding for making this welded connection.
(d) List *three* defects that may be found when resistance *spot* welding, and in each case state the probable cause of the defect.

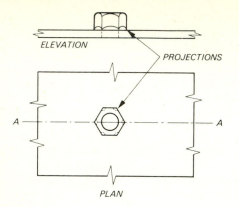

ELEVATION

PROJECTIONS

A ——————————————— A

PLAN

Question 9

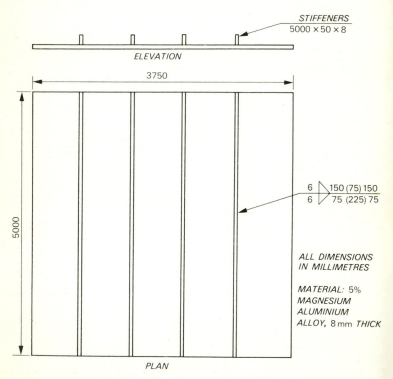

STIFFENERS
5000 × 50 × 8

ELEVATION

3750

5000

6 ▷ 150 (75) 150
6   75 (225) 75

ALL DIMENSIONS
IN MILLIMETRES

MATERIAL: 5%
MAGNESIUM
ALUMINIUM
ALLOY, 8 mm THICK

PLAN

Question 10

10 As part of a welding sequence stiffeners require to be welded to the aluminium alloy plate shown in the figure.
(a) Select a suitable process for welding the stiffeners.
(b) Sketch and label the essential components of the welding circuit for the process selected.
(c) Using a graph to illustrate your answer, give a brief outline of the output characteristics of the power source needed for the process used.
(d) What information is indicated by the weld symbol shown in the drawing?

11 The following thermal cutting equipment is available for use.

  Oxy-Propane manual cutting equipment.

  Arc plasma Profile cutting machine.

  Oxy-Acetylene cutting equipment complete with lance.

  Oxy-Acetylene cutting equipment complete with dispenser powder unit.

(a) From the cutting equipment available select with reasons which cutting process should be used for cutting each of the following materials to the shape or form specified. *Do not* use the same process for cutting both materials.
  (1) 250 mm diameter discs to be cut from 20 mm thick austenitic stainless steel plate.
  (2) The removal of excess material from the grey iron casting shown in the figure.
(b) Outline the principles of the thermal cutting processes used.
(c) State *two* aspects of safety that are particularly appropriate to each process used.

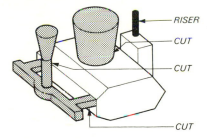

NOTE: SHADED PARTS ARE EXCESS MATERIAL

GREY IRON CASTING

Question 11

12 Alloy steel of the following % composition has to be manual metal arc welded.

Carbon (C)	Chromium (Cr)	Molybdenum (Mo)	Manganese (Mn)	Silicon (Si)	Nickel (Ni)	Venadium (V)
0.16	1.5	0.5	0.8	0.4	0.0	0.0

Remainder – Iron with acceptable limits of impurities.

(*a*) State *three* advantages of using alloying elements in steel.

(*b*) (1) Using the carbon equivalent formula given below determine the carbon equivalent of the alloy steel outlined above.

$$\text{Carbon Equivalent (\%)} = C\% + \frac{Mn\%}{20} + \frac{Ni\%}{15} + \frac{Cr\% + Mo\% + V\%}{10}$$

(2) Indicate how the information produced may be used to determine the welding procedure used.

(*c*) Explain why pre-heating this alloy may reduce the occurrence of under-bead cracking.

(*d*) Welds made on low-alloy steels are generally heat treated by normalizing.

(1) What is meant by normalizing?

(2) State *three* advantages produced by normalizing welded fab-rications.

13 State *five* factors that will influence dilution during fusion welding.

14 (*a*) Give *three* reasons for pre-heating air hardenable steels before or during oxy-fuel gas cutting.

(*b*) Give the main constituents in an air hardenable steel.

15 Briefly explain how a friction weld is produced.

16 (*a*) State *three* advantages of using fully automatic welding processes in preference to manual welding processes.

(*b*) Under what circumstances may manual welding be preferred to fully automatic welding?

17 (*a*) Explain how heat is produced during cutting when using the oxygen lance.

(*b*) State *two* safety precautions that should be taken when using the oxygen lance for cutting.

18 List *five* problems which may be encountered when fusion welding aluminium.

19 Explain briefly why single-pass welds made by the arc plasma process may have the form shown in the macrograph in the figure.

Question 19

20 One thousand circular containers as shown in the figure have been fabricated from low-carbon steel by using the metal arc gas-shielded welding process with carbon dioxide shielding gas.

(*a*) Name the main type of stress, in *each* case, that weld *A* and *B* would be subjected to during hydraulic testing.

(*b*) During testing some welds were found to contain porosity. State *three* probable causes of this defect.

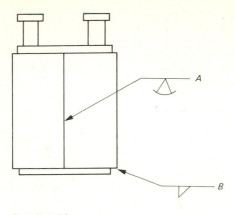

Question 20

21 500 000 shear connectors are to be welded to the low-carbon steel box girder
shown in the figure, using the drawn arc stud welding process.
  (a) In what respect does the purpose built power source used for this type of
      equipment differ from the power source used for metal arc welding?
  (b) Sketch in section, and label the main parts of, a ceramic ferrule used for
      stud welding.
  (c) If the sheer connectors are to be positioned at 100 mm centres, explain
      how this could be best achieved.
  (d) (1) Describe a method of testing the stud welds.
      (2) Welds having the section shown in 'd' 2 below failed during testing
          due to the defects shown. State *four* possible causes of these defects.

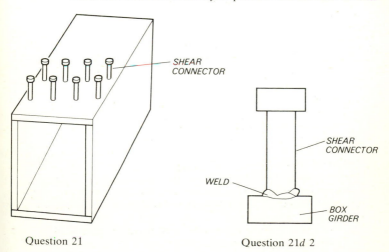

Question 21   Question 21*d* 2

22 A hinge bracket is shown in the figure.
  (*a*) Using graph paper, draw, to a scale of 1 : 4, the elevation in the direction of arrow *A*.
  (*b*) Calculate the total length of welding required to fabricate the bracket.
  (*c*) Why would ultrasonic testing *not* be recommended for testing the fillet welds between the base and the vertical plates?

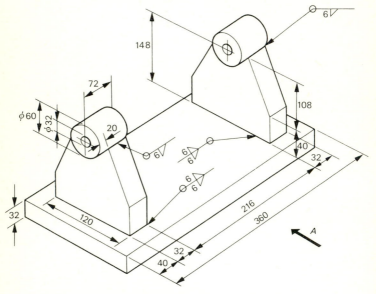

ALL DIMENSIONS IN mm

Question 22

23 (*a*) The figure shows a time and current graph making an electric resistance spot weld. The following two parts of this question refer to this diagram.
      (1) Calculate the weld time as a percentage of the welding cycle.
      (2) Explain why it is generally necessary for the forging time to be longer than the squeeze time.
  (*b*) A resistance spot welding machine was set up for welding low-carbon steel. If austenitic stainless steel of the same thickness is to be welded, what alterations would require to be made to
      (1) the welding current,
      (2) the welding time?
      Explain why these alterations are necessary.
  (*c*) Spot welding of sheet metal may also be carried out by fusion welding.
      (1) Select a fusion welding process and describe how a spot weld is made using this process.
      (2) What additional equipment would be necessary for making spot welds using standard equipment?

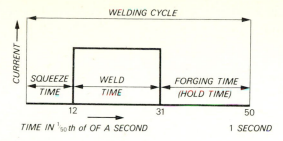

Question 23

24 The figure shows a low-alloy steel rotary shear blade, used for trimming steel plate. The cutting edge has become worn due to severe abrasion during service and is to be repaired by using the manual metal arc welding process.
   (a) The general composition of three electrodes is shown below. Select the most suitable electrode for use in the building up of the cutting edge, and give a reason for your selection.
      (1) Austenitic stainless steel.
      (2) Medium carbon low-alloy steel.
      (3) Pure nickel.
   (b) Outline the welding procedure that should be used to carry out this repair.
   (c) Explain the influence that the dilution of the weld metal by the parent metal would have on the weld's mechanical properties.
   (d) Under what circumstances would the oxy-acetylene welding process be preferred to arc welding processes for hard surfacing?

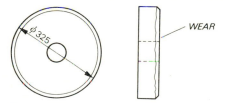

Question 24

25 The figure shows a low-carbon steel shaft to be fabricated by welding using the submerged arc welding process.
   (a) With the aid of a sketch, explain how the weld area is protected from atmospheric contamination.
   (b) State *three* advantages of using a backing strip for this joint.
   (c) Sketch in detail the weld preparation that would be used for the butt weld shown at A.
   (d) Why is the filler used in submerged arc welding copper coated?
   (e) When submerged arc welding, explain why it is an advantage to use a multipower source which has one electrode using alternative current and the other direct current.

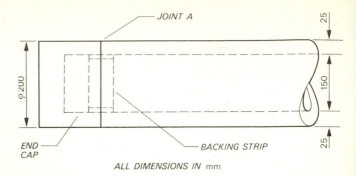

Question 25

26 Power sources used for electron beam welding may be rated as 30 kV.
   (a) Explain the term 30 kV.
   (b) Give *two* reasons why electron beam welding is generally carried out in a vacuum.
27 (a) State *four* of the variables involved in the production of friction welds.
   (b) State the range or temperature necessary for the production of friction welds in low-carbon steel.
28 Sketch in section, and label the parts of, the head of an arc plasma torch for welding.
29 (a) State *three* advantages of introducing iron powder into the flame when oxy-acetylene powder cutting.
   (b) State *two* hazards that the operator should guard against when oxy-acetylene powder cutting.
30 Low-carbon steel plate 25 mm thick has to be surfaced by the submerged arc welding process using stabilized austenitic stainless steel filler. List *five* variables that would influence the degree of dilution found in the weld.
31 Explain why it is recommended that materials containing sulphur should be removed from the weld area before welding nickel and nickel alloys.
32 State *five* factors that would need to be considered before deciding the pre-heating temperature to be used on low-alloy steel plate.
33 The grey cast iron angle plate shown in the figure has to be repaired by manual metal-arc welding and then machined flush. Rutile covered low-carbon steel and nickel alloy electrodes are available for making the weld.
   (a) Select the most suitable electrode to carry out this repair.
   (b) State *two* advantages and *one* disadvantage of the electrode selected.
34 (a) With the aid of a graph, explain why a constant potential power source is often preferred for automatic arc welding processes.
   (b) List *four* variables, under the control of the welder, that can influence weld quality when using the submerged arc welding process.
   (c) Explain why the quality of submerged arc welds made in low-alloy steels is particularly high compared with the quality of welds made by other arc welding processes.

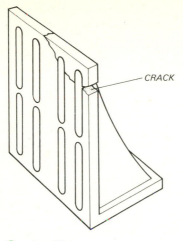

CRACK

Question 33

(d) Inspection authorities may specify that welds made by the submerged arc welding process should be heat treated. Outline a suitable heat treatment.

35 The aluminium–magnesium alloy components shown in the figure are to be welded using the metal arc gas-shielded welding process.

(a) Explain why an argon/carbon dioxide gas mixture would be unsuitable as the gas shield to be used for welding the components.

(b) State *three* factors that would influence the weld profile.

(c) Explain why the oxy-acetylene fusion welding process would be unsuitable for welding the components.

(d) Outline the influence that welding would have on the metal's mechanical properties and grain structure.

(e) If 1000 components are to be fabricated, calculate the total length of welding carried out to complete the contract. Each joint is to be made by means of a single-run weld deposit.

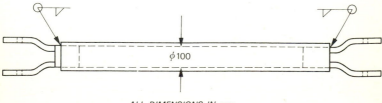

$\phi 100$

ALL DIMENSIONS IN mm

Question 35

36 (*a*) Calculate the volume of metal deposited in the welded joint shown in the figure if the length of the joint is 10 m, and the total weld reinforcement is taken as one quarter of the total volume of the joint gap.

(*b*) Outline the influence of *each* of the following factors on the cost of welded fabrication:
(1) current density,
(2) joint set-up.

(*c*) Outline how *four* factors in the welding procedure can influence welding costs.

(*d*) Sketch an example of a non-load bearing fillet weld, indicating by an arrow the direction of the applied load on the component when in service.

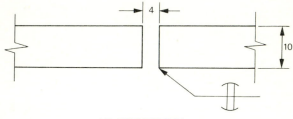

ALL DIMENSIONS IN mm

Question 36

37 A fabricated tank is to be manufactured from stabilized austenitic stainless steel. The weld preparations and holes are to be thermally cut using the arc plasma process.

(*a*) Explain why the arc plasma process would be used in preference to oxygen and fuel gas for cutting this material.

(*b*) Sketch in detail the design of the welded butt joint shown at *A* in the figure.

(*c*) State *two* problems which may be encountered during the arc welding of this joint.

(*d*) Explain why the depth of penetration would be greater in welds made in austenitic stainless steel, compared with similar welds made in low-carbon steel, assuming the energy input to be the same.

(*e*) Explain, with the use of a graph, why a drooping characteristic power source should be used when manual metal arc welding, rather than a constant potential power source.

38 Component parts for an oil rig, having the form shown in the figure are to be fabricated and joined by welding.

(*a*) Calculate the carbon equivalent of the steel given the following
Material composition: Carbon, 0.22%; Silicon, 0.5%; Manganese, 1.5%; Niobium, 0.1%; Vanadium, 0.1%; Sulphur, 0.05%; Phosphorus, 0.05%; remainder iron

$$\text{Carbon equivalent} = C\% + \frac{Mn\%}{6} + \frac{Cr\% + Mo\% + V\%}{5} + \frac{Ni\% + - Cu\%}{15}$$

(b) Sketch in detail the butt joint preparation shown at *B*.

(c) Outline a welding procedure that could be used to make the joint.

(d) Since this component will be subjected to fatigue conditions during service, state *three* precautions which should be taken when welding has been completed.

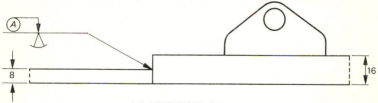

ALL DIMENSIONS IN mm

Question 37

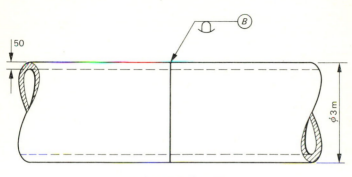

ALL DIMENSIONS IN mm

Question 38

MULTIPLE CHOICE QUESTIONS

Note. The following are examples of the multiple choice type of question but may not be representative of the entire scope of the examination either in content or difficulty.

1 Because of the possibility of explosions, acetylene line fittings should *not* be made from
(a) steel
(b) copper
(c) aluminium
(d) cast iron.

2 One reason why low-carbon steel may be successfully welded by oxy-acetylene without the use of a flux, is that the oxide
(a) is under the surface

(*b*) has a higher melting point than the parent metal

(*c*) has a lower melting point than the parent metal

(*d*) melts at the same temperature as the parent metal.

3 An undesirable property of an aluminium flux residue is that it

(*a*) is corrosive

(*b*) obstructs the vision of the molten pool

(*c*) decreases fluidity

(*d*) requires great heat to melt it.

4 When a low-alloy steel has a hard and brittle structure it may be rendered soft and malleable by

(*a*) recrystallization

(*b*) cold working

(*c*) lowering its temperature

(*d*) hot quenching.

5 What happens to the mechanical properties of steel if the carbon content is increased to 0.5%?

(*a*) The material becomes softer.

(*b*) Malleability is increased.

(*c*) The tensile strength is increased.

(*d*) Ductility is increased.

6 The main reason for pre-heating medium and high-carbon steels before cutting by the oxy-fuel gas technique is to

(*a*) improve the quality of cut

(*b*) increase the cutting speed

(*c*) refine the grain structure

(*d*) prevent hardening and cracking.

7 Which one of the following factors restricts the use of town gas as an oxy-fuel cutting gas?

(*a*) Its low calorific value.

(*b*) Its tendency to cause rapid melting.

(*c*) Its unsuitability for cutting plates less than 12 mm thick.

(*d*) Its relatively high cost.

8 A suitable filler wire for brazing pure aluminium would consist of

(*a*) aluminium bronze

(*b*) aluminium alloy containing 10/13% silicon

(*c*) aluminium alloy containing 5% magnesium

(*d*) pure aluminium.

9 Columnar growth takes place when a metal is

(*a*) cold

(*b*) losing heat

(*c*) being heated

(*d*) being rolled.

10 Difficulty may be encountered when welding aluminium because

(*a*) the weld metal expands during solidification

(*b*) its coefficient of expansion is low compared to steel

(*c*) no colour change takes place to indicate its melting points

(*d*) its thermal conductivity is low compared to steel.

11 One purpose of a microscopic examination of a weld is to establish the
   (a) strength of the weld
   (b) number of alloying elements
   (c) grain size
   (d) number of runs used.

12 Which one of the following components is employed to control amperage in an a.c. arc-welding circuit?
   (a) Rheostat.
   (b) Choke.
   (c) Voltmeter.
   (d) Resistor.

13 When carrying out welds in low-carbon steel, using the carbon dioxide welding process, one purpose of the inductance control is to reduce
   (a) porosity
   (b) penetration
   (c) undercut
   (d) spatter.

14 One purpose of a reactor (choke) when manual metal arc welding is to
   (a) change alternating current to direct current
   (b) allow the correct amperage to be selected
   (c) allow the desired arc voltage to be selected
   (d) enable the correct polarity to be chosen.

15 Which shielding gas is generally recommended when butt-welding 6 mm nickel alloy sheet by the metal arc gas-shielded process?
   (a) Argon.
   (b) CO_2.
   (c) Hydrogen.
   (d) Nitrogen.

16 When TIG welding using a.c. output, which one of the following is essential in the circuit to stabilize the arc?
   (a) A surge injector.
   (b) An open circuit voltage of 100 volts.
   (c) A flow meter.
   (d) An amperage regulator.

17 In manual metal arc welding the flux coating to give deep penetration characteristics would contain
   (a) iron oxide
   (b) manganese
   (c) cellulose
   (d) calcium carbonate.

18 Which element is used as a deoxidant in copper filler rods?
   (a) Aluminium.
   (b) Tin.
   (c) Sulphur.
   (d) Phosphorus.

19 An oxygen cylinder regulator being used in a flame-cutting supply may freeze up if the

(a) gas withdrawal rate is exceeded
(b) cylinder content is too low
(c) cylinder is on its side
(d) needle valve on the regulator is not fully open.

20 To test a component part for a vibrational loading, a suitable mechanical test would be
(a) impact
(b) tensile
(c) compressive
(d) fatigue.

21 The principal advantage of arc-on-gas welding is that it
(a) allows controlled penetration of initial bead
(b) requires less operator skill
(c) entirely eliminates distortion
(d) improves surface finish.

22 One reason why a grey cast iron casting should be slowly cooled after welding is to keep it
(a) soft
(b) spheroidal
(c) hard
(d) brittle.

23 An iron casting has a crack in it. Before oxy-acetylene fusion welding it may be necessary to drill the ends of the crack. One reason for this is to
(a) balance out any shrinkage stresses
(b) stop the crack from spreading
(c) prevent the ends of the crack from being carburized
(d) prevent grain growth.

24 Which one of the following metals may require the studding techniques to be used when being repaired by manual metal arc welding?
(a) Low-carbon steel.
(b) Aluminium.
(c) Nickel.
(d) Cast iron.

25 During the deposition of a manual metal arc electrode, a certain percentage of the core wire is lost. This is due to
(a) voltage drop across the arc
(b) short arc length
(c) spatter
(d) excessive build-up.

26 Which one of the following can be welded by d.c. using the tungsten arc gas-shielded process?
(a) Copper.
(b) Commercial pure aluminium.
(c) Silicon-aluminium.
(d) Magnesium alloys.

27 Backing bars for manual metal arc welding of low-carbon steel should be made from

(a) copper
(b) low-carbon steel
(c) tool steel
(d) cast iron.

28 Peening may be carried out when manual metal arc welding cast-iron in order to
(a) reduce the effects of contraction
(b) make the bond more firmly adhering
(c) refine the grain structure
(d) speed up the welding.

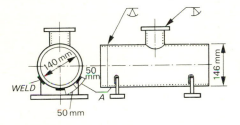

Question 29
The figure shows a component fabricated from stabilized austenitic stainless steel sheet 3 mm thick.

29 The fillet welds on the support brackets in the figure should have 5 mm leg length with 50 mm intermittent welds as shown. The symbol at '*A*' to communicate this information should be

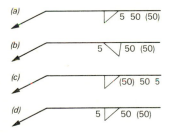

30 What is the volume of deposited metal in a fillet weld indicated by the symbol in the figure neglecting reinforcement?
(a) 14 000 mm³.
(b) 14 900 mm³.
(c) 15 000 mm³.
(d) 15 100 mm³.

Question 30

ANSWER KEY

1 – b	11 – c	21 – a
2 – c	12 – b	22 – a
3 – a	13 – d	23 – b
4 – a	14 – b	24 – d
5 – c	15 – a	25 – c
6 – d	16 – a	26 – a
7 – a	17 – c	27 – b
8 – b	18 – d	28 – a
9 – b	19 – a	29 – d
10 – c	20 – d	30 – c

FABRICATION AND WELDING ENGINEERING TECHNICIANS – PART I

Science

1 (a) What is the relationship between potential difference, current and resistance for a conductor of electricity?

(b) What is the effect of temperature change on the electrical resistance of metals?

(c) (1) For what purpose is a pyrometer used?
   (2) Name two types of pyrometer.

(d) Give a formula to express work done when a force of F newtons moves through a distance d metres.

(e) (1) What is meant by 'oxidation'?
   (2) Name *one* element present in low-carbon steels which serves to minimize oxidation of the alloy during welding.

(f) Explain what is meant by (1) heat, (2) temperature.

(g) Explain simply, the difference between (1) normalizing and (2) annealing.

2 A 100 Ω resistor is connected across a 440 V supply. Calculate (a) the current flow, (b) the resistance that should be used for a current flow of 10 A, (c) the increase in power consumption with the resistor calculated in (b).

3 A fabrication has a volume of 1 m³ at room temperature. Find the new volume of the fabrication after it has been subjected to a temperature rise of 180 °C. (The coefficient of linear expansion for the material is 0.000 011 per degree Celsius.)

4 (a) 'Heat treatment' is the name given to a group of processes designed to affect the properties of metals. Name *four* of these processes.

(b) Outline a suitable procedure for carrying out *one* of the heat treatment processes named in (a). Indicate the temperature range of the process by reference to any given metal.

(c) Give the relationship between force and extension when a metal is loaded within the limit of proportionality.

(*d*) (1) What is meant by the term 'allotropic form of an element'?

(2) Name *two* allotropic forms of *one* common element.

(*e*) Explain the meaning of 'Young's Modulus of Elasticity'.

(*f*) Explain the meaning of 'coefficient of linear expansion'.

(*g*) In a chemical reaction explain what is meant by (1) reduction, (2) oxidation.

(*h*) Explain, with the aid of sketches, how a bi-metal strip may be used *either* to control an electrically operated heating system *or* to activate a fire alarm.

5 A duct of elliptical cross-section, major axis 500 mm and minor axis 360 mm, is to be replaced by a rectangular duct having its cross-sectional length and breadth in the same proportion as those of the elliptical axes. If the replacement duct (rectangular) is to have the same cross-sectional area as the original (elliptical) duct, calculate the length and breadth of the section.

6 A heated steel ingot, of mass 12 kg and specific heat capacity 500 J/kg °C, is immersed in 15 litres of water contained in a vessel of mass 10 kg, specific heat capacity 840 J/kg °C, and at an initial temperature of 15 °C. If the final temperature is 85 °C, calculate the initial temperature of the ingot, neglecting heat losses.

7 (*a*) What is meant by the transfer of heat by radiation?

(*b*) With the aid of a sketch graph, explain the term 'latent heat'.

(*c*) Explain, with the aid of diagrams, the difference between (1) tensile stress, (2) shear stress.

(*d*) Explain why the efficiency of a machine can never reach 100%.

(*e*) Explain what is meant by the terms (1) yield stress (2) factor of safety.

(*f*) A safety (fail safe) device is used in most electrical power circuits. Briefly describe *one* of these devices and explain its operation.

(*g*) What is meant by the term 'solid solution'?

(*h*) Explain, with the aid of a diagram, the principle of operation of any *one* type of pyrometer.

Fabrication processes and welding technology

1 (*a*) Give *two* main differences which distinguish the hose assemblies used for

(1) fuel gas

(2) oxygen.

(*b*) Name *two* devices used for protecting electrical circuits against excess currents.

(*c*) Give *two* reasons for using self-secured joints.

(*d*) Give *three* reasons for swaging metal plate.

(*e*) Sketch the scale of a micrometer barrel and thimble showing a reading of 7.24 mm.

(*f*) State *four* factors which may influence good quality brazed joints.

(*g*) What is the purpose of a folding allowance when folding sheet metal?

(*h*) Give *five* factors to be considered when austenitic stainless steel is suggested as an alternative to aluminium for fabrication purposes.

2 Write explanatory notes on safety requirements in fabrication engineering for the following:

(*a*) lifting tackle,

(*b*) compressed air equipment,

(*c*) handling of gas cylinders,

(*d*) protective clothing,

(*e*) ventilation.

3 Compare the ability to be worked during fabrication operations of *one* ferrous metal with *one* non-ferrous metal.

4 Name *five* materials used for templates and give in each case *two* examples of the type of work for which they are suitable.

5 Compare the cutting processes (*a*) oxy-fuel gas and (*b*) arc, with respect to (1) applications and (2) limitations in fabrication work.

6 (*a*) Longitudinal and circumferential grooves are sometimes provided on horizontal bending rolls. Explain why these grooves are provided.

  (*b*) Compare *two* advantages and *two* disadvantages or the following types of power-driven plate bending rolls:

    (1) horizontal

    (2) vertical.

7 What are the relative merits of

  (*a*) an a.c. transformer welding set,

  (*b*) a d.c. motor generator welding set?

8 (*a*) What precautions should be taken when setting up a single-vee butt joint preparation prior to welding?

  (*b*) What information could be obtained by visual inspection of the joint after welding?

9 Describe what is meant by

  (*a*) self-secured joint,

  (*b*) pitch of holes,

  (*c*) edge distance of holes,

  (*d*) torque spanner,

  (*e*) impact wrench.

10 Sketch a cross section of each of the joint preparations required for the following butt joints, stating in each case *one* advantage of the preparation.

  (*a*) close-square

  (*b*) open-square

  (*c*) open-single-bevel

  (*d*) open-single-vee

  (*e*) close-double-vee.

11 (*a*) Give *two* reasons for depositing a fillet weld in multi-runs.

  (*b*) What is the difference between a welding return lead and an earth return on a metal arc welding set?

  (*c*) Give *four* factors which govern the amount of springback in a piece of metal formed into a cylinder by the use of plate bending rolls.

  (*d*) Make a diagram of the vernier scale and the relevant portion of the main scale of a vernier protractor showing a reading of 32°15′.

  (*e*) What is the difference between chip forming and non-chip forming processes as applied to metal cutting?

    (*f*) Give *two* reasons for ensuring that an acetylene cylinder must be in the upright position when in use.

    (*g*) What is meant by 'the rating of a fuse' in an electrical circuit?

    (*h*) State briefly what is meant by:
        (1) hardness
        (2) toughness.

12 Give *one* advantage and *two* limitations of using *each* of the following metals in fabrication engineering:
    (*a*) cast iron,
    (*b*) high yield steel,
    (*c*) austenitic stainless steel,
    (*d*) copper,
    (*e*) aluminium.

13 Tee section bars of nominal length 1 m are to be made by manual metal arc welding together low-carbon steel of rectangular cross-section 100 mm × 6 mm and length 1 m.

    Describe, with the aid of sketches, *three* different methods of minimizing distortion of the tee sections.

14 (*a*) Give *five* operating advantages of the electro-hydraulic press brake when compared with the mechanical press brake.

    (*b*) Explain, with the aid of sketches, the essential differences between the following for making right-angle bends in plate:
        (1) air bend dies,
        (2) bottoming dies.

15 (*a*) Describe the flame brazing process and give *one* example of its application.

    (*b*) State *four* fundamental differences between brazing and braze-welding.

16 (*a*) Explain the purpose of the following as applied to butt weld joint preparations
        (1) root gap
        (2) root face
        (3) included angle of bevel.

    (*b*) What is the relationship between throat thickness and leg length of a mitre fillet weld?

    (*c*) How is the effective length of a fillet weld determined?

17 The figure overleaf shows a cross-section of a butt joint in low-carbon steel in which the sealing run side is yet to be completed. Describe briefly:

    (*a*) *three* different hand guided processes which would efficiently remove the unfused portion of the joint down to sound weld metal to permit completion of the weld,

    (*b*) *one* manual metal arc welding method which could be used to complete the weld soundly without the removal of the unfused portion of the joint.

18 (*a*) State *three* functions of the flanges on a plate girder.

    (*b*) Name and sketch *three* different methods of stiffening metal plate work without the attachment of additional members.

19 Describe, with the aid of sketches, the following types of structural stanchion (column):

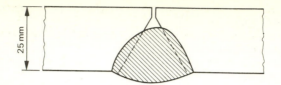

Question 17

(a) compound,
(b) laced,
(c) battened.

20 Compare and contrast the cutting of thin sheet metal by (1) power-operated rotary shears and (2) power-operated nibbling machines under the headings:
(a) cost,
(b) limitations of cut,
(c) quality of cut.

21 (a) Give *two* reasons for 'pre-setting' the butting edges of sheet metal prior to rolling into a cylinder.
(b) A resistance seam weld has been completed in the form of a pillow test piece. Name *two* tests which may be made on this weld.
(c) State *three* essential differences between plumber's (coarse) solder and tinman's solder.
(d) What is meant by the term 'grades of filter' in connection with eye protection in arc welding?
(e) Give *three* reasons why a fuse should be connected in an electrical circuit.
(f) Give *three* reasons why a change in size of electrode may be necessary when arc-welding metal of uniform thickness and composition.
(g) Give *four* observations that may be revealed by the macro-examination of a cross-section of a sound multi-run fillet weld.
(h) A large triangular plate is to be marked off so that the largest possible diameter circle may be flame cut from it. Describe a marking-off procedure to find the centre of the circle.

22 (a) State *five* advantages and *five* disadvantages when using rubber die pads in the press forming of thin sheet metal.
(b) Give *five* reasons for using a hollow false-wired edge on thin sheet metal.

23 Comment upon the following in connexion with safe electrical practice
(a) colour coding,
(b) maintenance of equipment,
(c) earthing,
(d) insulation,
(e) emergency switches.

24 (a) Distinguish, with the aid of labelled, plan-view sketches, between staggered-intermittent and chain-intermittent fillet welds.
(b) Describe, with the aid of sketches, *two* different methods to minimize angular distortion when plate-type flanges are welded to pipe ends.

25 Give any *five* essential properties that a filler metal must possess when used for brazing.

26 Sketch cross-sections of *each* of the following defective metal arc welds in plate material:
   (*a*) unequal leg lengths in a tee fillet,
   (*b*) undersize throat thickness in a tee fillet,
   (*c*) excessive root penetration in a single-bevel butt,
   (*d*) incomplete penetration in a double-bevel butt,
   (*e*) undercut in an asymmetrical double-vee butt.

27 Give *one* example where *each* of the following properties are advantageous in fabrication technology:
   (*a*) tensile strength,
   (*b*) ductility,
   (*c*) malleability,
   (*d*) hardness,
   (*e*) toughness.

28 How does (1) hot forming and (2) cold forming of a low-carbon steel affect
   (*a*) the strucutre of the metal,
   (*b*) extent of forming,
   (*c*) quality of finish?

29 (*a*) Give *seven* advantages when oxy-flame cutting low-carbon steel plates in stack formation.
   (*b*) How may the stack be prepared for cutting without the use of bolts or clamps?
   (*c*) Give *five* requirements for the efficient formation of the stack for cutting purpose .

FABRICATION AND WELDING ENGINEERING TECHNICIANS – PART II

1 A cylindrical low pressure vessel, shown in the figure overleaf, with hemispherical ends is required for the storage of mixed acids. In use it will stand on one end and a flanged inlet will be fitted into the top end. The tank is to be made from austenitic stainless steel of thickness 2 mm. The capacity of the tank, excluding the volume of the inlet, is 3 m³.
   (*a*) Determine the height and surface area of the storage vessel (excluding the flanged inlet), given that the diameter of the vessel is 1 metre.
   (*b*) (1) Describe the methods you would choose to cut and form the hemispherical ends.
       (2) State *four* factors that would influence the choice of the most suitable cutting and forming processes.
   (*c*) (1) Assuming that the joints are welded by the TIG (tungsten inert gas) process, indicate on a sketch the position of the joints and annotate the sketch by use of weld symbols to BS 499.
       (2) Detail the weld procedure required to give full and uniform penetration in the joints.

(d) List the appropriate quality control checks to ensure that the finished tank will be within dimensional tolerances and that the joints will be pressure tight.

(e) Discuss the effects on manufacture and control of increasing the metal thickness of 4 mm to order to accommodate higher pressure.

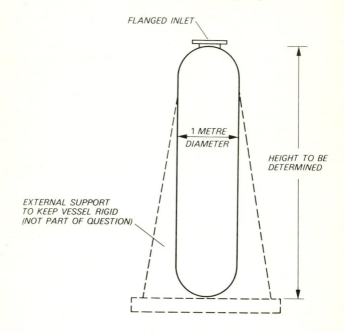

Question 1

2 The portal frame steel trestle, shown in the figure below, has been designed to support a temporary road fly-over. The columns (items *A* and *B*) are to be made from rolled universal beams and are welded, at the lower ends, to plates of thickness 25 mm through which bolts will be placed to locate and secure the trestle.

The cross beam (item *C*) is to be fabricated from plate. The flanges are of thickness 18 mm whilst the web and stiffeners are of thickness 12 mm.

The joints between the columns and the cross beam are to be made on site. Twenty-five trestles are required and a carbon–manganese steel (0.16% carbon and 0.82% manganese) is to be used throughout.

(a) (1) Discuss *two* different methods that could be used to make the beam to column joint on site.

(2) Discuss critically the factors which should be considered in making a choice between these two methods.

(b) Draw up a suitable procedure sheet which could be used by the planning department to programme the manufacture of items *A*, *B* and *C*, excluding site work.

(*c*) (1) Detail the welding processes you would use to fabricate item *C* and explain, with the aid of sketches, what distortion you would expect to occur during welding.

(2) State how this distortion could be controlled.

(*d*) List the inspection checks that should be made on the fabricated sections *A*, *B* and *C* before they are despatched to site.

(*e*) Describe the changes that would need to be made in the manufacturing procedures if a second batch of trestles is to be made from a high yield strength steel that may be sensitive to heat affected zone cracking during welding.

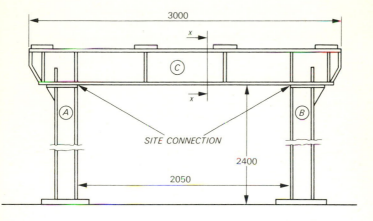

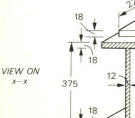

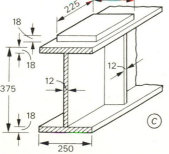

VIEW ON
x—x

ALL FILLET WELDS
ARE TO HAVE LEG
LENGTH OF 9 mm

ALL DIMENSIONS IN
MILLIMETRES

DO NOT SCALE

Question 2

3 The gearbox end unit, shown in the figure, is to be a welded fabrication manufactured in large numbers, from a heat-treatable aluminium–magnesium–silicon alloy.

After welding the components are to be heat-treated as follows:
  (1) heat to 500 °C for 2 hours,
  (2) water quench,
  (3) heat to 175 °C for 10 hours,
  (4) air cool.

The bores of the bosses are to be machined after fabrication.

(a) Choose, giving *four* reasons, *one* method for profiling items A and B from the plate material and show on a sketch, to a reasonable scale, how the items would be set out on a plate to achieve minimum scrap loss. Assume that the plate measures 2500 mm by 1250 mm.

(b) Describe *two* cleaning treatments recommended for the preparation of surfaces prior to welding and comment on any safety hazards associated with these techniques.

(c) Describe a typical manufacturing sequence, paying particular attention to the choice of method of welding and suitable assembly procedure.

(d) Explain why it is necessary to heat treat the unit after fabrication and state the changes in properties produced at each stage of the heat-treatment.

(e) Describe, with reasons, an alternative method of manufacturing the component, bearing in mind that a large number is required.

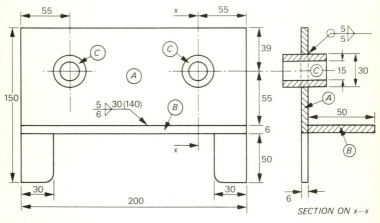

ALL DIMENSIONS IN MILLIMETRES

Question 3

4 Fifty press frames, one of which is shown in the figure, are to be fabricated from steel plate to BS 4860 grade 50D (0.16% carbon, 1.5% manganese). Flat plates of suitable thicknesses are available in sizes which allow each individual

item to be cut in one piece. The welds are to be deposited by manual metal arc welding.

(a) (1) Describe, in detail, *one* method of forming item A to the shape shown.

(2) Detail the aspects of the forming procedure which determine the minimum radius (R) that can be achieved with the method described in (1) and discuss the relevant factors which control the dimensional accuracy of the finished component.

(b) (1) Choose *one* suitable thermal cutting method for profiling the side panels (items B) and explain the reasons for your choice.

(2) List the factors which govern the quality and accuracy of the cut edges. How does plate composition affect quality?

(c) (1) It has been found from previous experience that the volume of weld metal deposited by an electrode is not constant and in checks on the deposition of 100 electrodes, the following results were obtained:

Volume of weld deposited by *one* electrode (mm³)	Number of electrodes
9000 – 9249	2
9250 – 9749	1
9500 – 9749	4
9750 – 9999	3
10 000 – 10 249	7
10 250 – 10 499	15
10 500 – 10 749	24
10 750 – 10 999	27
11 000 – 11 249	17

Plot this data in the form of a histogram and determine the arithmetic mean.

(2) The fillet weld joining the outer member of the frame (item C) to the side plate (item B) can be deposited with a flat face and a leg length accuracy of ± 1 mm as shown in the figure.
Calculate:

(1) the average number of electrodes required to weld the joint,

(2) the maximum number of electrodes that might be needed to make one joint.

(d) (1) Identify and discuss the cause of *two* defects which are most likely to occur in the welded joints.

(2) Describe, in detail, *one* non-destructive test which could be satisfactorily used to examine the quality of the completed welds.

(e) Draw up a brief fabrication procedure sheet for the complete press frame and discuss the role of supervision at each stage of fabrication in achieving dimensional accuracy of the finished unit.

5 The lightweight vat lid shown in the figure is to be fabricated in large numbers from low-carbon steel sheet of thickness 1 mm and preformed Z-section stiffeners (items A) of the same thickness.

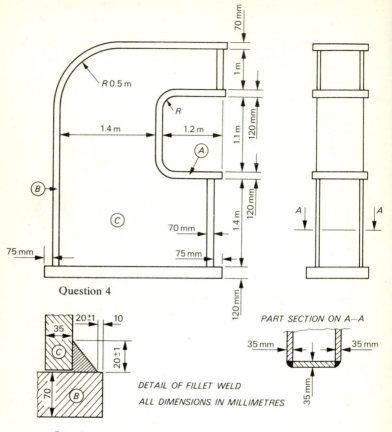

Question 4

DETAIL OF FILLET WELD

ALL DIMENSIONS IN MILLIMETRES

PART SECTION ON A—A

Question 4

(a) (1) Calculate the dimension X required to give a volume of 0.58 m³ contained within the inner skin of the lid.

(2) Determine the surface area of the inner skin (item B) and give the size of sheet required to form this item in one piece.

(b) Describe *one* procedure and the associated techniques for the forming of the inner and outer skins (items B and C), paying particular attention to the degree of accuracy that can be achieved.

(c) (1) Define the mechanical properties that are important in assessing the suitability of a steel for forming by pressing and bending.

(2) Outline *two* tests by which it would be possible to check the acceptability of sheets which are to be used to form the inner and outer skins of the lid.

(*d*) Describe *two* methods of joining the Z-section stiffeners (items *A*) to the inner and outer skins (items *B* and *C*), bearing mind that access is limited when the second sheet is being joined to the stiffeners.

(*e*) Describe *two* methods by which the lid could be protected against rusting. For each method discuss *two* advantages for its use.

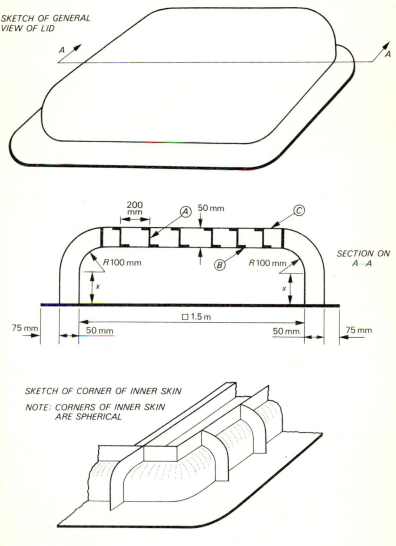

SKETCH OF GENERAL VIEW OF LID

SECTION ON A—A

SKETCH OF CORNER OF INNER SKIN

NOTE: CORNERS OF INNER SKIN ARE SPHERICAL

Question 5

6 A carbon-manganese steel pipe (A) of diameter 1.2 m is part of an existing chemical processing plant. It is planned, during a shut-down period, to connect a stainless steel branch pipe (B) into the system as shown in the figure. The branch is to be set-on and full penetration welds are required. The pipes have the following compositions

Composition (%)	Carbon	Manganese	Chromium	Nickel	Titanium
M in pipe (A)	0.18	0.85	nil	nil	nil
Branch pipe (B)	0.11	2.00	19.5	10.5	0.45

    (a) (1) Describe, in detail, *one* method by which the hole in the main pipe can be cut, bearing in mind that the pipe cannot be moved, and outline *one* technique for preparing the end of the branch pipe.

        (2) Discuss the factors which will determine the accuracy of fit that can be obtained on assembly of the joint.

    (b) Identify the safety hazards that could be involved in cutting the pipes and discuss the provisions that should be made to protect the personnel involved.

    (c) (1) Assuming that an electrode or filler wire of the same composition as the branch pipe is used to make the weld, what is the maximum proportion of main pipe material that can be melted into the weld pool if the chromium content of the solidified weld metal is to be not less than 16%?

     Calculate the weld metal composition at this dilution.

        (2) Explain why titanium is added to the stainless steel for the branch pipe.

    (d) Describe a suitable weld procedure for the branch joint paying particular attention to the choice of welding process and the metallurgical problems involved in making a stainless steel to carbon steel joint.

    (e) (1) Discuss the tests that an insurance surveyor would require before allowing the joint to be made on site.

        (2) Outline the difficulties which would be involved in inspecting the completed joint and suggest, with reasons for your choice, a suitable non-destructive testing technique.

7 The component shown in the figure is part of a machine tool assembly which is to be produced in batches of 1000 units. Item A is to be made from a steel, containing 0.45% carbon and 1.0% manganese, which can be hardened by heat-treatment. The remaining parts of the component are to be made in steel containing 0.15% carbon and 0.8% manganese. All the joints are to be welded with fillet welds of leg length 6 mm.

    (a) (1) Describe *one* thermal and *one* mechanical method of cutting the individual items, excluding the stub tube (item B).

        (2) Explain the effect of the composition of item A on the quality of the cut produced by these processes and show how this could influence the choice of the most suitable method.

    (b) Outline a suitable fabrication procedure for the component, and discuss

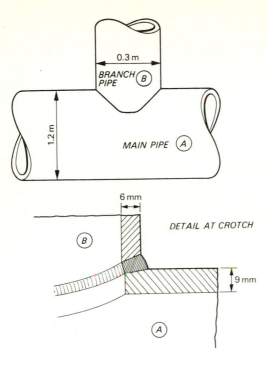

0.3 m

BRANCH
PIPE (B)

1.2 m

MAIN PIPE (A)

6 mm

DETAIL AT CROTCH

(B)

9 mm

(A)

Question 6

in detail those aspects of fabrication which affect the accuracy of the finished unit.

(c) (1) Describe briefly, with the aid of a diagram, a suitable welding technique for the deposition of the fillet welds, giving reasons for your choice.

(2) Detail the changes in welding procedure which must be made when welding onto item A and specify an inspection procedure which could be used to verify the quality of the welds.

(d) (1) Describe a suitable heat-treatment process for the hardening of the machined face of item A and comment on the metallurgical changes produced.

(2) Outline those aspects of the heat-treatment that must be taken into account to achieve uniformity of the finished product.

(e) (1) Identify the changes which would be needed to make the unit suitable for production by means of casting.

(2) Discuss the *four* most important factors which would be taken into account when choosing between fabrication and casting for the production of the 1000 units.

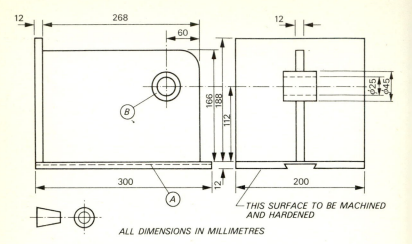

THIS SURFACE TO BE MACHINED AND HARDENED

ALL DIMENSIONS IN MILLIMETRES

Question 7

8 The tee-section duct shown in the figure is to be produced from steel sheet of thickness 0.9 mm at a rate of 750 per week. Both the internal and external surfaces of the ducts are to have a galvanized finish. The dimensional tolerances at the ends are based on the need for a sliding fit into neighbouring units in a ventilation system carrying moist air. The joints of the completed ducting must be leak tight.

(*a*) Describe, with the aid of sketches, how the square to circular transformer can be fabricated. State, with reasons, the most suitable method of cutting the material for the transformer.

(*b*) Calculate the total mass of the unit to the nearest 0.1 kg, assuming that the corner joints are not flanged and that the 0.9 mm thick sheet is of mass 7.02 kg/m².

(*c*) (1) Describe *one* welding technique and *one* mechanical method suitable for making the corner joints (*A*).

(2) Detail *two* advantages and *two* disadvantages of each method and state, with reasons, which would be preferred in practice.

(*d*) (1) Discuss the health hazards that can be encountered when welding and cutting steel sheet and comment on any special risks associated with the use of galvanized material. Explain briefly how workshop personnel can be protected against these hazards.

(2) Outline *one* workshop test that could be used to check the duct for leak tightness.

(*e*) (1) Comment on the reactions that may occur in service at areas where the galvanizing has been damaged and explain what may happen if copper pipes and fittings were introduced into the galvanized duct system.

(2) Name *one* suitable metallic or non-metallic material which could be

used as an alternative to galvanized steel sheet in ventilation, fume removal or hot air distribution systems. Give *two* advantages for the use of this alternative material and explain why galvanized sheet is still preferred for the majority of applications.

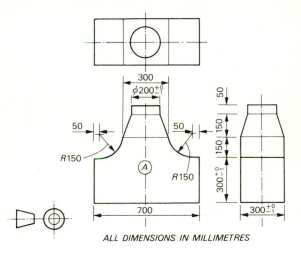

ALL DIMENSIONS IN MILLIMETRES

Question 8

9 The shell and the dished end of the cylindrical vessel shown in the figure are of thickness 30 mm and the base plate has a thickness of 50 mm. The dished end can be purchased ready formed. All the joints must be pressure tight and the welds must be free from significant defects.

The vessel is to be fabricated from *either* (1) aluminium–5%, magnesium–1%, manganese alloy *or* (2) austenitic stainless steel.

With reference to *one* of these materials

(*a*) Describe *one* thermal and *one* mechanical method of cutting the holes at positions *A* and *B*. State *two* advantages of each method and suggest, with reasons, which technique would be preferred.

(*b*) (1) Outline how the cylindrical shell (*C*) can be fabricated. Comment in particular on the factors governing the accuracy of fit between the shell and the dished end, and the alignment of the holes at *A* and *B*, which must be diametrically opposed to within $\pm \frac{1}{2}°$.

(2) Explain why an intermediate heat-treatment may be necessary when cold forming the plate and state the metallurgical changes produced by the heat-treatment.

(*c*) Detail the procedure for preparing and cleaning the plate edges at *D* and *E*. Identify those stages at which hazards are likely to arise and recommend suitable safety precautions necessary to protect the personnel involved.

(d) (1) Bearing in mind the size of the shell, select and describe in detail a suitable process for the welding of the joints at D and E.

(2) What safety precautions must be taken during the welding operation?

(e) List the defects which are likely to occur in the longitudinal weld in C and describe one non-destructive test which could be used to check that the weld is free from internal defects.

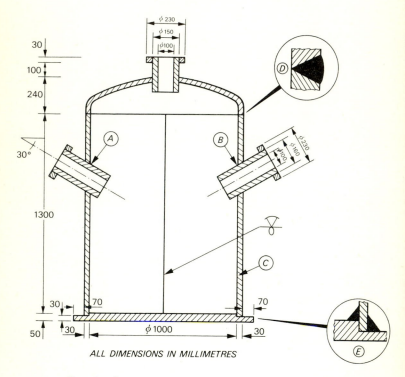

ALL DIMENSIONS IN MILLIMETRES

Question 9

10 (a) State *three* factors which are likely to influence brittle fracture in low-carbon steels.

(b) Name *one* grade of steel which could be used to reduce brittle fracture to a minimum and state *one* welding process which would be suitable for use on this material.

(c) State why metals such as tin, lead and zinc are never used for creep resistance applications.

(d) If an internal defect is present in a welded butt joint subjected to alternating stress, what type of failure would be most likely to occur? State *one* means by which the likelihood of failure could be reduced.

(e) Show, by means of a simple diagram, the likely mode of variation in (1) strength and (2) ductility of a low-carbon steel subjected to a temperature rise from 15 °C to 500 °C.

11 (a) What common characteristic do certain alloys have which enables them to respond to precipitation hardening?

(b) Why is solution treatment necessary in precipitation hardening?

(c) State the effects of fusion welding on the tensile strength and ductility of precipitation-hardened alloys.

(d) State *two* likely effects of fusion welding on the corrosion resistance of precipitation-hardened alloys.

(e) State, giving *one* reason for its suitability, *one* method of jointing an aluminium–4% copper heat-treatable alloy bracket to an existing component. (Solution treatment is not possible after joining.)

12 (a) State *three* properties necessary to ensure that two different metals may be effectively fusion welded to each other.

(b) In the heat-treatment of ferrous metals why is allotropy of iron important?

(c) Why are some steels hardened by quenching in oil and others by cooling in air? Name *one* type of steel in each case.

(d) Name *one* type of alloy steel most suitable for case-hardening, and give *one* reason for its suitability.

(e) Outline, with the aid of simple diagrams, the principal changes that take place in the grain structure of normalized low-carbon steel when it is (1) annealed and (2) cold-worked.

13 (a) With the aid of a simple outline diagram, show how a plasma stream may be generated for use as a heat source in welding.

(b) State *two* operations other than welding that may utilize arc plasma as a source of energy.

(c) Explain what is meant by the term 'laser beam'.

(d) Describe briefly a type of torch suitable for the fusion-welding of thermoplastics.

(e) State *two* operations necessary to complete a solid-phase weld.

14 State the information that can be obtained from *each* of the following types of test:

(a) notch sensitivity,

(b) dye penetrant,

(c) creep,

(d) controlled thermal severity,

(e) Erichsen cupping test.

15 How is industrial organization influenced by *each* of the following:

(a) variation of management structure to suit method of production,

(b) productivity,

(c) industrial training policy,

(d) trade unions,

(e) quality of product?

16 In a random sample of metal strips cut on a guillotine the width of the strips (w) and the frequency (f) at which these occur are tabulated below

Width of strip (w) mm	29.1	29.2	29.3	29.4	29.5	29.6
Frequency (f)	60	90	150	140	110	50.

(a) On graph paper draw a frequency polygon to illustrate these results.
   Horizontal scale 2 cm = 0.1 mm between 29.0 mm and 29.6 mm.
   Vertical scale 1 cm   = 20 components.

(b) Calculate the arithmetic mean width of the strips.

(c) Find the standard deviation.

(d) What information does the standard deviation convey to an inspector?

(e) If an acceptable range is $\left.\begin{array}{l}+ \text{ two} \\ - \text{ one}\end{array}\right\}$ standard deviations, how many of the strips are to be rejected?

17 (a) Upon what factor does the loudness of a musical note depend?

(b) What is the difference between a longitudinal and a transverse vibration?

(c) State how sound may be totally eliminated or regulated.

(d) Give *one* industrial use of an echo-sounder. State briefly how the echo-sounder works.

(e) Describe how an ultrasonic detector works in determining the position of a defect in a welded component.

18 (a) An electric heater with an effective resistance of 10 Ω is coupled to a 250 V supply. If electrical energy costs 0.9 p/kWh, how much does it cost to operate the heater for 40 hours?

(b) the figure shows a circuit diagram, calculate:
   (1) the current in the branch *ABD*
   (2) the current in the branch *ACD*
   (3) the reading of the voltmeter *V*.
   (4) the value of the resistance *X*.

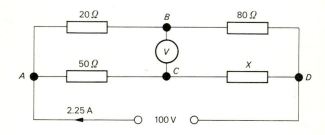

Question 18(*b*)

19 (a) Explain the meaning of the term 'mass effect' in the heat-treatment of plain carbon steels.

(b) When is it necessary to determine the 'carbon equivalent' for an alloy steel? How is the information obtained applied in practice?

(c) How is the critical cooling rate affected by an increase in carbon content of the steel?

(*d*) How is the grain size of a carbon steel affected by:
    (1) normalizing,
    (2) annealing.

(*e*) Explain why a steel of 0.55% carbon content needs pre-heating before welding. Suggest a suitable temperature for this operation.

20 (*a*) How is the formation of chromium carbide prevented when austenitic stainless steel is fusion welded?

(*b*) Explain how a malleable cast iron fitting could be joined by welding to a low-carbon steel casing.

(*c*) Describe briefly the changes in structure which occur when cold rolled copper is welded.

(*d*) Name *four* elements which may be found in materials used for the hard-facing of ferrous alloys.

(*e*) What is the meaning of sacrificial corrosion and how is this made use of on some fabrications?

21 (*a*) What is the function of a suppressor on a tungsten arc gas-shielded welding set?

(*b*) What is the function of a contactor in a gas-shielded welding circuit?

(*c*) The following gases are used for gas-shielded metal arc welding. In *each* case state *one* material and the mode of metal transfer for which each gas is most suitable:
    (1) argon,
    (2) argon + 15% nitrogen,
    (3) argon + 5% oxygen.

22 (*a*) A d.c. welding power supply has an open circuit voltage of 70 V and it is rated to supply 160 A. During normal loading the generator terminal voltage is reduced by 50%. To prevent the current exceeding 150% of the normal when striking the arc, a resistor of 0.04 Ω is permanently connected in series with the generator terminals and the electrodes. Determine the voltage across the electrodes and the load current, when;
    (1) the set is switched on,
    (2) the electrode touches the work,
    (3) the arc is established.

(*b*) Show, by means of a diagram, how a Wheatstone Bridge may be used to measure a resistance.

(*c*) Show, by means of a diagram, where a voltmeter and an ammeter should be fitted into an electrical circuit.

23 (*a*) State the underlying principles involved in;
    (1) magnetic crack detection,
    (2) radiographic examination,
    (3) dye-penetrant inspection.

(*b*) What is the difference between microscopic and macroscopic examination of metal?

(*c*) State the conditions necessary to ensure a satisfactory bond in a cold pressure weld.

24 Write a brief account on *each* of the following:

(*a*) photo-copying and its application to the marking out of templates,

(*b*) safety requirements with regard to lifting equipment,
(*c*) methods of inspection of materials and consumables,
(*d*) the use of sub-assembly drawings,
(*e*) interstage component inspection.

25 (*a*) The figure shows a rectangular pyramid to be fabricated by welding together four separate plates. Calculate:
(1) the total length of weld required
(2) the angle that the edge *AE* makes with the horizontal
(3) the cutting angle at the apex for plate *X*.

(*b*) The connection between the absolute pressure *P* and the volume *V* of a gas is given by the formula $PV^n = C$. When *P* equals 12 units and *V* equals 3 units the value of *C* is 62. Calculate the value of the index *n*.

(*c*) A batch of four plates contains two which are defective and two which are acceptable. What is the probability of selecting one defective and one acceptable plate in any order?

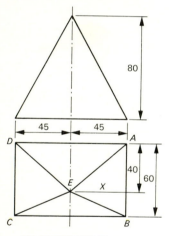

ALL DIMENSIONS IN mm

Question 25

26 (*a*) Show, by means of a diagram, the relationship between the Celsius and Kelvin temperature scales.
(*b*) State the circumstances where a joint would be secured by means of an adhesive.
(*c*) A thin walled cylinder of internal diameter 800 mm has walls of thickness 12 mm. If the internal pressure is 120 N/mm², find the circumferential stress in the walls.
(*d*) Give a brief explanation why fusion welding should not be used for securing joints on heat-treated aluminium–copper alloys.
(*e*) State the relationship between heat, electrical and mechanical energy and explain how these may be calculated.

27 (a) Explain the meaning of the term 'ruling section' in the heat-treatment of steel.

(b) Why does low-carbon steel corrode when exposed to a damp atmosphere?

(c) Iron is said to be an 'allotropic element'. Explain what this means.

(d) State *four* properties of ceramic materials.

(e) State *four* properties of titanium.

28 (a) State *two* consumable electrode and *two* non-consumable electrode arc welding processes and give *one* application for *each*.

(b) Explain briefly how the electroslag welding process is carried out.

(c) What is the difference between electroslag welding and electrogas welding?

(d) State *four* essential conditions to be satisfied in order to make an acceptable soft-soldered joint.

(e) A tin-lead solder has a composition of 63% tin and 27% lead. Indicate *two* special properties possessed by this solder.

29 (a) When calculating the effective area of the butt welded joint carrying a tensile load, as shown in the figure below, it is required to deduct twice the plate thickness from the total length of the weld. Give *two* reasons why this should be done.

(b) In the figure below, the butt welded joint, on a plate of width 500 mm, is required to support a tensile load of 720 kN. If the maximum permissible stress in the weld is 80 N/mm², show that the plate thickness t is obtained from the equation

$$t^2 - 250t + 4500 = 0.$$

(c) Calculate the plate thickness t required to satisfy the above equation.

(d) If the plates for the butt welded joint, shown in the figure, are to be welded in large quantities by manual methods, the production rate is 12 per hour at a production cost of £4.00 per hour. If a fixture is made for a

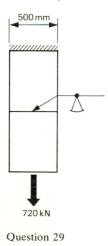

Question 29

cost of £150, output can be increased to 20 per hour at a cost of £3.00 per hour.

(1) Plot a graph to show the production cost and output by each method. (Scale: horizontal 25 mm = £50.00; vertical 25 mm = 250 components.)

(2) Determine, from the graph, the minimum number of plates to be welded in order that a fixture may be justified.

30 (*a*) Show, by means of a diagram, the path of a parallel beam of light after refraction through a converging lens. Indicate on the diagram;

(1) the principal focus,

(2) the focal length.

(*b*) Show, by means of a diagram, how light rays in a microscope are employed to illuminate the surface of a metallurgical specimen.

(*c*) How does the use of an etchant enable the grain boundaries of a polished metal surface to be clearly observed when viewed through a microscope?

(*d*) Describe briefly any *two* methods of non-destructive testing for inspection of welded joints.

31 (*a*) What effect does the cold working of low-carbon steel have upon,

(1) mechanical properties,

(2) grain structure?

(*b*) Sketch a typical stress–strain diagram for a tensile test on a fully annealed low-carbon steel specimen.

(*c*) Indicate on the diagram sketched in (*b*) the expected changes if the steel specimen had been subjected to cold working.

(*d*) What influence does the grain size have on the mechanical properties of a metal?

(*e*) Name *three* atomic structures found in metals. Give *one* example of *each*.

32 (*a*) State *four* main factors which influence the brittle fracture of low-carbon steels.

(*b*) Why is it necessary to pre-heat martensitic stainless steel before welding?

(*c*) What effect do high service temperatures have on the properties of a material?

33 Write brief notes on each of the following topics.

(*a*) Methods of inspection of materials as received from suppliers.

(*b*) Function of an approval authority, such as Lloyds.

(*c*) Pressure testing of vessels containing liquids.

(*d*) Compilation of material and parts lists on fabrication drawings.

34 (*a*) Explain what is meant by;

(1) electrical resistivity

(2) three-phase supply.

(*b*) Calculate the resistance of 100 m of copper cable of cross-sectional area 1.5 mm^2. (Resistivity of copper is $1.78 = 10^{-8}$ Ω/m^3.)

(*c*) In a hydraulic press, oil flows at the rate of 6 litres per minute against a piston of diameter 56 mm at a pressure of 1.8 N/mm^2. Calculate:

(1) the force on the piston

(2) the power output, in watts,

(3) the efficiency, if the power input meter reads 1 A at 240 V.

DRAWING AND DEVELOPMENT

1 A pictorial sketch having only the main dimensions on it and showing a mixing
   tank with stiffening bracing on the legs is shown in the figure.
   (*a*) Draw, to a scale of 1 to 20, a fully-dimensioned working drawing, in third
      angle projection, of the mixing tank.
      (Include the motor support platform and drain, but omit the legs and
      bracing.)
   (*b*) Draw, to a scale of 1 to 20, a fully-dimensioned working drawing, in first
      angle projection, of the assembled cross-bracing and supporting legs.
2 Develop, to a scale of 1 to 5, a template for the drain which is to be formed
   from one piece of material.
3 Cleat **A** is to be cut from plate and bent to shape.
   (*a*) Determine, geometrically and to a scale of 1 to 10, the correct bend
      angle θ.
   (*b*) Draw, full size, a fully-dimensioned template for this cleat.
4 The motor support platform is to be profile cut in one piece and bent to shape.
   Develop, to a scale of 1 to 5, a profile template for this platform and include
   the elongated holes.
5 Develop, to a scale of 1 to 2, a pattern for the capping plates **C** which are
   manually metal arc welded to the top of each leg.
6 Draw, full size, a true view of the gusset plate **B** locating the members in a
   suitable position on the gusset plate.
7 Develop, full size, by any suitable method, a template for one of the spherical
   corner pieces in the tank.
8 A pictorial sketch, having only the main dimensions on it, of a transfer hopper
   and support is shown in the figure.
   (*a*) Draw to a scale of 1 : 20, a front and side elevation of the hopper only in
      the direction of arrows *A* and *B*. Indicate *five* main dimensions and add
      five weld symbols.
   (*b*) Draw to a scale of 1 : 20, a front elevation and plan of the supporting
      framework only in the direction of arrows *A* and *C*. Indicate *five* main
      dimensions.
9 Develop, to a scale of 1 : 20, a template for the hopper body **F**.
10 Draw, full size, a fully dimensioned template for cleat **A** which is to be cut
   from plate and bent to shape.
11 Draw, to a scale of 1 : 20, a template for the hopper platform. Fully dimension
   this template in order that it may be marked out direct from your drawing.
12 Make sketches in good proportion to illustrate;
   (*a*) a cross section, showing a possible weld preparation suitable for the neck
      flange to neck joint detail,
   (*b*) a *fully dimensioned* template for part **E** of the hopper outlet elbow,
   (*c*) a cross section, through a bolt, of the platform flange to platform joint,
   (*d*) a pictorial view of leg top and capping plate.
   Use symbols to indicate the welds.
13 Develop, to a scale of 1 : 10, a template for the oblique conical hopper top **G**.
   Mark clearly on your template the position of the platform flange.

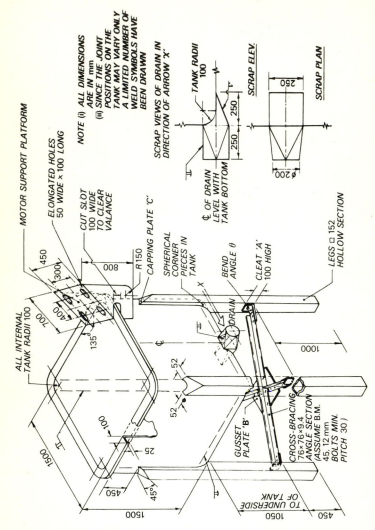

NOTE (i) ALL DIMENSIONS ARE IN mm
(ii) SINCE THE JOINT POSITIONS ON THE TANK MAY VARY ONLY A LIMITED NUMBER OF WELD SYMBOLS HAVE BEEN DRAWN

SCRAP VIEWS OF DRAIN IN DIRECTION OF ARROW 'X'

TANK RADII 100

250
250
250

℄ OF DRAIN LEVEL WITH TANK BOTTOM

SCRAP ELEV.

250
200 ∅

SCRAP PLAN

MOTOR SUPPORT PLATFORM

ELONGATED HOLES 50 WIDE × 100 LONG

CUT SLOT 100 WIDE TO CLEAR VALANCE

CAPPING PLATE 'C'

SPHERICAL CORNER PIECES IN TANK

BEND ANGLE θ

CLEAT 'A' 100 HIGH

LEGS □ 152 HOLLOW SECTION

450
300
008
R 150
700
400
135°

ALL INTERNAL TANK RADII 100

100

1000

DRAIN

X

℄

52
52

GUSSET PLATE 'B'

CROSS-BRACING 76×76×9.4 ANGLE SECTION (ASSUME B.M. 45, 12 mm BOLTS MIN. PITCH 30)

1500
1500
450
45°
25

TO UNDERSIDE OF TANK

1050
450

Questions 1–7

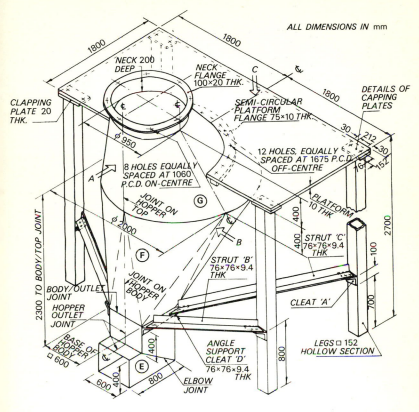

ALL DIMENSIONS IN mm

NECK 200 DEEP

NECK FLANGE 100×20 THK.

C

SEMI-CIRCULAR PLATFORM FLANGE 75×10 THK.

DETAILS OF CAPPING PLATES

CLAPPING PLATE 20 THK.

ϕ 950

8 HOLES EQUALLY SPACED AT 1060 P.C.D. ON-CENTRE

A

JOINT ON HOPPER TOP

G

ϕ 2000

F

JOINT ON HOPPER BODY

BODY/OUTLET JOINT

HOPPER OUTLET JOINT

BASE OF HOPPER BODY

□ 600

E

600 400

800

ELBOW JOINT

2300 TO BODY/TOP JOINT

400

ANGLE SUPPORT CLEAT 'D' 76×76×9.4 THK

STRUT 'B' 76×76×9.4 THK

B

12 HOLES, EQUALLY SPACED AT 1675 P.C.D. OFF-CENTRE

30   212

30

152

6

PLATFORM 10 THK

400

400

STRUT 'C' 76×76×9.4 THK

CLEAT 'A'

LEGS □ 152 HOLLOW SECTION

2700

100

700

800

Questions 8–14

14 Draw in detail, to a scale of 1 : 10, bracing struts **B** and **C**.

15 A pictorial sketch, showing only the main dimensions, of a storage hopper platform and support is given in the figure.

   (*a*) Draw to a scale of 1 : 20, a front elevation and plan, in the direction of arrows *A* and *B* of the *hopper only* with the domed top bolted onto it. Add five main dimensions, five weld symbols and indicate whether your drawing is in first or third angle projection.

   (*b*) Draw to a scale of 1 : 20, a front elevation and plan, in the direction of arrows *A* and *B* of the *platform* and *supporting structure* only. Add five main dimensions and indicate whether your drawing is in first of third angle projection.

16 Develop, to a scale of 1 : 20, a template for the hopper body **C**. Mark clearly on your template the position of the four brackets **E**.

17 Draw, full size, a fully-dimensioned template for the bracket **E** which is to be cut from plate and bent to shape. Indicate bend angle θ.

18 Draw, to a scale of 1 : 20, a template for the platform **F.** Fully dimension this template in order that direct marking may be used.

19 Develop, to a scale of 1 : 10, a template for *one* segment on the domed top marked **G.** Assume the metal thickness remains constant during forming.

20 Draw, full size, a fully-dimensioned template for gusset plate **D,** locating the cross-bracing in a suitable position.

21 Draw in detail, to a scale of 1 : 10, a fully-dimensioned template suitable for making the angle supports **S.**

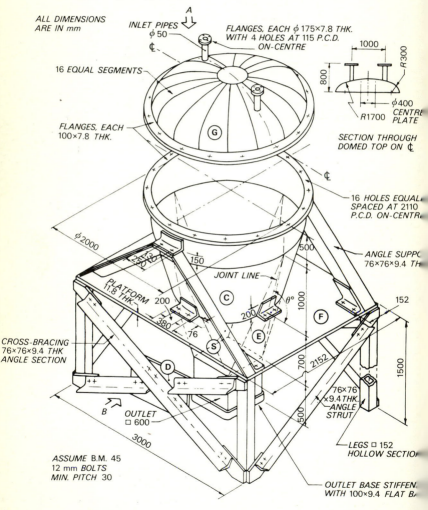

ALL DIMENSIONS ARE IN mm

INLET PIPES
φ 50

FLANGES, EACH φ 175×7.8 THK. WITH 4 HOLES AT 115 P.C.D. ON-CENTRE

16 EQUAL SEGMENTS

FLANGES, EACH 100×7.8 THK.

SECTION THROUGH DOMED TOP ON ₵

16 HOLES EQUAL SPACED AT 2110 P.C.D. ON-CENTR

ANGLE SUPPO 76×76×9.4 TH

JOINT LINE

PLATFORM 11.8 THK. 200

CROSS-BRACING 76×76×9.4 THK ANGLE SECTION

OUTLET □ 600

76×76 ×9.4 THK. ANGLE STRUT

LEGS □ 152 HOLLOW SECTIO

OUTLET BASE STIFFEN WITH 100×9.4 FLAT B

ASSUME B.M. 45 12 mm BOLTS MIN. PITCH 30

Questions 15–21

Index